Lecture Notes in Computer Science 14449

Founding Editors

Gerhard Goos
Juris Hartmanis

Editorial Board Members

The series Lecture Notes in Computer Science (LNCS), including its subseries Lecture Notes in Artificial Intelligence (LNAI) and Lecture Notes in Bioinformatics (LNBI), has established itself as a medium for the publication of new developments in computer science and information technology research, teaching, and education.

LNCS enjoys close cooperation with the computer science R & D community, the series counts many renowned academics among its volume editors and paper authors, and collaborates with prestigious societies. Its mission is to serve this international community by providing an invaluable service, mainly focused on the publication of conference and workshop proceedings and postproceedings. LNCS commenced publication in 1973.

Biao Luo · Long Cheng · Zheng-Guang Wu ·
Hongyi Li · Chaojie Li
Editors

Neural Information Processing

30th International Conference, ICONIP 2023
Changsha, China, November 20–23, 2023
Proceedings, Part III

Springer

Editors
Biao Luo (iD)
Central South University
Changsha, China

Long Cheng (iD)
Chinese Academy of Sciences
Beijing, China

Zheng-Guang Wu (iD)
Zhejiang University
Hangzhou, China

Hongyi Li (iD)
Guangdong University of Technology
Guangzhou, China

Chaojie Li (iD)
UNSW Sydney
Sydney, NSW, Australia

ISSN 0302-9743   ISSN 1611-3349 (electronic)
Lecture Notes in Computer Science
ISBN 978-981-99-8066-6   ISBN 978-981-99-8067-3 (eBook)
https://doi.org/10.1007/978-981-99-8067-3

This Springer imprint is published by the registered company Springer Nature Singapore Pte Ltd.
The registered company address is: 152 Beach Road, #21-01/04 Gateway East, Singapore 189721, Singapore

Paper in this product is recyclable.

Preface

Welcome to the 30th International Conference on Neural Information Processing (ICONIP2023) of the Asia-Pacific Neural Network Society (APNNS), held in Changsha, China, November 20–23, 2023.

The mission of the Asia-Pacific Neural Network Society is to promote active interactions among researchers, scientists, and industry professionals who are working in neural networks and related fields in the Asia-Pacific region. APNNS has Governing Board Members from 13 countries/regions – Australia, China, Hong Kong, India, Japan, Malaysia, New Zealand, Singapore, South Korea, Qatar, Taiwan, Thailand, and Turkey. The society's flagship annual conference is the International Conference of Neural Information Processing (ICONIP). The ICONIP conference aims to provide a leading international forum for researchers, scientists, and industry professionals who are working in neuroscience, neural networks, deep learning, and related fields to share their new ideas, progress, and achievements.

ICONIP2023 received 1274 papers, of which 256 papers were accepted for publication in Lecture Notes in Computer Science (LNCS), representing an acceptance rate of 20.09% and reflecting the increasingly high quality of research in neural networks and related areas. The conference focused on four main areas, i.e., "Theory and Algorithms", "Cognitive Neurosciences", "Human-Centered Computing", and "Applications". All the submissions were rigorously reviewed by the conference Program Committee (PC), comprising 258 PC members, and they ensured that every paper had at least two high-quality single-blind reviews. In fact, 5270 reviews were provided by 2145 reviewers. On average, each paper received 4.14 reviews.

We would like to take this opportunity to thank all the authors for submitting their papers to our conference, and our great appreciation goes to the Program Committee members and the reviewers who devoted their time and effort to our rigorous peer-review process; their insightful reviews and timely feedback ensured the high quality of the papers accepted for publication. We hope you enjoyed the research program at the conference.

October 2023

Biao Luo
Long Cheng
Zheng-Guang Wu
Hongyi Li
Chaojie Li

Organization

Honorary Chair

Weihua Gui   Central South University, China

Advisory Chairs

Jonathan Chan   King Mongkut's University of Technology
Thonburi, Thailand
Zeng-Guang Hou   Chinese Academy of Sciences, China
Nikola Kasabov   Auckland University of Technology, New Zealand
Derong Liu   Southern University of Science and Technology,
China
Seiichi Ozawa   Kobe University, Japan
Kevin Wong   Murdoch University, Australia

General Chairs

Tingwen Huang   Texas A&M University at Qatar, Qatar
Chunhua Yang   Central South University, China

Program Chairs

Biao Luo   Central South University, China
Long Cheng   Chinese Academy of Sciences, China
Zheng-Guang Wu   Zhejiang University, China
Hongyi Li   Guangdong University of Technology, China
Chaojie Li   University of New South Wales, Australia

Technical Chairs

Xing He   Southwest University, China
Keke Huang   Central South University, China
Huaqing Li   Southwest University, China
Qi Zhou   Guangdong University of Technology, China

Local Arrangement Chairs

Wenfeng Hu   Central South University, China
Bei Sun   Central South University, China

Finance Chairs

Fanbiao Li   Central South University, China
Hayaru Shouno   University of Electro-Communications, Japan
Xiaojun Zhou   Central South University, China

Special Session Chairs

Hongjing Liang   University of Electronic Science and Technology,
                              China
Paul S. Pang   Federation University, Australia
Qiankun Song   Chongqing Jiaotong University, China
Lin Xiao   Hunan Normal University, China

Tutorial Chairs

Min Liu   Hunan University, China
M. Tanveer   Indian Institute of Technology Indore, India
Guanghui Wen   Southeast University, China

Publicity Chairs

Sabri Arik   Istanbul University-Cerrahpaşa, Turkey
Sung-Bae Cho   Yonsei University, South Korea
Maryam Doborjeh   Auckland University of Technology, New Zealand
El-Sayed M. El-Alfy   King Fahd University of Petroleum and Minerals,
                              Saudi Arabia
Ashish Ghosh   Indian Statistical Institute, India
Chuandong Li   Southwest University, China
Weng Kin Lai   Tunku Abdul Rahman University of
                              Management & Technology, Malaysia
Chu Kiong Loo   University of Malaya, Malaysia

Qinmin Yang   Zhejiang University, China
Zhigang Zeng   Huazhong University of Science and Technology,
                                   China

Publication Chairs

Zhiwen Chen   Central South University, China
Andrew Chi-Sing Leung   City University of Hong Kong, China
Xin Wang   Southwest University, China
Xiaofeng Yuan   Central South University, China

Secretaries

Yun Feng   Hunan University, China
Bingchuan Wang   Central South University, China

Webmasters

Tianmeng Hu   Central South University, China
Xianzhe Liu   Xiangtan University, China

Program Committee

Rohit Agarwal   UiT The Arctic University of Norway, Norway
Hasin Ahmed   Gauhati University, India
Harith Al-Sahaf   Victoria University of Wellington, New Zealand
Brad Alexander   University of Adelaide, Australia
Mashaan Alshammari   Independent Researcher, Saudi Arabia
Sabri Arik   Istanbul University, Turkey
Ravneet Singh Arora   Block Inc., USA
Zeyar Aung   Khalifa University of Science and Technology,
                                   UAE
Monowar Bhuyan   Umeå University, Sweden
Jingguo Bi   Beijing University of Posts and
                                   Telecommunications, China
Xu Bin   Northwestern Polytechnical University, China
Marcin Blachnik   Silesian University of Technology, Poland
Paul Black   Federation University, Australia

Anoop C. S.	Govt. Engineering College, India
Ning Cai	Beijing University of Posts and Telecommunications, China
Siripinyo Chantamunee	Walailak University, Thailand
Hangjun Che	City University of Hong Kong, China
Wei-Wei Che	Qingdao University, China
Huabin Chen	Nanchang University, China
Jinpeng Chen	Beijing University of Posts & Telecommunications, China
Ke-Jia Chen	Nanjing University of Posts and Telecommunications, China
Lv Chen	Shandong Normal University, China
Qiuyuan Chen	Tencent Technology, China
Wei-Neng Chen	South China University of Technology, China
Yufei Chen	Tongji University, China
Long Cheng	Institute of Automation, China
Yongli Cheng	Fuzhou University, China
Sung-Bae Cho	Yonsei University, South Korea
Ruikai Cui	Australian National University, Australia
Jianhua Dai	Hunan Normal University, China
Tao Dai	Tsinghua University, China
Yuxin Ding	Harbin Institute of Technology, China
Bo Dong	Xi'an Jiaotong University, China
Shanling Dong	Zhejiang University, China
Sidong Feng	Monash University, Australia
Yuming Feng	Chongqing Three Gorges University, China
Yun Feng	Hunan University, China
Junjie Fu	Southeast University, China
Yanggeng Fu	Fuzhou University, China
Ninnart Fuengfusin	Kyushu Institute of Technology, Japan
Thippa Reddy Gadekallu	VIT University, India
Ruobin Gao	Nanyang Technological University, Singapore
Tom Gedeon	Curtin University, Australia
Kam Meng Goh	Tunku Abdul Rahman University of Management and Technology, Malaysia
Zbigniew Gomolka	University of Rzeszow, Poland
Shengrong Gong	Changshu Institute of Technology, China
Xiaodong Gu	Fudan University, China
Zhihao Gu	Shanghai Jiao Tong University, China
Changlu Guo	Budapest University of Technology and Economics, Hungary
Weixin Han	Northwestern Polytechnical University, China

Xing He	Southwest University, China
Akira Hirose	University of Tokyo, Japan
Yin Hongwei	Huzhou Normal University, China
Md Zakir Hossain	Curtin University, Australia
Zengguang Hou	Chinese Academy of Sciences, China
Lu Hu	Jiangsu University, China
Zeke Zexi Hu	University of Sydney, Australia
He Huang	Soochow University, China
Junjian Huang	Chongqing University of Education, China
Kaizhu Huang	Duke Kunshan University, China
David Iclanzan	Sapientia University, Romania
Radu Tudor Ionescu	University of Bucharest, Romania
Asim Iqbal	Cornell University, USA
Syed Islam	Edith Cowan University, Australia
Kazunori Iwata	Hiroshima City University, Japan
Junkai Ji	Shenzhen University, China
Yi Ji	Soochow University, China
Canghong Jin	Zhejiang University, China
Xiaoyang Kang	Fudan University, China
Mutsumi Kimura	Ryukoku University, Japan
Masahiro Kohjima	NTT, Japan
Damian Kordos	Rzeszow University of Technology, Poland
Marek Kraft	Poznań University of Technology, Poland
Lov Kumar	NIT Kurukshetra, India
Weng Kin Lai	Tunku Abdul Rahman University of Management & Technology, Malaysia
Xinyi Le	Shanghai Jiao Tong University, China
Bin Li	University of Science and Technology of China, China
Hongfei Li	Xinjiang University, China
Houcheng Li	Chinese Academy of Sciences, China
Huaqing Li	Southwest University, China
Jianfeng Li	Southwest University, China
Jun Li	Nanjing Normal University, China
Kan Li	Beijing Institute of Technology, China
Peifeng Li	Soochow University, China
Wenye Li	Chinese University of Hong Kong, China
Xiangyu Li	Beijing Jiaotong University, China
Yantao Li	Chongqing University, China
Yaoman Li	Chinese University of Hong Kong, China
Yinlin Li	Chinese Academy of Sciences, China
Yuan Li	Academy of Military Science, China

Yun Li	Nanjing University of Posts and Telecommunications, China
Zhidong Li	University of Technology Sydney, Australia
Zhixin Li	Guangxi Normal University, China
Zhongyi Li	Beihang University, China
Ziqiang Li	University of Tokyo, Japan
Xianghong Lin	Northwest Normal University, China
Yang Lin	University of Sydney, Australia
Huawen Liu	Zhejiang Normal University, China
Jian-Wei Liu	China University of Petroleum, China
Jun Liu	Chengdu University of Information Technology, China
Junxiu Liu	Guangxi Normal University, China
Tommy Liu	Australian National University, Australia
Wen Liu	Chinese University of Hong Kong, China
Yan Liu	Taikang Insurance Group, China
Yang Liu	Guangdong University of Technology, China
Yaozhong Liu	Australian National University, Australia
Yong Liu	Heilongjiang University, China
Yubao Liu	Sun Yat-sen University, China
Yunlong Liu	Xiamen University, China
Zhe Liu	Jiangsu University, China
Zhen Liu	Chinese Academy of Sciences, China
Zhi-Yong Liu	Chinese Academy of Sciences, China
Ma Lizhuang	Shanghai Jiao Tong University, China
Chu-Kiong Loo	University of Malaya, Malaysia
Vasco Lopes	Universidade da Beira Interior, Portugal
Hongtao Lu	Shanghai Jiao Tong University, China
Wenpeng Lu	Qilu University of Technology, China
Biao Luo	Central South University, China
Ye Luo	Tongji University, China
Jiancheng Lv	Sichuan University, China
Yuezu Lv	Beijing Institute of Technology, China
Huifang Ma	Northwest Normal University, China
Jinwen Ma	Peking University, China
Jyoti Maggu	Thapar Institute of Engineering and Technology Patiala, India
Adnan Mahmood	Macquarie University, Australia
Mufti Mahmud	University of Padova, Italy
Krishanu Maity	Indian Institute of Technology Patna, India
Srimanta Mandal	DA-IICT, India
Wang Manning	Fudan University, China

Piotr Milczarski	Lodz University of Technology, Poland
Malek Mouhoub	University of Regina, Canada
Nankun Mu	Chongqing University, China
Wenlong Ni	Jiangxi Normal University, China
Anupiya Nugaliyadde	Murdoch University, Australia
Toshiaki Omori	Kobe University, Japan
Babatunde Onasanya	University of Ibadan, Nigeria
Manisha Padala	Indian Institute of Science, India
Sarbani Palit	Indian Statistical Institute, India
Paul Pang	Federation University, Australia
Rasmita Panigrahi	Giet University, India
Kitsuchart Pasupa	King Mongkut's Institute of Technology Ladkrabang, Thailand
Dipanjyoti Paul	Ohio State University, USA
Hu Peng	Jiujiang University, China
Kebin Peng	University of Texas at San Antonio, USA
Dawid Połap	Silesian University of Technology, Poland
Zhong Qian	Soochow University, China
Sitian Qin	Harbin Institute of Technology at Weihai, China
Toshimichi Saito	Hosei University, Japan
Fumiaki Saitoh	Chiba Institute of Technology, Japan
Naoyuki Sato	Future University Hakodate, Japan
Chandni Saxena	Chinese University of Hong Kong, China
Jiaxing Shang	Chongqing University, China
Lin Shang	Nanjing University, China
Jie Shao	University of Science and Technology of China, China
Yin Sheng	Huazhong University of Science and Technology, China
Liu Sheng-Lan	Dalian University of Technology, China
Hayaru Shouno	University of Electro-Communications, Japan
Gautam Srivastava	Brandon University, Canada
Jianbo Su	Shanghai Jiao Tong University, China
Jianhua Su	Institute of Automation, China
Xiangdong Su	Inner Mongolia University, China
Daiki Suehiro	Kyushu University, Japan
Basem Suleiman	University of New South Wales, Australia
Ning Sun	Shandong Normal University, China
Shiliang Sun	East China Normal University, China
Chunyu Tan	Anhui University, China
Gouhei Tanaka	University of Tokyo, Japan
Maolin Tang	Queensland University of Technology, Australia

Qiang Xiao	Huazhong University of Science and Technology, China
Hao Xiong	Macquarie University, Australia
Dongpo Xu	Northeast Normal University, China
Hua Xu	Tsinghua University, China
Jianhua Xu	Nanjing Normal University, China
Xinyue Xu	Hong Kong University of Science and Technology, China
Yong Xu	Beijing Institute of Technology, China
Ngo Xuan Bach	Posts and Telecommunications Institute of Technology, Vietnam
Hao Xue	University of New South Wales, Australia
Yang Xujun	Chongqing Jiaotong University, China
Haitian Yang	Chinese Academy of Sciences, China
Jie Yang	Shanghai Jiao Tong University, China
Minghao Yang	Chinese Academy of Sciences, China
Peipei Yang	Chinese Academy of Science, China
Zhiyuan Yang	City University of Hong Kong, China
Wangshu Yao	Soochow University, China
Ming Yin	Guangdong University of Technology, China
Qiang Yu	Tianjin University, China
Wenxin Yu	Southwest University of Science and Technology, China
Yun-Hao Yuan	Yangzhou University, China
Xiaodong Yue	Shanghai University, China
Paweł Zawistowski	Warsaw University of Technology, Poland
Hui Zeng	Southwest University of Science and Technology, China
Wang Zengyunwang	Hunan First Normal University, China
Daren Zha	Institute of Information Engineering, China
Zhi-Hui Zhan	South China University of Technology, China
Baojie Zhang	Chongqing Three Gorges University, China
Canlong Zhang	Guangxi Normal University, China
Guixuan Zhang	Chinese Academy of Science, China
Jianming Zhang	Changsha University of Science and Technology, China
Li Zhang	Soochow University, China
Wei Zhang	Southwest University, China
Wenbing Zhang	Yangzhou University, China
Xiang Zhang	National University of Defense Technology, China
Xiaofang Zhang	Soochow University, China
Xiaowang Zhang	Tianjin University, China

Xinglong Zhang	National University of Defense Technology, China
Dongdong Zhao	Wuhan University of Technology, China
Xiang Zhao	National University of Defense Technology, China
Xu Zhao	Shanghai Jiao Tong University, China
Liping Zheng	Hefei University of Technology, China
Yan Zheng	Kyushu University, Japan
Baojiang Zhong	Soochow University, China
Guoqiang Zhong	Ocean University of China, China
Jialing Zhou	Nanjing University of Science and Technology, China
Wenan Zhou	PCN&CAD Center, China
Xiao-Hu Zhou	Institute of Automation, China
Xinyu Zhou	Jiangxi Normal University, China
Quanxin Zhu	Nanjing Normal University, China
Yuanheng Zhu	Chinese Academy of Sciences, China
Xiaotian Zhuang	JD Logistics, China
Dongsheng Zou	Chongqing University, China

Contents – Part III

Human Centred Computing

Theory and Algorithms

Efficient Lightweight Network with Transformer-Based Distillation for Micro-crack Detection of Solar Cells

Xiangying Xie[1,2], Xinyue Liu[1], QiXiang Chen[3], and Biao Leng[1(✉)]

[1] School of Computer Science and Engineering, Beihang University,
Beijing 100191, China
{liuxinyue7,lengbiao}@buaa.edu.cn
[2] State Grid Digital Technology Holding Co., Ltd., Beijing 100053, China
xiexiangying@sgec.sgcc.com.cn
[3] AnHui Jiyuan Software Co., Ltd., Hefei 230088, China
chenqixiang@sgit.sgcc.com.cn

Abstract. Micro-cracks on solar cells often affect the power generation efficiency, so this paper proposes a lightweight network for cell image micro-crack detection task. Firstly, a Feature Selection framework is proposed, which can efficiently and adaptively decide the number of layers of the feature extraction network, and clip unnecessary feature generation process. In addition, based on the design of the Transformer layer, Transformer Distillation is proposed. In Transformer Distillation, the designed Transformer Refine module excavates the distillation information from the two dimensions of features and relations. Using a combination of Feature Selection and Transformer Distillation, the lightweight networks based on ResNet and ViT can achieve much better effects than the original networks, with classification accuracy rates of 88.58% and 89.35% respectively.

Keywords: Knowledge distillation · Defect detection · Image classification · Deep learning

1 Introduction

Photovoltaic power generation technology which is recently widely used generates electricity with the help of solar energy which is one of the renewable energy sources. The realization of photovoltaic power generation is inseparable from its core component, solar cells. Due to problems such as external force and aging, solar cells may appear micro-cracks in actual use, thereby reducing power generation efficiency. Therefore, it is necessary to regularly detect the micro-cracks to maintain the normal function of solar cells.

Figure 1 shows some examples of electroluminescence images of solar cells, these high-resolution images are often used for micro-crack detection. As shown in Fig. 1, according to different manufacturing processes, solar cells are divided into two types: monocrystalline cells and polycrystalline cells. In addition, it can

© The Author(s), under exclusive license to Springer Nature Singapore Pte Ltd. 2024
B. Luo et al. (Eds.): ICONIP 2023, LNCS 14449, pp. 3–15, 2024.
https://doi.org/10.1007/978-981-99-8067-3_1

be seen that the defective cells contain obvious stripe cracks compared to the normal cells.

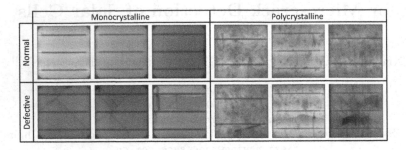

Fig. 1. Some examples from the solar cell micro-crack dataset elpv.

The efficiency of traditional artificial micro-crack detection is low, while the cost is high. In recent years, deep learning networks have been gradually introduced into micro-crack detection and have achieved excellent results. However, considering that the micro-crack detection task is a binary classification task, and the information contained in the solar cell images is relatively simple, we believe that the current micro-crack detection models based on the common convolutional neural networks are too redundant for the micro-crack detection task. Therefore, we propose Incomplete Small Network (ISN), relying on the Feature Selection (FS) framework to adaptively select a network of appropriate size for the micro-crack detection task. Through experiments, we prove that the smaller networks achieve better detection results than the existing original networks.

The current mainstream knowledge distillation algorithms are implemented based on the teacher-student framework. Among them, the selected teacher network with the better effect generally has more parameters than the student network. In the micro-crack detection task, we find that the classification effect of the larger network is worse than that of the smaller network. The traditional logit-based knowledge distillation methods and the selection rules of the teacher and student network are not applicable to the distillation task of the micro-crack detection network. Therefore, we innovatively propose a Transformer Distillation (TD) method, which can integrate feature-based distillation and relation-based distillation into one distillation module, extract instructive information from the teacher network, and transfer it to the student network.

In summary, the main contributions of this paper are as follows:

1. Aiming at the characteristics of the micro-crack detection task, Feature Selection module is proposed to obtain a lightweight network Incomplete Small Network, which is better than the original ResNet [8] and ViT [6] for the classification of micro-cracked solar cells.
2. We propose a new distillation method, Transformer Distillation, which uses the Transformer Refine (TR) module to integrate the teacher's network features, so that the student network can learn effective information for the micro-crack detection task.

3. Related experiments are conducted on the public solar cell dataset, which proves the advantages of the proposed lightweight network, and verifies the effectiveness of Transformer Distillation on CNN and Transformer networks.

2 Related Work

2.1 Image Classification and Defect Detection

Image classification is an important research content in the field of computer vision, and its task is to predict the corresponding category label for the image. In recent years, with the development of deep learning, image classification algorithms based on deep convolutional networks such as ResNet [8] and Efficientnet [13] have become mainstream research directions.

Defect detection mainly has two methods: Photoluminescence (PL) [10] and Electroluminescence (EL) [7]. In the defect detection task of solar cells, the EL method is widely used. Traditional solar cell defect detection methods usually use Fourier image reconstruction [15], filtering [1], clustering [16], and so on. The defect detection task of solar cells is essentially an image classification method, so with the development of image classification methods based on deep learning, more and more defect detection methods gradually introduce deep learning for classification [4,5].

2.2 Knowledge Distillation

Knowledge distillation is one of the model compression methods. The initial knowledge distillation methods [9,12,17] transfer the information in a wider or deeper teacher network to a smaller student network with poorer performance. According to the different information transmitted, knowledge distillation algorithms can be divided into logit-based, feature-based, and relation-based. Recently, some methods improve the traditional distillation architecture, such as teachers and networks learning from each other [19], introducing teacher assistance [11], self-distillation [18] and other methods are proposed.

3 Methodology

In this section, our goal is to get a lightweight network for solar cell micro-crack detection. In Sect. 3.1, we first introduce the advantages of small models in micro-crack detection task for solar cells. In Sect. 3.2, we innovatively propose a knowledge distillation method named Transformer Distillation, and discuss the applicability of this method to knowledge distillation between networks of different sizes and self-distillation within small networks.

3.1 Smaller is Better

The objective of the micro-crack detection task is to identify whether there are micro-cracks in the solar cell image and to determine whether the detected cell image is normal (represented by class 0) or micro-cracked (represented by class 1). Therefore, micro-crack detection can be regarded as a binary image classification problem. In the existing micro-crack detection methods based on deep learning, the neural networks such as VGGNet, ResNet and Vision Transformer (ViT) are usually used to extract the features of the detected solar cell image, and then the fully connected layers are applied to binary classification on the obtained features to get the final classification result.

Based on the deep convolutional network ResNet and ViT commonly used in academic and industrial fields, we conduct micro-crack detection experiments on the elpv dataset which is an academic public solar cell dataset. The networks we use for micro-crack detection include ResNets of different sizes (ResNet18, ResNet34, ResNet50, and ResNet101) and ViTs of different sizes (Vit-Tiny, Vit-Small, and Vit-Base). We find that for the simple binary classification task of crack detection, the classification effect of the smaller network is better than that of the larger network, as detailed in Sect. 4.3.

In solar cell images, micro-cracks often appear in the shape of lines. Therefore, the information contained in the deep features of the neural network is too abstract, while the low-level features containing rich texture information are more conducive to the detection of micro-cracks. Considering the importance of low-level features for micro-crack detection task, we further propose a low-level feature-based micro-crack detection method using lightweight neural networks, in which a framework for Feature Selection (FS) is designed to judge which layer features are the most suitable for micro-crack detection.

Figure 2 (a) and (b) show the Feature Selection frameworks we propose for ResNet18 and Vit-Tiny, respectively. The main body of the FS framework corresponding to ResNet18 is the same as the original network structure, including 5 stages which have multiple convolutional layers and the residual structure. Similarly, the main body of the FS framework corresponding to ViT-Tiny is the same as the original ViT-Tiny, including 12 layers of transformer layers with Multi-Head Attention and Multi-Layer Perceptron (MLP).

In addition, after the output of each layer in the FS framework, a linear layer is added as the classifier to perform binary classification on the solar cell images according to the output of this layer. The Feature Selection mechanism is to select the layer suitable for micro-crack detection according to the accuracy of the classifier. We will delete the deep features with poor classification effects, only keep the layer with the best classification effect and lower layers, and directly use the classifier corresponding to the optimal layer in the FS framework as the classifier of the final micro-crack detection model.

A series of experiments prove that for ResNet18, compared to using 5 stages for feature extraction, the output of Stage 4 is better for micro-crack detection. ViT-related experiments have similar results, and the 6-layer ViT structure can perform better in micro-crack detection tasks. The detailed experimental results

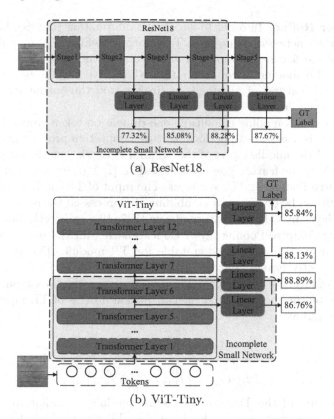

(a) ResNet18.

(b) ViT-Tiny.

Fig. 2. The Feature Selection framework for ResNet18 and ViT-Tiny.

can be found in Sect. 4.3. With the help of FS framework, we can boldly delete deep features that are not conducive to micro-crack detection, and obtain an Incomplete Small Network (ISN) that uses low-level features for classification. Compared with the original ResNet and ViT, ResNet-ISN and ViT-ISN we propose greatly reduce parameters, better effect, and are more suitable for real-time micro-crack detection tasks in industrial scenarios.

3.2 Transformer Distillation

In this section, we propose a novel distillation method Transformer Distillation (TD). First, use Transformer Refine (TR) module to process the output of intermediate layers in the teacher network to obtain the processed features. Afterwards, let the student network learn the information contained in the features output by Transformer Refine module to realize the transfer of information from the teacher network to the student network. In addition, we respectively propose multiple options for the teacher network and discuss the implementation ways of Transformer Distillation for distillation between networks of different sizes and self-distillation.

Transformer Refine. In order to enrich the information that can be obtained from the teacher network, we design a Transformer Refine (TR) module to refine the intermediate feature information of the teacher network. The information processed by TR module is from the intermediate layers of the teacher network like the output features of all stages in ResNet and the output features of all transformer layers in ViT.

The input of TR module is multiple one-dimension token embeddings of the same size, so it is necessary to perform feature adaptation processing on the output features of the middle layers of the original network. For the output of each stage in ResNet, the feature size is converted to $[1, 100]$ by using the combination of AdaptivePooling and Linear layer. The input of TR module with a size of $[n_t^{ResNet}, 100]$ can be obtained by combining the processed features. For the output of each layer in ViT, the feature of size $[197, 192]$ is directly processed into $[1, 192]$ using AdaptivePooling layer. Concatenating all the processed features gives an input of size $[n_t^{ViT}, 192]$ suitable for TR module. The above n_t^{ResNet} and n_t^{ViT} respectively represent the number of stages of the ResNet teacher and the number of layers of the ViT teacher. Let $f_{i,t}$ represent the output feature of layer i in the teacher network, and the calculation process of TR input $Input_{TR}$ can be expressed as:

$$f'_{i,t} = \text{FeatureAdaptation}(f_{i,t}), \tag{1}$$

$$Input_{TR} = \{f'_{1,t}, f'_{2,t}, ..., f'_{n_t,t}\}. \tag{2}$$

The structure of the Transformer Refine module is similar to that of the traditional Transformer encoder. Each layer in TR module includes Multi-head Self-Attention (MSA), Feed-Forward Network (FFN), and residual connection, where MSA is based on the self-attention mechanism with Query -Key-Value, FFN is a two-layer MLP. The overall structure is shown in Fig. 3. The calculation process of the entire Transformer Refine module is as follows:

$$z_0 = Input_{TR}, \tag{3}$$

$$z_i^* = \text{MSA}(\text{Norm}(z_{i-1})) + z_{i-1}, \tag{4}$$

$$z_i = \text{FFN}(\text{Norm}(z_i^*)) + z_i^*, \tag{5}$$

where z_i^* and z_i denote the output feature of MSA and FFN in the ith layer in TR module.

After further processing the tokens generated by the intermediate features of the teacher network with the help of TR module, the output features not only contain the original token information, but also refine the relationship information between tokens. Therefore, TR module can obtain the relationship information between the intermediate features of the teacher network. Considering this, using the TR module as the distillation medium, it is possible to adaptively extract the feature information transmitted from the teacher network to the student network, and enrich the types of information transmitted.

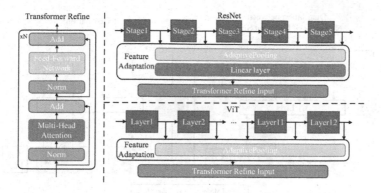

Fig. 3. The left side shows the implementation of Transformer Refine module. The upper and lower sections on the right show the generation processes of Transformer Refine module inputs.

Distillation Process. Different from the previous distillation scenarios, in the micro-crack detection task, the detection effect of the deep network is worse than that of the shallow network according to Sect. 3.1. We cannot get a better teacher network for training the student network, which is contrary to the premise assumed in the previous application of distillation. Therefore, traditional feature-based knowledge distillation and logit-based knowledge distillation are difficult to apply to the micro-crack detection task.

Although the effect of deep network for micro-crack detection is not good, our intuition is that the low-level features in the deep network and the information contained in the feature generation process from the lower layers to the deeper layers are still helpful for the training of our proposed Incomplete Small Network. In this regard, we propose Transformer Distillation.

In the previous section, Transformer Refine module we propose can adaptively process the deep and low-level features in the network. The outputs of TR module not only contain the original information of the corresponding layers, but also include the relationship information between the features of the multi-layer features. For example, among the five TR features obtained by processing the output of ResNet containing five stages through TR module, the TR feature corresponding to Stage 1 not only contains the information in the original Stage 1 output, but also contains the information in the deeper stages and the information of the feature generation process.

The proposed Transformer Distillation method based on TR module can be divided into two implementation methods according to the size selection of the teacher network: distillation between networks of different sizes and self-distillation, as shown in Fig. 4. The specific network selection methods of the two types are introduced below.

Distillation between Networks of Different Sizes (DNDS). In this type of distillation, a deeper network is used as the teacher network, and our proposed Incomplete Small Network is used as the student network. Specifically,

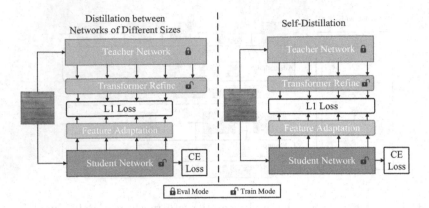

Fig. 4. The distillation framework of Transformer Distillation.

for ResNet, ResNet18 with five stages is used as the teacher network, and ResNet-ISN with four stages is used as the student network. For ViT, the original 12-layer ViT-Tiny is used as the teacher network, and the 6-layer ViT-ISN is used as the student network.

Self-Distillation (SD). In self-distillation, Incomplete Small Network is used as the teacher and student network at the same time, and offline distillation is performed. For the self-distillation of ResNet and ViT, ResNet-ISN and ViT-ISN are used as both teacher and student networks, respectively.

The loss function of Transformer Distillation training is calculated using the processed features of teacher and student networks. First, use TR module to process the intermediate features of the teacher network to get the output tokens z_t. Then, use the feature adaptation processing method in TR module to process the intermediate features of the student network to get the embeddings z_s with the same size as the tokens. Finally, select the feature z_t' of the layer corresponding to z_s in z_t, and use the $L1$ distance as the loss function to calculate the loss value of distillation:

$$Loss_{TD} = \frac{1}{n_s} \sum_{i=1}^{n_s} |z_t^i - z_s^i|, \tag{6}$$

where $Loss_{TD}$ represents the loss value of Transformer Distillation, n_s is the number of layers in student network, z_t^i and z_s^i denote the ith token in z_t and z_s.

We use the cross-entropy loss $Loss_{CE}$ to calculate the classification loss value in micro-crack detection. Therefore, the overall loss value for optimization is:

$$Loss = Loss_{CE} + \lambda \times Loss_{TD}, \tag{7}$$

where λ is the weight of the Transformer Distillation loss.

4   Experiments

4.1   Dataset and Evaluation Metrics

In order to verify the effectiveness of our proposed lightweight micro-crack detection network and Transformer Distillation method, we conduct a series of experiments on the public elpv dataset [2–4]. The elpv dataset is a dataset of solar cell images obtained from electroluminescence images of photovoltaic modules, containing 2,624 polycrystalline and monocrystalline solar cell image samples. We regard all the images in the dataset with a micro-crack probability greater than 0 as the micro-cracked cell images and label them as class 1. Correspondingly, the images with a micro-crack probability equal to 0 are regarded as normal cell images without micro-cracks and are represented by label 0.

We divide the elpv dataset into a training set and a test set for the training and testing process of the network. The training set contains 1968 pictures, and the test set contains 656 pictures. The specific division method is shown in Table 1.

Table 1. The Division Method of The Dataset (0, 33%, 67% and 100% indicates the probability that the cell has micro-crack)

	Train				Test				
Label	0	1			0	1			Sum
Probability	0%	33%	67%	100%	0%	33%	67%	100%	
Polycrystal	683	132	37	301	237	46	13	101	1550
Monocrystal	438	87	41	249	150	30	15	64	1074
Sum	1121	219	78	550	387	76	28	165	2624

Common evaluation metrics such as accuracy, precision, recall, F1-score, and AUC are used to evaluate the experimental results.

4.2   Experiment Settings

The used initial parameters of the networks in this paper refer to the pre-trained networks on ImageNet dataset. We use 4 GeForce GTX 1080Ti GPUs to train the networks for 100 epochs using the Adam optimizer with the learning rate set to $1e^{-4}$ and the batch size of 32.

Data augmentation is first performed on the images in the training set. We choose the rotation method to expand the data quantity to 4 times the original, so as to ensure that the integrity of the micro-cracks in the images will not be affected when the data is enhanced. In addition, during the training process, the images are normalized. The hyperparameter λ is set to 0.1 and 0.2 in the experiments of ResNet and ViT respectively.

4.3   Feature Selection Experiments

Smaller Networks Perform Better. We conduct micro-crack detection experiments using different sizes of ResNet and ViT on the elpv dataset, and the results are shown in Table 2. It can be seen that ResNet18 and ViT-Tiny can get better classification results, which proves that the simpler the network is more suitable for the simple micro-crack detection task.

Table 2. Classification results of ResNet and ViT on the elpv dataset.

Type	ResNet				ViT		
Size	18	34	50	101	Tiny	Small	Base
Params(M)	11.18	21.29	23.51	42.50	5.52	9.25	36.19
Accuracy(%)	87.67	86.45	86.91	87.21	85.84	85.69	84.78
Precision(%)	85.93	86.94	88.33	90.79	86.12	82.84	84.84
Recall(%)	83.70	78.89	78.52	76.67	78.15	82.22	76.67
F1-score(%)	84.80	82.71	83.14	83.13	81.94	82.53	80.54
AUC(%)	87.07	85.31	85.64	85.62	84.68	85.17	83.55

Feature Selection Experiments. Table 3 and Table 4 respectively show the experimental results of our proposed Feature Selection framework on ResNet18 and ViT-Tiny. The results in the tables correspond to the classifiers of different stages or Transformer layers of the networks. Experimental results show that ResNet with 4 stages and ViT with 6 layers perform better than the original ResNet with 5 stages and ViT with 12 layers, achieving accuracy of 88.18% and 88.89% respectively. It also proves that the feature layers in the original networks are redundant, and the Feature Selection framework can help to judge and retain useful feature layers more accurately.

Table 3. Experimental results of Feature Selection framework on ResNet18.

Stage	2	3	4	5
Params(M)	0.16	0.68	**2.78**	11.18
Accuracy(%)	77.32	85.08	**88.28**	87.67
Precision(%)	74.69	86.13	**87.84**	85.93
Recall(%)	67.78	75.93	82.96	**83.70**
F1-score(%)	71.07	80.71	**85.33**	84.80
AUC(%)	75.88	83.70	**87.48**	84.80

4.4   Transformer Distillation Experiments

Settings of Related Parameters. For the selection of the number of layers n_{layer}, the number of heads n_{head}, and the dimension of the features d_{model} in

Table 4. Experimental results of Feature Selection framework on ViT-Tiny.

Layer	1	2	3	4	5	6	7	8	9	10	11	12
Params(M)	0.63	1.08	1.52	1.97	2.41	2.86	3.30	3.75	4.19	4.64	5.08	5.52
Accuracy(%)	68.80	71.08	78.08	86.45	86.76	**88.89**	88.13	87.52	87.37	87.21	87.06	85.84
Precision(%)	74.81	76.32	77.88	86.35	87.04	**92.64**	88.71	89.17	87.85	86.33	85.71	86.12
Recall(%)	36.30	42.96	65.19	79.63	79.63	79.26	81.48	79.26	80.37	81.85	**82.22**	78.15
F1-score(%)	48.88	54.98	70.97	82.85	83.17	**85.43**	84.94	83.92	83.95	84.03	83.93	81.94
AUC(%)	63.88	66.83	76.13	85.42	85.68	**87.43**	87.13	86.27	86.31	86.40	86.33	84.68

Transformer Refine module, we conduct a series of experiments. In this section, all the experiments are based on Self-Distillation method in Transformer Distillation and performed on ResNet18. The obtained results are shown in Table 5. It can be seen that the distillation effect is the best when n_{layer} is 3, n_{head} is 4, and d_{model} is 100, and this setting will also be maintained in subsequent experiments.

Table 5. Experimental results of settings of parameters in Transformer Refine module.

	n_{layer}			n_{head}			d_{model}		
	$n_{head} = 4\ d_{model} = 100$			$n_{layer} = 3\ d_{model} = 100$			$n_{layer} = 3\ n_{head} = 4$		
	2	3	4	2	4	5	80	100	120
Accuracy(%)	86.30	**87.82**	86.91	86.45	**87.82**	87.21	86.61	**87.82**	87.06
Precision(%)	88.46	88.31	**88.98**	87.55	**88.31**	86.90	87.92	**88.31**	87.76
Recall(%)	76.67	**81.11**	77.78	78.15	**81.11**	81.11	78.15	**81.11**	79.63
F1-score(%)	82.14	**84.56**	83.00	82.58	**84.56**	83.91	82.75	**84.56**	83.50
AUC(%)	84.84	**86.81**	85.53	85.20	**86.81**	86.29	85.33	**86.81**	85.94

Ablation Study. For the proposed two distillation methods Distillation between Networks of Different Sizes (DNDS) and Self-Distillation (SD) in Sect. 3.2, the results of our ablation experiments are shown in Table 6. For ResNet, DNDS can improve the classification accuracy of the student network by 0.3%, while SD is not feasible. But for ViT, DNDS can not improve the effect much, while SD can effectively improve the accuracy by about 0.5%.

Overall Results. Table 7 shows the classification effect of the networks obtained by combining the Feature Selection and Transformer Distillation methods proposed in this paper. The obtained ResNet and ViT can achieve accuracy rates of 88.58% and 89.35% respectively. In addition, Table 7 also compares the proposed TD method with other mainstream distillation methods. It can be seen that in the micro-crack detection task, Transformer Distillation can effectively further improve the detection effect on the basis of small networks.

Table 6. Ablation experiment results of the proposed distillation methods.

		DNDS	SD	Accuracy(%)	Precision(%)	Recall(%)	F1-score(%)	AUC(%)
ResNet	ISN	✗	✗	88.28	87.84	82.96	85.33	87.48
	TD	✓	✗	**88.58**	87.64	**84.07**	**85.82**	**87.90**
		✗	✓	87.67	88.89	80.00	84.21	86.51
ViT	ISN	✗	✗	88.89	92.64	79.26	85.43	87.13
	TD	✓	✗	88.89	90.53	81.48	85.77	87.77
		✗	✓	**89.35**	91.67	**81.48**	**86.27**	**88.16**

Table 7. Comparison results with mainstream distillation methods.

		Accuracy(%)	Precision(%)	Recall(%)	F1-score(%)	AUC(%)
ResNet (DNDS)	Student	88.28	87.84	82.96	85.33	87.48
	Teacher	87.67	85.93	83.70	84.80	87.07
	KD [9]	86.30	90.54	74.44	81.71	84.51
	FitNets [12]	86.91	88.66	78.15	83.07	85.59
	TD(Ours)	**88.58**	**87.64**	**84.07**	**85.82**	**87.90**
ViT (SD)	Student	88.89	92.64	79.26	85.43	87.13
	Teacher					
	DeiT [14]	87.52	90.87	77.41	83.60	85.99
	TD (Ours)	**89.35**	**91.67**	**81.48**	**86.27**	**88.16**

5 Conclusion

In this paper, we propose a lightweight network named Incomplete Small Network for the solar cell image micro-crack detection task, using the Feature Selection framework to adaptively select the depth of the feature extraction network, which improves the classification effect while reducing the number of network parameters. In addition, we innovatively propose Transformer Distillation and design a Transformer Refine module to extract instructive information from the teacher network to improve the effect of distillation. The classification results on the public dataset of solar cell images show that our proposed lightweight network and distillation method are effective, and can be well adapted to the industrial scene of micro-crack detection with high-efficiency and real-time requirements.

References

1. Anwar, S.A., Abdullah, M.Z.: Micro-crack detection of multicrystalline solar cells featuring an improved anisotropic diffusion filter and image segmentation technique. EURASIP J. Image Video Process. **2014**, 1–17 (2014)
2. Buerhop-Lutz, C., et al.: A benchmark for visual identification of defective solar cells in electroluminescence imagery. In: 35th European PV Solar Energy Conference and Exhibition, vol. 12871289, pp. 1287–1289 (2018)

3. Deitsch, S., et al.: Segmentation of photovoltaic module cells in uncalibrated electroluminescence images. Mach. Vis. Appl. **32**(4), 84 (2021)
4. Deitsch, S., et al.: Automatic classification of defective photovoltaic module cells in electroluminescence images. Sol. Energy **185**, 455 468 (2019)
5. Demirci, M.Y., Beşli, N., Gümüşçü, A.: Efficient deep feature extraction and classification for identifying defective photovoltaic module cells in electroluminescence images. Expert Syst. Appl. **175**, 114810 (2021)
6. Dosovitskiy, A., et al.: An image is worth 16x16 words: transformers for image recognition at scale. arXiv preprint arXiv:2010.11929 (2020)
7. Fuyuki, T., Kitiyanan, A.: Photographic diagnosis of crystalline silicon solar cells utilizing electroluminescence. Appl. Phys. A **96**, 189–196 (2009)
8. He, K., Zhang, X., Ren, S., Sun, J.: Deep residual learning for image recognition. In: Proceedings of the IEEE Conference on Computer Vision and Pattern Recognition, pp. 770–778 (2016)
9. Hinton, G., Vinyals, O., Dean, J.: Distilling the knowledge in a neural network. arXiv preprint arXiv:1503.02531 (2015)
10. Johnston, S., et al.: Correlating multicrystalline silicon defect types using photoluminescence, defect-band emission, and lock-in thermography imaging techniques. IEEE J. Photovoltaics **4**(1), 348–354 (2013)
11. Mirzadeh, S.I., Farajtabar, M., Li, A., Levine, N., Matsukawa, A., Ghasemzadeh, H.: Improved knowledge distillation via teacher assistant. In: Proceedings of the AAAI Conference on Artificial Intelligence, vol. 34, pp. 5191–5198 (2020)
12. Romero, A., Ballas, N., Kahou, S.E., Chassang, A., Gatta, C., Bengio, Y.: FitNets: hints for thin deep nets. arXiv preprint arXiv:1412.6550 (2014)
13. Tan, M., Le, Q.: EfficientNet: rethinking model scaling for convolutional neural networks. In: International Conference on Machine Learning, pp. 6105–6114. PMLR (2019)
14. Touvron, H., Cord, M., Douze, M., Massa, F., Sablayrolles, A., Jégou, H.: Training data-efficient image transformers & distillation through attention. In: International Conference on Machine Learning, pp. 10347–10357. PMLR (2021)
15. Tsai, D.M., Wu, S.C., Li, W.C.: Defect detection of solar cells in electroluminescence images using Fourier image reconstruction. Sol. Energy Mater. Sol. Cells **99**, 250–262 (2012)
16. Tseng, D.C., Liu, Y.S., Chou, C.M.: Automatic finger interruption detection in electroluminescence images of multicrystalline solar cells. Math. Prob. Eng. **2015** (2015)
17. Zagoruyko, S., Komodakis, N.: Paying more attention to attention: improving the performance of convolutional neural networks via attention transfer. arXiv preprint arXiv:1612.03928 (2016)
18. Zhang, L., Song, J., Gao, A., Chen, J., Bao, C., Ma, K.: Be your own teacher: improve the performance of convolutional neural networks via self distillation. In: Proceedings of the IEEE/CVF International Conference on Computer Vision, pp. 3713–3722 (2019)
19. Zhang, Y., Xiang, T., Hospedales, T.M., Lu, H.: Deep mutual learning. In: Proceedings of the IEEE Conference on Computer Vision and Pattern Recognition, pp. 4320–4328 (2018)

MTLAN: Multi-Task Learning and Auxiliary Network for Enhanced Sentence Embedding

Gang Liu[1,2], Tongli Wang[1(✉)], Wenli Yang[1], Zhizheng Yan[1], and Kai Zhan[3]

[1] College of Computer Science and Technology, Harbin Engineering University,Harbin, China
watoli@hrbeu.edu.cn
[2] Modeling and Emulation in E-Government National Engineering Laboratory, Harbin Engineering University, Harbin, China
[3] PwC Enterprise Digital, PricewaterhouseCoopers, Sydney, Australia

Abstract. The objective of cross-lingual sentence embedding learning is to map sentences into a shared representation space, where semantically similar sentence representations are closer together, while distinct sentence representations exhibit clear differentiation. This paper proposes a novel sentence embedding model called MTLAN, which incorporates multi-task learning and auxiliary networks. The model utilizes the LaBSE model for extracting sentence features and undergoes joint training on tasks related to sentence semantic representation and distance measurement. Furthermore, an auxiliary network is employed to enhance the contextual expression of words within sentences. To address the issue of limited resources for low-resource languages, we construct a pseudo-corpus dataset using a multilingual dictionary for unsupervised learning. We conduct experiments on multiple publicly available datasets, including STS and SICK, to evaluate both monolingual sentence similarity and cross-lingual semantic similarity. The empirical results demonstrate the significant superiority of our proposed model over state-of-the-art methods.

Keywords: Cross-lingual · Sentence embedding · Multi-task learning · Contrastive learning · Auxiliary network

1 Introduction

Sentence embedding is a fundamental technique for using deep learning to solve natural language processing problems, also known as a distributed representation of sentences. By using an encoder model that encodes words or sentences into fixed-length, dense vectors, it is widely used in downstream tasks such as text classification and information retrieval. Recently, self-supervised learning has become the paradigm for sentence embedding encoding.

B. Luo et al. (Eds.): ICONIP 2023, LNCS 14449, pp. 16–27, 2024.
https://doi.org/10.1007/978-981-99-8067-3_2

Traditional methods mainly include monolingual mapping [1] and training on pseudo-corpora [2]. In recent years, Facebook introduced XLM [3], XLM-R [4], which are translation language models trained on bilingual sentence pairs. Additionally, researchers have explored multilingual joint pre-training, training a cross-lingual word vector model using monolingual or cross-lingual loss functions as training objectives. Examples of such models include RankCSE [5], and miCSE [6].

However, traditional linear transformation methods suffer from poor scalability and heavy reliance on corpus resources. Moreover, the sentence vector space obtained by encoding sentences using pre-trained language models often exhibits anisotropic properties, meaning that the learned word embeddings form a conical space in the vector space. This can lead to degradation of representations due to factors such as word frequency and anisotropy in the corpus. To address this issue, we propose the application of multi-task learning to sentence embedding tasks, utilizing the LaBSE model for transfer learning to extract sentence features.

We jointly train on sentence semantic representation and distance measurement tasks to enforce encoding distances in the Euclidean space that reflect semantic similarity between sentences. This is achieved through multi-task joint optimization of contrastive loss on the encoded sentence vectors and Euclidean distance loss. Furthermore, to tackle the issues of incomplete extraction of sentence semantic features by pre-trained language models and insufficient data for low-resource languages, we construct pseudo-corpora datasets using a multilingual dictionary for unsupervised learning. We also employ an auxiliary network based on a Bert encoder and LaBSE decoder for training.

To evaluate the effectiveness of our model, we conducted experiments on monolingual sentence similarity using the SentEval benchmark, including STS2012-STS2016 [7], STS-B [8], and SICK-R [9] datasets. In addition, we conducted cross-lingual sentence retrieval experiments on the publicly available dataset Tatoeba [10]. The results demonstrate that our model achieves state-of-the-art performance in both monolingual sentence embedding and cross-lingual sentence embedding tasks.

The contributions of this paper can be summarized as follows:

- Multi-task learning approach: By utilizing contrastive learning methods, the model is optimized to learn sentence semantic features. Additionally, the Euclidean distance measurement task is employed to constrain the positional information of encoded sentences. The model achieves joint optimization of contrastive loss and Euclidean distance loss through multi-task learning.
- Construction of a training set that aligns sentences in different languages: A training set is designed to align sentences from different languages, enabling pre-trained models to further capture alignment information between languages. This allows for the capturing of interaction information between multilingual and monolingual contexts.
- Auxiliary network: We introduce an auxiliary network using unsupervised learning methods. The encoder and decoder of the auxiliary network are

employed to predict masked pseudo-corpus sentences, thereby strengthening the model's encoding capacity and representation ability for cross-lingual sentences.

2   Related Work

In the current landscape of sentence embedding methods, there are many excellent approaches ranging from unsupervised learning to supervised learning. As in the Quick-Thought [11] in the field of unsupervised learning. In supervised learning, there is the InferSent [12] network model. With the rise of pre-training models such as BERT, there have been several models based on pre-training methods. The mainstream sentence embedding models in recent years are depicted in Fig. 1.

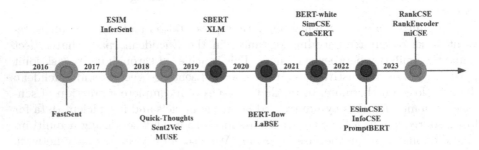

Fig. 1. A timeline of the sentence embedding models. The green ones are the classic models, the blue ones represent recent state-of-the-art models that incorporate new methods such as pre-training and contrastive learning. The red one is the latest sentence embedding model in 2023 (Color figure online)

LaBSE (Language-agnostic BERT Sentence Embedding) [13], introduced in 2020, is a multilingual BERT-based sentence embedding model that leverages the Masked Language Modeling (MLM) and Translation Language Modeling (TLM) methods. It was pre-trained on a massive corpus of 17 billion monolingual sentences and 6 billion bilingual sentence pairs. In the preprocessing stage, positive and negative training samples are constructed from the Natural Language Inference (NLI) dataset. The LaBSE model is then used for transfer learning to extract sentence features from these samples.

Yu Zhang [14] pointed out that multi-task learning (MTL) is a learning paradigm in machine learning, whose purpose is to use the useful information contained in multiple related tasks to help improve the generalization performance of all tasks. Therefore, we choose contrastive learning task and distance metric task to represent the semantic features and sentence encoding position information of sentences.

3   Method

The overall architecture of the MTLAN model is depicted in Fig. 2, consisting of four functional modules: yellow, blue, green, and purple, corresponding to pseudo data augmentation, feature extraction network (M_0), auxiliary network (M_1, M_2), and multi-task learning, respectively. The red portion represents the sentence embedding vectors output by the model.

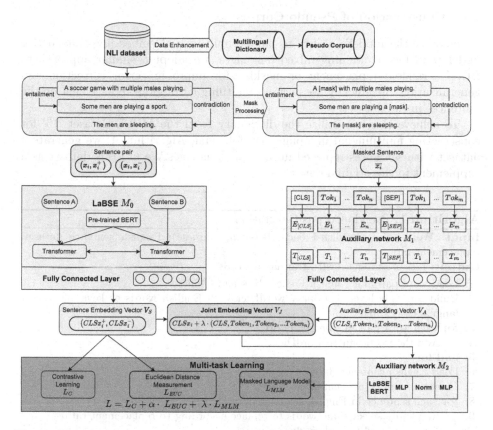

Fig. 2. The overall architecture of the MTLAN model. Yellow: pseudo-corpus data augmentation, blue: feature extraction network, green: auxiliary network, purple: multi-task learning, red: output sentence embeddings (Color figure online)

Firstly, we perform data augmentation by constructing a pseudo dataset using a multilingual dictionary, as shown in the yellow module. Next, we extract positive and negative sample pairs (x_i, x_i^+) and (x_i, x_i^-) from the NLI dataset, and utilize the pre-trained LaBSE model and fully connected layer to extract sentence embeddings V_S, as shown in the blue module.

Simultaneously, we apply masked processing to the sentences in the NLI dataset and train the auxiliary network. Initially, we extract auxiliary embedding vectors V_A using the frozen parameters of the BERT model M_1. Then, we concatenate V_A with the sentence embedding vectors V_S obtained from LaBSE to obtain V_J. Subsequently, we continue training using M_2, which reuses the BERT weights from LaBSE, as depicted in the green module.

Lastly, the multi-task learning component jointly optimizes the losses of contrastive learning, Euclidean distance measurement, and Masked Language Model, as shown in the purple module.

3.1 Construction of Pseudo Corpus

To enhance the model's ability to handle multilingual embeddings and better understand text containing mixed languages, we adopt a similar approach to BERT's random replacement or masking for improving error correction and generalization. Specifically, we perform multilingual random replacement on the sentences in the dataset.

To achieve this, we utilize the dictionary from the MUSE dataset [15] for constructing multilingual mappings. By following Algorithm. 1, we generate a dataset with sentences replaced in multiple languages.We use this dataset as a supplement to the original dataset.

Algorithm 1. Pseudo Corpus Generation

Input: Word pairs (English-French, Russian, Chinese); English sentence dataset; replacement ratio

Output: List of sentences with translated words

  1: Load word pairs for English-French, Russian, Chinese
  2: Build a double-layer dictionary wordDict with English words as keys and target languages as subkeys
  3: **for** each word pair in word pairs **do**
  4:      Add the word pair to wordDict
  5: **end for**
  6: Load English sentence dataset
  7: Create an empty list sentences
  8: **for** each sentence in English sentence dataset **do**
  9:      Randomly select some words to replace according to replacement ratio
10:      **for** each word to replace **do**
11:         **if** word not in wordDict or word is a stopword **then**
12:           Continue
13:         **end if**
14:         Randomly select a target language from wordDict[word]
15:         Randomly select a translation word from wordDict[word][target language]
16:         Replace word with translation word
17:      **end for**
18:      Add the replaced sentence to sentences
19: **end for**

3.2 Sentence Feature Extraction

We extract sentence pairs that entail relationships from the dataset, forming positive samples as pairs $\left(x_i, x_i^+\right)$ and negative samples as pairs $\left(x_i, x_i^-\right)$. These pairs are then inputted to the pre-trained LaBSE model for sentence encoding. The LaBSE model utilizes the CLS vector to represent the semantics of the entire sequence $[Tok_1], [Tok_2], ..., [Tok_n]$.

Subsequently, the CLS vector is passed through a fully connected layer for fine-tuning, resulting in the sentence embeddings V_S for both positive and negative samples. To simplify the description, we refer to the module responsible for sentence feature extraction as M_0.

3.3 Auxiliary Network

The auxiliary network is divided into two parts: an encoder and a decoder, denoted as M_1 and M_2, respectively. The encoder M_1 consists of a 12-layer BERT model and a 1-layer fully connected network. The parameters of the BERT model are initialized with the pre-trained LaBSE model parameters. During the training process, the parameters are frozen to improve training efficiency and enhance the sentence embedding capability of the M_0 module. The sentences in the dataset are masked and inputted into M_1 to obtain the sentence embedding, V_A.

The decoder M_2 is composed of a BERT network, two fully connected layers, and a normalization layer. The BERT network in M_2 is reused from the BERT network in $M_0 - LaBSE$. Therefore, during the training of M_2, the model parameters of M_0 are also updated through back-propagation. The input of M_2, V_J, is a fusion of the output embedding vectors V_S from M_0 and the output embedding vector V_A from M_1, as shown in Eq. 1. This allows V_J to retain the information of the original sentence and the representation information of each token.

$$V_J = V_S + \lambda * V_A = CLSx_i + \lambda * (CLS, Token_1, Token_2, ...Token_n) \quad (1)$$

The output of M2 is the probability of each word in the masked sentence, and the prediction loss is calculated through cross-entropy in the multi-task learning module.

3.4 Multi-task Learning

The multi-task learning part involves jointly updating the losses for contrastive learning, Euclidean distance metric, and Masked Language Model (MLM).

Firstly, we have the contrastive learning loss. We obtain the output Vs from the M0 module and then calculate the batch-wise contrastive loss L_C using the contrastive loss function as shown in Eq. 2.

$$L_C = - \log \frac{e^{\cos(z_i, z_i^+)}}{\sum\limits_{i=1}^{N} e^{\cos(z_i, z_i^+)} + e^{\cos(z_i, z_i^-)}} \quad (2)$$

where z_i represents the sentence embedding vector obtained by encoding the input x_i through the M_0 network.

Next, we introduce the Euclidean distance metric loss. To enforce distinguishability among semantically similar sentences, we add the Euclidean distance loss on top of the contrastive loss. The loss is calculated using Eq. 3.

$$L_{EUC} = -\frac{1}{N} \bullet \sum_{i=1}^{N} \frac{e^{euc(z_i, z_i^+)}}{e^{euc(z_i, z_i^+)} + e^{euc(z_i, z_i^-)}} \tag{3}$$

Finally, we have the Masked Language Model loss. The input V_J is fed to the M_2 network to obtain the predicted probabilities $Prob_{MLM}$ for each masked word, and then the cross-entropy loss is computed using the masked word labels $Label_{MLM}$, as shown in Eq. 4.

$$L_{MLM} = crossEntropy(M_2(Detach(V_J)), Label_{MLM}) \tag{4}$$

Putting it all together, the final objective function L for optimizing the MTLAN model is calculated using Eq. 5.

$$L = L_C + \alpha * L_{EUC} + \lambda * L_{MLM} \tag{5}$$

4   Experiments

We conducted experiments and analysis on the MTLAN model using publicly available datasets such as Tatoeba [11], XNLI [16], and the SentEval evaluation tool. The experiments mainly include:

- Monolingual Sentence Similarity: We used the STS dataset as a test set to evaluate the model's effectiveness in representing sentences and performing inference-based sentence similarity tasks.
- Cross-lingual Sentence Retrieval: We utilized the Tatoeba dataset for cross-lingual sentence retrieval tasks to assess the model's ability to represent sentences across different languages.
- Sentence Embedding: We employed MTLAN for cross-lingual sentence encoding and visualized the results using t-SNE dimensionality reduction technique to evaluate the quality of the sentence representations.
- Ablation Study: We conducted ablation experiments by removing the multi-task learning module and the auxiliary network module separately to validate the effectiveness of the model.

4.1   Evaluation Indicators

In the task of sentence semantic similarity identification, we evaluated the results using the STS2012-STS2016 [17] dataset. The evaluation of English dataset similarity is performed based on the Spearman's rank correlation coefficient.

In the cross-language sentence retrieval experiment, we used accuracy and F1 score as evaluation metrics and compared them with other baseline models.

4.2 Monolingual Sentence Similarity

We use the MTLAN model to perform semantic similarity prediction tasks on the STS2012-2016 dataset and the STS-B and STS-R dataset, and compare it with the best-performing models in recent years. The experimental results are shown in Table 1.

Table 1. Monolingual (English) sentence similarity experiments, results in terms of Spearman's rank correlation coefficient (%)

	STS12	STS13	STS14	STS15	STS16	STS-B	SICK-R	Avg.
BERTbase [18] (Jacob D, et al. 2019)	39.7	59.38	49.67	66.03	66.19	53.87	62.06	56.7
ConSERTbase [19] (Yuan Y, et al. 2021)	64.64	78.49	69.07	79.72	75.95	73.97	67.31	72.74
SimCSE-BERTbase [20] (Gao T, et al. 2021)	68.4	82.41	74.38	80.91	78.56	76.85	72.23	76.25
InfoCSE-BERTlarge [21] (Wu X, et al. 2022)	71.89	**86.17**	77.72	86.2	81.29	83.16	74.84	80.18
LaBSE [13] (Fang F, et al. 2022)	65.08	67.98	64.03	76.59	72.98	69.66	69.74	69.44
PromptBERT [22] (Jiang T, et al. 2022)	75.48	85.59	80.57	85.99	81.08	84.56	**80.52**	81.97
miCSE [6] (Tassilo K, et al. 2023)	71.71	83.09	75.46	83.13	80.22	79.7	73.62	78.13
RankEncoder [23] (Yeon S, et al. 2023)	74.88	85.59	78.61	83.5	80.56	81.55	75.78	80.07
RankCSE_listNet [5] (Jiduan L, et al. 2023)	73.23	85.08	77.5	85.67	**82.99**	84.2	72.98	80.24
MTLAN (ours. 2023)	**82.12**	83.34	**80.94**	**87.11**	81.81	**84.58**	79.93	**82.83**

The MTLAN model achieved the best performance on the STS2012 dataset, obtaining a Spearman's rank correlation coefficient of 82.12% after inference. This represents a 6.64% improvement over PromptBERT on STS2012. The MTLAN model also outperformed other models on the STS2014, STS2015, and STS-B datasets. On average, the MTLAN model achieved the highest score of 82.83%. These results demonstrate that the MTLAN model excels in the task of single-sentence encoding, exhibiting both strong performance and stability.

4.3 Cross-Lingual Sentence Retrieval

Due to the inclusion of the LaBSE model in the feature extraction network M0, which has been trained on 109 languages, the MTLAN model is capable of transfer learning for other languages as well. To evaluate the cross-lingual sentence representation capabilities of the MTLAN model, tes is conducted on both the Tatoeba dataset.

To assess the cross-lingual transfer ability of the MT-LaBSE model, we selected languages with rich and low resources from the multilingual Tatoeba dataset. The languages with rich resources include Hindi (hin), French (fra), German (deu), Afrikaans (afr), and Swahili (swh). The low-resource languages include Telugu (tel), Tagalog (tgl), Irish (gle), Georgian (kat), and Amharic (amh). The results of multilingual sentence retrieval accuracy are presented in Table 2.

Table 2. Accuracy comparison of cross-lingual sentence retrieval tasks on Tatoeba dataset (%)

	hin	fra	deu	afr	tel	tgl	gle	kat	amh	swh
LaSER [10]	94.7	**95.7**	**99.0**	89.4	79.7	-/-	52.0	35.9	42.0	42.4
DuEAM [24]	92.9	-/-	96.0	84.8	90.6	60.6	42.0	76.4	56.0	-/-
mSimCSEen [20]	94.4	93.9	98.6	85.6	92.9	70.0	54.8	89.2	79.5	42.1
MTLAN (ours.)	**97.8**	**95.7**	**99.0**	**96.1**	**97.2**	**95.2**	**92.6**	**93.8**	**92.9**	**85.8**

The experimental results demonstrate that MTLAN outperforms other models in terms of accuracy across different languages. This is attributed to the strong multilingual understanding ability of the feature extraction module in MTLAN, as well as the augmentation of pseudo-training data that enhances the model's generalization across multiple languages. Furthermore, the pre-training model used in MTLAN, LaBSE, is smaller in size compared to the one used in mSimCSE. This means that while MTLAN achieves higher accuracy, it also maintains a more lightweight model architecture.

4.4 Sentence Embedding

To visually demonstrate the effectiveness of sentence embeddings, we selected the first 200 parallel sentences from an NLI dataset in English and Urdu. Using the MTLAN model, we performed cross-lingual sentence encoding and visualized the results using t-SNE dimensionality reduction, as shown in Fig. 3. The left plot represents the processing results using the pre-trained LaBSE model directly on English (en) and Urdu (ur), while the right plot represents the results using the MTLAN model.

From the visualization, it is evident that the MTLAN results exhibit a higher degree of overlap between sentences from different languages, indicating that the sentence embeddings produced by MTLAN are capable of bringing semantically similar sentences closer together.

4.5 Ablation Study

The results of the ablation experiments are shown in Fig. 4. The MTLAN model without the auxiliary network is denoted as "w/o auxiliary network," and the model without multi-task learning but only keeping the feature extraction module M0 is denoted as "w/o multi-task learning." These experiments were conducted on the STS2012-2016, STS-B, and SICK-R datasets.

From the results of the ablation experiments, we can observe that removing the auxiliary network leads to a slight decline in the model's performance. However, removing the multi-task learning module results in a significant degradation of the model's capability. This indicates that both the auxiliary network and the multi-task learning module play crucial roles in improving the performance of the

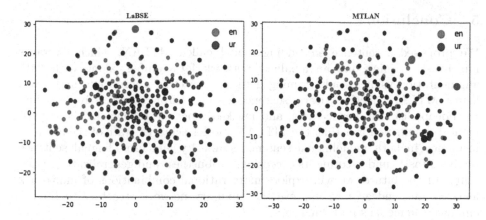

Fig. 3. Visualization of sentence embeddings. Left: LaBSE results, Right: MT-LaBSE results. The red dots correspond to the 20th parallel sentence pair in the selected NLI dataset for English and Urdu, the black dots correspond to the 110th pair, and the purple dots correspond to the 130th pair (Color figure online)

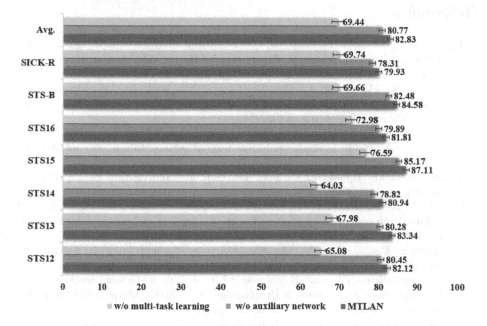

Fig. 4. Ablation study results of MTLAN model

MTLAN model. The results of the ablation experiments provide strong evidence for the effectiveness of the auxiliary network and multi-task learning module in the MTLAN model.

5   Conclusion

We proposed a sentence embedding model called MTLAN, which leverages multi-task learning and an auxiliary network for training, along with data augmentation using pseudo-training data. By selecting appropriate multi-task losses, MTLAN effectively learns sentence features through contrastive learning, Euclidean distance metric tasks, and masked language modeling. The experimental results confirm that the MTLAN model achieves state-of-the-art performance in both monolingual sentence similarity and cross-lingual sentence retrieval tasks, underscoring its exceptional robustness and generalization capability. In the future, we will explore more rational combinations of multi-task learning and endeavor to improve the auxiliary network structure to further enhance the model's performance.

Acknowledgments. This work is supported by Natural Science Foundation of Heilongjiang Province under grant number LH2021F015.

References

1. Pham, H., Luong, M.T., Manning, C.D.: Learning distributed representations for multilingual text sequences. In: Proceedings of the 1st Workshop on Vector Space Modeling for Natural Language Processing, pp. 88–94 (2015)
2. Qin, L., Ni, M., Zhang, Y., Che, W.: CoSDA-ML: multi-lingual code-switching data augmentation for zero-shot cross-lingual NLP, pp. 3853–3860 (2020). https://doi.org/10.24963/ijcai.2020/533
3. Conneau, A., Lample, G.: Cross-lingual language model pretraining 32 (2019)
4. Conneau, A., et al.: Unsupervised cross-lingual representation learning at scale, pp. 8440–8451 (2020)
5. Liu, J., et al.: RankCSE: Unsupervised sentence representations learning via learning to rank (2023)
6. Nie, Z., Zhang, R., Mao, Y.: On the inadequacy of optimizing alignment and uniformity in contrastive learning of sentence representations. In: The Eleventh International Conference on Learning Representations (2023)
7. Agirre, E., Cer, D., Diab, M., Gonzalez-Agirre, A.: Semeval-2012 task 6: a pilot on semantic textual similarity. In: *SEM 2012: The First Joint Conference on Lexical and Computational Semantics-Volume 1: Proceedings of the main conference and the shared task, and Volume 2: Proceedings of the Sixth International Workshop on Semantic Evaluation (SemEval 2012), pp. 385–393 (2012)
8. Cer, D., Diab, M., Agirre, E., Lopez-Gazpio, I., Specia, L.: Semeval-2017 task 1: Semantic textual similarity-multilingual and cross-lingual focused evaluation. arXiv preprint arXiv:1708.00055 (2017)
9. Marelli, M., Menini, S., Baroni, M., Bentivogli, L., Bernardi, R., Zamparelli, R., et al.: A sick cure for the evaluation of compositional distributional semantic models. In: LREC, pp. 216–223. Reykjavik (2014)
10. Artetxe, M., Schwenk, H.: Massively multilingual sentence embeddings for zero-shot cross-lingual transfer and beyond. Trans. Assoc. Comput. Linguist. **7**, 597–610 (2019)

11. Logeswaran, L., Lee, H.: An efficient framework for learning sentence representations (2018)
12. Conneau, A., Kiela, D., Schwenk, H., Barrault, L., Bordes, A.: Supervised learning of universal sentence representations from natural language inference data, pp. 670–680 (2017)
13. Feng, F., Yang, Y., Cer, D., Arivazhagan, N., Wang, W.: Language-agnostic BERT sentence embedding, pp. 878–891, May 2022
14. Zhang, Y., Yang, Q.: A survey on multi-task learning. IEEE Trans. Knowl. Data Eng. **34**(12), 5586–5609 (2021)
15. Lample, G., Conneau, A., Denoyer, L., Ranzato, M.: Unsupervised machine translation using monolingual corpora only (2018)
16. Conneau, A., et al.: XNLI: evaluating cross-lingual sentence representations. In: Proceedings of the 2018 Conference on Empirical Methods in Natural Language Processing, pp. 2475–2485 (2018)
17. Agirre, E., et al.: Semeval-2014 task 10: multilingual semantic textual similarity. In: Proceedings of the 8th International Workshop on Semantic Evaluation (SemEval 2014), pp. 81–91 (2014)
18. Devlin, J., Chang, M.W., Lee, K., Toutanova, K.: BERT: pre-training of deep bidirectional transformers for language understanding, pp. 4171–4186, June 2019
19. Yan, Y., Li, R., Wang, S., Zhang, F., Wu, W., Xu, W.: ConSERT: a contrastive framework for self-supervised sentence representation transfer, pp. 5065–5075, August 2021
20. Gao, T., Yao, X., Chen, D.: SimCSE: simple contrastive learning of sentence embeddings, pp. 6894–6910, November 2021
21. Wu, X., Gao, C., Lin, Z., Han, J., Wang, Z., Hu, S.: InfoCSE: information-aggregated contrastive learning of sentence embeddings, pp. 3060–3070, December 2022
22. Jiang, T., et al.: PromptBERT: improving BERT sentence embeddings with prompts, pp. 8826–8837, December 2022
23. Seonwoo, Y., et al.: Ranking-enhanced unsupervised sentence representation learning (2023)
24. Goswami, K., Dutta, S., Assem, H., Fransen, T., McCrae, J.P.: Cross-lingual sentence embedding using multi-task learning. In: Proceedings of the 2021 Conference on Empirical Methods in Natural Language Processing, pp. 9099–9113. Association for Computational Linguistics, Online and Punta Cana, Dominican Republic, November 2021

Correlated Online k-Nearest Neighbors Regressor Chain for Online Multi-output Regression

Zipeng Wu[1], Chu Kiong Loo[1], and Kitsuchart Pasupa[2](✉) ⓘ

[1] Faculty of Computer Science and Information Technology, University of Malaya, 50603 Kuala Lumpur, Malaysia
s2013980@siswa.um.edu.my, ckloo.um@um.edu.my
[2] School of Information Technology, King Mongkut's Institute of Technology Ladkrabang, Bangkok 10520, Thailand
kitsuchart@it.kmitl.ac.th

Abstract. Online multi-output regression is a crucial task in machine learning with applications in various domains such as environmental monitoring, energy efficiency prediction, and water quality prediction. This paper introduces CONNRC, a novel algorithm designed to address online multi-output regression challenges and provide accurate real-time predictions. CONNRC builds upon the k-nearest neighbor algorithm in an online manner and incorporates a relevant chain structure to effectively capture and utilize correlations among structured multi-outputs. The main contribution of this work lies in the potential of CONNRC to enhance the accuracy and efficiency of real-time predictions across diverse application domains. Through a comprehensive experimental evaluation on six real-world datasets, CONNRC is compared against five existing online regression algorithms. The consistent results highlight that CONNRC consistently outperforms the other algorithms in terms of average Mean Absolute Error, demonstrating its superior accuracy in multi-output regression tasks. However, the time performance of CONNRC requires further improvement, indicating an area for future research and optimization.

Keywords: Multi output regression · Online machine learning · k-Nearest Neighbors

1 Introduction

Online machine learning, characterized by its real-time learning and adaptive capabilities [14], is increasingly emerging as a critical component in a wide range of applications, including river flow prediction and water quality assessment. The development of robust and efficient online machine learning algorithms is important, considering their aptitude for handling continuous data streams and adapting to evolving data patterns.

B. Luo et al. (Eds.): ICONIP 2023, LNCS 14449, pp. 28–39, 2024.
https://doi.org/10.1007/978-981-99-8067-3_3

Despite the limited attention received by online multi-output regression, it is noteworthy that in an increasing number of applications, the need to predict multiple outputs rather than a single output is becoming more prevalent. Multi-output regression poses a greater challenge as it involves handling potential structural relationships between the outputs. Nevertheless, online multi-output regression holds significant potential in a variety of applications, e.g., Train Carriage Load prediction [22], water quality prediction [6].

Typically, multi-output regression can be categorized into global and local methods. Global methods output all output variables at once, while local methods often consist of multiple sub-models. However, existing multi-output regression algorithms are designed to learn a mapping from the input space to the output space on the entire training dataset, making them suitable for batch processing environments but challenging to apply in environments requiring online model updates.

Several online learning algorithms have been proposed [1,3,5,8,11,15,23] with Incremental Structured Output Prediction Tree (iSOUP-Tree) [15] and Passive-Aggressive (PA) [3] algorithms being notable examples. iSOUP-Tree is a tree-based method that simultaneously predicts all targets in multi-target regression tasks on data streams. On the other hand, PA is particularly suitable for single-output tasks and quickly adapts to data changes and updates model parameters analytically. It balances proximity to the current classifier and achieves a unit margin on the latest example. However, both algorithms face challenges in effectively capturing and exploiting correlations in multi-output regression tasks.

In this paper, we aim to address this problem by introducing a novel online multi-output regression algorithm called "Correlated Online k-Nearest Neighbors Regressor Chain" (CONNRC). Our proposed algorithm utilizes a maximum correlation chain structure to capture the associations among output variables while also leveraging the strengths of the k-nearest neighbors (kNN) algorithm.

The main contributions of this paper are as follows:

- We propose a novel online multi-output regression algorithm, CONNRC, which extends the kNN algorithm to handle online multi-output regression tasks. This algorithm leverages a maximum correlation chain structure to capture the association between output variables, thus addressing the challenge of handling potential structural relationships between the outputs in multi-output regression tasks.
- We comprehensively evaluate the proposed CONNRC algorithm on six real-world datasets. The evaluation includes a comparison with the existing online multi-output regression algorithm, iSOUP-Tree [15], and several classic online learning algorithms, including Hoeffding Adaptive Tree (HAT) [1], Adaptive Model Rules (AMRules) [5], Stochastic Gradient Trees (SGT) [8], and PA [3]. Our results demonstrate that CONNRC consistently outperforms other state-of-the-art algorithms in terms of average Mean Absolute Error (aMAE), signifying its superior accuracy in online multi-output regression tasks.

2   Related Work

The related work for this study can be broadly categorized into two main areas: online multi-output regression and classic online learning methods.

2.1   Online Multi-output Regression

Online multi-output regression is a crucial technique for modeling, predicting, and compressing multi-dimensional correlated data streams. The MORES method [11] dynamically learns the structure of regression coefficients and residual errors to improve prediction accuracy. It introduces three modified covariance matrices to extract necessary information from all seen data for training and sets different weights on samples to track the evolving characteristics of data streams. The iSOUP-Tree method [15] learns trees that predict all targets simultaneously. The iSOUP-optionTree extends the iSOUP-Tree through the use of option nodes and can be used as a base learner in ensemble approaches like online bagging and online random forest. The MORStreaming method [23] is another notable online multi-output regression method. It uses an instance-based model to make predictions and consists of two algorithms: an online algorithm based on topology networks to learn the instances and an online algorithm based on adaptive rules to learn the correlation between outputs automatically.

2.2   Classic Online Regression

Classic online learning methods have been extensively studied and applied in various domains. These methods typically focus on single-output regression tasks. The PA [3] is a well-known online learning algorithm for regression tasks. It is a margin-based online learning algorithm and has an analytical solution to update model parameters as new sample(s) arrive. The HAT [1] is a tree-based method for online regression. It uses an ADWIN concept-drift detector at each decision node to monitor possible changes in the data distribution. If a drift is detected in a node, an alternate tree begins to be induced in the background. When enough information is gathered, the node where the change was detected is swapped by its alternate tree. Lastly, SGT [8,12] is a tree-based method for regression. It directly minimizes a loss function to guide tree growth and update their predictions, differentiating it from other incremental tree learners that do not directly optimize the loss but a data impurity-related heuristic. AMRules [5] is an online regression algorithm that constructs rules with attribute conditions as antecedents and linear combinations of attributes as consequents. It employs a Page-Hinkley test to identify and adapt to changes in the data stream, and includes outlier detection to prevent model skewing due to anomalous examples.

3   Proposed Method

The Correlated Online k-Nearest Neighbors Regressor Chain (CONNRC) algorithm, inspired by the Regressor Chain (RC) [21] and its variants [13,17,19,20], is designed for online multi-output regression tasks. It uses a chain structure to leverage correlations among multiple output variables, improving prediction performance. While the kNN algorithm forms its basis, CONNRC enhances it by incorporating correlation information, addressing a limitation of traditional online kNN in online multi-output regression tasks.

Throughout this paper, we use the following notation: a vector of output variables $\boldsymbol{y} = (y_1, \ldots, y_t, \ldots, y_N)$, where $t = 1, \ldots, N$, with y_t representing the t-th sample and N being the total number of samples. $\mathbf{Y}$ denotes a matrix of output variables, $\mathbf{Y} = (\boldsymbol{y}_1, \ldots, \boldsymbol{y}_m)$, where $\boldsymbol{y}_i$ is the i-th vector of output variables and m is the total number of output variables. $\mathbf{X}$ represents the input feature space, which is a matrix with N samples and d feature variables. The correlation coefficient of $\boldsymbol{y}_i$ and $\boldsymbol{y}_j$ can be denoted as $c_{i,j}$, $c_{i,j} = \text{corr}(\boldsymbol{y}_i, \boldsymbol{y}_j)$. This is represented as $\mathbf{COE}_{m \times m} = \text{corr}(\mathbf{Y})$, where $\mathbf{COE}$ is the correlation matrix of size $m \times m$, with m being the number of output variables. By summing each row of $\boldsymbol{y}_i$, we obtain the cumulative correlation value c_i, which indicates the degree of association between $\boldsymbol{y}_i$ and all other variables in $\mathbf{Y}$. The correlation among output variables is calculated using the Spearman correlation coefficient, which has the advantage that there is no requirement for the distribution of the data, and the effect of the size of the absolute values can be ignored. $\mathbf{I}$ represents the index vector corresponding to c_l that is the order of meta-model in the chain structure. It is the descending rank of each output variable cumulative correlation value c_i, $\boldsymbol{I} = (I_1, \ldots, I_m)$. This information serves as a guide for the order in which the kNN models are updated, based on the assumption that the output variables with higher cumulative correlation are more informative and should be predicted first.

A kNN model is initialized for each output variable, and a window of recent data points is maintained. The maximum window size, w, is a parameter that can be adjusted based on the specific requirements of the task. The window always contains the most relevant data points for prediction, which is ensured by a mechanism that adds the current point to the window if its distance to the nearest neighbor is greater than a certain threshold, *min_distance_keep*, and removes the oldest point in the window if the window is full.

For each training example in the stream, $(\boldsymbol{x}, \boldsymbol{y})$, the kNN model for each output variable is updated. For the first output variable, the model kNN_1 is updated with the input features and the corresponding output value $(\boldsymbol{x}, y_1)$. For subsequent output variables, the model kNN_i is updated with the input features and the predicted values of the previous output variables $(\boldsymbol{x}, y_i)$.

After updating the model, the $k_{neighbors}$ nearest neighbors for the current point $(\boldsymbol{x}, y_i)$ are computed from the window. If the distance to the nearest neighbor is greater than a certain threshold, the current point is added to the window. If the window is full, the oldest point in the window is removed. This process ensures that the window always contains the most relevant data points for prediction. The pseudocode of CONNRC is illustrated in Algorithm 1.

Algorithm 1: Correlated Online k-Nearest Neighbors Regressor Chain

Data: Historical training dataset $D = (\mathbf{X}_{history}, \boldsymbol{y}_{history})$, and stream data $(\mathbf{X}, \boldsymbol{y})$

Result: Online Multi-output regression model $h_j, j = \{1, \ldots, m\}$

$\mathbf{COE}_{m \times m} = \mathrm{corr}(\mathbf{Y})$;

$c_l = \sum \mathbf{COE}(:, l), l = \{1, \ldots, m\}$;

$[\boldsymbol{c'}, \mathbf{I}] = \mathrm{rank}(c_l, \text{'decending'})$, where $\mathbf{I} = (I_1, \ldots, I_m)$;

for y_i *in* $\boldsymbol{y}_{1,m}$ **do**
  Let kNN_i be a kNN model for the output y_i;
  Initialize the window W for kNN_i with a maximum window size w;
end

for *all training examples* $(\boldsymbol{x}, \boldsymbol{y})$ *in stream* **do**
  if $i = 1$ **then**
    Update kNN_1 with $(\boldsymbol{x}, y_1)$;
  else
    $\boldsymbol{x} = \boldsymbol{x}_i \cup (y_1 \ldots y_{i-1})$;
    Update kNN_i with $(\boldsymbol{x}, y_i)$;
  end
  for *each* y_i *in* $\boldsymbol{y} = (y_{I_1}, \ldots, y_{I_m})$ **do**
    Compute the $k_{neighbors}$ nearest neighbors **for** $(\boldsymbol{x}, y_i)$ *from the window*
    do
      Calculating the Euclidean distance between $(\boldsymbol{x}, y_i)$ and all points in the window;
      Selecting the $k_{neighbors}$ points with the smallest distances;
    end
    if *the distance to the nearest neighbor* $> min\_distance\_keep$ **then**
      Add $(\boldsymbol{x}, y_i)$ to the window of kNN_i;
      if $\mathrm{size}(W) > w$ **then**
        Remove $W[0]$;
      end
    end
  end
end

In this algorithm, the historical dataset is denoted as $D = (\mathbf{X}_{history}, \boldsymbol{y}_{history})$. For the current experiment, we utilize the first 25% of the entire dataset as our training data.

The CONNRC algorithm has several advantages. First, it considers the correlation among multiple output variables by correlated chain structure, which can improve the prediction performance in multi-output regression tasks. Second, it is based on the simple yet effective kNN algorithm, which makes it easy to implement and understand. It is worth noting that CONNRC consists of two hyperparameters: the maximum window size (w) and the threshold for adding points to the window ($min\_distance\_keep$). These hyperparameters need to be tuned. Furthermore, the computation of the $k_{neighbors}$ nearest neighbors can be time-consuming, especially when the window size is large.

4 Experiment Framework

In this section, we present the experimental framework used to evaluate the performance of our proposed method, CONNRC, in comparison with several state-of-the-art online regression algorithms. All experiments were implemented in Python 3.9 and executed on a PC with an Intel Core i7 12700 processor (4.90 GHz) and 16 GB of RAM.

4.1 Datasets

Our proposed method, CONNRC, underwent evaluation on six real-world datasets, each exhibiting unique characteristics and posing specific challenges. These datasets were chosen based on their previous utilization in online multi-output regression research [4,15] and their role as benchmark datasets in the field of multi-output regression [21]. The datasets used in our evaluation are as follows:

- Water Quality Prediction [6]: It comprises 16 input attributes that are associated with physical and chemical water quality parameters. Additionally, it includes 14 target attributes that represent the relative presence of plant and animal species in Slovenian rivers.
- Supply Chain Management Prediction [9,16]: It is derived from the Trading Agent Competition in Supply Chain Management (TAC SCM) tournament from 2010. This consists of two sub-datasets. The dataset consists of 16 regression targets, where each target corresponds to either the mean price for the next day (SCM1D) [9] (as the first sub-dataset) or the mean price for a 20-day period in the future (SCM20D) [16] (as the second sub-dataset) for each product within the simulation.
- River flow Prediction [21]: This dataset consists of hourly flow observations collected from 8 sites within the Mississippi River network in the United States. The data was obtained from the US National Weather Service. Each row of the dataset contains the most recent observation for each of the eight sites and time-lagged observations from 6, 12, 18, 24, 36, 48, and 60 h in the past.
- Energy Building Prediction [18]: The focus of this dataset is the prediction of heating load and cooling load requirements for buildings. The prediction is based on eight building parameters, including glazing area, roof area, overall height, and other relevant factors.
- Sea Water Quality Prediction [10]: The Andromeda dataset focuses on predicting future values for six water quality variables, namely temperature, pH, conductivity, salinity, oxygen, and turbidity. The dataset pertains specifically to the Thermaikos Gulf of Thessaloniki in Greece. The predictions are made using a window of five days and a lead time of five days, indicating that the model aims to forecast the values of these variables five days ahead based on the observations from the previous five days.

These datasets represent a variety of application domains and provide a comprehensive basis for evaluating the performance of the proposed method.

4.2   Experiment Settings

To evaluate the performance of each multi-output method, we employed the pre-quential strategy introduced in [7]. In this strategy, a sample is used to update the model after it has been evaluated by this model. This approach allows us to simulate a real-time learning environment, which is a key characteristic of online machine learning. We also employed progressive validation, as suggested in [2]. This method is considered the canonical way to evaluate a model's performance, as it allows us to accurately assess how a model would have performed in a production scenario. In progressive validation, the dataset is transformed into a sequence of queries and responses. For each step, the model is tasked with predicting an observation or undergoing an update, and the samples are processed one after the other.

In our comparison, we assessed the performance of our proposed algorithm, CONNRC, in comparison to the state-of-the-art iSOUP-Tree, as well as several classic online learning algorithms: PA, AMRules, HAT, and SGT. It's important to note that our task involves multi-output regression, while these classical online learning algorithms were originally designed for single-output tasks. We trained separate models for each output variable to adapt them to the multi-output setting, denoting them with the prefix "MT-". The reason for choosing single-output classical online learning algorithms as baselines is that online multi-output regression currently receives little attention, and few open-source algorithms are available for this purpose. These baseline algorithms represent a diverse range of approaches to online regression, providing a comprehensive benchmark for evaluating the performance of our proposed method, CONNRC.

4.3   Performance Evaluation

To evaluate the overall performance of the model, we conducted performance evaluations at regular intervals, incrementally increasing the data sample by 25%. This approach allowed us to monitor the model's performance as the dataset size grew and gain a comprehensive understanding of its effectiveness across various sample sizes. In our evaluation, we considered several metrics to assess the performance of the model, including average Mean Absolute Error (aMAE), memory usage, and time. It is important to note that smaller values for all three metrics indicate better performance. The aMAE can be calculated by the following equation:

$$\text{aMAE} = \frac{1}{m} \sum_{m}^{t=1} \frac{1}{N} \sum_{N}^{i=1} \left| y_i^t - \widehat{y}_i^t \right|. \tag{1}$$

5   Experiment Results and Discussion

Figure 1 illustrates the accuracy performance of the algorithms based on the average Mean Absolute Error (aMAE). Our proposed method, CONNRC, consistently outperforms the other algorithms by achieving the lowest aMAE values across all sample sizes and datasets. The performance of CONNRC remains stable and promising at every regular checkpoint, whereas the performance of the other algorithms fluctuates. This underscores the effectiveness of CONNRC in capturing and leveraging the correlations among multiple outputs, resulting in enhanced prediction accuracy.

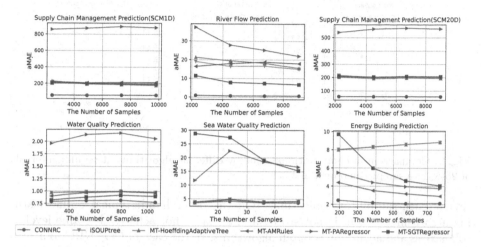

Fig. 1. Comparison of aMAE by different algorithms at regular intervals of 25% data sample incremental

While CONNRC demonstrates superior accuracy compared to other methods, it also exhibits the highest time consumption across all datasets, as depicted in Fig. 2. This can be attributed to the time complexity associated with searching for the nearest neighbors in the kNN algorithm, indicating a potential area for future improvement. Nevertheless, it is important to note that CONNRC still maintains reasonable efficiency in certain cases, suggesting a balance between accuracy and computational speed.

It is worth noting that although MT-PARegressor demonstrates the poorest accuracy in 4 out of 6 datasets, it stands out as the fastest algorithm. This highlights the potential trade-off between accuracy and computational efficiency when selecting an algorithm for specific applications.

Figure 3 illustrates the memory usage comparison among the algorithms. It is evident that MT-PARegressor exhibits the highest efficiency in terms of memory consumption across all datasets. In contrast, in most cases, CONNRC and MT-SGTRegressor tend to consume more memory. However, it is worth noting that CONNRC demonstrates reasonable efficiency, particularly in memory usage, on certain datasets, such as the River Flow dataset.

One key difference between the online kNN algorithm and other online learning algorithms lies in their approach to handling data. While both algorithms learn from each sample once, the kNN algorithm retains the data, allowing its performance to be comparable to batch algorithms that have the ability to

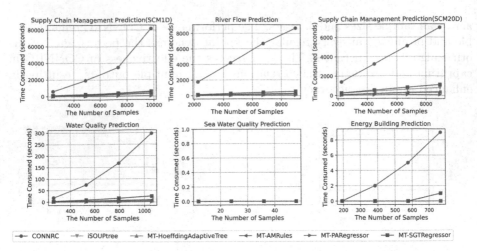

Fig. 2. Comparison of computational time by different algorithms at regular intervals of 25% data sample incremental

review each sample multiple times. On the other hand, typical online learning algorithms, which do not store data, often do not achieve the same level of robustness and accuracy as batch algorithms. In the case of CONNRC, our proposed method, we leverage the advantage of data retention inherited from kNN. Additionally, CONNRC incorporates a relevance chain structure to capture and exploit the correlations among structured multi-outputs. Moreover, the neighboring meta-models in CONNRC utilize the output of the preceding meta-model in the chain as latent information. Similar to knowledge distillation, this process

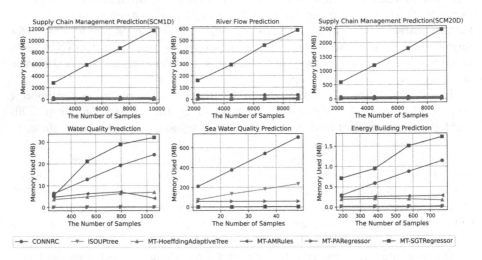

Fig. 3. Comparison of memory usage by different algorithms at regular intervals of 25% data sample incremental

enhances CONNRC's robustness against concept drift, allowing it to adapt to changing data patterns.

6   Conclusion and Future Work

This paper introduced and evaluated CONNRC, an online multi-output regression algorithm, using six different datasets. Our experimental results demonstrated that CONNRC consistently outperformed other algorithms in terms of accuracy, as measured by the aMAE metric. This indicates that CONNRC is capable of providing more accurate predictions in online multi-output regression tasks compared to the evaluated algorithms. However, it is important to note that the time performance of CONNRC was relatively slower compared to the other algorithms. This could be a potential limitation in scenarios where real-time predictions are crucial and time efficiency is a primary concern. While CONNRC demonstrated reasonable efficiency in terms of memory usage, it was not the most memory-consuming among the tested algorithms.

Despite these limitations, the superior accuracy performance of CONNRC underscores its potential for online multi-output regression tasks. The algorithm's ability to adapt to changes in the data stream and make accurate predictions is a significant advantage in many real-world applications, such as environmental monitoring, energy efficiency prediction, and water quality prediction.

For future work, we have plans to improve the efficiency of CONNRC, particularly its time performance. One potential approach is to enhance the nearest neighbor search in the kNN component of the algorithm, as it currently constitutes the most time-consuming part. Through optimization of this process, we aim to significantly reduce the prediction time, making CONNRC more suitable for real-time applications. Given the potential application scenarios involving massive streaming data, we will seek large datasets to validate the performance of our algorithms in such environments.

References

1. Bifet, A., Gavaldà, R.: Adaptive learning from evolving data streams. In: Adams, N.M., Robardet, C., Siebes, A., Boulicaut, J.-F. (eds.) IDA 2009. LNCS, vol. 5772, pp. 249–260. Springer, Heidelberg (2009). https://doi.org/10.1007/978-3-642-03915-7_22

2. Blum, A., Kalai, A., Langford, J.: Beating the hold-out: bounds for K-fold and progressive cross-validation. In: Ben-David, S., Long, P.M. (eds.) Proceedings of the Twelfth Annual Conference on Computational Learning Theory, COLT 1999, Santa Cruz, CA, USA, 7–9 July 1999, pp. 203–208. ACM (1999). https://doi.org/10.1145/307400.307439

3. Crammer, K., Dekel, O., Keshet, J., Shalev-Shwartz, S., Singer, Y.: Online passive-aggressive algorithms. J. Mach. Learn. Res. **7**, 551–585 (2006)

4. Duarte, J., Gama, J.: Multi-target regression from high-speed data streams with adaptive model rules. In: 2015 IEEE International Conference on Data Science

and Advanced Analytics, DSAA 2015, Campus des Cordeliers, Paris, France, 19–21 October 2015, pp. 1–10. IEEE (2015). https://doi.org/10.1109/DSAA.2015.7344900

5. Duarte, J., Gama, J., Bifet, A.: Adaptive model rules from high-speed data streams. ACM Trans. Knowl. Discov. Data 10(3), 30:1–30:22 (2016). https://doi.org/10.1145/2829955

6. Dzeroski, S., Demsar, D., Grbovic, J.: Predicting chemical parameters of river water quality from bioindicator data. Appl. Intell. 13(1), 7–17 (2000). https://doi.org/10.1023/A:1008323212047

7. Gama, J.: Knowledge Discovery from Data Streams. CRC Press (2010)

8. Gouk, H., Pfahringer, B., Frank, E.: Stochastic gradient trees. In: Lee, W.S., Suzuki, T. (eds.) Proceedings of The 11th Asian Conference on Machine Learning, ACML 2019, 17–19 November 2019, Nagoya, Japan. Proceedings of Machine Learning Research, vol. 101, pp. 1094–1109. PMLR (2019)

9. Groves, W., Gini, M.: Improving prediction in TAC SCM by integrating multivariate and temporal aspects via PLS regression. In: David, E., Robu, V., Shehory, O., Stein, S., Symeonidis, A. (eds.) AMEC/TADA -2011. LNBIP, vol. 119, pp. 28–43. Springer, Heidelberg (2013). https://doi.org/10.1007/978-3-642-34889-1_3

10. Hatzikos, E.V., Tsoumakas, G., Tzanis, G., Bassiliades, N., Vlahavas, I.P.: An empirical study on sea water quality prediction. Knowl. Based Syst. 21(6), 471–478 (2008). https://doi.org/10.1016/j.knosys.2008.03.005

11. Li, C., Wei, F., Dong, W., Wang, X., Liu, Q., Zhang, X.: Dynamic structure embedded online multiple-output regression for streaming data. IEEE Trans. Pattern Anal. Mach. Intell. 41(2), 323–336 (2019). https://doi.org/10.1109/TPAMI.2018.2794446

12. Mastelini, S.M., de Leon Ferreira de Carvalho, A.C.P.: Using dynamical quantization to perform split attempts in online tree regressors. Pattern Recognit. Lett. 145, 37–42 (2021). https://doi.org/10.1016/j.patrec.2021.01.033

13. Melki, G., Cano, A., Kecman, V., Ventura, S.: Multi-target support vector regression via correlation regressor chains. Inf. Sci. 415, 53–69 (2017). https://doi.org/10.1016/j.ins.2017.06.017

14. Montiel, J., et al.: River: machine learning for streaming data in Python. J. Mach. Learn. Res. 22, 110:1–110:8 (2021)

15. Osojnik, A., Panov, P., Dzeroski, S.: Tree-based methods for online multi-target regression. J. Intell. Inf. Syst. 50(2), 315–339 (2018). https://doi.org/10.1007/s10844-017-0462-7

16. Pardoe, D., Stone, P.: The 2007 TAC SCM prediction challenge. In: Ketter, W., La Poutré, H., Sadeh, N., Shehory, O., Walsh, W. (eds.) AMEC/TADA -2008. LNBIP, vol. 44, pp. 175–189. Springer, Heidelberg (2010). https://doi.org/10.1007/978-3-642-15237-5_13

17. Read, J., Martino, L.: Probabilistic regressor chains with monte Carlo methods. Neurocomputing 413, 471–486 (2020). https://doi.org/10.1016/j.neucom.2020.05.024

18. Tsanas, A., Xifara, A.: Accurate quantitative estimation of energy performance of residential buildings using statistical machine learning tools. Energy Build. 49, 560–567 (2012)

19. Wu, Z., Lian, G.: A novel dynamically adjusted regressor chain for taxi demand prediction. In: 2020 International Joint Conference on Neural Networks, IJCNN 2020, Glasgow, United Kingdom, 19–24 July 2020, pp. 1–10. IEEE (2020). https://doi.org/10.1109/IJCNN48605.2020.9207160

20. Wu, Z., Loo, C.K., Pasupa, K., Xu, L.: An interpretable multi-target regression method for hierarchical load forecasting. In: Tanveer, M., Agarwal, S., Ozawa, S., Ekbal, A., Jatowt, A. (eds.) Neural Information Processing - 29th International Conference, ICONIP 2022, Virtual Event, 22–26 November 2022, Proceedings, Part VII. CCIS, vol. 1794, pp. 3–12. Springer, Singapore (2022). https://doi.org/10.1007/978-981-99-1648-1_1

21. Xioufis, E.S., Tsoumakas, G., Groves, W., Vlahavas, I.P.: Multi-target regression via input space expansion: treating targets as inputs. Mach. Learn. **104**(1), 55–98 (2016). https://doi.org/10.1007/s10994-016-5546-z

22. Yu, H., Lu, J., Liu, A., Wang, B., Li, R., Zhang, G.: Real-time prediction system of train carriage load based on multi-stream fuzzy learning. IEEE Trans. Intell. Transp. Syst. **23**(9), 15155–15165 (2022). https://doi.org/10.1109/TITS.2021.3137446

23. Yu, H., Lu, J., Zhang, G.: MORStreaming: a multioutput regression system for streaming data. IEEE Trans. Syst. Man Cybern. Syst. **52**(8), 4862–4874 (2022). https://doi.org/10.1109/TSMC.2021.3102978

Evolutionary Computation for Berth Allocation Problems: A Survey

Xin-Xin Xu[1], Yi Jiang[2(✉)], Lei Zhang[3], Xun Liu[3], Xiang-Qian Ding[1], and Zhi-Hui Zhan[2(✉)]

[1] School of Computer Science and Technology, Ocean University of China, Qingdao 266100, China
[2] School of Computer Science and Engineering, South China University of Technology, Guangzhou 510006, China
vladimir_jiangyi@qq.com, zhanapollo@163.com
[3] Qingdao Port International Co., Ltd., Qingdao 266000, China

Abstract. Berth allocation problem (BAP) is to assign berthing spaces for incoming vessels while considering various constraints and objectives, which is an important optimization problem in port logistics. Evolutionary computation (EC) algorithms are a class of meta-heuristic optimization algorithms that mimic the process of natural evolution and swarm intelligence behaviors to generate and evolve potential solutions to optimization problems. Due to the advantages of strong gobal search capability and robustness, the EC algorithms have gained significant attention in many research fields. In recent years, many studies have successfully applied EC algorithms in solving BAPs and achieved encouraging performance. This paper aims to survey the existing literature on the EC algorithms for solving BAPs. First, this survey introduces two common models of BAPs, which are continuous BAP and discrete BAP. Second, this paper introduces three typical EC algorithms (including genetic algorithm, particle swarm optimization, and ant colony optimization) and analyzes the existing studies of using these EC algorithms to solve BAPs. Finally, this paper analyzes the future research directions of the EC algorithms in solving BAPs.

Keywords: Berth allocation problem · evolutionary computation · genetic algorithm · particle swarm optimization · ant colony optimization

1 Introduction

Berth allocation problem (BAP) has been the subject of extensive research due to its vital role in improving smart port logistics and increasing efficiency. Berth allocation is the process of providing berthing places to arriving boats while taking into account various constraints and objectives. For example, minimizing vessel wait times, minimizing service times, and avoiding congestion. The BAPs often involve multiple objectives, constraints, and dynamic factors, making it challenging to find an optimal solution using traditional optimization methods. At present, researchers have explored various optimization techniques, in order to effectively tackle the more challenging BAPs [1–3].

© The Author(s), under exclusive license to Springer Nature Singapore Pte Ltd. 2024
B. Luo et al. (Eds.): ICONIP 2023, LNCS 14449, pp. 40–51, 2024.
https://doi.org/10.1007/978-981-99-8067-3_4

Evolutionary computation (EC) refers to a set of meta-heuristic optimization techniques inspired by principles of natural evolution and the swarm intelligence behaviors of animals [4–6]. These techniques mimic the process of natural selection to generate and evolve potential solutions to optimization problems. The EC technique generally includes two branches, evolutionary algorithms and swarm intelligence. The typical algorithms of evolutionary algorithms are genetic algorithm (GA) [7–10] and differential evolution (DE) [11–14], while the typical algorithms of swarm intelligence are particle swarm optimization (PSO) [15–18] and ant colony optimization (ACO) [19–21]. EC algorithms generally have good global search ability, robustness to uncertainty, and are well-suited for solving complex and nonlinear optimization problems. Nowadays, multiple EC algorithms have achieved considerable success in complex optimization [22–24] and real-world optimization problems [25, 26].

Due to their ability to explore a wide search space and handle uncertainty in diverse problems, the EC algorithms are able to effectively address the challenging BAPs. Therefore, in recent years, EC has gained significant attention in the context of berth allocation, offering promising approaches to address the complexities and uncertainties associated with this problem domain [27, 28].

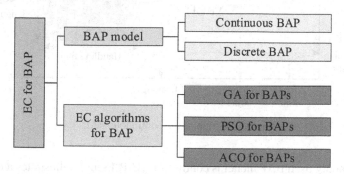

Fig. 1. The architecture of existing studies of the EC for BAP.

In order to present the success achieved by the EC algorithm in solving BAPs and to discuss how to implement a well-performing EC algorithm to effectively handle the more challenging BAPs, this paper presents a comprehensive survey of EC for BAPs. The diagram of the studies of the EC for BAP is shown in Fig. 1, and the outline of this paper follows this diagram. First, this survey describes two commonly-used BAP models, which are continuous BAPs and discrete BAPs. Second, we introduce three typical EC algorithms, which are GA, PSO, and ACO. Then, this survey analyzes the existing studies of using these three EC algorithms to solve BAPs. Finally, we analyze the future research directions of the EC algorithm in solving BAPs.

The rest of this paper is organized as follows: Section 2 gives the description of two commonly used BAP models, including the continuous BAPs and the discrete BAPs. Section 3 introduces three typical EC algorithms, i.e., GA, PSO, and ACO, and analyzes the related studies of these EC algorithms for BAPs. Section 4 discusses several potential future research directions. Section 5 draws the conclusion.

2 BAP Models

2.1 Continuous BAPs

The BAP's complexity and practical importance have prompted researchers to investigate numerous optimization strategies. The BAP's major objective is to optimize the allocation of berths to vessels while taking into account various objectives and constraints in order to ensure efficient port operations and improve overall performance.

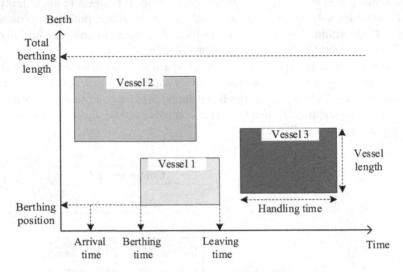

Fig. 2. Diagram of the continuous BAPs.

A commonly used BAP model is continuous BAP. Figure 2 illustrates a diagram of the continuous BAP with three vessels. In continuous BAP, the vessels can be placed arbitrarily within the total berthing length subject to collision avoidance constraints. To be more specific, a general continuous BAP model can be simply described based on the problem parameters, decision variable, objective function, and constraints as follows:

Problem Parameters:

N: Number of vessels.
TBL: Total berthing length.
$AT_1, AT_2, ..., AT_N$: Arrival time of the N vessels.
$HT_1, HT_2, ..., HT_N$: Handling time of the N vessels.
$VL_1, VL_2, ..., VL_N$: Vessel length of the N vessels.

Decision Variables:

$P_1, P_2, ..., P_N$: Berthing position of the N vessels.
$BT_1, BT_2, ..., BT_N$: Berthing time of the N vessels.

Objective Function: In order to meet the requirement of the common BAP model, minimizing the total waiting time of all the vessels is adopted as the objective function, which is shown as

$$\min f = \min \sum_{i=1}^{N} (BT_i - AT_i) \tag{1}$$

Constraints: All the vessels should be located within the total berth length:

$$P_i + VL_i \le N, \ i = 1, 2, ..., N \tag{2}$$

Every vessel can not overlap with other vessels:

$$Overlap\_area = 0 \tag{3}$$

2.2 Discrete BAPs

Another commonly used BAP model is discrete BAP. Figure 3 illustrates a diagram of the discrete BAP with three vessels and two berths. In discrete BAP, the port is divided into several discrete berths, and each vessel can only be assigned to a suitable berth. To be more specific, a general discrete BAP model can be simply described based on the problem parameters, decision variable, objective function, and constraints as follows:

Problem Parameters:

N: Number of vessels.
B: Number of berths.
$BL_1, BL_2, ..., BL_B$: Berth length of the berths.
$AT_1, AT_2, ..., AT_N$: Arrival time of the N vessels.
$HT_1, HT_2, ..., HT_N$: Handling time of the N vessels.$VL_1, VL_2, ..., VL_N$: Vessel length of the N vessels.

Decision Variables:

$IB_1, IB_2, ..., IB_N$: Index of the assigned berth of the N vessels.
$BT_1, BT_2, ..., BT_N$: Berthing time of the N vessels.

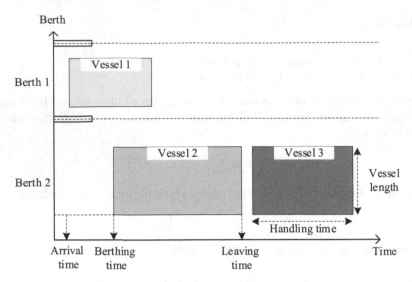

Fig. 3. Diagram of the discrete BAP.

Objective Function: In order to meet the requirement of the common BAP model, minimizing the total waiting time of all the vessels is adopted as the objective function, which is shown as

$$\min f = \min \sum_{i=1}^{N} (BT_i - AT_i) \tag{4}$$

Constraints: The vessel length should be smaller than the length of the assigned berth:

$$\forall IB_i = k, VL_i \leq BL_k \tag{5}$$

A vessel can not be assigned to a berth that is already occupied by another vessel:

$$\forall IB_i = IB_j, \ BT_i + HT_i \leq BT_j \lor BT_i \geq BT_j + HT_j \tag{6}$$

3   EC Algorithms for BAPs

The EC algorithms are able to effectively address the challenging BAPs due to their ability to explore a wide search space and handle uncertainty in diverse problems. Therefore, in recent years, EC has gained significant attention in the context of berth allocation, offering promising approaches to address the complexities and uncertainties associated with this problem domain. Based on the classes of the EC algorithms for solving BAPs, the existing related works can be classified into three taxonomies: GA for BAPs, PSO for BAPs, and ACO for BAPs. The following part of this section first describes the general process of GA, PSO, and ACO, and then introduces the existing studies using these three EC algorithms for solving BAPs.

3.1   GA for BAPs

GA is a population-based metaheuristic optimization method that searches for optimum or near-optimal solutions to complicated problems by mimicking the process of genetic evolution [7]. GA maintains a population containing a population of individuals, each of which represents a potential solution to the optimization problem. The general process of GA includes four operations, which are initialization, selection, crossover, and mutation. At the beginning of the GA, the initialization process is executed. Commonly, the uniform initialization strategy is adopted to uniformly place the individuals into the search space as:

$$x_d = \text{rand}() \times (U_d - L_d) + L_d \tag{7}$$

where x_d indicates the dth dimension of the individual x, U_d and L_d indicate the upper bound and lower bound search space of the dth dimension, respectively.

After initialization, the selection, crossover, and mutation operations are executed iteratively to search for the global optimum. Firstly, the selection operation is carried out to select the parental individuals for crossover. The goal of the selection operation is to select elite individuals as parents in order to generate better offspring. The commonly-used selection strategies include tournament selection [29], roulette selection [30], and a recent-proposed promote selection [31]. Then, the crossover operation is carried out on each two parental individuals to generate offspring. After that, the mutation operation is executed on the generated offspring to enhance the exploration ability of the off-spring. Simulated binary crossover [32] and polynomial mutation [33] are respectively a commonly-used crossover strategy and a commonly-used mutation strategy.

Because of its capacity to handle broad search space and multiple objectives [34–36], the GA has been widely used for BAPs. The advantages of GA for BAP mainly exist in solving continuous BAP models, as the GA has good search capacity in continuous optimization problems. For example, Ganji et al. [37] discussed the disadvantages of the traditional optimization algorithm on large-scale BAPs and adopted a GA to solve the BAPs. In experimental verification, the GA showed promising performance on both small-scale and large-scale BAPs. Chen et al. [38] focused on the study of the continuous BAP model and designed a GA to solve the continuous BAP. Li et al. [39] established a mixed integer nonlinear programming BAP model and proposed a genetic-harmony search algorithm for this model. The proposed genetic-harmony search algorithm combined the advantages of GA and the harmony search algorithm, which enhanced the effectiveness of solving BAPs. Hu et al. [40] modified the GA and proposed an improved GA for BAPs. In this improved GA, the prior experience of the BAP is used to help better generate the initialization solution. In the experimental studies, the improved GA showed more encouraging performance than the traditional method of solving BAPs. Ji et al. [41] modeled the continuous BAP into a bi-objective optimization problem and designed an enhanced non-dominated sorting GA II to solve the bi-objective BAP. Experimental results indicated that the enhanced non-dominated sorting GA II can effectively solve the bi-objective BAP and show generally better performance than the five compared algorithms.

3.2 PSO for BAPs

The idea of PSO is imitating the search for food by birds to make the solutions gradually approach the global optimum [15]. PSO maintains a population containing multiple particles. Each particle has three properties, which are the position, the velocity, and the personal best position. Herein, the position indicates the potential solution, the velocity means the searching direction of the particle, and the personal best position represents the best position that the particle has visited. Also, the whole population has a global best position to indicate the best position visited by all the particles.

The general process of PSO includes five operations, which are initialization, velocity update, position update, personal best update, and global best update. First, the initialization is executed to initialize the position and the velocity of each particle. Then, in the main loop of PSO, the velocity update is executed, whose mathematic formula is shown as:

$$v_i = w \cdot v_i + r_1 \cdot c_1 \cdot (pb_i - x_i) + r_2 \cdot c_2 \cdot (gb - x_i) \tag{8}$$

where v_i and x_i respectively denote the velocity and the position of the ith particle; r_1 and r_2 are two random numbers from $[0, 1]$; pb_i and gb denote the personal best position of the ith particle and the global best position of the whole swarm, respectively. c_1 and c_2 are two user-defined parameters. Then, the position is updated in the position update process via

$$x_i = x_i + v_i \tag{9}$$

In the personal best update process and global best update process, the personal best position of each particle and the global best position are updated if the current particle reaches a better position than the personal best position or the global best position.

Also, the PSO has achieved promising performance in solving BAPs, as the PSO has the advantages of fast convergence and ease of implementation. The advantages of PSO for BAP mainly exist in solving continuous BAP models, as the PSO has a fast convergence speed in continuous optimization problems. For example, Tengecha et al. [42] considered the constraints of worker performance and truck deployment and designed a new BAP model. To effectively solve the new BAP model, a PSO algorithm with a local search strategy was proposed. In the experiments, the real-world application scenario in the Dar es Salaam port is adopted as the test scenario of the proposed PSO variant. The experimental results indicate the encouraging performance of the PSO. Aiming to minimize the berth time of the ships in the port, Zhu et al. [43] proposed a PSO-based heuristic algorithm. Yang et al. [44] proposed an improved coded PSO to effectively minimize vessel berthing time. Experimental results indicate the improved coded PSO is effective.

3.3 ACO for BAPs

The idea of ACO is imitating the search for food process of ants [19]. ACO maintains a population containing multiple ants. The ants communicate through pheromone trails to find the shortest paths between their nest and food sources.

The general process of PSO includes three operations, which are initialization, solution construction, and pheromone update. ACO begins by initializing a colony of virtual ants and the pheromone information. In initialization, randomly put at a beginning point during startup, which in the context of the BAP can represent an unallocated vessel or a potential berth assignment. Then, the solution construction process is carried out to generate a solution for each ant. Specifically, each ant constructs a solution iteratively by building a sequence of berth assignments for the vessels. At each step, an ant chooses the next berth assignment based on a probabilistic rule that combines pheromone trails and heuristic information. The commonly-used state transition rule of the ants on the point i to select the next point j is shown as:

$$
j \in J = \begin{cases} \arg \max \left[\tau_{i,j} \times \left(\eta_{i,j} \right)^{\beta} \right], & \text{if rand}() \leq q_0 \\ \text{roulette wheel selection}, & \text{otherwise} \end{cases} \tag{10}
$$

where $\tau_{i,j}$ and $\eta_{i,j}$ respectively indicate the pheromone and the heuristic information between point i and point j. q_0 is a user-defined parameter. After solution construction, the pheromone update is carried out. The pheromone update includes the local pheromone update and the global pheromone update. The local pheromone update is carried out on each ant to update the pheromone of all the visited paths by this ant. The global pheromone update is carried out on the global best ant to update the pheromone of the paths visited by the best ant.

Due to its features, the ACO is widely adopted in solving discrete BAPs. In the context of the BAP, the ACO algorithm can be adapted to handle various constraints, dynamic environments, and uncertainty by incorporating problem-specific heuristics and adaptive mechanisms. For example, Wang et al. [45] proposed an adaptive ant colony system (AACS) to effectively deal with the discrete BAP with two berths. In the AACS, three well-performing strategies were combined to enhance the algorithm's performance. The three strategies are adaptive heuristic information strategy, variable-range receding horizon control strategy, and partial solution memory strategy. The adaptive heuristic information strategy adjusts the heuristic information based on the feedback of the optimizing results. The variable-range receding horizon control strategy divides the whole BAP into several subproblems in order to reduce computational complexity. The partial solution memory strategy was proposed to speed up the convergence of the AACS. Based on the three well-performing strategies, the AACS has shown promising performance in the experiments. Also, Li et al. [46] proposed a berth and yard collaboration allocation model and designed a cascade multi-stage integrated heuristic algorithm to solve this problem. This cascade multi-stage integrated heuristic algorithm combined the advantages of the tabu search and ACO. The experimental results have shown that the proposed algorithm was effective.

4   Potential Research Avenue

In the previous section of this paper, the existing studies of the EC algorithms for BAPs are reviewed. EC algorithms have achieved great success in the research field of BAP. Also, the EC algorithms still have great potential and can show promising performance in future research of BAPs. This section gives three future research directions of the EC algorithms for BAPs from three different aspects, including the BAP modeling aspect, the algorithm designing aspect,and the properties extending aspect.

4.1 BAP Modeling

In the existing research of EC algorithms for BAPs, the BAP models are often remote to the real-world BAPs [47, 48]. For example, the real-world BAP usually has many objectives and constraints, and the coming sequence of the vessels can dynamically change. Therefore, to further promote the development of the research on BAP, designing a more accurate model that is close to the real-world scenario is important. Also, in real-world applications, the algorithms should consider the berth allocation in multiple ports. As considering multiple ports or multi-port berth allocation scenarios introduces new challenges, researchers should consider inter-port transportation, transshipment, and coordination aspects.

4.2 Algorithm Designing

Improving the design and performance of EC algorithms for BAPs is a future research avenue with good potential. For example, researchers can investigate innovative evolutionary operators, adaptive parameter control methods, and population initialization procedures that are suited to the scenarios of the BAP. The efficiency and efficacy of EC algorithms in discovering high-quality solutions may be considerably enhanced by creating problem-specific operators and strategies. Furthermore, the integration of EC algorithms with other metaheuristics or hybrid techniques has the potential to yield excellent research results. Hybrid algorithms can make use of the complementary qualities of several techniques, resulting in improved exploration and exploitation capabilities for dealing with complicated BAP cases.

4.3 Properties Extending

In the existing research of EC algorithms for BAPs, the single-objective and simple BAP model is mainly adopted. However, the BAP in the real-world scenario can be more complex. Therefore, including complicated and dynamic restrictions in the BAP model, such as vessel compatibility, handling time changes, environmental conditions, and berth availability fluctuations, is one prospective research area. Another fascinating research direction is the investigation of the multi-objective BAP model. Multiple competing objectives can be handled efficiently using EC algorithms. Researchers can look at Pareto-based techniques to balance these objectives and identify trade-offs between competing aims, resulting in more robust and adaptable berth allocation systems.

5 Conclusion

This paper provides a comprehensive survey of the existing studies on EC algorithms for BAPs. Specifically, first, two commonly-used mathematical models of the BAPs are introduced. These two BAP models are the continuous BAPs and the discrete BAPs. In continuous BAP, the vessels can be placed arbitrarily within the total berthing length subject to collision avoidance constraints. In discrete BAP, the port is divided into several discrete berths, and each vessel can only be assigned to a suitable berth. Then, three representative EC algorithms, i.e., GA, PSO, and ACO are introduced. According to the types of EC algorithms, the existing studies on EC for BAP are introduced based on three classes, which are the studies of GA for BAP, the studies of PSO for BAP, and the studies of ACO for BAP. Finally, we analyze several future research directions of the EC algorithm for BAPs. We hope this paper can highlight the research directions and inspire the researchers.

Acknowledgment. This work was supported in part by the National Key Research and Development Program of China under Grant 2020YFB1710803.

References

1. Rodrigues, F., Agra, A.: Berth allocation and quay crane assignment/scheduling problem under uncertainty: a survey. Eur. J. Oper. Res. **303**(2), 501–524 (2022)
2. Martin-Iradi, B., Pacino, D., Ropke, S.: The multiport berth allocation problem with speed optimization: exact methods and a cooperative game analysis. Transp. Sci. **56**(4), 972–999 (2022). https://doi.org/10.1287/trsc.2021.1112
3. Yin, D., Niu, Y., Yang, J., Yu, S.: Static and discrete berth allocation for large-scale marine-loading problem by using iterative variable grouping genetic algorithm. J. Marine Sci. Eng. **10**(9), 1294 (2022)
4. Zhan, Z.H., Shi, L., Tan, K.C., Zhang, J.: A survey on evolutionary computation for complex continuous optimization. Artif. Intell. Rev. **55**(1), 59–110 (2022)
5. Zhan, Z.H., et al.: Matrix-based evolutionary computation. IEEE Trans. Emerg. Top. Comput. Intell. **6**(2), 315–328 (2022)
6. Chen, Z.G., Zhan, Z.H., Kwong, S., Zhang, J.: Evolutionary computation for intelligent transportation in smart cities: a survey. IEEE Comput. Intell. Mag. **17**(2), 83–102 (2022)
7. Holland, J.: Genetic algorithm. Sci. Am. **267**(1), 66–83 (1992)
8. Liu, S., Chen, Z., Zhan, Z.H., Jeon, S., Kwong, S., Zhang, J.: Many-objective job shop scheduling: a multiple populations for multiple objectives-based genetic algorithm approach. IEEE Trans. Cybern. **53**(3), 1460–1474 (2023)
9. Wu, S.H., Zhan, Z.H., Zhang, J.: SAFE: scale-adaptive fitness evaluation method for expensive optimization problems. IEEE Trans. Evol. Comput. **25**(3), 478–491 (2021)
10. Jiang, Y., Zhan, Zi.-H., Tan, K.C., Zhang, Jun: A bi-objective knowledge transfer framework for evolutionary many-task optimization. IEEE Trans. Evol. Comput. **27**(5), 1514–1528 (2023). https://doi.org/10.1109/TEVC.2022.3210783
11. Storn, R., Price, K.: Differential evolution–a simple and efficient heuristic for global optimization over continuous spaces. J. Global Optim. **11**, 341–359 (1997)
12. Jiang, Y., Chen, C., Zhan, Z.H., Li, Y., Zhang, J.: Adversarial differential evolution for multimodal optimization problems, In: Proceedings of IEEE Conference on Evolutionary Computation, pp. 1–8. (2022)
13. Zhan, Z.H., Wang, Z.J., Jin, H., Zhang, J.: Adaptive distributed differential evolution. IEEE Trans. Cybern. **50**(11), 4633–4647 (2020)
14. Jiang, Y., Zhan, Z.H., Tan, K., Zhang, J.: Knowledge learning for evolutionary computation. IEEE Trans. Evol. Comput. (2023).https://doi.org/10.1109/TEVC.2023.3278132

15. Kennedy, J., Eberhart, R.: Particle swarm optimization. In: Proceedings of IEEE International Conference on Neural Networks, pp. 1942–1948 (1995)
16. Jian, J., Chen, Z., Zhan, Z.H., Zhang, J.: Region encoding helps evolutionary computation evolve faster: a new solution encoding scheme in particle swarm for large-scale optimization. IEEE Trans. Evol. Comput. **25**(4), 779–793 (2021)
17. Liu, X.F., Zhan, Z.H., Gao, Y., Zhang, J., Kwong, S., Zhang, J.: Coevolutionary particle swarm optimization with bottleneck objective learning strategy for many-objective optimization. IEEE Trans. Evol. Comput. **23**(4), 587–602 (2019)
18. Liu, X.-F., Fang, Y., Zhan, Z.-H., Zhang, J.: Strength learning particle swarm optimization for multiobjective multirobot task scheduling. IEEE Trans. Syst. Man Cybern. Syst. **53**(7), 4052–4063 (2023). https://doi.org/10.1109/TSMC.2023.3239953
19. Dorigo, M., Gambardella, L.: Ant colony system: A cooperative learning approach to the traveling salesman problem. IEEE Trans. Evol. Comput. **1**(1), 53–66 (1997)
20. Wu, L., Shi, L., Zhan, Z.H., Lai, K., Zhang, J.: A buffer-based ant colony system approach for dynamic cold chain logistics scheduling. IEEE Trans. Emerg. Top. Comput. Intell. **6**(6), 1438–1452 (2022)
21. Shi, L., Zhan, Z.H., Liang, D., Zhang, J.: Memory-based ant colony system approach for multi-source data associated dynamic electric vehicle dispatch optimization. IEEE Trans. Intell. Transp. Syst. **23**(10), 17491–17505 (2022)
22. Jiang, Y., Zhan, Z.H., Tan, K., Zhang, J.: Optimizing niche center for multimodal optimization problems. IEEE Trans. Cybern. **53**(4), 2544–2557 (2023)
23. Li, J.Y., Zhan, Z.H., Zhang, J.: Evolutionary computation for expensive optimization: a survey. Mach. Intell. Res. **19**(1), 3–23 (2022)
24. Jiang, Y., Zhan, Z.H., Tan, K., Zhang, J.: Block-level knowledge transfer for evolutionary multi-task optimization. IEEE Trans. Cybern. (2023). https://doi.org/10.1109/TCYB.2023.3273625
25. Wang, C., et al.: A scheme library-based ant colony optimization with 2-opt local search for dynamic traveling salesman problem. Comput. Model. Eng. Sci. **135**(2), 1209–1228 (2022)
26. Zhan, Z.H., Li, J.Y., Zhang, J.: Evolutionary deep learning: a survey. Neurocomputing **483**, 42–58 (2022)
27. Karafa, J., Golias, M., Ivey, S., Saharidis, G., Leonardos, N.: The berth allocation problem with stochastic vessel handling times. Int. J. Adv. Manuf. Technol. **65**, 473–484 (2013)
28. Cheong, C., Tan, K., Liu, D., Lin, C.: Multi-objective and prioritized berth allocation in container ports. Ann. Oper. Res. **180**(1), 63–103 (2010)
29. Boeh, R., Hanne, T., Dornberger, R.: A comparison of linear rank and tournament for parent selection in a genetic algorithm solving a dynamic travelling salesman problem, In: Proceedings of International Conference on Soft Computing and Machine Intelligence, pp. 97–102. (2022)
30. Zhu, Y., Yang, Q., Gao, X., Lu, Z.: A ranking weight based roulette wheel selection method for comprehensive learning particle swarm optimization, In: Proceedings of IEEE International Conference on Systems, Man, and Cybernetics, pp. 1–7 (2022)
31. Chen, J.C., Cao, M., Zhan, Z.H., Liu, D., Zhang, J.: A new and efficient genetic algorithm with promotion selection operator. In: Proceedings of IEEE International Conference on Systems, Man, and Cybernetics, pp. 1532–1537 (2020)
32. Agrawal, R., Deb, K., Agrawal, R.: Simulated binary crossover for continuous search space. Complex Syst. **9**(2), 115–148 (1995)
33. Deb, K., Goyal, M.: A combined genetic adaptive search (GeneAS) for engineering design. Comput. Sci. Inform. **26**(4), 30–45 (1996)
34. Li, J.Y., Zhan, Z.H., Li, Y., Zhang, J.: Multiple tasks for multiple objectives: a new multiobjective optimization method via multitask optimization. IEEE Trans. Evol. Comput. (2023). https://doi.org/10.1109/TEVC.2023.3294307

35. Yang, Q.T., Zhan, Z.H., Kwong, S., Zhang, J.: Multiple populations for multiple objectives framework with bias sorting for many-objective optimization. IEEE Trans. Evol. Comput. (2022). https://doi.org/10.1109/TEVC.2022.3212058
36. Zhan, Z.H., Li, J.Y., Kwong, S., Zhang, J.: Learning-aid evolution for optimization. IEEE Trans. Evol. Comput. (2022). https://doi.org/10.1109/TEVC.2022.3232776
37. Ganji, S., Babazadeh, A., Arabshahi, N.: Analysis of the continuous berth allocation problem in container ports using a genetic algorithm. J. Mar. Sci. Technol. 15(4), 408–416 (2010)
38. Chen, L., Huang, Y.: A dynamic continuous berth allocation method based on genetic algorithm. In: Proceedings of IEEE International Conference on Control Science and Systems Engineering, pp. 770–773 (2017)
39. Li, S., Li, G., Zhu, Y.: Research on continuous berth allocation problem based on genetic-harmony search algorithm. IOP Conf. Ser. Mater. Sci. Eng. 782(3), 032071 (2020)
40. Hu, X., Ji, S., Hua, H., Zhou, B., Hu, G.: An improved genetic algorithm for berth scheduling at bulk terminal. Comput. Syst. Sci. Eng. 43(3), 1285–1296 (2022)
41. Ji, B., Huang, H., Yu, S.: An enhanced NSGA-II for solving berth allocation and quay crane assignment problem with stochastic arrival times. IEEE Trans. Intell. Transp. Syst. 24(1), 459–473 (2023)
42. Tengecha, N., Zhang, X.: An efficient algorithm for the berth and quay crane assignments considering operator performance in container terminal using particle swarm model. J. Marine Sci. Eng. 10(9), 1232 (2022)
43. Zhu, S., Tan, Z., Yang, Z., Cai, L.: Quay crane and yard truck dual-cycle scheduling with mixed storage strategy. Adv. Eng. Inform. 54, 101722 (2022)
44. Yang, Y., Yu, H., Zhu, X.: Study of the master bay plan problem based on a twin 40-foot quay crane operation. J. Marine Sci. Eng. 11(4), 807 (2023)
45. Wang, R., et al.: An adaptive ant colony system based on variable range receding horizon control for berth allocation problem. IEEE Trans. Intell. Transp. Syst. 23(11), 21675–21686 (2022)
46. Li, B., Jiang, X.: A Joint operational scheme of berths and yards at container terminals with computational logistics and computational intelligence, In: Proceedings of IEEE Int. Conference on Computer Supported Cooperative Work in Design, pp. 1095–1101 (2022)
47. Sun, X., Qiu, J., Tao, Y., Yi, Y., Zhao, J.: Distributed optimal voltage control and berth allocation of all-electric ships in seaport microgrids. IEEE Trans. Smart Grid 13(4), 2664–2674 (2022)
48. Jiang, X., Zhong, M., Shi, J., Li, W., Sui, Y., Dou, Y.: Overall scheduling model for vessels scheduling and berth allocation for ports with restricted channels that considers carbon emissions. J. Marine Sci. Eng. 10(11), 1757 (2022)

Cognitive Neurosciences

Privacy-Preserving Travel Time Prediction for Internet of Vehicles: A Crowdsensing and Federated Learning Approach

Hongyu Huang[1], Cui Sun[1], Xinyu Lei[2], Nankun Mu[1], Chunqiang Hu[3], Chao Chen[1], Huaqing Li[4], and Yantao Li[1(✉)]

[1] College of Computer Science, Chongqing University, Chongqing, China
yantaoli@cqu.edu.cn
[2] Department of Computer Science, Michigan Technological University, Chongqing, USA
[3] School of Big Data and Software Engineering, Chongqing University, Chongqing, China
[4] College of Electronic and Information Engineering, Southwest University, Chongqing, China

Abstract. Travel time prediction (TTP) is an important module task to support various applications for Internet of Vehicles (IoVs). Although TTP has been widely investigated in the existing literature, most of them assume that the traffic data for estimating the travel time are comprehensive and public for free. However, accurate TTP needs real-time vehicular data so that the prediction can be adaptive to traffic changes. Moreover, since real-time data contain vehicles' privacy, TTP requires protection during the data processing. In this paper, we propose a novel Privacy-Preserving TTP mechanism for IoVs, ℙTPrediction, based on crowdsensing and federated learning. In crowdsensing, a data curator continually collects traffic data from vehicles for TTP. To protect the vehicles' privacy, we make use of the federated learning so that vehicles can help the data curator train the prediction model without revealing their information. We also design a spatial prefix encoding method to protect vehicles' location information, along with a ciphertext-policy attribute-based encryption (CP-ABE) mechanism to protect the prediction model of the curator. We evaluate ℙTPrediction in terms of MAE, MSE, RMSE on two real-world traffic datasets. The experimental results illustrate that the proposed ℙTPrediction shows higher prediction accuracy and stronger privacy protection comparing to the existing methods.

Keywords: Internet of Vehicles · Travel Time Prediction · Privacy Protection · Crowdsensing · Federated Learning

1 Introduction

In recent years, we have experienced the rapid development of Internet of Vehicles (IoVs) technology. Various applications based on IoVs have been imple-

B. Luo et al. (Eds.): ICONIP 2023, LNCS 14449, pp. 55–66, 2024.
https://doi.org/10.1007/978-981-99-8067-3_5

mented [1–3], e.g. real-time trajectory planning [4], online car hailing [5], and driving safety warning [6]. To support these services, the travel time, i.e., how long it needs for a vehicle to pass through a region or a series of road segments, is a very important information. Obviously, it is not easy to accurately obtain the travel time in IoVs. Many infrastructures, such as cameras and pressure circuits, have been deployed to monitor the traffic condition. Based on the monitored information from infrastructures, there are some works [7,8] which predict the travel time by investigating the regulations. However, we notice two limitations. First, these devices are usually very expensive and not penetrative, and thus they are only deployed on some main streets and intersections in well-developed cities. Second, these methods may not be sensitive to the occurrence of new patterns or even unexpected accidents.

Driven by the quick development of the sensing technology, IoV-based crowd-sensing has become a promising method to collect real-time and comprehensive traffic data. In an IoV-based crowdsensing system [9], a data curator issues an sensing task through a crowdsensing platform. The platform then publishes the task to crowd workers who are responsible for sensing and submitting data to the curator. Based on the mechanism, the data curator can sense any arbitrary region on demand and the real-time data are benefit for more accurate travel time prediction (TTP). However, when collecting data in a crowdsensing paradigm, the curator must consider the privacy of vehicles or even itself. Specifically, we consider the following three privacy requirements: 1) Task privacy. Since the data collection task is region-associated, the data curator may not intend to release tasks to vehicles that are outside of the targeted region. 2) Location privacy. As mentioned above, the interactions between data curator and vehicles are through a public crowdsensing server. In this case, vehicles may not plan to reveal their locations to the server to obtain the region-associated tasks. 3) Data privacy. In traditional crowdsensing schemes, participants (workers) usually submit their original data to a server. However, the original sensed data could contain the participants' private information, such as preferences, attributes, and trajectories. So the participants have concerns that the data curator can infer their privacy when predicting the travel time.

In order to deal with the aforementioned issues, we propose ℙℙTPrediction, a novel Privacy-Preserving Travel Time Prediction for IoVs based on crowdsensing and federated learning in this paper. Specifically, we first utilize an elaborately-designed prefix based spatial encoding method to decrypt the location and shape of the customized region. Then, we construct an access policy tree using these prefixes. The privacy-preserving matching process is based on the ciphertext-policy attributed-based encryption (CP-ABE) scheme [10]. Finally, to protect the data privacy, we resort to the federated learning (FL) [11]. In the FL, vehicles do not need to submit their original data to the curator. Instead, vehicles' sensed data are stored locally and privately. They only train the local model and submit updates. To validate the prediction performance of ℙℙTPrediction, we compare it with the stacked autoencoder (SAE), LSTM, and GRU models in terms of MAE, MSE, RMSE and security analysis under the same configuration on two real-world taxi trajectory datasets. The experimental results illustrate that the proposed PTPrediction show higher prediction accuracy and stronger privacy protection comparing with the existing models. The main contributions of this work are summarized as follows:

- We propose PTPrediction, a novel Privacy-Preserving Travel Time Prediction for IoVs based on the crowdsensing and federated learning technology.
- We are among the first to make use of the federated learning based on the LSTM neural network for TTP, and design a spatial prefix encoding method to protect vehicles' location information.
- We evaluate the performance of PTPrediction on two real-world taxi trajectory datasets, and the experimental results show that the proposed PTPrediction outperforms the existing models.

The rest of this paper is organized as follows: In Sect. 2, we discuss the related literature about crowdsensing and TTP. Then, we introduce the system architecture and formally define our problem in Sect. 3. Next, we propose PTPrediction in detail in Sects. 4. In Sect. 5, we present our experimental results and analysis. Finally, we conclude this paper in Sect. 6.

2   Related Work

For the location privacy, the common solution is masking or perturbation so that the adversary cannot access the exact location directly. For example, cloaking region techniques [12] obfuscate an extract location to a region which is large enough to thwart inferring attacks. Wang et al. [13] propose a personalized location privacy protection, where each worker uploads fuzzy distance and personal privacy level to the server instead of uploading the real location or task distance, which accomplishes the privacy-preserving task assignment for mobile crowdsensing. In terms of task privacy issues, Yuan et al. [14] establish a secure channel between the requester and the targeted participants by leveraging the CP-ABE encryption and symmetric encryption scheme.

To predict the travel time in a given target region, some considerable methods are based on machine models. Jang et al. [15] design a recurrent neural network (RNN) attention mechanism to capture temporal information to accurately predict bus travel time. In addition to RNN, the long short-term memory (LSTM) neural network shows good performance in TTP because it can avoid the vanishing gradient of RNN during back propagation process. To the protect privacy, the federated learning (FL) [16] has been proposed to alleviate privacy concerns. For example, Liu et al. [17] incorporate the FL and GRU to protect privacy while accomplish accurate the traffic flow prediction. Wang et al. [18] integrate the swarm learning (SL) into a FL framework, which aggregate the local SL model with neighboring vehicles to obtain the global model.

3   Problem Formulation

In this section, we first introduce the architecture of PTPrediction, and then present the privacy model and assumptions. Finally, we formally propose our design goals.

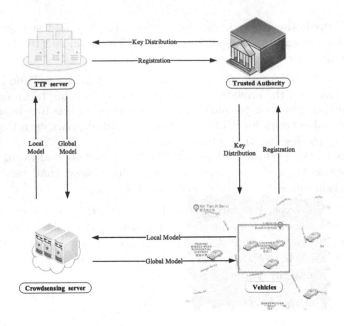

Fig. 1. Architecture of PTPrediction for IoVs.

3.1   System Architecture

The architecture of PTPrediction for IoVs is shown in Fig. 1, which consists of four entities: a TTP server, a spatial crowdsensing (CS) server, a trusted authority (TA), and certain vehicles. We introduce the four entities in the following:

TTP Server: TTP server provides the TTP service to its subscribers by collecting the real-time traffic data.

Vehicles: The vehicles can sense and record the traveling time in a certain region, and then send the data to nearby sensors by radios. Vehicles can be regarded as participants to fulfill crowdsensing tasks.

Crowdsensing (CS) Server: The TTP server exploits a CS server to publish tasks to participants (vehicles) located in the required region.

Trusted Authority (TA): The TA is required to secure secret connections between the TTP server and vehicles.

From a high-level perspective, PTPrediction works sequentially in the following four steps:

Step 1: Each vehicle and TTP server register at the TA, which is responsible for generating and distributing the corresponding key pairs.

Step 2: TTP server sends the encrypted global model Enc_gm and the region-related ciphertext to the CS server, which publishes them to all vehicles.

Step 3: For a vehicle decrypting Enc_gm by using its location information successfully, it will obtain the global model. Then, the vehicle trains the global model with its local dataset. After the training, the vehicle obtains the local model Enc_lm and then submits it to the TTP server.

Step 4: Upon receiving Enc_lm, the TTP server decrypts it to train a new global model. If the model does not converge, the TTP server will encrypt it in a new Enc_gm and send it to vehicles that answer Enc_lm before.

Please note that steps 2 to 4 repeat until the global model reaches a convergence.

3.2 Privacy Model and Assumptions

Similar to most of previous works, we assume that TA can guarantee the correctness of the TTP server and vehicles, so we do not consider the malicious participants in this system. In addition, we assume that TA is fully trusted. The privacy threats only come from a semi-honest CS server, which indicates that the CS server has two sides. On the one side, it correctly and honestly operates the crowdsensing. However, on the other side, the CS server may infer the participants' privacy using the collected data.

3.3 Design Goals

Based on the above system architecture, privacy model and assumptions, we aim to design a TTP system which should meet the objectives of privacy, accuracy, and efficiency.

Privacy: We try to build a secret channel between the TTP server and the targeted vehicle when there are a middleman (CS server) and many potential eavesdroppers (untargeted vehicles). The content of tasks and answers is only visible to the TTP server and the targeted vehicle.

Accuracy: The task is assigned to vehicles who are in a specified region, and other vehicles out of the given region cannot access to the content of the tasks. In the process of federated learning, the TTP server should accurately predict the travel time in the specified region using the collected model updates.

Efficiency: In IoVs, vehicles can obtain the task information released by the TTP server in time. When predicting the traffic time, we need to reduce the overhead as much as possible, and thus the system can effectively learn the machine model to predict the traffic time.

4 Methodology

4.1 Attribute Set and Access Policy Generation

In this section, we utilize a spatial encoding method [19] to access the policy of regions.

Location-induced Attribute Set: We use a two-tuple (u, v) to represent the longitude and latitude. Then, we give a definition to construct the attribute set of a vehicle.

Definition 1. (The prefix family set): Given an integer u, its prefix family set, denoted as $F(u)$, consists of $w + 1$ binary strings, each of which has $w + 1$ bits. These strings are '$b_0 b_1 b_2 \cdots b_w$', '$b_0 b_1 b_2 \cdots b_{w-1}*$', '$b_0 b_1 b_2 \cdots b_{w-2} * *$', ..., and '$b_0 * \cdots *$', where b_i is "0" or "1", and "*" is the wildcard character.

Suppose that there is a vehicle locating at (u, v). We first transform u into a binary string of $w + 1$ bits, where the highest bit is '0' and the following bits are potential padding '0's. Then, we replace the lowest bit with '*' iteratively until we get '0*...*'. We group these strings into a set and denote it as $F(u)$. Similarly, we obtain $F(v)$ for the longitude v where each string starts with '1'. Finally, we formulate the attribute set of the location (u, v) as $\mathbb{A}_{(u,v)} = F(u) \cup F(v)$.

Region-induced Access Policy Generation: To customize a target region, we use multiple rectangles to assemble and surround it. Each rectangle is defined as a range, e.g. $[a, b]$, which denotes its interval on the longitude and latitude. Then, we give a definition for constructing the access policy of a region.

Definition 2. (The minimum prefixes set): For a given range $[a, b]$ of integers, its minimum prefixes set, denoted as $R([a, b])$, is a set of strings which can cover all strings of $F(u)$ where $u \in [a, b]$.

Please note that 'a string S covers another string S_1' indicating that S is the prefix of S_1. For example, '01**' covers '010*' and '0110'. It is not straightforward to construct $R([a, b])$. We first construct a set $P_x = F(a) \cup F(a+1) \cup \cdots \cup F(b)$. Then, we remove strings which represent the integer outside the range from P_x. Finally, we remove strings which are covered by the other strings from P_x, and obtain the remaining strings $R([a, b])$. In order to decide whether a number u is in a range $[a, b]$, we simply check whether $|F(u) \cap R([a, b])| > 0$. If yes, then we know that $u \in [a, b]$.

4.2 CP-ABE-based Crowdsensing

In our crowdsensing system, the privacy-preserving interactions between the TTP server and vehicles are based on a CP-ABE-HP [20] scheme. This scheme includes four algorithms, i.e. *ABE.Setup*, *ABE.KeyGen*, *ABE.Enc*, and *ABE.Dec*. Specifically, from the aspect of data flow, we explain how our scheme works, which is also illustrated in Fig. 2.

System Setup: The TTP server and all vehicles should register at the TA to obtain an authorized credential. Afterwards, the TA invokes the *ABE.Setup* and distributes public key PK to the TTP server and master key MSK to all vehicles. Each vehicle periodically generates an attribute set $\mathbb{A}$ according to its real-time location. Then, the TA invokes the *ABE.KeyGen(MSK, $\mathbb{A}$)* and distributes the corresponding SK to each vehicle.

Task Encryption: The TTP server first generates a CS task $ST = (gm, r)$, where gm denotes the global model and r denotes the targeted region. Then, it constructs an access policy $\mathbb{P}$ according to the r. Next, it generates a symmetric key Ks and invokes the *Sym.Enc(Ks, gm)* to encrypt the gm. Afterwards, it invokes the *ABE.Enc(PK, Ks, $\mathbb{P}$)* to encrypt Ks and produces a region-specified

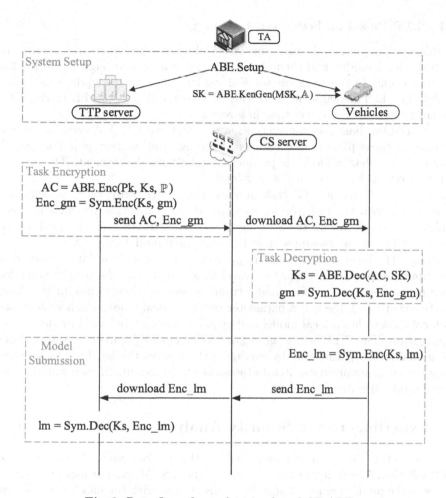

Fig. 2. Data flow of crowdsensing-based CP-ABE.

ciphertext AC. It updates $ST = (AC, Enc\_gm)$, which is then published through the CS server. Now, all vehicles know the ST from the CS server.

Task Decryption: The vehicles invokes $ABE.Dec(AC, SK)$ to decrypt the AC. If $\mathbb{A}$ satisfies $\mathbb{P}$ specified for ST, the algorithm will decrypt AC and output the symmetric key Ks. Then, the participant invokes the $Sym.Dec(Ks, Enc\_gm)$ to decrypt the task content gm, namely global model.

Model Submission: The matched vehicles train a local model lm using the global model and local data without compromising. The vehicle invokes the $Sym.Enc(Ks, lm)$ to encrypt the lm and outputs the ciphertext Enc_lm. Then, the vehicle sends Enc_lm to the CS server through a secure communication channel. After receiving Enc_lm from the CS server, the TTP server invokes $Sym.Dec(Ks, Enc\_lm)$ to decrypt to obtain multiple local models lm. The TTP server will aggregate lm to obtain a gm to predict the travel time.

4.3   TTP Based on Federated Learning

In this section, we mainly introduce a vehicle-based LSTM neural network to train the local model, and then obtain the global model through FL to predict the travel time. In this work, we use FedAvg to aggregate and update the global model. The LSTM neural network can be integrated into the FL in the TTP system to forecast the travel time in a region.

The FedAvg based on the encryption protocol using LSTM network process consists of three phases: initialization, training, and aggregation. For the ith round, a new shared model is produced for better enhancement. The specific iteration on the FL procedure is as follows:

Initialization: Given a TTP task in a FL framework, only when vehicles can be matched with the task region, they are allowed to access the region. These vehicles will check-in with the TTP server to participate to train the local model, and the TTP server transmits an initialized global model w_0 to them.

Training: The input-output pair (x_i, y_i) is a vector of a local data, where x_i is the input sample vector (e.g., the travel time data), and the output value for training sample x_i is y_i. The model training process aims to minimize a local loss function $f_i(w)$, the w is a parameter vector of local model. Each vehicle uses its local data to learn local model and acquires the updated local model.

Aggregation: The TTP server aggregates local model parameters and updates the global model parameters by averaging the sum of the local model weights. Only the client involved can iterate the local model weights in each round of the global model training.

5   Experiments and Security Analysis

In this section, we evaluate and analyze the performance of the proposed $\mathbb{PT}$Prediction. We compare the accuracy of FedLSTM, on the aspect of errors between the prediction result and the ground truth, with the baseline models on the real-world taxi trajectory datasets. Furthermore, we analyze the security of $\mathbb{PT}$Prediction and costs of both TTP server and vehicles.

5.1   Experimental Settings

Implementation. We implement the CP-ABE-HP scheme using the Python cryptographic platform charm library. We employ the 3DES algorithm as the sysmetric cryptography algorithm. We conduct our experiments using Tensorflow 2.4.0 framework based python 3.6.7 on ubuntu-16.03.1-desktop-amd64 under vmware 12.0.0.

Datasets. In the experiment, we exploit both the Shanghai dataset and Chongqing dataset. In the Shanghai dataset, over 4,000 taxis in Shanghai metropolitan region uploaded their GPS locations and timestamps for every one to five minutes. The time period of this dataset is about five months. We use the data of the first four months as the training set to learn the global model. The data of the fifth month is used to simulate a real-time scenario for testing.

Table 1. Structure of FedLSTM for TTP.

Metrics	Hidden layers	Hidden units	MAE	MSE	RMSE
FedLSTM	1	50	16.43	423.87	20.59
		100	15.31	369.68	19.23
		150	16.33	417.32	20.43
	2	50, 50	15.52	363.87	19.08
		100, 100	**14.32**	**313.72**	**17.71**
		150, 150	14.85	332.94	18.25
	3	50, 50, 50	15.58	357.10	18.90
		100, 100, 100	14.77	337.83	18.38
		150, 150, 150	15.00	357.14	18.90

Table 2. Comparison of MAE, MSE and RMSE among models of FedLSTM, LSTM, GRU and SAE on Shanghai dataset.

Metrics	MAE	MSE	RMSE
FedLSTM model	**14.32**	**313.72**	**17.71**
LSTM model	14.28	312.71	17.68
GRU model	14.58	321.04	17.92
FedGRU model	14.51	325.03	18.03
SAE model	15.85	382.42	19.42

Baseline Models and Metrics. We compare FedLSTM with several baseline models on Shanghai datasets. These neural network models are simply expressed as follows:

LSTM: Long-short-term-memory network, which is mainly designed to solve the problem of gradient disappearance.

GRU: Gated-recurrent-unit network, is a variant model of RNN by intention to address the vanishing problem.

SAE: Stacked autoencoder is a deep neural network model composed of multiple layers of sparse autoencoders.

We use Mean Absolute Error (MAE), Mean Square Error (MSE), and Root Mean Square Error (RMSE) as our evaluation metrics to compare and analyze the performance of the TTP.

Parameter Settings. In the framework of deep learning, appropriate parameter selection is a crucial factor to better reflect the performance of the model. We tune different hyper-parameter settings to investigate their impacts on the performance of FedLSTM.

5.2   Results and Discussion

As shown in Table 1, we list the values of MAE, MSE and RMSE under different combinations of parameters. In our experiment, we can observe that all three

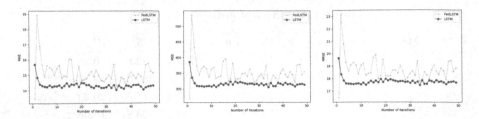

Fig. 3. Model performance on three metrics.

metrics achieve the lowest values when FedLSTM has two hidden layers with 100 units each. The number of hidden layers and unit of the FedLSTM model must not be too large or too small from the perspective of experimental results.

Using the same configuration mentioned above, i.e., two hidden layers with 100 units each, we compare the FedLSTM with three centralized models: LSTM, GRU and SAE on the Shanghai dataset. From Table 2, we observe that LSTM shows the best performance, followed by FedLSTM, GRU and SAE. We notice that both LSTM models, i.e. LSTM and FedLSTM, have better performance than GRU and SAE. The LSTM model has better performance in dealing with big data, and our experiment has more trajectory data. Since the FedLSTM has the same procedure with LSTM on each vehicle, it can also outperform GRU and SAE.

In order to evaluate the accuracy of the TTP, we present the changes of three errors in terms of learning iterations between the TTP server and vehicles. As shown in Fig. 3, we find that all three metrics are positively correlated, and as the number of iterations increases, the MAE, MSE and RMSE of LSTM tend to FedLSTM.

5.3 Security Analysis

We illustrate the security of our crowdsensing system. Specifically, the location privacy protection and model privacy protection can be satisfied through $\mathbb{PT}$Prediction.

Location Privacy Protection: The TTP server generates the localized policy tree $\mathbb{P}$ according to the target region through the CP-ABE-HP scheme. At the same time, the policy tree $\mathbb{P}$ is encrypted through a symmetric key. Only when the attributes of the vehicle meet the policy tree $\mathbb{P}$, the symmetric key can be obtained to decrypt the AC, and thus the CS server does not get any information from the TTP server. The vehicles generate the localized attribute sets $\mathbb{A}$ according to vehicle's location information, and thus the location information of the vehicle is not revealed to any other party.

Model Privacy Protection: The global model gm is encrypted by the TTP server. Only when the attribute set generated by the vehicle can match the policy tree, the ciphertext AC can be decrypted and the symmetric key Ks can be obtained. No other party can decrypt the model task without obtaining the symmetric key Ks. Therefore, the information of the model task will not be revealed. The vehicle performs local training to get the lm sent to the TTP server over a secure communication channel. The CS server is only a transfer model function, and can not obtain the information of the model.

6 Conclusion

In this paper, we present PTPrediction, which integrates FL and crowdsensing into a TTP system, to predict the travel time in a customized region in a privacy-preserving way. We first exploit the CP-ABE-HP scheme to accomplish privacy-preserving task distribution. Then, to predict the travel time, we utilize the LSTM neural network because it is suitable to learn from the time-series data. We incorporate the LSTM into the federated setting, namely FedLSTM, so that every vehicle can learn the model locally and protects the data privacy. Finally, we evaluate the performance of FedLSTM on two real-world taxi trajectory datasets. Experimental results show that FedLSTM has fast convergence speed and satisfied prediction accuracy. We notice that even comparing with centralized learning models, i.e., LSTM, GRU and SAE, the performance of FedLSTM is still competitive.

Acknowledgment. This work was supported in part by the National Natural Science Foundation of China under Grants 62072061, 62072065, 62172066, 62173278, 62272073 and U20A20176, in part by the Natural Science Foundation under Grant CNS-2153393, in part by the National Key R&D Program of China under Grant 2020YFB1805400, in part by the Chongqing Science Fund for Distinguished Young Scholars under Grant CSTB2023NSCQ-JQX0025, and in part by the Regional Innovation and Cooperation Project of Sichuan Province under Grant 2023YFQ0028.

References

1. Huang, H., Yang, Y., Li, Y.: PSG: local privacy preserving synthetic social graph generation. In: Gao, H., Wang, X. (eds.) CollaborateCom 2021. LNICST, vol. 406, pp. 389–404. Springer, Cham (2021). https://doi.org/10.1007/978-3-030-92635-9_23
2. Anthi, E., Williams, L., Słowińska, M., Theodorakopoulos, G., Burnap, P.: A supervised intrusion detection system for smart home IoT devices. IEEE Internet Things J. **6**(5), 9042–9053 (2019)
3. Huang, H., Zhao, H., Hu, C., Chen, C., Li, Y.: Find and dig: a privacy-preserving image processing mechanism in deep neural networks for mobile computation. In: 2021 International Joint Conference on Neural Networks (IJCNN), pp. 1–8 (2021)
4. Ordóñez, M.D., et al.: IoT technologies and applications in tourism and travel industries. In: Internet of Things-The Call of the Edge, pp. 341–360. River publishers (2022)
5. Atiqur, R.: Automated smart car parking system for smart cities demand employs internet of things technology. Int. J. Inf. Commun. Technol. ISSN **2252**(8776), 8776 (2021)

6. Kim, H., Kim, B., Jung, D.: Effect evaluation of forward collision warning system using IoT log and virtual driving simulation data. Appl. Sci. **11**(13), 6045 (2021)
7. Luo, S., Zou, F., Zhang, C., Tian, J., Guo, F., Liao, L.: Multi-view travel time prediction based on electronic toll collection data. Entropy **24**(8), 1050 (2022)
8. Chen, M.Y., Chiang, H.S., Yang, K.J.: Constructing cooperative intelligent transport systems for travel time prediction with deep learning approaches. IEEE Trans. Intell. Transp. Syst. **23**(9), 16590–16599 (2022)
9. Wang, S., Sun, S., Wang, X., Ning, Z., Rodrigues, J.J.: Secure crowdsensing in 5g internet of vehicles: when deep reinforcement learning meets blockchain. IEEE Consum. Electron. Mag. **10**(5), 72–81 (2020)
10. Xu, R., Wang, Y., Lang, B.: A tree-based CP-ABE scheme with hidden policy supporting secure data sharing in cloud computing. In: 2013 International Conference on Advanced Cloud and Big Data, pp. 51–57. IEEE (2013)
11. Zhan, Y., Li, P., Qu, Z., Zeng, D., Guo, S.: A learning-based incentive mechanism for federated learning. IEEE Internet Things J. **7**(7), 6360–6368 (2020)
12. Ardagna, C.A., Cremonini, M., di Vimercati, S.D.C., Samarati, P.: An obfuscation-based approach for protecting location privacy. IEEE Trans. Dependable Secure Comput. **8**(1), 13–27 (2009)
13. Wang, Z., Hu, J., Lv, R., Wei, J., Wang, Q., Yang, D., Qi, H.: Personalized privacy-preserving task allocation for mobile crowdsensing. IEEE Trans. Mob. Comput. **18**(6), 1330–1341 (2018)
14. Yuan, D., Li, Q., Li, G., Wang, Q., Ren, K.: PriRadar: a privacy-preserving framework for spatial crowdsourcing. IEEE Trans. Inf. Forensics Secur. **15**, 299–314 (2019)
15. Jang, V.: Bus dynamic travel time prediction: using a deep feature extraction framework based on RNN and DNN. Electronics **9**(11), 1876 (2020)
16. Yang, Q., Liu, Y., Chen, T., Tong, Y.: Federated machine learning: concept and applications. ACM Trans. Intell. Syst. Technol. (TIST) **10**(2), 1–19 (2019)
17. Liu, Y., James, J., Kang, J., Niyato, D., Zhang, S.: Privacy-preserving traffic flow prediction: a federated learning approach. IEEE Internet Things J. **7**(8), 7751–7763 (2020)
18. Wang, Z., Li, X., Wu, T., Xu, C., Zhang, L.: A credibility-aware swarm-federated deep learning framework in internet of vehicles. Digit. Commun. Netw. (2023)
19. Huang, W., Lei, X., Huang, H.: PTA-SC: privacy-preserving task allocation for spatial crowdsourcing. In: 2021 IEEE Wireless Communications and Networking Conference (WCNC), pp. 1–7. IEEE (2021)
20. Nishide, T., Yoneyama, K., Ohta, K.: Attribute-based encryption with partially hidden encryptor-specified access structures. In: Bellovin, S.M., Gennaro, R., Keromytis, A., Yung, M. (eds.) ACNS 2008. LNCS, vol. 5037, pp. 111–129. Springer, Heidelberg (2008). https://doi.org/10.1007/978-3-540-68914-0_7

A Fine-Grained Domain Adaptation Method for Cross-Session Vigilance Estimation in SSVEP-Based BCI

Kangning Wang[1,2], Shuang Qiu[2,3]([envelope]), Wei Wei[2], Ying Gao[2], Huiguang He[2,3], Minpeng Xu[1,4], and Dong Ming[1,4]

[1] Academy of Medical Engineering and Translational Medicine, Tianjin University, Tianjin, China

[2] Laboratory of Brain Atlas and Brain-Inspired Intelligence, State Key Laboratory of Multimodal Artificial Intelligence Systems, Institute of Automation, Chinese Academy of Sciences, Beijing, China
shuang.qiu@ia.ac.cn

[3] School of Artificial Intelligence, University of Chinese Academy of Sciences, Beijing, China

[4] College of Precision Instruments and Optoelectronics Engineering, Tianjin University, Tianjin, China

Abstract. Brain-computer interface (BCI), a direct communication system between the human brain and external environment, can provide assistance for people with disabilities. Vigilance is an important cognitive state and has a close influence on the performance of users in BCI systems. In this study, a four-target BCI system for cursor control was built based on steady-state visual evoked potential (SSVEP) and twelve subjects were recruited and carried out two long-term BCI experimental sessions, which consisted of two SSVEP-based cursor-control tasks. During each session, electroencephalogram (EEG) signals were recorded. Based on the labeled EEG data of the source domain (previous session) and a small amount of unlabeled EEG data of the target domain (new session), we developed a fine-grained domain adaptation network (FGDAN) for cross-session vigilance estimation in BCI tasks. In the FGDAN model, the graph convolution network (GCN) was built to extract deep features of EEG. The fined-grained feature alignment was proposed to highlight the importance of the different channels figured out by the attention weights mechanism and aligns the feature distributions between source and target domains at the channel level. The experimental results demonstrate that the proposed FGDAN achieved a better performance than the compared methods and indicate the feasibility and effectiveness of our methods for cross-session vigilance estimation of BCI users.

Keywords: Vigilance Estimation · Domain Adaptation · Brain-Computer Interface (BCI) · Electroencephalogram (EEG)

B. Luo et al. (Eds.): ICONIP 2023, LNCS 14449, pp. 67–80, 2024.
https://doi.org/10.1007/978-981-99-8067-3_6

1 Introduction

Vigilance, as a vital cognitive state, denotes the ability of organisms to maintain long-term attention to stimuli [1]. It plays an important role in human-computer interaction and affects the performance of users. Previous studies show that the decline of the vigilance state is one of the main factors leading to the decrease of the performance of users in human-computer interactions [2, 3]. In recent years, vigilance estimation, an important method to recognize human vigilance states, has received substantial attention.

Brain-computer interface (BCI), an important kind of human-computer interaction system, is a communication system that allows a direct connection between the human brain and external devices, which can be used to provide assistance for (disabled) people [4]. Our previous study found that the vigilance state of BCI users has a significant impact on their BCI performance [5]. During the BCI control tasks, the classification accuracy of BCI dropped significantly at the low-vigilance level and led to undesirable performance declines of the task [5]. Therefore, it is necessary to estimate the vigilance state of users during BCI tasks to achieve warning of users' low-vigilance state and further maintain their BCI performance.

In the past years, various modalities have been used as indicators of the vigilance state, including facial expressions [6], speech signals [7], and physiological signals [2, 3]. Thereinto, Electroencephalogram (EEG), as a high-temporal resolution and effective physiological signal, has been widely used to estimate the vigilance state [3, 8, 9]. The differential entropy (DE) and power spectral density (PSD) extracted from EEG signals have been found to be effective features for vigilance estimation [8–10]. Based on DE or PSD features, several approaches based on traditional machine learning methods, such as support vector regression (SVR) [8], extreme learning machine (ELM) [9], continuous conditional neural field (CCNF) [2, 10], and continuous conditional random field (CCRF) [2, 10], have been combined and applied for drivers' vigilance estimation. Recently, some deep learning-based vigilance estimation methods have been proposed. Ko et al. developed a deep convolutional neural network (CNN) method based on DE to estimate the vigilance state of drivers [11]. Khessiba et al. combined a one-dimensional CNN with long short-term memory (LSTM) to improve the performance of vigilance estimation [12]. Most recently, Zhang et al. proposed the LSTM-CapsAtt architecture, consisting of an LSTM network with a capsule attention mechanism, for vigilance estimation [13].

Although some progress has been made in vigilance estimation, few studies have focused on cross-session or cross-subject vigilance estimation tasks, which are two general application scenarios in practice. It is still challenging to develop a model that is generalized to different sessions from the same subject due to the nonstationary characteristic of EEG [14]. The EEG data distribution of one subject was different across sessions [15], which caused the domain shift across sessions. To solve this problem, unsupervised domain adaptation methods have been introduced as an efficient solution. By using unlabeled target data, domain adaptation methods can improve the learning performance in the target domains (new sessions) through the transfer of knowledge from the source domains (previous sessions) [14]. In 2018, Li et al. utilized adversarial domain adaptation networks (DANN) to build cross-subject vigilance estimation models without using any label information from new subjects [15]. In 2021, Luo et al. proposed a Wasserstein-distance-based multi-source adversarial domain adaptation method to overcome the domain shift problem and achieved valid performances in both cross-subject emotion recognition and vigilance estimation tasks [16]. In the existing EEG-based domain adaptation vigilance estimation methods, most of them regard the source domain and the target domain samples as a whole in domain alignment, without considering the complex EEG distributions of different brain regions or channels. Different brain regions or channels respond differently to different vigilance states [10]. Thus, traditional domain adaptation methods which treat all brain regions or channels equally during domain alignment may lead to suboptimal results. In this paper, we proposed a channel-based fine-grained domain adaptation method for cross-session vigilance estimation in the BCI task.

The aim of the present study is to develop a cross-session vigilance estimation method for BCI users. We first designed a 4-target brain-controlled cursor platform based on steady-state visual evoked potential (SSVEP), which is the typical visual evoked paradigm-based BCI system. Then, twelve subjects were recruited and underwent two long-term BCI experimental sessions (90-min continuous cursor-control BCI tasks), when EEG signals were recorded during each session. And, we developed a Fine-Grained Domain Adaptation Network (FGDAN) based on the graph convolution network (GCN) for cross-session vigilance estimation. In this method, we built a GCN network to extract deep EEG features. After obtaining the extracted deep features of each EEG channel, we designed a fine-grained feature alignment module to alleviate the domain differences between the source and target domains on each channel and learn discriminative features of each channel. The experimental results show that the proposed FGDAN achieved better performance than the compared methods and suggest that an effective method for cross-session vigilance estimation is established in our study.

2 Experiment

2.1 Subjects

We recruited twelve subjects, aged between 21 and 30 years, to participate in our experiments (nine males and three females). All of the subjects had normal or corrected-to-normal vision and were asked to avoid consuming alcohol or caffeine before the experiment. The experiment was conducted in accordance with the standards of the Declaration of Helsinki and was approved by the Institutional Review Board of the Institute of Automation, Chinese Academy of Sciences. All subjects provided and signed written informed consent in advance.

2.2 Vigilance Experiment

This experiment consisted of two sessions that were conducted on two different days. Each session contains one long-term BCI task. The long-term BCI task is a cursor-control task based on SSVEP. In each session, the BCI task was performed continuously for 90 min. The cursor-control task was conducted based on our designed SSVEP-based BCI system. The task was to control the cursor to move along the path step-by-step in a specific scenario. During the task, flickering targets continued to flicker until the end of the session, and the subjects controlled the cursor by focusing their gaze on different flickering targets. We designed an online SSVEP-based BCI system with four flickering targets corresponding to four cursor control commands: moving-downward, moving-left, moving-upward, and moving-right. The duration of a command (step) was 1.2 s, and in each step, the position of the cursor was updated once.

To evoke SSVEP, the flickering targets of SSVEP-based BCI were modulated by the sampled sinusoidal stimulation method [17]. The stimulus frequencies were 8 Hz (down), 9 Hz (left), 10 Hz (up), and 11 Hz (right). The stimulus interface was presented on a 19-inch LED screen with a resolution of 1280×1024 pixels and a refresh rate of 60 Hz. Each stimulus was presented within a 100×100 pixel square. To detect the frequency of SSVEP, filter bank canonical correlation analysis (FBCCA), a zero-training algorithm, was utilized [18]. FBCCA is a widely used method in the frequency detection of SSVEP and has satisfactory performance. It mainly includes three steps [18]. First, it decomposes the EEG data into multiple subbands. Second, it calculates the correlation coefficients between each subband and the reference signals. Third, the frequency of the reference signals that has the maximum weighted sum of the correlation coefficients was denoted as the stimulus frequency, and the command corresponding to this frequency is the final classification result. More details of SSVEP decoding are presented in our previous study [5].

2.3 Data Acquisition

During the experiment, EEG data and eye movement data were collected. EEG data were recorded using a Neuroscan SynAmps2 amplifier (Neuroscan, Inc., Australia) with a 64-channel electrode cap, and the EEG electrodes in the cap were placed according to the international 10–20 system. The reference and ground electrodes were placed at the M1 (left mastoid) channel and AFz channel respectively and all electrode impedances were kept below 10 kΩ during the experiment. Thereinto, the data of eight channels (O1, O2, Oz, PO3, PO4, POz, PO7, and PO8) were also used for the online frequency detection of SSVEP. The signals of eye movement were measured by an EyeLink eye tracker (SR Research Ltd., Canada) with a computer to process images. The sampling rate of EEG and eye movement signals was 1000 Hz.

2.4 Data Preprocessing and Feature Extraction

First, the EEG data of the 62 channels were re-referenced to the average signals of both mastoids (M1 and M2). The Automatic Artifact Removal (AAR) toolbox [19] was used to remove the artifacts and noise caused by the movement of eyes and muscle. Then, a 4-order ButterWorth bandpass filter was applied to filter the EEG data with the 1–50 Hz passing band. Finally, the data were epoched into non-overlapping segments, 4 s each, and downsampled to 200 Hz.

The DE feature has been widely used to estimate the vigilance state [10, 11, 13]. In this study, DE features of the delta (1–4 Hz), theta (4–8 Hz), alpha (8–13 Hz), beta (13–30 Hz), and gamma (30–50 Hz) bands were calculated in each segment. DE can be calculated according to [10]:

$$h(X) = \tfrac{1}{2}\log(2\pi e\sigma^2),\tag{1}$$

where the variable X obeys the Gaussian distribution $N(\mu, \sigma^2)$. After preprocessing and feature extraction of the acquired EEG data, there were 62×5 (channels $\times$ bands) DE features for each segment.

2.5 Vigilance Labeling

In this study, we adopted the eye tracking-based PERCLOS index as the label (ground truth) of the vigilance state, which is one of the most widely accepted indices [2, 10]. PERCLOS denotes the PERcentage of eye CLOSure over time. According to the eye movement information related to blink, saccade, and fixation measured from the eye tracker, the PERCLOS values were calculated from the percentage of the durations of blinks and eye closures over a specified time interval as follows [10]:

$$PERCLOS = \tfrac{blink+CLOS}{blink+fixation+saccade+CLOS}\tag{2}$$

where $CLOS$ denotes the duration of the eye closures, and $blink$, $fixation$, and $saccade$ represent the blink duration, fixation duration, and saccade duration, respectively. In this study, the specified time interval was set as 4 s (same length as the EEG segment). The PERCLOS values (ranging from 0 to 1) were further smoothed (sliding average).

2.6 Evaluation

In the cross-session vigilance estimation situation, for each subject, EEG data from one session were set as the source data (domain), and EEG data from another session were set as the target data (domain). In this study, the root mean square error (RMSE) and correlation coefficient (COR), which are widely used in vigilance estimation, were used to evaluate the vigilance estimation performance. In general, the more accurate the method is, the lower the RMSE is and the higher the COR is. Analysis of Variance (ANOVA) was applied to analyze the effect of different methods on the vigilance estimation. The paired-samples t-test was conducted to evaluate the difference between two compared results and the significance level was set at 0.05.

3 Fine-Grained Domain Adaptation Network

In this study, we focus on the cross-session vigilance estimation problem and the source domain (existing session) and target domain (new session) share the same label space. An EEG sample (segment) was organized into a DE feature matrix $X \in \mathbb{R}^{N \times F_{\text{fb}}}$ as a source/target domain sample, where N is the number of EEG channels and F_{fb} is the number of frequency bands. Its corresponding PERCLOS index value was used as the label y, and note that the label of target domain data was only used to evaluate the performance of the methods. We defined X as an undirected and weighted graph and $A \in \mathbb{R}^{N \times N}$ is the adjacency matrix which denotes the connection weights between nodes. In this study, each node corresponds to an EEG channel, and the feature at each node is a DE feature vector extracted from the F_{fb} bands of the corresponding channel.

3.1 Overview

In this study, we proposed a Fine-Grained Domain Adaptation Network (FGDAN) for cross-session vigilance estimation in SSVEP-based BCI tasks. The framework of the proposed FGDAN is illustrated in Fig. 1. Specifically, FGDAN consists of three components: a GCN-based feature extractor, a vigilance label regressor, and a fine-grained feature alignment module. A GCN-based feature extractor was used as the backbone network to extract deep features while the channel information of the EEG data was preserved. After obtaining the extracted deep features of each channel, the fine-grained feature alignment module was designed to reduce the distribution discrepancy between the source and target domains on each channel and at the same time to learn discriminative features of each channel. The vigilance label regressor outputs the final predicted results of the vigilance state through the fully connected (FC) layers.

3.2 GCN-Based Feature Extractor

Adaptive Graph Structure. To capture the dynamic functional relationships between EEG channels, an adaptive graph structure learning mechanism [20] was employed to construct distinct graph structures (adjacency matrices) for different EEG samples. Each element in the adjacency matrix A is calculated as follows:

$$A_{m,n} = \frac{\exp(\text{ReLU}(\omega^T |x_m - x_n|))}{\sum_{n=1}^{N} \exp(\text{ReLU}(\omega^T |x_m - x_n|))}, \tag{3}$$

where $A_{m,n}$ $(m, n \in \{1, 2, \ldots, N\})$ is the m-th row and n-th col element of A, represents the connection relationship (edge weights) between nodes m and n and is determined by the current EEG sample. $x_m \in \mathbb{R}^{F_{\text{fb}} \times 1}$ and $x_n \in \mathbb{R}^{F_{\text{fb}} \times 1}$ are F_{fb} features of nodes m and n in X. ReLU($\cdot$) is the rectified linear unit (ReLU) activation function. $\omega = (\omega_1, \omega_2, \ldots, \omega_{\text{fb}})^T \in \mathbb{R}^{F_{\text{fb}} \times 1}$ is the learnable parameter. $|\cdot|$ denotes the absolute value operator. The learnable parameter ω is updated by minimizing the following loss function:

$$\mathcal{L}_{\text{graph}} = \sum_{m,n=1}^{N} \|x_m - x_n\|_2^2 A_{m,n} + \lambda_1 \|A\|_F^2, \tag{4}$$

where $\|\cdot\|_2$ and $\|\cdot\|_F$ are the l_2-norm and F-norm operators, respectively, and λ_1 is the regularization parameter $(\lambda_1 > 0)$. The larger the distance between x_m and x_n, the smaller the edge weights $A_{m,n}$ between nodes m and n is. The learned adjacency matrix A provides a flexible way to represent the functional relationships between EEG channels.

GCN. In this study, a GCN based on spectral graph theory is utilized to extract the spatial representations of EEG by aggregating information from neighboring nodes. In

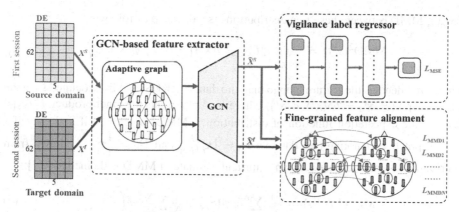

Fig. 1. The framework of FGDAN.

spectral graph analysis, a graph G is represented by its Laplacian matrix $L = D - A$, where $D \in \mathbb{R}^{N \times N}$ is the degree matrix. The graph convolution operation $*_G$ on signal $x_g \in \mathbb{R}^N$ approximated by the K-1 order Chebyshev expansion [21] is defined as:

$$g_\theta *_G x_g = \sum_{k=0}^{K-1} \theta_k T_k \left(\tilde{L} \right) x_g, \tag{5}$$

where g_θ denotes the graph convolution kernel; $\theta \in \mathbb{R}^K$ is the vector of Chebyshev coefficients; $\tilde{L} = \frac{2}{\lambda_{\max}} L - I_N$, where $\lambda_{\max}$ is the maximum eigenvalue of L and $I_N \in \mathbb{R}^{N \times N}$ is the identity matrix. $T_k(\cdot)$ denotes the Chebyshev polynomials and is recursively defined as $T_k(x) = 2x T_{k-1}(x) - T_{k-2}(x)$, where $T_0(x) = 1$ and $T_1(x) = x$. By the graph convolution operation, the information from 0 to K-1 order neighbors is aggregated into the center node. After the graph convolution operations on each EEG sample in the spatial dimension, we obtained $\widehat{X} \in \mathbb{R}^{N \times N_f}$, where N_f is the number of GCN filters.

3.3 Fine-Grained Feature Alignment Module

In EEG-based vigilance estimation, the importance and contribution of each EEG channel are different [10]. However, the coarse-grained domain adaptation aligns all the channels of the whole brain without focus, which probably degrades adaptation performance. To deal with this problem, the fine-grained feature alignment module, with separated alignment heads for each EEG channel, was designed to align the distributions (reduce the distribution discrepancy) across the source domain and target domain in this study. The fine-grained feature alignment module raises different attention to align the distributions of each channel across two domains and is composed of two components: the feature alignment and the attention-based dynamic weights.

Feature Alignment. We applied the Maximum Mean Discrepancy (MMD) [22], a widely used discrepancy metric between the distributions of different domains, to help reduce the distribution divergence between source and target domains. Given two sets of data, MMD embeds each sample to a Reproducing Kernel Hilbert Space (RKHS) and calculates the distance between these embeddings of two sets [22]. Given two distributions

s and t, MMD between these two distributions is calculated as follows:

$$\text{MMD}^2(s, t) = \sup_{\|\phi\|_{\mathcal{H}} \leq 1} \|E_{\mathbf{x}^s \sim s}[\phi(\mathbf{x}^s)] - E_{\mathbf{x}^t \sim t}[\phi(\mathbf{x}^t)]\|_{\mathcal{H}}^2, \tag{6}$$

where $\phi(\cdot)$ denotes the feature map to map the data to RKHS, and is associated with the kernel map $k(\mathbf{x}^s, \mathbf{x}^t) = \langle \phi(\mathbf{x}^s), \phi(\mathbf{x}^t) \rangle$ where $\langle \cdot, \cdot \rangle$ represents the inner product. $E_{\mathbf{x}^s \sim s}[\cdot]$ denotes the expectation function of distribution s. $\text{MMD}^2(s, t) = 0$ if and only if $s = t$. In practice, denoting two datasets $\mathcal{D}_s = \{\mathbf{x}_i^s\}_{i=1}^{N_s}$ and $\mathcal{D}_t = \{\mathbf{x}_j^t\}_{j=1}^{N_t}$ drawn from distributions s and t respectively, an empirical estimate of MMD is defined as [22]:

$$\text{MMD}^2(\mathcal{D}_s, \mathcal{D}_t) = \|\frac{1}{N_s} \sum_{i=1}^{N_s} \phi(\mathbf{x}_i^s) - \frac{1}{N_t} \sum_{j=1}^{N_t} \phi(\mathbf{x}_j^t)\|_{\mathcal{H}}^2. \tag{7}$$

And the $\text{MMD}^2(\mathcal{D}_s, \mathcal{D}_t)$ is commonly used as an MMD loss term $\mathcal{L}_{\text{MMD}}$ to add to the final loss function.

In this study, we use Eq. (7) to calculate the discrepancy of each channel between the source and target domains. We define the fine-grained feature alignment loss $\mathcal{L}_{\text{DA}}$ as:

$$\mathcal{L}_{\text{DA}} = \sum_{n=1}^{N} \alpha_n' \mathcal{L}_{\text{MMD}}^n, \tag{8}$$

where N denotes the number of EEG channels, $\alpha_{n'}$ denotes the weights of $\mathcal{L}_{\text{MMD}}$ of each channel and will be presented next. During the process of training, $\mathcal{L}_{\text{DA}}$ is minimized to narrow the source domain and the target domain in the feature space, which helps make better predictions for the target domain.

Dynamic Weights. Inspired by the success of SENet [23], we naturally utilize the deep features of each channel ($\hat{X}$) to generate the attention weights. Specifically, given an EEG sample $X \in \mathbb{R}^{N \times F_{\text{fb}}}$ from the source domain or target domain, we can obtain the output feature map $\hat{X} \in \mathbb{R}^{N \times N_f \times 1}$ of the GCN-based feature extractor, where N and $N_f \times 1$ denote the channel dimension and the spatial (feature) dimensions respectively. Then, global average pooling was adopted to generate channel-wise statistics. Formally, a statistic $z \in \mathbb{R}^N$ is calculated by shrinking $\hat{X}$ from its spatial dimensions $N_f \times 1$, and the c-th element of z is calculated as follows:

$$z_c = \frac{1}{N_f \times 1} \sum_{i=1}^{N_f} \hat{X}(c). \tag{9}$$

Next, we employed a simple gating mechanism to generate the attention weights:

$$\alpha = Sigmoid(W_2 \text{ReLU}(W_1 z)), \tag{10}$$

where $\alpha = (\alpha_1, \alpha_2, \ldots, \alpha_N) \in \mathbb{R}^N$, Sigmoid($\cdot$) is the Sigmoid activation function, $W_1 \in \mathbb{R}^{\frac{N}{r} \times N}$ and $W_2 \in \mathbb{R}^{N \times \frac{N}{r}}$ are the learnable parameters where r is the reduction ratio ($r = 2$ in this study). Finally, for a pair of EEG samples X^s and X^t drawn from source and target domains respectively, we obtained the attention weights vector α^s and α^t, and the final attention weights vector α' is defined as:

$$\alpha' = \tfrac{1}{2}(\alpha^s + \alpha^t), \tag{11}$$

where $\alpha' = (\alpha'_1, \alpha'_2, \ldots, \alpha'_N) \in \mathbb{R}^N$. Combining MMD loss with the attention weights mechanism, the fine-grained feature alignment module aligns the distributions of each channel that are more transferable for the vigilance estimation task.

3.4 Vigilance Label Regressor

The vigilance label regressor consists of four FC layers with Sigmoid activation functions to extract the representations and ensure that the predicted results range from 0 to 1, as with the labels of vigilance (PERCLOS index values). The mean square error $\mathcal{L}_{\mathrm{mse}}$ is used as the prediction loss function, and the final loss function $\mathcal{L}_{\mathrm{loss}}$ is defined as follows:

$$\mathcal{L}_{\mathrm{loss}} = \mathcal{L}_{\mathrm{mse}} + \mathcal{L}_{\mathrm{graph}} + \lambda_2 \mathcal{L}_{\mathrm{DA}}, \tag{12}$$

where $\mathcal{L}_{\mathrm{mse}} = \frac{1}{N_{\mathrm{m}}} \sum_{i_{\mathrm{m}}=1}^{N_{\mathrm{m}}} (y_{i_{\mathrm{m}}} - \hat{y}_{i_{\mathrm{m}}})^2$, with $\hat{y}$ denoting the predicted labels, and λ_2 is the trade-off parameter.

3.5 Implementation Details and Compared Methods

We implemented FGDAN using TensorFlow libraries on an Nvidia 1080Ti GPU. The learning rate is set as 0.0001, the number of training epoch is set as 100, the parameters λ_1 and λ_2 are set as 0.001, and the model is trained using the Adam optimizer with a batch size of 20.

To evaluate the proposed FGDAN model, we conduct the same experiments using another 12 methods, including baseline vigilance estimation methods (without domain adaptation): SVR [8], ELM [9], CCNF [10], CCRF [10], 4D-CRNN [24], CDCN [25], GraphSleepNet [20], LSTM_CapsAtt [13], and domain adaptation methods: DDC [26], DAN [27], DANN [28], RSD [29]. We directly ran (or reproduced) the codes of the compared methods on our data to ensure a convincing comparison with the proposed method. To allow fair comparisons, we used the same backbone network (our GCN-based feature extractor) for all domain adaptation methods.

4 Results and Discussions

4.1 Comparison with Vigilance Estimation Methods

For each subject, when data from one session were set as the training set, data from another session were set as the test set. Two-way repeated-measures ANOVA was conducted to analyze the effects of different sessions and different methods on the vigilance estimation. The results of ANOVA show that there is a significant main effect of different methods on the vigilance estimation ($p < 0.05$), no significant main effect of different sessions on the vigilance estimation ($p > 0.05$), and no significant interaction between session and method ($p > 0.05$). Thus, there was no significant difference between the results of the two sessions. For the subsequent analysis, the average of the two results obtained by swapping the training set and test set was used as the final result of one subject.

The cross-session vigilance estimation results of our baseline methods and compared methods are listed in Table 1. As shown in Table 1, our FGDAN (without the feature alignment module) achieves the best performance on both RMSE and COR compared to other baseline methods. The t-test shows that the RMSE value of FGDAN is significantly lower than those of SVR, ELM, CCRF, and CDCN (all $p < 0.05$) and the COR value is significantly higher than those of SVR, ELM, and 4D-CRNN (all $p < 0.05$). Additionally, the COR value of FGDAN tends to be significantly higher than that of CCNF ($p = 0.097$) and the RMSE value tends to be significantly lower than that of LSTM_CapsAtt ($p = 0.076$). Compared with the GraphSleepNet model, our method achieves a higher COR value and a lower RMSE value, but the results are not significant. These results demonstrate that the backbone module (feature extractor) of the proposed FGDAN is superior to the compared methods, which shows an effective feature extraction ability. It is helpful for domain adaptation-based vigilance estimation.

Table 1. The results of baseline methods (mean $\pm$ STD). '*' and '**' represent $p < 0.05$ and $p < 0.01$ respectively when compared with our method (the same below).

Methods	RMSE	COR
SVR	0.180 ± 0.039**	0.703 ± 0.162**
ELM	0.178 ± 0.031**	0.688 ± 0.198*
CCNF	0.151 ± 0.030	0.746 ± 0.184
CCRF	0.160 ± 0.029*	0.752 ± 0.140
4D-CRNN	0.149 ± 0.028	0.699 ± 0.225*
CDCN	0.160 ± 0.033*	0.758 ± 0.163
GraphSleepNet	0.154 ± 0.043	0.758 ± 0.158
LSTM_CapsAtt	0.164 ± 0.049	0.760 ± 0.156
FGDAN (baseline)	**0.146 ± 0.025**	**0.772 ± 0.137**

4.2 The Results of Domain Adaptation Methods

Table 2 presents the comparison cross-session vigilance estimation results between our FGDAN and the compared domain adaptation methods. For each subject, data from one session were set as the source domain, data from another session were set as the target domain. The training set was consisted of the labeled source domain data and 3 min unlabeled target domain data, and the remaining target domain data were set as the test set. The one-way repeated-measures ANOVA reveals a significant main effect of different methods on the cross-session vigilance estimation ($p < 0.05$). Our FGDAN achieves the highest value of COR and the smallest value of RMSE and outperforms all the compared methods in the vigilance estimation. Specifically, FGDAN achieves a significantly lower RMSE value than that of DDC, DAN, DANN, and RSD (all $p <$ 0.05) and a significantly higher COR value than that of DAN, DANN, and RSD (all $p < 0.05$). And FGDAN achieves a higher COR value than that of DDC ($p = 0.131$). These results show that a lower mean error and a higher consistency between the labels and predicted values are achieved in our method and demonstrate that the proposed FGDAN is superior to the compared domain adaptation methods. Through the design of fine-grained feature alignment, our FGDAN achieves higher performance than the compared method. Compared with other methods, the proposed fined-grained feature alignment highlights the importance of the different channels figured out by the attention weights mechanism and aligns the feature distributions across domains at the channel level. Therefore, the distributions between the source domain and the target domain achieved better alignment through our method, and FGDAN performed better on the test set.

Table 2. The results of the domain adaptation methods (mean $\pm$ STD).

Methods	RMSE	COR
DDC	0.145 ± 0.029**	0.799 ± 0.122
DAN	0.141 ± 0.024**	0.798 ± 0.128*
DANN	0.145 ± 0.027*	0.783 ± 0.151**
RSD	0.153 ± 0.033*	0.785 ± 0.130**
FGDAN	**0.134 ± 0.028**	**0.822 ± 0.115**

4.3 Ablation Study

To evaluate the effectiveness of the fine-grained feature alignment in our FGDAN, we conducted an ablation study. The results are listed in Table 3. FGDAN-1 was implemented using coarse-grained feature alignment, and FGDAN-2 was implemented using region-based feature alignment. The one-way repeated-measures ANOVA revealed a significant effect of different alignment components on the vigilance estimation ($p <$ 0.05). As shown in Table 3, FGDAN outperforms the others with a higher COR and a lower RMSE. The t-test shows that the RMSE value of FGDAN is significantly lower than those of Source only and FGDAN-1 (all $p < 0.01$) and the COR value is significantly higher than those of Source only ($p < 0.01$). Also, the RMSE value of FGDAN tends to be significantly lower than that of FGDAN-2 ($p = 0.092$) and the COR value is higher than those of FGDAN-1 ($p = 0.164$) and FGDAN-2 ($p = 0.180$). FGDAN-1, FGDAN-2, and FGDAN are based on whole brain-based feature alignment, region-based feature alignment, and channel-based feature alignment, respectively. It can be seen from the results of Table 3 that the performances of the models gradually increase with the granularity of feature alignment from coarser to finer. This indicates that the proposed fine-grained feature alignment can effectively improve the cross-session vigilance estimation performance compared to coarse-grained feature alignment and verifies the effectiveness of our proposed feature alignment mechanism.

Table 3. The results of the ablation study (mean ± STD).

Methods	RMSE	COR
Source only	0.146 ± 0.025**	0.772 ± 0.137**
FGDAN-1	0.144 ± 0.029**	0.803 ± 0.111
FGDAN-2	0.137 ± 0.028	0.808 ± 0.135
FGDAN	**0.134 ± 0.028**	**0.822 ± 0.115**

4.4 Visualization

To show the effectiveness of our method in an intuitive way, we further utilized t-distributed Stochastic Neighbor Embedding (t-SNE) [30] to embed the features into two dimensions. Figure 2 presents an example of visualization from the test data of one subject. Thereinto, the high-vigilance level was set as PERCLOS < 0.5, and the low-vigilance level was set as PERCLOS ≥ 0.5. The output embeddings of FGDAN are more separable than the output embeddings of Source only and FGDAN-1. This further demonstrates the effectiveness of our fine-grained feature alignment.

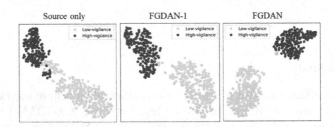

Fig. 2. An example of t-SNE visualization.

5 Conclusions

In this study, we proposed a fined-grained domain adaptation method (FGDAN) for the cross-session vigilance estimation of users in the SSVEP-based BCI task. The proposed fined-grained feature alignment highlights the importance of the different channels figured out by the attention weights mechanism and aligns the feature distributions across domains at the channel level. Comprehensive experiments were conducted and the experimental results show that our proposed FGDAN achieved a better performance of cross-session vigilance estimation than the compared methods in the BCI task. In conclusion, the experimental results demonstrate the feasibility and efficiency of our proposed approach.

Acknowledgements. This work was supported by the Beijing Natural Science Foundation [grant numbers 7222311 and J210010], the National Natural Science Foundation of China [grant numbers U21A20388 and 62206285] and China Postdoctoral Science Foundation (grant number 2021M703490).

References

1. Oken, B.S., Salinsky, M.C., Elsas, S.: Vigilance, alertness, or sustained attention: physiological basis and measurement. Clin. Neurophysiol. **117**(9), 1885–1901 (2006)
2. Zheng, W., et al.: Vigilance estimation using a wearable EOG device in real driving environment. IEEE Trans. Intell. Transp. Syst. **21**(1), 170–184 (2020)
3. Sauvet, F., et al.: In-flight automatic detection of vigilance states using a single EEG channel. IEEE Trans. Biomed. Eng. **61**(12), 2840–2847 (2014)
4. Wolpaw, J., et al.: Brain-computer interfaces for communication and control. Clin. Neurophysiol. **113**(6), 767–791 (2002)
5. Wang, K., et al.: Vigilance estimating in SSVEP-based BCI using multimodal signals. In: 2021 43rd Annual International Conference of the IEEE Engineering in Medicine & Biology Society (EMBC), pp. 5974–5978 (2021)
6. Du, R., Liu, R., Wu, T., Lu, B.: Online vigilance analysis combining video and electrooculography features. In: 2012 International Conference on Neural Information Processing (ICONIP), pp. 447–454 (2012)
7. Krajewski, J., Batliner, A., Golz, M.: Acoustic sleepiness detection: framework and validation of a speech-adapted pattern recognition approach. Behav. Res. Methods **41**(3), 795–804 (2009)
8. Shi, L., Jiao, Y., Lu, B.: Differential entropy feature for EEG-based vigilance estimation. In: 2013 35th Annual International Conference of the IEEE Engineering in Medicine and Biology Society (EMBC), pp. 6627–6630 (2013)
9. Shi, L., Lu, B.: EEG-based vigilance estimation using extreme learning machines. Neurocomputing **102**, 135–143 (2013)
10. Zheng, W., Lu, B.: A multimodal approach to estimating vigilance using EEG and forehead EOG. J. Neural Eng. **14**, 026017 (2017)
11. Ko, W., Oh, K., Jeon, E., Suk, H.: VIGNet: a deep convolutional neural network for EEG-based driver vigilance estimation. In: 2020 8th International Winter Conference on Brain-Computer Interface (BCI), pp. 1–3 (2020)
12. Khessiba, S., Blaiech, A.G., Khalifa, K.B., Abdallah, A.B., Bedoui, M.H.: Innovative deep learning models for EEG-based vigilance detection. Neural Comput. Appl. **33**, 6921–6937 (2020)
13. Zhang, G., Etemad, A.: Capsule attention for multimodal EEG-EOG representation learning with application to driver vigilance estimation. IEEE Trans. Neur. Syst. Rehabil. **29**, 1138–1149 (2021)
14. Jayaram, V., Alamgir, M., Altun, Y., Scholkopf, B., Grosse-Wentrup, M.: Transfer learning in brain-computer interfaces. IEEE Comput. Intell. M. **11**(1), 20–31 (2016)
15. Li, H., Zheng, W., Lu, B.: Multimodal vigilance estimation with adversarial domain adaptation networks. In: 2018 International Joint Conference on Neural Networks (IJCNN), pp. 1–6 (2018)

16. Luo, Y., Lu, B.: Wasserstein-distance-based multi-source adversarial domain adaptation for emotion recognition and vigilance estimation. In: 2021 IEEE International Conference on Bioinformatics and Biomedicine (BIBM), pp. 1424–1428 (2021)

17. Manyakov, N.V., Chumerin, N., Robben, A., Combaz, A., Van Vliet, M., Van Hulle, M.M.: Sampled sinusoidal stimulation profile and multichannel fuzzy logic classification for monitor-based phase-coded SSVEP brain-computer interfacing. J. Neural Eng. **10**, 036011 (2013)

18. Chen, X., Wang, Y., Gao, S., Jung, T.-P., Gao, X.: Filter bank canonical correlation analysis for implementing a high-speed SSVEP-based brain-computer interface. J. Neural Eng. **12**, 046008 (2015)

19. Gomez-Herrero, G., et al.: Automatic removal of ocular artifacts in the EEG without an EOG reference channel. In: Proceedings of the 7th Nordic Signal Processing Symposium, pp. 130–133 (2006)

20. Jia, Z., et al.: GraphSleepNet: adaptive spatial-temporal graph convolutional networks for sleep stage classification. In: Proceedings of the Twenty-Ninth International Joint Conference on Artificial Intelligence (IJCAI 2020), pp. 1324–1330 (2021)

21. Defferrard, M., Bresson, X., Vandergheynst, P.: Convolutional neural networks on graphs with fast localized spectral filtering. In: Proceedings of the 30th International Conference on Neural Information Processing Systems, pp. 3844–3852 (2016)

22. Gretton, A., et al.: A kernel two-sample test. J. Mach. Learn. Res. **13**(25), 723–773 (2012)

23. Hu, J., Shen, L., Albanie, S., Sun, G., Wu, E.: Squeeze-and-excitation networks. IEEE Trans. Pattern Anal. Mach. Intell. **42**(8), 2011–2023 (2020)

24. Shen, F., Dai, G., Lin, G., Zhang, J., Zeng, H.: EEG-based emotion recognition using 4D convolutional recurrent neural network. Cogn. Neurodyn. **14**, 815–828 (2020)

25. Gao, Z., Wang, X., Yang, Y., Li, Y., Ma, K., Chen, G.: A channel-fused dense convolutional network for EEG-based emotion recognition. IEEE T. Cogn. Dev. Syst. **13**, 945–954 (2021)

26. Tzeng, E., et al.: Deep domain confusion: maximizing for domain invariance. arXiv preprint arXiv:1412.3474 (2014)

27. Long, M., Cao, Y., Wang, J., Jordan, M.: Learning transferable features with deep adaptation networks. In: Proceedings of the 32nd International Conference on Machine Learning, pp. 97–105 (2015)

28. Ganin, Y., et al.: Domain-adversarial training of neural networks. J. Mach. Learn. Res. **17**(59), 1–35 (2016)

29. Chen, X., Wang, S., Wang, J., Long, M.: Representation subspace distance for domain adaptation regression. In: Proceedings of the 38th International Conference on Machine Learning, pp. 1749–1759 (2021)

30. Maaten, L., Hinton, G.: Visualizing data using t-SNE. J. Mach. Learn. Res. **9**, 2579–2605 (2008)

RMPE:Reducing Residual Membrane Potential Error for Enabling High-Accuracy and Ultra-low-latency Spiking Neural Networks

Yunhua Chen[1], Zhimin Xiong[1], Ren Feng[1], Pinghua Chen[1], and Jinsheng Xiao[2(✉)]

[1] School of Computers, Guangdong University of Technology, Guangzhou, China
yhchen@gdut.edu.cn
[2] School of Electronic Information, Wuhan Universigy, Wuhan, China
xiaojs@whu.edu.cn

Abstract. Spiking neural networks (SNNs) have attracted great attention due to their distinctive properties of low power consumption and high computing efficiency on neuromorphic hardware. An effective way to obtain deep SNNs with competitive accuracy on large-scale datasets is ANN-SNN conversion. However, it requires a long time window to get an optimal mapping between the firing rates of SNNs and the activation of ANNs due to conversion error. Compared with the source ANN, the converted SNN usually suffers a huge loss of accuracy at ultra-low latency. In this paper, we first analyze the residual membrane potential error caused by the asynchronous transmission property of spikes at ultra-low latency, and we deduce an explicit expression for the residual membrane potential error (RMPE) and the SNN parameters. Then we propose a layer-by-layer calibration algorithm for these SNN parameters to eliminate RMPE. Finally, a two-stage ANN-SNN conversion scheme is proposed to eliminate the quantization error, the truncation error, and the RMPE separately. We evaluate our method on CIR-FARs and ImageNet, and the experimental results show that the proposed ANN-SNN conversion method has a significant reduction in accuracy loss at ultra-low-latency. When T is ≤ 64, our method requires about half the latency of other methods of similar accuracy on ImageNet. The code is available at https://github. com/JominWink/SNN_Conversion_Phase.

Keywords: Spike Neural Networks · Rate Coding · ANN-SNN Conversion

1 Introduction

Spiking neural networks (SNNs) have attracted great attention in recent years due to their inherent properties of low power consumption and high computational efficiency [1]. Each neuron in the SNN will generate a spike only when its accumulated membrane potential exceeds a threshold, or it will remain inactive. Therefore, when SNNs are deployed on neuromorphic computing hardware, they will be able to achieve ultra-low power consumption, latency, and high computing capability that artificial neural networks (ANN) cannot achieve [2]. This makes SNNs very valuable in applications

© The Author(s), under exclusive license to Springer Nature Singapore Pte Ltd. 2024
B. Luo et al. (Eds.): ICONIP 2023, LNCS 14449, pp. 81–93, 2024.
https://doi.org/10.1007/978-981-99-8067-3_7

where low latency and low power consumption are critical, such as high-speed object detection [3,4] and tracking, wearable devices, etc.

Despite the significant advantages in terms of power consumption and computational efficiency, it is difficult to train SNNs due to the non-differentiable nature of spiking processes. The surrogate gradient (SG) methods [5] and spike time-dependent plasticity (STDP) algorithms [6] have been proposed to solve the non-differentiable problem. Since the existing mainstream frameworks, such as TensorFlow, PyTorch, etc., are not optimized for the training of SNNs, SNNs need to be calculated on time steps, which are more sensitive to gradient vanishing or explosion problems. Therefore, it is still difficult to train deeper SNNs with complex network structures. Cao et al. [7] proposed to convert a trained deep ANN into a deep SNN based on ANN-SNN conversion, which can make full use of existing deep learning algorithms and network models and the performance of SNNs have been improved [8]. However, since spiking neurons with complex spatio-temporal dynamics are quite different from artificial neurons in terms of information transfer and processing, there are conversion errors during the conversion process, resulting in decreased accuracy of the converted SNN.

Early researches [9–12] mainly eliminate conversion errors by optimizing parameters such as weights, thresholds, and initial membrane potentials of SNNs. Although ANN-SNN conversion errors have been reduced, since they mainly deal with the truncation error between ANN activation values and SNN output values, it is often necessary to set a long inference time window to achieve accuracy close to ANN, resulting in high latency. Recent researches [10,11,13] divide the ANN-SNN conversion error into quantization error and truncation error and optimize corresponding parameters. Based on the assumption that the SNN inputs conform to a uniform distribution, they further reduce the network latency of SNNs by adjusting the optimal shift of the initial membrane potential. Bu et al. [1] further divide the transformation error into clipping error, quantization error, and unevenness error. They proposed a quantization clipping shift activation function to train the thresholds of SNN in ANN, which has achieved state-of-the-art performance. In this work, they first assume that the remaining membrane potential $v^l(T)$ is in $[0, V_{th})$, and then assume that the input spike trains conform to a uniform distribution so that the unevenness error can be treated as quantization error.

However, assuming that the SNN inputs conform to a uniform distribution is oversimplified or even wrong, as shown in [14], leading to a huge loss of accuracy at ultralow latency (e.g., T = 4). Besides, restricting the value of the residual membrane potential $v^l(T)$ to $[0, V_{th})$ is not consistent with the actual situation. After analyzing the residual membrane potential of SNN neurons at extremely few time steps, we deduce the relationship between residual membrane potential error and initial membrane potential and weights, and propose a layer-wise calibration algorithm to calibrate the initial membrane potential and weights to reduce residual membrane potential errors. Furthermore, to solve the problem of accuracy fluctuations caused by the direct use of quantization clipping shift activation functions for ANN training, we propose a quantization clipping activation function with a trainable threshold and use it to fine-tune the pretrained ANN to obtain the quantized clipped ANN (QC-ANN), as a pre-trained SNN. Which forms a two-stage ANN-SNN conversion scheme, where quantization, clipping, and residual membrane potential errors are eliminated separately.

2 The Spike Response of if Neurons and Its Conversion Error with ReLU

2.1 Spike Neuron and ANN-SNN Conversion

We employ the Integrate-and-Fire (IF) neuron and the soft-reset [15] mechanism to avoid extra information loss, i.e., when a neuron's membrane potential exceeds the threshold V_{th}^l, it fires a spike, and its membrane potential subtracts the threshold. This process can be formulated as:

$$v^l(t) = v^{l-1}(t) + W^l x^{l-1} - s^l(t) V_{th}^l \tag{1}$$

where $s^l(t)$ denotes the output spikes of all neurons in layer l at moment t, the element of which equals 1 if there is a spike and 0 otherwise. $H(\cdot)$ is the Heaviside step function. The principle of ANN-SNN conversion based on rate coding is to make the firing rate in SNN approximate the activation value in ANN. Ideally, the two are approximately equal after T time steps, as shown below.

$$a_i^l \approx r_i^l = \frac{1}{T} \sum_{t=0}^{T} s_i^l(t) \tag{2}$$

As for batch normalization, we follow [16] and merge the convolutional layer and the subsequent BN layer to form a new convolutional layer as follows.

$$W \leftarrow W \frac{\gamma}{\sigma}, b \leftarrow \beta + (b - \mu) \frac{\gamma}{\sigma} \tag{3}$$

where W and b are the weights and bias parameters of the previous layer, γ, σ, β and μ are the parameters of the BN layer, and γ, β are the hyper-parameters to be trained.

2.2 The Spike Response of if Neurons

The product of the spike firing rate of SNN and the threshold value is denoted as $\phi^{l-1}(T) = \frac{\sum_{t=1}^{T} s^{l-1}(t)}{T} V_{th}^{l-1}$. Assuming that the inputs of ANN and SNN in network layer l are the same, i.e., $a^{l-1} = \frac{\sum_{t=1}^{T} s^{l-1}(t)}{T} V_{th}^{l-1}$, and the initial membrane potential in layer l is $v^l(0) = 0$. According to the rate coding property, each neuron fires at most one spike per time step, i.e., $\sum_{t=1}^{T} s^l(t) \in \{0, 1, 2, ..., T\}$. Ideally, according to Eq. (2), the number of spikes is equal to the membrane potential increment rounded down, as in $\sum_{t=1}^{T} s^l(t) = \lfloor \frac{W^l \phi^{l-1}(T)T}{V_{th}^l} \rfloor$, so:

$$\phi^l(T) = clip \left(\frac{V_{th}^l}{T} \left\lfloor \frac{W^l \phi^{l-1}(T)T}{V_{th}^l} \right\rfloor, 0, V_{th}^l \right)$$

$$\approx V_{th}^l clip \left(\frac{1}{T} \left\lfloor \frac{W^l \phi^{l-1}(T)T}{V_{th}^l} \right\rfloor, 0, 1 \right) \tag{4}$$

Here the *clip* function sets the upper bound 1 and the lower bound 0. $\lfloor \cdot \rfloor$ denotes the *floor* function, as shown in Eq. (4), the spike response of IF neuron in the SNN can be expressed as a step function as shown in Fig. 1(a).

2.3   ANN-SNN Conversion Errors

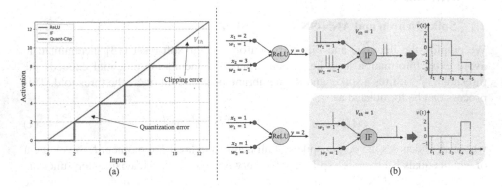

Fig. 1. Diagram of ANN-SNN conversion error.(a) Conversion error between ANN activation value and SNN output spike (V_{th}^l=10, T=5). (b) Diagram of the misfiring of spikes caused by the asynchronous transmission of information in SNN.

The Clipping Error and Quantization Error. The forward propagation equation of ANN is shown in the blue curve in Fig. 1(a). Since the output of SNN is discrete and in the form of a *floor* function, while the output of ANN is continuous, there is unavoidable information loss in ANN-SNN conversion, where we need to map the ANN activation value a^l in $[0, a_{max}^l]$ to SNN output in $[0, V_{th}^l]$. If $V_{th}^l \leq a_{max}^l$, as shown in Fig. 1 and Eq. (4), there is an error of information representation inequality, which is called *clipping error*. Since the output spike $\sum_{t=1}^{T} s^l(t) \in \{0, 1, 2, ..., T\}$ is discrete, Eq. (4) is essentially a quantization function with a quantization factor of $\frac{V_{th}^l}{T}$. As shown in Fig. 1(a), a *quantization error* will inevitably arise when quantization the activation value a^l to $\phi^l(T)$. From Eq. (4), the quantization error and the clipping error are mainly determined by the threshold. As the threshold increases monotonically, the clipping error decreases and the quantization error increases, and vice versa. Therefore, the early researches mainly focus on adjusting the threshold to minimize the sum of the above two errors.

The Residual Membrane Potential Error (RMPE). As shown in Fig. 1(b), in the first row, it is assumed that the data $x_1 = 2$ and $x_2 = 3$ are input to the ANN, and the corresponding weights are $w_1 = 1$ and $w_2 = -1$, and the activation value of the ANN neuron output is 0. In the second row, it is assumed that the data $x_1 = 1$ and $x_2 = 1$ are input to the ANN, the corresponding weights are $w_1 = 1$ and $w_2 = 1$, and the activation value of the ANN neuron output is 2. According to rate coding, the number of spikes fired in SNN corresponds to the ANN activation value and is represented by a certain distribution. Here we set the time step and threshold to $T = 5$, $V_{th} = 1$, respectively.

In the first row of Fig. 1(b), the neuron fires spikes at time step t_1 and t_2 when its membrane potential reaches the threshold, as a result, a total of two spikes are fired, which is greater than the correct number of the expected spikes. In the second row of Fig. 1(b), in the finite time step t_5, the membrane potential contains enough information to fire two spikes, but only one spike is fired, which is less than the number of the expected spikes. From the above examples, we can see that in the time window containing more time steps, the distribution of spikes with positive and negative weights approximately conforms to a uniform distribution, and the residual membrane potential error generated by the early spikes can be compensated by the later spikes. In the time window with fewer time steps, the distribution of spikes with positive and negative weights does not conform to a uniform distribution, and the residual membrane potential cannot be compensated by the later spikes and can only be discarded. This is the key to the large accuracy loss in the rate-based ANN-SNN conversion at ultra-low latency. As shown in Fig. 1(b), the error can be reflected by the residual membrane potential in time window T, therefore we term this error as the *residual membrane potential error*.

The Parameters Corresponding to RMPE. Here, we analyze the RMPE related parameters based on the relationship between the input and output firing rates of spiking neurons. Assuming that the number of error spikes generated by RMPE is M and the firing threshold $V_{th} = 1$. Equation (5) can be obtained by adding Eq. (1) from 1 to T and dividing by T.

$$r^l(T) = \frac{W^l \sum_{t=1}^{T} s^{l-1}(t)}{T} \pm \frac{W^l M}{T} - \frac{v^l(T) - v^l(0)}{T} \qquad (5)$$

$$v^l(T) = \sum_{t=1}^{T} W^l s^{l-1}(t) + v^l(0) \pm W^l M - clip\left(\left\lfloor \sum_{t=1}^{T} W^l s^{l-1}(t) \right\rfloor, 0, T\right) \qquad (6)$$

Eq. (5) describes the relationship between the input and output firing rates of the spiking neurons, similar to the forward process in ANN. When the simulation time T is long enough, $\frac{v^l(T)-v^l(0)}{TV_{th}^l} \approx 0$ and $\frac{(W^l M)}{T} \approx 0$, however, bigger T leads to larger inference latency. From Eq.(5), we can derive Eq.(6), which exhibits an explicit relationship between RMPE and parameters W^l and $v^l(0)$. In fact, the distribution of the input spikes can be tuned to be as close to a uniform distribution as possible by adjusting the initial membrane potential $v^l(0)$, thus reducing the number of misfired spikes M, and the residual membrane potential $v^l(T)$ can be restricted to $[0, V_{th}^l)$ by optimizing both $v^l(0)$ and W^l.

3   Method

3.1   The Framework of the Training Algorithm

Based on the above analysis, we propose a two-phase ANN-SNN conversion scheme, where quantization error, truncation error, and residual membrane potential error are eliminated separately. As shown in Fig.(2), in the first phase, the source ANN is trained based on ReLU. Then, the QC-ANN is fine-tuned using a quantization clipping activation function with trainable thresholds to minimize the quantization and clipping errors. In the second phase, the neurons in QC-ANN are replaced with IF neurons and the parameters of QC-ANN are transferred to SNN to get a converted SNN, which will be fine-tuned layer-by-layer to reduce the residual membrane potential error, so that the ANN-SNN conversion at ultra-low latency can also achieve high accuracy.

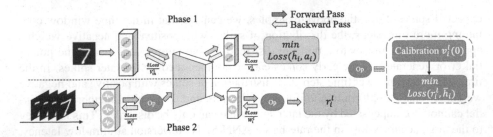

Fig. 2. Diagram of the two-stage training algorithm.

3.2 Training in Phase I

Quantization Clipping Activation Functions. Based on the previous analysis of the spike response of IF neurons (Eq. (4)), the ReLU activation function is quantized, and the quantization clipping activation function $\bar{h}$ is obtained as follows:

$$a^l = \bar{h}(z^l) = \lambda^l clip\left(\frac{1}{L}\left\lfloor\frac{z^l L}{\lambda^l}\right\rfloor, 0, 1\right) \tag{7}$$

where the hyper-parameter L denotes the quantization steps of the ANN, and the trainable parameter λ^l corresponds to the firing threshold for the l-th layer in the SNN. To directly train the QC-ANN, we use the straight-through estimator for the $floor$ function, i.e.,

$$\frac{\partial\lfloor x\rfloor}{\partial x} = 1 \tag{8}$$

When we train QC-ANN, we will train the threshold of each layer of SNN, so that the curve of quantization clipping activation function is as close as possible to the spike response curve of IF neurons, thus the sum of quantization error and clipping error is minimized.

3.3 Training in Phase II

After training QC-ANN in Phase 1, the weights and thresholds from the trained QC-ANN are transferred to SNN, and the QC-neurons are replaced with IF neurons to obtain a pre-trained SNN. In Phase 2, as shown in Fig.(2)(lower part of dotted line), we propose a layer-by-layer calibration algorithm to calibrate the initial membrane potential $v^l(0)$, and to optimize the weight W^l based on fine-tuning learning to reduce RMPE.

Layer-by-layer Calibration of the Initial Membrane Potentials. We calibrate the initial membrane potentials to constrict the residual membrane potentials to $[0, V_{th}^l)$, and to adjust the distribution of the spike trains. To fit the activation value of QC-ANN $\bar{h}(x)$, we use N samples to calibrate the initial membrane potentials. Unlike Li et al. [11] who use the computed simplified mean to calibrate the bias, we assume that the number of misfired spikes is not considered, the error function between the ideal activation value $\bar{h}(x)$ and the firing rate of SNN can be deduced according to Eq.(5), as follows:

$$\min_{v_i^l(0)} \left\{ \left| \frac{\phi_i^{l-1}(T)}{V_{th}^l} - \frac{v_i^l(T) - v_i^l(0)}{TV_{th}^l} - \bar{h}(z_i^l) \right| \right\} \tag{9}$$

In an ideal state, Eq.(16) can be set equal to 0, to explicitly derive the expression of initial membrane potential $v_i^l(0)$, as follows:

$$v_i^l(0) = \frac{T}{N} \sum_{i-1}^{N} |\bar{h}(z_i^l) - \phi_i^l(T)| \tag{10}$$

According to Eq. (10), the error between the ideal output value of SNN and the actual activation value of QC-ANN can be used to calibrate the initial membrane potential.

Thus, the increment in membrane potentials of an IF neuron in the whole time window T (i.e., the ideal value of SNN output) can be used to calibrate the initial membrane potential, control the value range of the residual membrane potential and adjust the distribution of the input spike trains, thereby reducing the RMPE.

Fine-tuning of SNN Weights. To eliminate RMPE, we use the minimum mean square error of the target activation values a^l and the actual spike firing rates r^l as the loss function to calibrate the weights under different time steps. Here, the hierarchical optimization method similar to Li et al. [11] and Deng et al. [13] is adopted for fine-tuning of SNN weights, i.e.,

$$\arg \min_{W^l} Loss = \arg \min_{W^l} \{ (r^l - \bar{h}^l(x))^2 \} \tag{11}$$

The actual spike firing rate r^l needs to be accumulated on time step T. We use surrogate gradient $f(v)$ to solve the non-differentiability of the spike function in the back-propagation through time (BPTT) algorithm:

$$\frac{\partial H(v)}{\partial v} \approx f(v) = \gamma max\{0, 1 - |v|\} \tag{12}$$

where γ is a constant representing the maximum value of the gradient. To reduce the computational cost, a mini-batch of samples is used for fine-tuning. Based on the above two-stage conversion scheme, high-accuracy ANN-SNN conversion can be achieved with ultra-low latency (T≤4).

4   Experiments

4.1   Implementation Details

We use VGG and ResNet network structures to conduct experiments on CIFARs and ImageNet. We use SGD optimizer with a momentum parameter of 0.9 and a cosine attenuation scheduler to adjust the learning rate. In phase 1, the learning rate of source ANN based on ReLU training is set to 0.1, and the learning rate of training QC-ANN

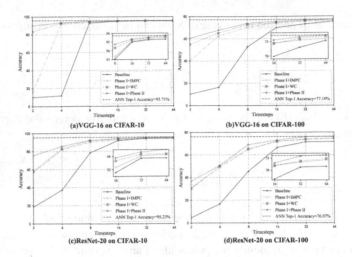

Fig. 3. Accuracy curves of different methods at different time steps on different data sets.

is set to 1e-5. In phase 2, the learning rates for CIFAR-10, CIFAR-100, and ImageNet are set to 1e4, 1e5, and 1e4, respectively. In our experiment, the quantization step L is uniformly set to 8. Our experiment is implemented based on Pytorch deep learning framework. Hardware platform: CPU: Intel Xeon E5-2698, GPU: Tesla V100.

Table 1. The top-1 accuracy of SNN under different simulation time steps.

Method	VGG-16(77.10)		ResNet-20(76.07)	
	4	16	4	16
Phase I	16.23	69.85	16.56	66.34
Phase I+OpI [10]	42.24	70.67	19.04	67.32
Phase I+Opt [13]	58.19	71.5	24.34	67.16
Phase I+LMPC	**62.42**	**74.48**	**48.47**	**71.99**

4.2 Ablation Study

This section verifies the validity of Phase I quantization and Phase II calibration methods. Under four conditions, on the CIFAR datasets, Test the accuracy of VGG-16 and ResNet-20 network structures under Phase I(QC-ANN), LMPC(layer-by-layer membrane potential calibration), LWC(layer-by-layer weight calibration), and Phase I+Phase II from T=2 to T=64.

As shown in Fig. 3, When T≤8, the Phase I conversion effect is the worst because QC-ANN only balances clip and quantization errors and ignores RMPE under ultra-low latency. The LMPC and LWC calibration methods improved the conversion accuracy, indicating that RMPE was reduced when the clip and quantization errors were balanced. As can be seen from the Phase I + Phase II curves, the method has the best accuracy at all time steps, indicating not only balanced clipping and quantization errors but also a significant reduction in the RMPE.

To further verify the effectiveness of the layer-by-layer calibration of the initial membrane potential, as shown in Table 1, the accuracy of the initial membrane potential calibration method proposed in this paper is compared with the existing methods at different time steps. Compared with OpI method [10], which uniformly sets the initial membrane potential as $\frac{1}{2}V_{th}$, and Opt method [13], which assigns $\frac{1}{2T}V_{th}$ offset to each time step, LMPC membrane potential calibration method proposed in this paper achieves higher accuracy under ultra-low latency.

To more explicitly illustrate the effect of calibrating RMPE, we counted the number of neurons in each layer with a residual membrane potential $v^l(T) \notin [0, V_{th}^l)$ at the end of time window T. As shown in Fig. 4, different calibration algorithms can reduce the proportion of SNN neurons with RMPE to a certain extent, indicating the effectiveness of the calibration algorithm in the second phase.

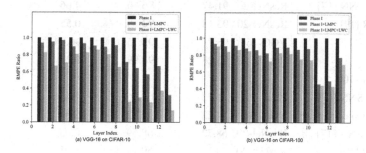

Fig. 4. The proportion of neurons with RMPE for different calibration algorithms.

4.3 Comparison with Other Works

First, we compare the accuracy loss corresponding to the best accuracy with other methods when the time step is limited to $T \leq 64$ as shown in Table 2.

Then, to further illustrate the effectiveness of our work, we compare top-1 accuracy of different time steps with some of the best models as shown in Table 3. For CIFAR-10 and CIFAR-100, our conversion method achieves a performance loss of less than 1% under 32 time steps.

For CIFAR-10 and CIFAR-100, our method has higher conversion accuracy than other methods under T≤32. When T = 4, the accuracy of VGG-16 and ResNet-20 trained on CIFAR-10 can reach 93.28% and 85.43%, and the accuracy of VGG-16 and ResNet-20 trained on CIFAR-100 can reach 67.69% and 50.41%, respectively.

For ImageNet, our method is twice as fast as the SNN QCFS method, achieving a performance of 68.80% with 16 time steps. To prove that our model does not require too much reasoning latency (T≤128), the reasoning accuracy of different time steps are also listed in Table 3 and compared with other works. The results show that the method proposed in this paper is superior to previous methods.

4.4 Energy Estimation

Spike neural networks have considerable potential in neuromorphic chips, and one of the benefits is reduced energy consumption. To verify the computational efficiency of the model proposed in this paper, the firing rate of the SNN converted from VGG-16 on the ImageNet dataset was statistically analyzed.

Table 2. Comparison of the conversion loss at the best accuracy.

Method	Network	ANN Acc(%)	SNN Acc(%)	Conversion Loss(%)
CIFAR-10				
RMP-SNN [17]	VGG-16	93.63	90.35	3.28
SpikeConverter [18]	VGG-16	93.63	93.71	−0.08
RMP-SNN [17]	ResNet-20	91.47	86.60	4.66
TTBR [19]	ResNet-20	93.18	92.68	0.55
Ours	VGG-16	95.71	95.75	−0.04
Ours	ResNet-20	95.23	94.99	0.24
CIFAR-100				
RMP-SNN [17]	VGG-16	71.22	63.76	7.46
SpikeConverter [18]	VGG-16	71.22	71.22	0.00
RMP-SNN [17]	ResNet-20	68.72	68.69	0.03
TTBR [19]	ResNet-20	70.15	69.14	1.01
Ours	VGG-16	77.10	77.10	0.00
Ours	ResNet-20	76.07	76.38	−0.29

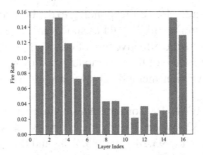

Fig. 5. The firing rate of the SNN converted from VGG-16 on the ImageNet dataset.

Figure 5 shows the firing rate of each layer when time step T is 64, and the average firing rate is 0.0616, which reflects the sparsity of spike activity. To further estimate the energy loss on the chip, the energy estimation equation in the existing work [20] was adopted in this paper to estimate the energy consumption. In this work, except for the first layer of neurons performs multiplication, the rest only perform addition. Therefore, the energy consumption ratio of SNN to ANN is:

$$\frac{Energy_{SNN}}{Energy_{ANN}} = \frac{c * \alpha + (1 - \frac{c}{b}) * b * \beta}{a * \alpha} \tag{13}$$

where α represent the energy cost for multiplication which is $4.6pJ$ and β for addition which is $0.9pJ$ [21]. a, b, and c represent the number of operations in ANN, SNN, and the first layer of SNN. Based on Eq.(13), we calculated that our model only needs 70.06% of ANN's energy consumption when T is 64.

Table 3. Comparison with other works. (* : indicates the quantization step L=8.)

Method	ANN Acc	T = 2	T = 4	T = 8	T = 16	T = 32	T = 64	T >= 128
VGG-16 on CIFAR-10								
Calibration [11]	95.72%	–	–	–	–	93.71%	95.14%	95.79%
OpI. [10]	94.57%	–	–	90.96%	93.38%	94.20%	94.45%	94.55%
Para.Calibration [22]	95.60%	–	86.57%	91.41%	93.64%	94.81%	95.60%	95.60%
QCFS* [1]	95.52%	83.93%	91.77%	94.45%	95.22%	95.56%	95.74%	95.79%
Ours	95.71%	89.97%	93.28%	94.73%	95.32%	95.53%	95.75%	95.75%
ResNet-20 on CIFAR-10								
Calibration [11]	95.46%	-	-	-	-	94.78%	95.30%	95.42%
OpI. [10]	92.74%	-	-	66.24%	87.22%	91.88%	92.57%	92.73%
QCFS* [1]	93.34%	58.67%	75.70%	87.79%	92.14%	93.04%	93.34%	93.24%
Ours	95.23%	75.46%	85.43%	91.98%	93.76%	94.62%	94.99%	95.23%
VGG-16 on CIFAR-100								
Calibration [11]	77.89%	–	–	–	–	73.55%	76.64%	77.87%
OpI. [10]	76.31%	–	–	60.49%	70.72%	74.84%	75.97%	76.31%
Para.Calibration [22]	77.93%	–	55.60%	64.13%	72.23%	75.53%	76.55%	77.79%
QCFS* [1]	76.28%	52.46%	62.09%	70.71%	74.83%	76.41%	76.73%	76.74%
Ours	77.10%	60.24%	67.69%	73.78%	75.48%	76.55%	77.10%	77.11%
ResNet-20 on CIFAR-100								
Calibration [11]	77.16%	–	–	–	–	76.32%	77.29%	77.73%
OpI. [10]	70.43%	–	–	23.09%	52.34%	67.18%	69.96%	70.51%
QCFS* [1]	69.69%	19.96%	34.14%	55.37%	67.33%	69.82%	70.49%	70.55%
Ours	76.07%	36.65%	50.41%	68.88%	72.95%	75.84%	76.38%	76.36%
VGG-16 on ImageNet								
Calibration [11]	75.36%	–	–	–	–	63.64%	70.69%	73.32%(T = 128)
OpI. [10]	74.85%	–	–	6.25%	36.02%	64.70%	72.47%	74.24%(T = 128)
Para.Calibration [22]	75.36%	–	–	–	65.02%	69.04%	72.52%	74.11%(T = 128)
QCFS* [1]	74.29%	–	–	–	50.97%	68.47%	72.85%	73.97%(T = 128)
Ours	74.49%	37.46%	58.04%	63.63%	68.80%	71.33%	72.60%	73.25%(T = 128)

5   Conclusion

In this paper, we analyzed the error caused by the RMPE of IF neurons under ultra-low latency, derived its relationship with the SNN parameters and proposed a layer-by-layer calibration algorithm for these SNN parameters to eliminate RMPE. Our proposed ANN-SNN conversion scheme can be divided into two phases: In the first phase, the pre-trained ANN is fine-tuned using a quantized clipping activation function with a trainable threshold to minimize the quantization error and clipping error. In the second phase, the weights of the SNN is fine-tuned layer-by-layer to reduce RMPE. As a result, the converted deep SNN can achieve high accuracy with ultra-low inference latency.

Acknowledgment. This work was supported by the National Key Research and Development Program of China (No.2021YFB2501104) and the Natural Science Foundation of Guangdong Province, China (No. 2021A1515012233).

References

1. Bu, T., Fang, W., Ding, J., Dai, P., Yu, Z., Huang, T.: Optimal ANN-SNN conversion for high-accuracy and ultra-low-latency spiking neural networks. In: International Conference on Learning Representations (2022)
2. Pfeiffer, M., Pfeil, T.: Deep learning with spiking neurons: opportunities and challenges. Front. Neurosci. **12**, 774 (2018)
3. Xiao, J., Guo, H., Zhou, J., Zhao, T., Yu, Q., Chen, Y.: Tiny object detection with context enhancement and feature purification. Expert Syst. Appl. **211**, 118665–118674 (2023)
4. Xiao, J., Wu, Y., Chen, Y., Wang, S., Wang, Z., Ma, J.: LSTFE-net: Long short-term feature enhancement network for video small object detection. In: Proceedings of the IEEE/CVF Conference on Computer Vision and Pattern Recognition, pp. 14613–14622 (2023)
5. Lee, J.H., Delbruck, T., Pfeiffer, M.: Training deep spiking neural networks using backprop-agation. Front. Neurosci. **10**, 508 (2016)
6. Tavanaei, A., Maida, A.: BP-STDP: approximating backpropagation using spike timing dependent plasticity. Neurocomputing **330**, 39–47 (2019)
7. Cao, Y., Chen, Y., Khosla, D.: Spiking deep convolutional neural networks for energy-efficient object recognition. Int. J. Comput. Vision **113**(1), 54–66 (2015)
8. Chen, Y., Mai, Y., Feng, R., Xiao, J.: An adaptive threshold mechanism for accurate and efficient deep spiking convolutional neural networks. Neurocomputing **469**, 189–197 (2022)
9. Diehl, P.U., Neil, D., Binas, J., Cook, M., Liu, S.C., Pfeiffer, M.: Fast-classifying, high-accuracy spiking deep networks through weight and threshold balancing. In: International Joint Conference on Neural Networks, pp. 1–8 (2015)
10. Bu, T., Ding, J., yu, Z., Huang, T.: Optimized potential initialization for low-latency spiking neural networks. In: Proceedings of the AAAI Conference on Artificial Intelligence, vol. 36, pp. 11–20, June 2022
11. Li, Y., Deng, S., Dong, X., Gong, R., Gu, S.: A free lunch from ANN: towards efficient, accurate spiking neural networks calibration. In: International Conference on Machine Learning, pp. 6316–6325 (2021)
12. Mueller, E., Hansjakob, J., Auge, D., Knoll, A.: Minimizing inference time: Optimization methods for converted deep spiking neural networks. In: International Joint Conference on Neural Networks, pp. 1–8 (2021)
13. Deng, S., Gu, S.: Optimal conversion of conventional artificial neural networks to spiking neural networks. ArXiv abs/2103.00476 (2021)

14. Datta, G., Beerel, P.A.: Can deep neural networks be converted to ultra low-latency spiking neural networks? In: Automation & Test in Europe Conference & Exhibition, pp. 718–723 (2022)
15. Rueckauer, B., Liu, S.C.: Conversion of analog to spiking neural networks using sparse temporal coding. In: 2018 IEEE International Symposium on Circuits and Systems, pp. 1–5 (2018)
16. Rueckauer, B., Lungu, I.A., Hu, Y., Pfeiffer, M., Liu, S.C.: Conversion of continuous-valued deep networks to efficient event-driven networks for image classification. Front. Neurosci. **11**, 682 (2017)
17. Han, B., Srinivasan, G., Roy, K.: RMP-SNN: residual membrane potential neuron for enabling deeper high-accuracy and low-latency spiking neural network. In: IEEE Conference on Computer Vision and Pattern Recognition, pp. 13558–13567 (2020)
18. Liu, F., Zhao, W., Chen, Y., Wang, Z., Jiang, L.: Spikeconverter: an efficient conversion framework zipping the gap between artificial neural networks and spiking neural networks, vol. 36, pp. 1692–1701 (2022)
19. Meng, Q., Yan, S., Xiao, M., Wang, Y., Lin, Z., Luo, Z.Q.: Training much deeper spiking neural networks with a small number of time-steps. Neural Netw. **153**, 254–268 (2022)
20. Rathi, N., Roy, K.: DIET-SNN: a low-latency spiking neural network with direct input encoding and leakage and threshold optimization. IEEE Trans. Neural Netw. Learn. Syst. **34**(6), 3174–3182 (2023). https://doi.org/10.1109/TNNLS.2021.3111897
21. Horowitz, M.: 1.1 computing's energy problem (and what we can do about it). In: IEEE International Solid-State Circuits Conference Digest of Technical Papers, pp. 10–14 (2014)
22. Li, Y., Deng, S.W., Dong, X., Gu, S.: Converting artificial neural networks to spiking neural networks via parameter calibration. ArXiv abs/2205.10121 (2022)

An Improved Target Searching and Imaging Method for CSAR

Yuxiao Deng$^{(\boxtimes)}$, Chuandong Li$^{(\boxtimes)}$, Yawei Shi, Huiwei Wang, and Huaqing Li

College of Electronic and Information Engineering, Southwest University, BeiBei, Chongqing 400715, China
dengyx0829@163.com, {cdli,tomson,hwwang,huaqingli}@swu.edu.cn

Abstract. Circular Synthetic Aperture Radar (CSAR) has attracted much attention in the field of high-resolution SAR imaging. In order to shorten the computation time and improve the imaging effect, in this paper, we propose a fast CSAR imaging strategy that searches the target and automatically selects the area of interest for imaging. The first step is to find the target and select the imaging center and interest imaging area based on the target search algorithm, the second step is to divide the full-aperture data into sub-apertures according to the angle, the third step is to approximate the sub-apertures as linear arrays and imaging them separately, and the last step is to perform sub-image fusion to obtain the final CSAR image. This method can greatly reduce the imaging time and obtain well-focused CSAR images. The proposed algorithm is verified by both simulation and processing real data collected with our mmWave imager prototype utilizing commercially available 77-GHz MIMO radar sensors. Through the experimental results we verified the performance and the superiority of the our algorithm.

Keywords: Circular Synthetic Aperture Radar · Sub-Apertures Divide · 2D-Multiple Signal Classification

1 Introduction

Circular synthetic aperture radar (CSAR) is a SAR system in which the antenna moves along a circular track line while illuminating the internal plane region of the scan path. By acquiring multi-angle measurements over 360°, CSAR can provide higher image resolution and more object information [1,2] than linear SAR (LSAR) systems. Combined with 77 Ghz millimeter wave frequency-modalized continuous wave (FMCW) signals, CSAR has shown considerable advantages in wall structure detection, indoor personnel detection, life monitoring, security equipment and other aspects [3].

B. Luo et al. (Eds.): ICONIP 2023, LNCS 14449, pp. 94–106, 2024.
https://doi.org/10.1007/978-981-99-8067-3_8

With the wide application of CSAR in various fields, the research on its imaging algorithm is also widely developed. The currently widely used BP algorithm can be applied to almost any imaging geometry, but its computational complexity is high and time consuming [3]. The wavefront reconstruction algorithm is used to effectively avoid complex operations and save operation time, however, it has good resolution for the center of the imaging scene but poor imaging effect for those far away from the center of the imaging scene [13]. Jia G W, Buchroithner M F, Chang W G, et al. in [14] introduced frequency domain imaging algorithm of CSAR based on Fourier transform and combines it with subaperture, but the theory of its conversion to frequency domain is relatively complex and computation-intensive. Compressed sensing algorithm is used to describe CSAR imaging, but how to avoid the defocusing of imaging caused by the change of scattering coefficient in full-aperture CSAR is not involved [15]. Subaperture algorithm is used to image CSAR under BP algorithm, but the theoretical basis of subaperture division is not elaborated [16].

In order to shorten the computation time and improve the imaging effect, in this paper, we propose a fast CSAR imaging algorithm that searches the target and automatically selects the area of interest for imaging. Our method can automatically search the targets and select regions of interest for imaging. The searching target part is based on Ranging and 2D Direction of Arrival (DOA) algorithm. The imaging part is based on sub-aperture (SA) processing. We approximate the sub-aperture as straight line and use the improved synthetic aperture algorithm in LSAR to perform sub-aperture imaging. After the sub-apertures are respectively imaged, we select the strong scattering points for aperture fusion to generate the final image. Our algorithm not only effectively avoids the defocus caused by the change of the target scattering coefficient under the full aperture of the circular array, but also solves the problem of low imaging efficiency of the BP algorithm.

The rest of this paper is arranged as follows: Section 2 introduces the geometry and imaging model, and Sect. 3 proposes our fast imaging strategy of CSAR. In Section 4 we introduce the experimental platform we built to verify our algorithm, and use both simulation and processing real data collected with our mmWave imager prototype utilizing commercially available 77-GHz MIMO radar sensors to verify the effectiveness of our proposed algorithm. Finally, Sect. 4.2 and Sect. 5 gives the imaging results and conclusion.

2   Imagine Model of CSAR

2.1   The Geometry Model

Figure 1 shows the geometry of CSAR radar imaging system. The radar motions along a uniform circular in the plane with radius R, and the beam angle is 120°. The position of radar is $(x_a, y_a, z_a) = (Rcos\theta, Rsin\theta, z_0)$, where $\theta \in (0, 2\pi)$ represents the slow time orientation point. According to the imaging geometry, the observation scene is a circular area with a radius of $R_0 = z_0 \times tan60°$.

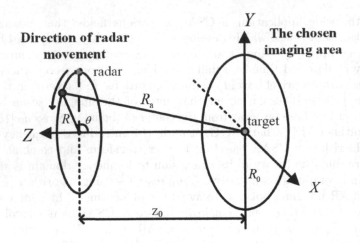

Fig. 1. The geometry of the CSAR imaging

2.2   The Signal Model

We use FMCW signal as the transmission signal, and its transmission waveform can be expressed as

$$s(t) = \sum_{n=1}^{N} p(t - n \cdot T_r)$$

$$p(t) = rect\left(\frac{t}{T_w}\right) e^{j\pi K_r t^2} e^{j\pi f_c t} \tag{1}$$

where $rect()$ is a rectangular signal; K_r is the frequency modulation slope of the transmitted FMCW pulse signal, T_r is the pulse repetition period, T_w is the pulse width, f_c is the carrier frequency, N is the total number of transmitted pulses. Then the corresponding single-point target SAR echo signal can be written as

$$s_r(t) = \sum_{n=1}^{N} \sigma \cdot w \cdot p(t - n \cdot T_r - \tau_n) \tag{2}$$

where σ is the radar cross-sectional area of the point target, w is the weighting of the two-way amplitude of the main lobe of the antenna pattern, τ_n is the electromagnetic wave propagating between the radar and the target when the SAR transmits the nth pulse The round-trip time can be expressed as $\tau_n = 2R(t)/c$.

3   Processing Strategy for CSAR Imagine

The main flow of our proposed algorithm is shown in the Fig. 2.

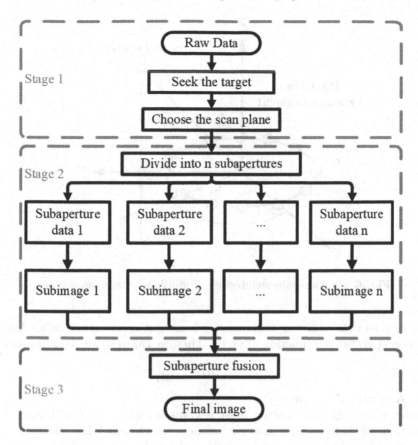

Fig. 2. Our algorithm flow

3.1   Stage 1: Target Searching

According to the geometric model, it is not difficult to find that the radar array can be equivalent to an omni-directional antenna array uniformly distributed on a circle after the radar element performs a circular motion of go-stop-go in a plane. We assumed that D mutually independent target object signals are incident on the array at time t, recorded as

$$(\mathbf{s}_1(t) \cdots \mathbf{s}_D(t)) \tag{3}$$

Their azimuth and elevation angles are respectively $\phi_i, \theta_i (i = 1, 2, \cdots, D)$ Then the echo signal can be expressed as

$$\mathbf{x}(t) = \sum_{k=1}^{D} \mathbf{s}_k(t)\mathbf{a}(\theta_k, \varphi_k) + \mathbf{n}(t) \tag{4}$$

Where the steering vector in the direction of (θ, ϕ) can be expressed as (Fig. 3)

$$\mathbf{a}(\theta_k, \phi_k) = \left[e^{j\xi \cos(\phi - \gamma_0)} \ldots e^{j\xi \cos(\phi - \gamma_{x+1})} \right]^T \tag{5}$$

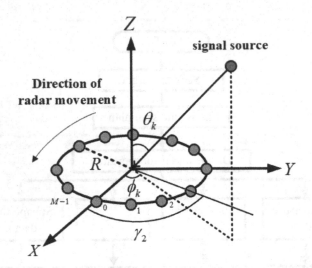

Fig. 3. The hardware architecture of mmWave imager prototype

It can be seen that when the frequency and antenna array structure are known, the steering vector is a function of (θ, ϕ). Then we can rewrite $\mathbf{X}(t)$ into matrix form

$$\mathbf{X}(t) = \mathbf{A}\mathbf{S}(t) + \mathbf{n}(t) \tag{6}$$

where $\mathbf{A} = [\mathbf{a}(\theta_1, \phi_1), ..., \mathbf{a}(\theta_D, \phi_D)]$.

Assume that the array output noise is Gaussian white noise with zero mean and variance $\sigma^2\mathbf{I}$. Inter-array noise is independent of each other, and noise and signal are also independent of each other. Therefore, the output covariance matrix of the array can be expressed as

$$\mathbf{R} = E[\mathbf{X}(t)\mathbf{X}^H(t)] = \mathbf{A}\mathbf{P}\mathbf{A}^H + \sigma^2\mathbf{I} \tag{7}$$

Construct the beamformer by phase mode excitation method, we can have

$$\mathbf{F}_e^H \mathbf{a}(\theta, \phi) = \mathbf{a}_e(\theta, \phi) \tag{8}$$

where $\mathbf{a}_e(\theta, \phi)$ is the matrix of the central Hermitian column. Assume the highest order of the phase mode that can be excited by a uniform circular array is M, so we have $\mathbf{M}' = 2M+1$ modes. In order to eliminate the influence of residual error in the far-field pattern formula, we assumed that the number of array elements $N > 2M + 6$.

Let a permutation matrix be Q, $\alpha_i = 2\pi i / M', i \in [-M, M], M' = 2M + 1$, we can have

$$\mathbf{W} = \frac{1}{\sqrt{M'}} [\mathbf{v}(\alpha_{-M}) \cdots \mathbf{v}(\alpha_0) \cdots \mathbf{v}(\alpha_M)] \tag{9}$$

The azimuth angle change of $\mathbf{a}_e(\theta, \phi)$ is similar to that of a uniform linear array, while the elevation angle changes in the form of symmetrical amplitude attenuation [15]. We can have:

$$\mathbf{R} = \sum_{i=1}^{D} \lambda_i \mathbf{e}_i \mathbf{e}_i^H + \sum_{i=D+1}^{2M+1} \delta^2 \mathbf{e}_i \mathbf{e}_i^H \tag{10}$$

Then we can get the signal subspace $\mathbf{S} = [\mathbf{s}_1 \cdots \mathbf{s}_D]$ and the noise subspace $\mathbf{G} = [\mathbf{g}_{D+1} \cdots \mathbf{g}_{2M+1}]$. The spatial spectrum can be obtained as

$$P(\theta, \phi) = \frac{1}{\mathbf{b}^T(\theta, \phi)\mathbf{G}\mathbf{G}^T\mathbf{b}(\ theta, \phi)} \tag{11}$$

Finally, the spectral peak search can be performed to obtain the DOA estimated value. According to the DOA estimated result, we can calculate the offset distance between the target and the array element center, We can choose the area that can contain all targets as the imaging region of interest, and select the imaging center according to the offset distance [17].

3.2 Stage 2: Subaperture Imaging

According to the angle, we can divide the full aperture into eight sub-apertures. According to the radar imaging geometric model, in a certain sub-aperture, the instantaneous slant distance between the platform and the target after moving t_a can be expressed as (Fig. 4)

$$R(t_n) = \sqrt{(V_a t_a - x_T)^2 + y_T^2 + z_0^2} = \sqrt{R_B^2 + (V_a t_a - x_T)^2} \tag{12}$$

where

$$R_B = \sqrt{z_0^2 + y_T^2} \tag{13}$$

For the convenience of theoretical analysis, we call t a fast time variable, and t_a a slow time variable. If the sampling interval of the SAR fast time is further set as ΔT, and the number of samples along the distance during a single pulse is M, then after the SAR sends N pulses, a $N \times M$ dimensional number matrix. And the n-th pulse and the m-th distance sampling baseband sign can be expressed as

$$s_r(n, m) = \sigma' \exp\left\{jK_r\pi\left[m - \frac{2R(n)}{c}\right]^2\right\} \exp\left[-j\frac{4\pi}{\lambda}R(n)\right] \tag{14}$$

where $\sigma' = \sigma w$, $R(n) = \sqrt{R_B^2 + (V_a n - x_T)^2}$ is the instantaneous distance between the SAR and the target at the slow time nT_r; n_0 is the sampling moment at the slow time.

$$s(t, \theta) = \int_r \int_x f(x, y)p(t - \tau)dxdy \tag{15}$$

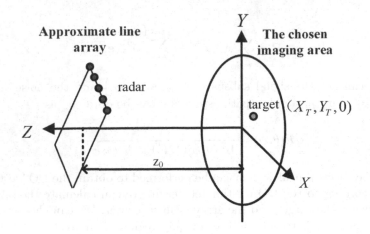

Fig. 4. Subsection imaging of the CSAR

where

$$\tau = \frac{2\sqrt{(x - R\cos\theta)^2 + (y - R\sin\theta)^2 + Z_0^2}}{c} \tag{16}$$

Perform Fourier transform on the echo signal to transform the signal into the wavenumber domain, and complete the distance motion correction and azimuth focusing through the stolt transform in the wavenumber domain [18]. Define the wave number corresponding to the carrier frequency as $K_c = 4\pi f_c/c$, the wave number corresponding to the fundamental frequency $f_m - f_c$ is $K_b = 4\pi(f_m - f_c)/c$ the wave number corresponding to the spatial frequency is $K_n = 2\pi f_a/V_a$ can be obtained by the stationary phase method

$$s_r(n, m) = \sigma' \exp\left\{jK_r\pi\left[m - \frac{2R(n)}{c}\right]^2\right\}\exp\left[-j\frac{4\pi}{\lambda}R(n)\right] \tag{17}$$

The two-dimensional Fourier transform of the range-to-matching micro-output signal is

$$S_r(K_n, K_m) = \sigma' \exp\left[-jR_B\sqrt{K_m^2 - K_n^2} - jK_n x_T\right] \tag{18}$$

where $K_m = K_c + K_b$, K_b is the frequency spectrum of the output envelope of the FM signal pulse pressure. Therefore, if the reference function is defined as

$$S_0(K_n, K_m) = \exp\left[jR_B\sqrt{K_m^2 - K_n^2}\right] \tag{19}$$

then use it with

$$S_r(K_n, K_m) = \sigma' \exp\left[-jR_B\sqrt{K_m^2 - K_n^2} - jK_n x_T\right] \tag{20}$$

Multiplying and performing two-dimensional Fourier transform can obtain the sub-aperture two-dimensional image of the target.

3.3   Stage 3: Subimages Combination

The target distribution function in the spatial domain of the imaging region and its spatial frequency domain spectrum observed by sub-apertures are expressed as

$$f(x_1, y_1) \leftrightarrow F(k_{x1}, k_{y1})$$

$$f(x_2, y_2) \leftrightarrow F(k_{x2}, k_{y2}) \tag{21}$$

$$\vdots$$

$$f(x_n, y_n) \leftrightarrow F(k_{xn}, k_{yn})$$

The frequency spectra are integrated into a coordinate system. In the process of spectrum synthesis, in order to perform inverse Fourier transform, we need to interpolate the rotated unevenly distributed spectral domain data into uniformly distributed Cartesian coordinate data [18].

$$\begin{pmatrix} k_{xn} \\ k_{yn} \end{pmatrix} = \begin{pmatrix} \cos(n-1)\theta - \sin(n-1)\theta \\ \sin(n-1)\theta \cos(n-1)\theta \end{pmatrix} \begin{pmatrix} k_{x1} \\ k_{y1} \end{pmatrix} \tag{22}$$

The spatial spectrum of the observed n observation paths is rotated at a corresponding angle, and then the spectrum is fused

$$F(k_x, k_y) = F(k_{x1}, k_{y1}) \bigcup \cdots \bigcup F_{\text{rotate}}(k_{xn}, k_{yn}) \tag{23}$$

After the spatial spectrum synthesis of all path observations is completed, the target area imaging can be obtained by inverse Fourier transform

$$f(x, y) = \mathcal{F}_{(k_x)}^{-1} \left| \mathcal{F}_{(k_y)}^{-1} [F(k_x, k_y)] \right| \tag{24}$$

4   Experiments and Discussion

4.1   System Overview

The imager prototype consists of a millimeter-wave radar and a custom circular orbit system powered by a rotating motor. The configuration of the Texas Instruments IWR1642-Boost, ESP32, and DAC1000 boards is shown in Fig. 6(b).

The IWR1642-Boost is an evaluation board for the single-chip IWR1642 mmWave sensor. It transmits FMCW signals with a frequency bandwidth of 3.6 GHz ranging from 77 GHz to 80.6 GHz. The DAC1000 is an add-on board used with the IWR1642-Boost to enable high-speed raw data capture for post-processing. The two-dimensional scanning platform designed as shown in Fig. 6(a) has a rotating motor and a communication network port. It is used to control the rotation angle and speed of the radar array element on the matlab user terminal. The radar hardware is stacked on a circular track, and the equivalent two-dimensional scanning can be realized through the circular array, and it is small in size and has a rotation radius of 0.2 m.

To complete the capture and image reconstruction process, the 2D imaging system including radar and scanner is connected to the computer environment through a serial interface. Figure 5 shows a simplified view of the main components and high-level system architecture of an imaging system. Both the radar and orbital systems are controlled through a MATLAB-based graphical user interface. At the end of each scan, the captured raw data was imported into a computer and image reconstruction was implemented in MATLAB.

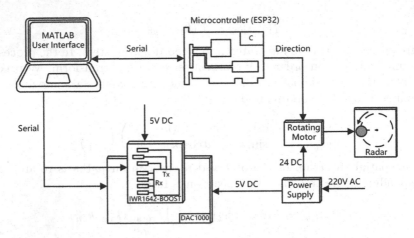

Fig. 5. The hardware architecture of mmWave imager prototype

Fig. 6. mmWave imager prototype. (a) Two-dimensional scanning platform (b) FMCW radar hardware stack.

Table 1. The parameters of mmWave imager prototype.

Parameter(units)	Value	Parameter(units)	Value
Max Range(m)	10.6496	Start Frequency(GHz)	77
Max Velocity(m/s)	13.0917	Slope(MHz/us)	70.48
Range resolution(m)	0.0416	Velocity resolution(m)	4.3639

In order to verify the effectiveness of our algorithm, we used our own mmWave imager prototype platform to collect real data to process. The main experimental parameters are shown in Table 1, and the actual test environment is shown in Fig. 7.

Besides, the full rotation aperture data is processed via traditional BP and full aperture WK algorithms to analyze and compare imaging qualities and speeds with our method.

Fig. 7. The experimental scene of mmWave imager prototype.

4.2   Measurements and Imaging Results

The imaging results of multiple point targets and single point targets at different positions by our method, traditional BP algorithm and full aperture WK algorithms are shown in the figures.

In Fig. 8 (a), (b), (c), it is obvious that our algorithm can clearly find the target point and image it accurately. Through the comparison of the three algorithms of Fig. 9 (a), (b), and (c) multiple point targets, we found that the full aperture WK imaging result has serious defocus, and our method and traditional BP algorithm has the focused imaging result of the precise position of multiple points.

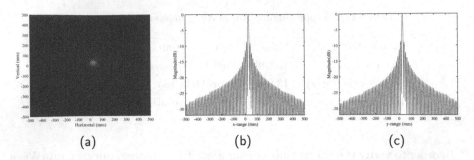

Fig. 8. CSAR imaging result of a point target. (a) CSAR result by our method. (b) x-range magnitude by our method. (c) y-range magnitude by our method.

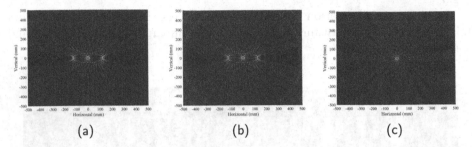

Fig. 9. Imaging results of the multiple points. (a) Result by our method. (c) Result by traditional BP algorithm. (d) Result by Full Aperture WKA.

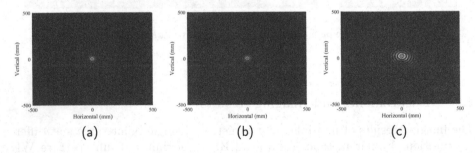

Fig. 10. Imaging results of the point located at (0, 0).(a) Result by our method. (c) Result by traditional BP algorithm. (d) Result by Full Aperture WKA.

In addition, Fig. 10 magnifies the details of the three focused images located in the center of the imaging plane obtained by different algorithms. For central point targets, by comparing peak sidelobe ratio (PSLR) and integrated sidelobe ratio (ISLR) and imaging time.

It is not difficult to find that our method is more efficient than other methods. Therefore, our method is more suitable for fast imaging (Tables 2 and 3).

Table 2. Image qualities of different algorithms.

Our method	PSLR/dB	−9.2330, −8.9128
	ISLR/dB	−4.5969, −4.5471
Traditional BP	PSLR/dB	−8.8821, −8.7651
	ISLR/dB	−4.3942, −4.3966
Full Aperture WKA	PSLR/dB	−7.8473, −7.8380
	ISLR/dB	−3.9364, −4.3301

Table 3. Run times of different algorithms.

Algorithms	Our method	Traditional BP	Full Aperture WKA
Run times/s	8.484805	85.344710	8.481291

5 Conclusion

In this paper, we have proposed a fast CSAR imaging method, which can search the objects in the scanning area and automatically select the interest imaging area, then get the imaging results quickly.

Based on the unique signal model and characteristics of FMCW SAR, we have studied the target imaging based on 2D-MUSIC algorithm and the fast imaging method based on sub-aperture. The running time of the program is greatly reduced by selecting the region of interest, so the problem of long imaging time of the traditional FMCW CSAR algorithm has been solved. In the imaging part, we used sub-apertures to image separately and then fuse them together, which effectively avoided the problem that the scattering coefficient of the target changes under the full aperture. Our method can obtain sharper images in less time.

Finally, we have verified the superiority of our proposed algorithm through both simulation and processing real data collected with our mmWave imager prototype utilizing commercially available 77-GHz MIMO radar sensors.

Acknowledgements. This work was supported by the National Natural Science Foundation of China (61873213).

References

1. Knaell, K.K., Cardillo, G.P.: Radar tomography for the generation of three-dimensional images. IEE Proc. Radar Sonar Navig. **142**(2), 54–60 (1995)
2. Soumekh, M.: Reconnaissance with slant plane circular SAR imaging. IEEE Trans. Image Process. **5**(8), 1252–1265 (1996)
3. Chen, L., An, D., Huang, X.: A backprojection-based imaging for circular synthetic aperture radar. IEEE J. Sel. Top. Appl. Earth Obs. Remote Sens. **10**(8), 3547–3555 (2017)

4. Ponce, O., Prats-Iraola, P., Pinheiro, M., et al.: Fully polarimetric high-resolution 3-D imaging with circular SAR at L-band. IEEE Trans. Geosci. Remote Sens. **52**(6), 3074–3090 (2013)
5. Ponce, O., Prats-Iraola, P., Scheiber, R., et al.: First airborne demonstration of holographic SAR tomography with fully polarimetric multicircular acquisitions at L-band. IEEE Trans. Geosci. Remote Sens. **54**(10), 6170–6196 (2016)
6. Gianelli, C.D., Xu, L.: Focusing, imaging, and ATR for the Gotcha 2008 wide angle SAR collection. In: Algorithms for Synthetic Aperture Radar Imagery XX, vol. 8746, pp. 174–181. SPIE (2013). https://doi.org/10.1117/12.2015773
7. Saville, M.A., Jackson, J.A., Fuller, D.F.: Rethinking vehicle classification with wide-angle polarimetric SAR. IEEE Aerosp. Electron. Syst. Mag. **29**(1), 41–49 (2014)
8. Frolind, P.O., Gustavsson, A., Lundberg, M., et al.: Circular-aperture VHF-band synthetic aperture radar for detection of vehicles in forest concealment. IEEE Trans. Geosci. Remote Sens. **50**(4), 1329–1339 (2011)
9. Cantalloube, H.M.J., Colin-Koeniguer, E., Oriot H.: High resolution SAR imaging along circular trajectories. In: IEEE International Geoscience and Remote Sensing Symposium, pp. 850–853. IEEE (2007). https://doi.org/10.1109/IGARSS.2007.4422930
10. Dupuis, X., Martineau, P.: Very high resolution circular SAR imaging at X band. In: 2014 IEEE Geoscience and Remote Sensing Symposium, pp. 930–933. IEEE (2014). https://doi.org/10.1109/IGARSS.2014.6946578
11. Lin, Y., Hong, W., Tan, W., et al.: Extension of range migration algorithm to squint circular SAR imaging. IEEE Geosci. Remote Sens. Lett. **8**(4), 651–655 (2011)
12. Chen, L., An, D., Huang, X., et al.: P-band ultra wideband circular synthetic aperture radar experiment and imaging. In: 2016 CIE International Conference on Radar (RADAR), pp. 1–3. IEEE (2016). https://doi.org/10.1109/RADAR.2016.8059352
13. Hao, J., Li, J., Pi, Y.: Three-dimensional imaging of terahertz circular SAR with sparse linear array. Sensors **18**(8), 2477 (2018)
14. Ao, D., Wang, R., Hu, C., et al.: A sparse SAR imaging method based on multiple measurement vectors model. Remote Sens. **9**(3), 297 (2017)
15. Liu, T., Pi, Y., Yang, X.: Wide-angle CSAR imaging based on the adaptive sub-aperture partition method in the terahertz band. IEEE Trans. Terahertz Sci. Technol. **8**(2), 165–173 (2017)
16. Zheng, Y., Cui, X., Wu, G., et al.: Polarimetric CSAR image quality enhancement using joint sub-aperture processing. In: 2022 7th International Conference on Signal and Image Processing (ICSIP), pp. 458–462. IEEE (2022). https://doi.org/10.1109/ICSIP55141.2022.9886812
17. Chu, L., Ma, Y., Yang, S., et al.: Imaging algorithm for circular SAR based on geometric constraints. In: 2022 International Conference on Computer Network, Electronic and Automation (ICCNEA), pp. 303–306. IEEE (2022). https://doi.org/10.1109/ICCNEA57056.2022.00073
18. Lou, Y., Liu, W., Xing, M., et al.: A novel motion compensation method applicable to ground cartesian back-projection algorithm for airborne circular SAR. IEEE Trans. Geosci. Remote Sens. (2023). https://doi.org/10.1109/TGRS.2023.3276051

Block-Matching Multi-pedestrian Tracking

Chao Zhang$^{(\boxtimes)}$ (iD)

Beihang University, No. 37, Beijing, China
chaozlimex@gmail.com

Abstract. Target association is an extremely important problem in the field of multi-object tracking, especially for pedestrian scenes with high similarity in appearance and dense distribution. The traditional approach of combining IOU and ReID techniques with the Hungarian algorithm only partially addresses these challenges. To improve the model's matching ability, this paper proposes a block-matching model that extracts local features using a Block Matching Module (BMM) based on the Transformer model. The BMM extracts features by dividing them into blocks and mines effective features of the target to complete target similarity evaluation. Additionally, a Euclidean Distance Module (EDM) based on the Euclidean distance association matching strategy is introduced to further enhance the model's association ability. By integrating BMM and EDM into the same multi-object tracking model, this paper establishes a novel model called BWTrack that achieves excellent performance on MOT16, MOT17, and MOT20 while maintaining high performance at 7 FPS on a single GPU.

Keywords: Multi-object tracking · Block Matching Module · Euclidean Distance Module · Target association · Similarity evaluation

1 Introduction

Multi-object tracking is a key computer vision task that involves detecting objects and associating them across frames. Its objective is to track multiple objects of a specific category, providing their object number and position. Multi-object tracking has a wide range of applications, including unmanned driving and human-computer interaction [28], and plays an important role in scene perception and intelligent recognition.

Target association is a critical aspect of multi-object tracking, playing a vital role in this field. Currently, mainstream algorithms for target association involve two steps: utilizing loss metric algorithms for similarity measurement (e.g., IOU loss) and employing matching algorithms for cross-frame target association (e.g., Hungarian algorithm). Traditional DEEPSORT [44] algorithms use naive IOU as the loss metric function. Recent models such as [3,18,38] have enhanced the discriminative ability of models to handle complex interference scenarios by optimizing the proportionality factor or incorporating historical information for different loss calculations. Regarding matching algorithms, several trackers [18,33] have improved model accuracy and reduced ID switches by employing staged interference or introducing graph neural networks.

While these models have improved performance to some extent, they often involve complex design choices and numerous parameter optimizations tailored to specific scenarios. Some models have increased robustness by incorporating historical information into the model structure, but this has resulted in increased complexity and a notable impact on tracking efficiency.

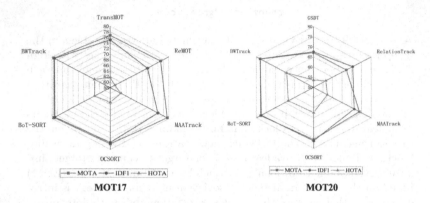

Fig. 1. Comparison of our multi-object tracking model with current state-of-the-art models on MOT17 and MOT20 benchmarks.

Regarding the mentioned multi-object tracking algorithms, we identify two key challenges: effective similarity measurement and robust target association. This paper proposes a rethinking of the model design from two perspectives. **(1) Effective similarity measurement.** In complex pedestrian tracking scenarios with occlusion and interference, robust target feature extraction is crucial. By partitioning the target and gradually extracting partition features in the network, the model's discriminative performance can be enhanced. **(2) Robust feature association.** Traditional methods heavily rely on historical information for the association, neglecting spatial information within frames. By introducing Euclidean distance matching methods based on frame-level feature association, the model's robustness is improved, reducing mismatching frequency.

In response to the above issues, this paper introduces BWTrack, a novel multi-object tracking model that addresses the aforementioned issues. It consists of two main components: **(1) Block feature matching module (BMM).** A redesigned model based on ViT [7] Block and traditional convolutional features. It performs step-by-step block feature extraction, enhancing local feature extraction capability. **(2) Euclidean Distance Module (EDM).** In order to mitigate interference from similar objects, we propose a new target-matching algorithm that utilizes frame-level information, deviating from the traditional reliance on historical information. Our core idea is to enhance the utilization of frame-level information.

We extensively validated our model on the MOT16, MOT17, and MOT20 datasets [6,9]. The experimental results, as depicted in Fig. 1, demonstrate that our model achieves state-of-the-art performance across these datasets. Notably, we achieved an IDF1 score of 76.6 on MOT20, surpassing BoT-SORT 3.5. Furthermore, our tracking model achieves a high tracking speed of 7 FPS, ensuring both accuracy and efficiency.

Our contributions can be summarized as follows:

1) We propose a novel target association matching model based on block matching, enhancing the model's local perception through hierarchical block feature extraction and convolutional feature fusion.
2) To address the limitation of traditional target tracking models in utilizing intra-frame information, we introduce the Euclidean distance matching algorithm, significantly improving the model's robustness by matching untracked and undetected targets.
3) We conduct extensive visualization and validation experiments, and the results show that our proposed model achieves state-of-the-art performance on the MOT16 [9], MOT17 [9], and MOT20 [6] datasets.

2 Related Work

In this section, we provide an overview of the current mainstream multi-object tracking process. We then delve into a detailed review of target matching and association algorithms in multi-object tracking.

2.1 MOT Algorithm

Mainstream multi-object tracking algorithms include object detection and object association modules. These algorithms can be categorized into two types: two-stage and one-stage. The details are described as follows.

Two-stage multi-object tracking models, as proposed in [1,4], adopt a separated target detection and target association module. The detection module predicts bounding boxes for specific targets, while effective target association matching is crucial. Traditional models like SORT [2] use the Kalman filter for trajectory prediction and IOU for inter-target loss measurement, filtering out low-confidence targets. However, these models struggle in densely distributed scenes and complex scenarios due to the limitations of IOU features. To address this, some models [27,33] have introduced ReID and graph neural networks to optimize target association by constructing a graph based on historical and detected target positions. These advancements improve performance in challenging scenarios involving trajectory intersection and complex occlusion.

One-stage multi-object tracking models, as proposed in [3,23,43], employ a fused target detection and association module. To improve computational efficiency, these models combine similarity measurement models with detection models, creating a multitask inference model. This model performs object detection and simultaneously extracts embedding features from target regions, speeding up the target-matching process. However, the performance of one-stage models is inherently limited due to the substantial differences between classification and regression tasks.

2.2 Similarity Metrics

Effective similarity measurement is vital for accurate target matching in multi-object tracking. Optimization in this area can be categorized into two main directions.

Optimization of traditional IOU similarity measurement includes extensions and alternative approaches. C-BIoU [18] addresses the limitations of IOU in measuring large offsets and evaluating differences between targets. It introduces a scaling factor scheme by incorporating a buffer scale for the template bounding box, enhancing model robustness. On the other hand, GHOST [38] divides motion trajectories into active and inactive categories. Active trajectories utilize naive similarity measurement methods like IOU or appearance similarity, while inactive trajectories rely on proxy distance for comparison.

Optimization of similarity measurement based on ReID models. ByteTrack [24] addresses low appearance similarity caused by target occlusion by introducing different confidence categories, extending the traditional single-confidence scheme. Targets surpassing a specified threshold undergo further filtering and association, enhancing target trajectory recovery, reducing ID switches, and improving model performance. CSTrack [8] tackles the task differences between detection and ReID models and reduces computational complexity. It proposes a scale-aware attention network within the one-stage multi-object tracking model. By employing spatiotemporal attention for multiple scales of target features, the challenge of feature alignment across different semantic tasks is significantly alleviated.

2.3 Target Association

Effective cross-frame target association is crucial for improving model performance. Mainstream methods for target association can be categorized into two aspects.

Based on similarity regularization scheme. Similarity regularization schemes have been extensively studied and optimized for loss measurement in target association. Traditional multi-object tracking models, such as [4,26], propose weighted fusion schemes for IOU similarity and ReID similarity. However, a drawback of this approach is the limited ability to effectively discriminate between positive and negative targets based on similarity. To address this limitation, BoT-SORT [1] introduces a staged target similarity measurement scheme. It first excludes low-confidence targets and then selects the final similarity using the minimum selection principle between the two similarity strategies. This scheme significantly enhances the discriminability between positive and negative labels, thereby improving overall model performance.

Based on graph neural network scheme. Scholars have explored the use of graph neural networks (GNNs) to address multi-object tracking tasks, leveraging the inherent characteristics of matching association and GNNs. SGT [33] integrates historical information and introduces an online long-term association model without a motion prediction component. By treating the similarity between detection targets and cross-frame targets as nodes and edges in a graph neural network, SGT establishes a model that achieves high tracking speed and significantly improves target association in challenging scenes through online feature updates.

3 Methodology

In this section, we present an overview of our multi-object tracking model, BWTrack. Our model is built upon BoT-SORT [1] and comprises an object detection model and an object association model. We then delve into the two key components proposed in this paper: the Block Matching Module (BMM) and the Euclidean Distance Module (EDM). Lastly, we provide a detailed account of the model training process, covering both the detection network and the ReID network.

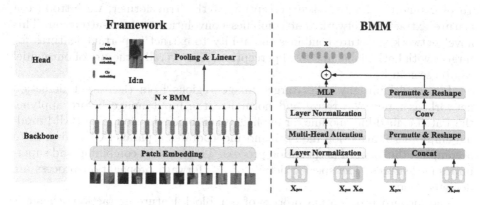

Fig. 2. The overall framework of our proposed model, which includes a ReID network consisting of a backbone and head. The core structure of our model is a Block-based Matching Module (BMM), which is detailed in the right half of the image.

3.1 Framework

Our tracking model comprises two main components: (1) the target detection network, based on Yolox [30], which extracts and predicts specific targets in video sequences, and (2) the target association model, inspired by C-BIoU [18]. We employ a matching scheme that combines IOU, ReID, and Euclidean distance, enhancing the robustness of traditional matching models against interference from similar targets. The overall model structure is as follows.

Target Detection Network. Our detection model takes an input of size $3 \times H \times W$. For the backbone, we utilize Darknet 53 [30] and generate separate feature pyramids $[D1, D2, D3]$. The head network is responsible for predicting the target category and position. In our experiments, we set $H = 800$, $W = 1440$, and use channel selection of $[256, 512, 1024]$ for the feature pyramid.

Target Association Model. Our ReID model is based on FastReid [32], but we redesign the backbone by incorporating a ViT-based model. We enhance the network blocks and employ a classification network for training. The final embedding output of the targets is obtained from the output features of the classification network's backbone.

3.2   Key Modules

BMM. BMM is the core module of our target association matching model. Traditionally, ResNeSt [22] is used as the backbone network for feature extraction in multi-object tracking models. However, in recent years, Transformer [15] has gained prominence in the NLP field and has been successfully applied to computer vision tasks, leading to the emergence of models like ViT. In this paper, we propose a novel BMM module based on the ViT model. It divides the feature extraction process into blocks, enabling explicit block feature extraction. To overcome the limited local perception of the Transformer, we introduce a feature extraction network that combines convolution and self-attention. This novel network structure combines the ability to extract important features for targets with both global and local perception. The overall structure of our model can be seen in Fig. 2.

The model initially takes target image regions from the target detection network as input. We resize and normalize the input image before applying the Patch embedding module for explicit token block division. The BMM module employs self-attention modules and incorporates a parallel CNN branch to accomplish the forward propagation process. This involves combining and superimposing features. The specific details of the forward propagation process are depicted below.

The forward propagation process of our block feature extraction scheme is described by Eqs. 1, 2, and 3, where X denotes the input patch embedding. MSA refers to multi-head self-attention, MLP represents MultiLayer Perceptron, and LN stands for Layer Normalization. The partitioned features are concatenated to achieve feature combinations.

$$
\begin{aligned}
X_{pre} &= X[:,: X.shape[1]//2,:] \\
X_{pos} &= X[:, X.shape[1]//2 :,:]
\end{aligned}
\tag{1}
$$

$$
\begin{aligned}
X_{pre} &= Dropout(MSA(LN(X_{pre}))) + X_{pre} \\
X_{pos} &= Dropout(MLP(LN(X_{pos}))) + X_{pos}
\end{aligned}
\tag{2}
$$

$$
X_{trans} = concat(X_{pre}, X_{pos})
\tag{3}
$$

For CNN feature extraction, we initially map the token to a two-dimensional image space and apply convolutional operations for feature extraction. Subsequently, we map the features obtained from the convolutional layer back to the embedding token space. We then perform linear superposition to fuse the features derived from both self-attention and convolutional operations. The detailed process is illustrated below.

$$X_{conv} = reshape(permute(X[: -1]))$$
$$X_{conv} = Conv(X_{conv}) \tag{4}$$
$$X_{conv} = reshape(permute(X_{conv}))$$

The one-dimensional token features and two-dimensional image features are mapped to each other using reshape and permute functions, as shown in Eq. 4. This enables the extraction of convolutional features.

The feature fusion is accomplished through straightforward linear operations, as depicted below.

$$X = X_{trans}$$
$$X[:, : -1, :] = X[:, : -1, :] + X_{conv} \tag{5}$$

As shown in Eq. 5, the extraction of target features is achieved by stacking N layers of BMM models, where we choose $N = 12$.

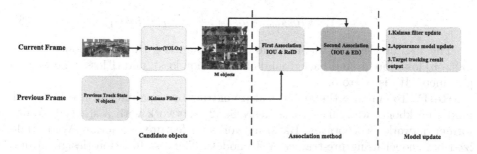

Fig. 3. The overall process of our proposed novel object association algorithm, which is based on IOU matching, ReID matching, and Euclidean distance matching.

EDM. In this paper, we propose the EDM algorithm for target association matching. It innovatively incorporates Euclidean distance into the traditional IOU and ReID matching strategies. By introducing Euclidean distance, we effectively address issues such as low target resolution caused by low IOU and identity switching due to high appearance similarity. This significantly enhances the model's performance. The overall process of our method is illustrated in Fig. 3. In the secondary matching module, we introduce the Euclidean distance matching strategy alongside the traditional IOU matching. The Euclidean distance loss measurement method is presented in Formula 6.

$$Dis_{euc} = 1/(1 + e^{-EUC(O_t, O_d)}) \tag{6}$$

In Formula 6, D_{euc} denotes the Euclidean distance loss, while EUC represents the distance matrix function used to compute the distances between untracked targets and undetected targets in the secondary matching target. The final loss is calculated using Formula 7, and the matching result is obtained through the utilization of the Hungarian algorithm.

$$Dis_{iou} = IOU(O_t, O_d)$$
$$Dis = min(Dis_{euc}, Dis_{iou}) \tag{7}$$

In Formula 7, IOU denotes the IOU distance matrix function used to calculate the distance between unmatched targets and untracked targets. The subsequent matching process selects the minimum value from the matching matrix as the final calculation.

3.3 Training

Detection. The target detection model in this paper utilizes the Yolox detection model, with Darknet 53 [30] serving as the backbone. We initialize the model training with a pre-trained model, derived from [1], to address the limited training dataset issue and expedite convergence speed. The designed loss function is presented below.

$$Loss = W_r * loss_{iou} + loss_{obj} + loss_{cls} + loss_{l1} \tag{8}$$

The loss function, as depicted in Eq. 8, comprises three components: iou loss, obj loss (foreground-background classification loss), and l1 loss. The weight parameter W_r is set to 5.

ReID. To enhance the global feature extraction capability of the FastReid model backbone, we replaced the ResNeSt [22] network with a stacked Transformer network incorporating BMM networks for feature extraction. We initialized our model using pre-trained ViT models. The loss function design in our paper follows the same approach as the Bot-SORT model. The specific details of the loss function are presented below.

$$Loss = W_c \times Loss_{crossentropy} + W_t \times Loss_{triplet} \tag{9}$$

The FastReid network in our approach employs a classification network for model training. The final loss is obtained by combining CrossEntropyLoss and TripletLoss, with the two losses weighted. In this paper, we set W_c and W_t to 1, as shown in Eq. 9.

4   Experiments

In this section, we begin by presenting the deployment details of our model. Next, we evaluate the performance of our tracking model, BWTrack, by comparing it with the current state-of-the-art model. Through this comparison, we showcase the superiority of our model. Additionally, we conduct comprehensive ablation experiments to assess the impact of each module. The experimental results highlight the significant enhancement of local perception and improved tracking robustness achieved by our model.

4.1 Model Deployment Details

The tracking model in this paper was trained and evaluated using Python 3.8, PyTorch 1.11, and CUDA 11.3. The CPU environment consisted of 12 vCPUs of Intel(R) Xeon(R) Platinum 8255C CPU @ 2.50 GHz and 40G of memory. For training, two RTX 3080 GPUs were utilized, while testing was performed on a single RTX 3080 GPU.

Model. The ReID model in this paper is initialized with a pre-trained ViT-Base model. The model's backbone consists of N layers of BMM, with N set to 12 in our experiments. Each BMM includes two channels: the self-attention channel for block feature extraction and the conv channel with n stacked fully convolutional layers for local view feature extraction. In our paper, n is set to 3. During actual deployment, we apply a dropout ratio of 0.1 to the self-attention branch. The model takes RGB images as input for both training and testing, with dimensions set to H = 384 and W = 128.

Train. The training dataset in this paper consists of images from MOT16, MOT17, and MOT20. During processing, we crop target image blocks from

Table 1. Comparison of the performance of our proposed BWTrack tracking model and current state-of-the-art tracking models on the MOT16 benchmark. The best performance for each metric is highlighted in bold.

Tracker	MOTA	IDF1	MT	ML	FP	FN	ID Sw	FPS
POI [19]	66.1	65.1	34.0	21.3	5061	55914	805	<5.2
DeepSORT-2 [45]	61.4	62.2	32.8	18.2	12852	56668	781	<6.7
RAN [29]	63.0	63.8	39.9	22.1	13663	53248	482	<1.5
IAT [5]	48.8	-	15.8	38.1	5875	86567	936	-
CNNMTT [36]	65.2	62.2	32.4	21.3	6578	55896	946	<5.2
Tracktor++ [25]	56.2	54.9	20.7	35.8	**2394**	76844	617	1.8
STPP [16]	50.5	-	19.6	39.4	5939	83694	638	-
TPM [12]	50.9	-	19.4	39.4	4866	84022	619	-
TubeTK POI [37]	66.9	62.2	39.0	16.1	11544	47502	1236	1.0
CTrackerV1 [11]	67.6	57.2	32.9	23.1	8934	48305	1897	6.8
JDE [43]	64.4	55.8	35.4	20.0	10642	52523	1544	18.8
FairMOTv2 [23]	74.9	72.8	**44.7**	**15.9**	10163	34484	1074	18.9
CSTrack [8]	75.6	73.3	42.8	16.5	9646	33777	1121	16.4
BoT-SORT [1]	**78.8**	79.1	-	-	11124	26976	495	8.0
BWTrack	**78.8**	**79.3**	-	-	11133	**26942**	**469**	7.0

the video sequences and assign the same category to video targets with the same ID. The model's total number of categories is equal to the number of IDs in the video sequence. For our experiments, we use a batch size of 32 and employ Adam as the optimizer. The initial learning rate is set to 0.00035, and WEIGHT_DECAY is set to 0.0005. We utilize the CosineAnnealingLR lr_scheduler, with DELAY_EPOCHS set to 30. Additionally, we employ a warm-up training method with WARMUP_FACTOR set to 0.1 and WARMUP_ITERS set to 2000. The training epoch in this paper is set to 60.

Testing and Evaluation Metrics. This work extensively evaluates MOT16, MOT17, and MOT20 datasets using the open-source code TrackEval. The evaluation criteria include MOTA, IDF1, ID Sw, and FPS. MOTA provides a comprehensive assessment of tracking accuracy for various target types. IDF1 evaluates the proportion of correctly recognized detections. ID Sw measures the frequency of ID switches.

4.2   Comparison with State-of-the-Arts

MOT16. MOT16 is a widely used dataset for evaluating the performance of multiple object-tracking algorithms. The dataset includes 14 video sequences captured from various real-world scenarios, such as crowded pedestrian areas. MOT16 provides annotations for object detection and tracking and includes various challenges such as occlusion, motion blur, and camera motion. As shown in Table 1, we compared our tracking model with current mainstream models on MOT16. Our multi-object tracking model surpasses BoT-SORT on the IDF1 metric, achieving a performance of 79.3.

MOT17. MOT17 is a commonly used dataset for evaluating multiple object-tracking algorithms. It features challenging real-world scenarios and provides annotations for object detection and tracking. The dataset comprises 14 video sequences with diverse challenges, including occlusion, motion blur, and crowded scenes. In Table 2, we compared our tracking model with mainstream models on MOT17. Our multi-object tracking model outperforms BoT-SORT in terms of the ID Sw metric, achieving a performance of 1439.

MOT20. MOT20 is a recently released dataset designed to evaluate multiple object-tracking algorithms. It consists of 8 video sequences captured in diverse real-world scenarios, including crowded pedestrian areas. The dataset provides annotations for object detection and tracking, presenting challenges such as occlusion, motion blur, and camera motion. Notably, MOT20 introduces a new challenge of detecting and tracking objects in crowded scenes with similar appearances, adding a realistic aspect compared to previous datasets. In Table 3, we compared our tracking model with mainstream models on MOT20. Our multi-object tracking model outperforms BoT-SORT in terms of the IDF1 metric, achieving a performance of 76.6.

Our experimental results highlight the robustness of our tracking model, particularly in crowded scenes. This is attributed to the BMM module, which enhances local perception, and the EDM module, which addresses limitations in

Table 2. Comparison of the performance of our proposed BWTrack tracking model and current state-of-the-art tracking models on the MOT17 benchmark. The best performance for each metric is highlighted in bold.

Tracker	MOTA	IDF1	HOTA	FP	FN	IDs	FPS
Tube TK [37]	63.0	58.6	48.0	27060	177483	4137	3.0
MOTR [21]	65.1	66.4	-	45486	149307	2049	-
CTracker [11]	66.6	57.4	49.0	22284	160491	5529	6.8
CenterTrack [49]	67.8	64.7	52.2	18498	160332	3039	17.5
QuasiDense [10]	68.7	66.3	53.9	26589	146643	3378	20.3
TraDes [46]	69.1	63.9	52.7	20892	150060	3555	17.5
MAT [31]	69.5	63.1	53.8	30660	138741	2844	9.0
SOTMOT [48]	71.0	71.9	-	39537	118983	5184	16.0
TransCenter [17]	73.2	62.2	54.5	23112	123738	4614	1.0
GSDT [42]	73.2	66.5	55.2	26397	120666	3891	4.9
Semi-TCL [34]	73.3	73.2	59.8	22944	124980	2790	-
FairMOT [23]	73.7	72.3	59.3	27507	117477	3303	25.9
RelationTrack [20]	73.8	74.7	61.0	27999	118623	**1374**	8.5
PermaTrackPr [40]	73.8	68.9	55.5	28998	115104	3699	11.9
CSTrack [8]	74.9	72.6	59.3	23847	114303	3567	15.8
TransTrack [14]	75.2	63.5	54.1	50157	86442	3603	10.0
FUFET [13]	76.2	68.0	57.9	32796	98475	3237	6.8
SiamMOT [35]	76.3	72.3	-	-	-	-	12.8
CorrTracker [41]	76.5	73.6	60.7	29808	99510	3369	15.6
TransMOT [27]	76.7	75.1	61.7	36231	93150	2346	9.6
ReMOT [47]	77.0	72.0	59.7	33204	93612	2853	1.8
MAATrack [39]	**79.4**	75.9	62.0	37320	**77661**	1452	189.1
OC-SORT [4]	78.0	77.5	63.2	**15129**	107055	1950	29.0
BoT-SORT [1]	79.3	79.0	**63.9**	22583	92392	1519	7.8
BWTrack	79.3	**79.1**	**63.9**	22683	92399	1439	6.4

appearance similarity for traditional Reid models. As a result, our model exhibits significant improvements in robustness.

The comparative data for this experiment is sourced from [1,8]. We retested the baseline model (BoT-SORT) data for accurate comparison.

Table 3. Comparison of the performance of our proposed BWTrack tracking model and current state-of-the-art tracking models on the MOT20 benchmark. The best performance for each metric is highlighted in bold.

Tracker	MOTA	IDF1	HOTA	FP	FN	IDs	FPS
FairMOT [23]	61.8	67.3	54.6	103440	**88901**	5243	13.2
TransCenter [17]	61.9	50.4	-	45895	146347	4653	1.0
TransTrack [14]	65.0	59.4	48.5	27197	150197	3608	7.2
CorrTracker [41]	65.2	69.1	-	79429	95855	5183	8.5
Semi-TCL [34]	65.2	70.1	55.3	61209	114709	4139	-
CSTrack [8]	66.6	68.6	54.0	25404	144358	3196	4.5
GSDT [42]	67.1	67.5	53.6	31913	135409	3131	0.9
SiamMOT [35]	67.1	69.1	-	-	-	-	4.3
RelationTrack [20]	67.2	70.5	56.5	61134	104597	4243	2.7
SOTMOT [48]	68.6	71.4	-	57064	101154	4209	8.5
MAATrack [39]	73.9	71.2	57.3	24942	108744	1331	14.7
OC-SORT [4]	75.7	76.3	62.4	**19067**	105894	**942**	18.7
BoT-SORT [1]	76.6	73.1	60.5	20720	98369	1952	1.8
BWTrack	**76.7**	**76.6**	**62.5**	20697	97939	1587	2.7

Fig. 4. Comparison of feature extraction between BMM, ResNeSt, and ViT models.

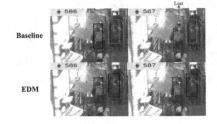

Fig. 5. Comparison of matching performance between EDM and the traditional baseline (BOT-SORT) strategy.

4.3 Ablation Study

To assess the effectiveness of each module, we conducted comprehensive exploratory experiments. These experiments involved visualizing a large number of results to demonstrate the actual tracking performance of our model.

Effect of Block Matching Module (BMM)
BMM vs ViT Block vs ResNeSt. We conducted comparative experiments on the BMM module introduced in this article, and the results can be found in Table 4. To ensure fairness in the comparison, we used the same detection model and association method for multiple multi-object tracking models, with modifica-

Table 4. Performance comparison of different backbone models on MOT17 and MOT20 datasets.

Backbone	MOT17			MOT20		
	MOTA	IDF1	IDs	MOTA	IDF1	HOTA
ResNeSt	79.3	79.0	1519	76.6	73.1	60.5
ViT	79.3	78.4	1464	76.7	76.1	62.4
BMM	79.3	78.9	1458	76.7	76.4	62.5

tions limited to the ReID model's backbone. The experiments demonstrate that our BMM module significantly enhances the IDF1 and HOTA metrics on the MOT20 test dataset, improving the model's robustness. However, while MOT20 focuses more on crowded scenes, MOT17 covers a wider range of scenarios. On the MOT17 dataset, although there is a slight decrease in performance according to the IDF1 metric, there is an improvement in the IDs metric, enhancing the stability of multi-object tracking to some extent.

What Features Does BMM Extract? To investigate the reasons behind the outstanding performance of BMM, we visualized the extracted features by comparing them with traditional ViT and ResNeSt models. Figure 4 showcases the experimental results, providing intuitive evidence. The traditional ResNeSt network exhibits poor focus, while the ViT network produces overly broad features. In contrast, the BMM network demonstrates a more focused feature extraction ability, leading to more accurate similarity measurements. The analysis indicates that the partitioning display and feature combination in the final stage enhance the model's ability to distinguish similar target scenes to a certain extent. Consequently, the proposed BMM module significantly contributes to improving the model's robustness.

Table 5. Performance comparison of different association methods on MOT17 and MOT20 datasets.

Method	MOT17			MOT20		
	MOTA	IDF1	IDs	MOTA	IDF1	HOTA
Baseline	79.3	78.9	1458	76.7	76.4	62.5
EDM	79.3	79.1	1439	76.7	76.6	62.5

Effect of Euclidean Distance Module (EDM)

EDM vs BOT-SORT Method. We conducted comparative experiments on the EDM module introduced in this article, and the results are presented in Table 5. To ensure a fair comparison, multiple multi-object tracking models were utilized with consistent detection and Reid models, with modifications limited to the association method. The experimental results demonstrate that our EDM module has the potential to enhance the IDF1 metric of the model to some extent.

Effect of Euclidean Distance Module(EDM)

Fig. 6. Tracking scenes from MOT17 train dataset using our BWtrack model. Different colored bounding boxes represent tracked targets, with the numbers in the top left corner indicating target IDs. The number in the top left corner of the image indicates the frame number.

What Tracking Scenes Does EDM Optimize? To investigate the reasons behind the exceptional performance of EDM, we visualized the outcomes of various association methods. Figure 5 provides intuitive evidence that EDM outperforms traditional association methods in densely populated and occluded scenes. In such scenarios, neither IOU nor traditional Reid matching strategies can effectively match targets, whereas EDM achieves successful target matching. This addresses the challenge of ineffective matching caused by high confidence levels of traditional IOU and Reid networks when dealing with similar targets. The analysis indicates that incorporating a Euclidean distance-matching strategy enhances the model's ability to distinguish densely populated and occluded scenes to some extent. As a result, the proposed EDM module significantly contributes to improving the model's robustness.

More Visualization Results

To visually showcase the tracking performance of our model, we conducted additional visualization experiments on the MOT17 train dataset. The results, depicted in Fig. 6, demonstrate the superior tracking performance of our model in occluded and densely crowded scenes.

Parameter Evaluation

To delve into the model's complexity, we conducted a comparative analysis of the parameter quantity. The measurement results can be found in Table 6. For the experiment, all Reid models utilized the same input (1, 3, 384, 128) and operated under identical software and hardware environments.

Table 6. Comparison of parameter quantity among different backbone models.

Backbone	FLOPs(G)	Params(M)
ResNeSt	7.0	29.5
ViT	16.5	87.1
BMM	53.2	278.3

5   Conclusion

In this paper, we proposed BWTrack, a novel multi-object tracking model that incorporates a Block Matching Module (BMM) and an Euclidean Distance Module (EDM) to enhance feature extraction and matching capabilities. The BMM utilizes block-based feature extraction to evaluate similarity in crowded scenes. The EDM further enhances matching capabilities using Euclidean distance. Our model achieved state-of-the-art performance on MOT16, MOT17, and MOT20 datasets while maintaining a high speed of 7 FPS on a single GPU. However, generalizing our model's performance to other datasets or real-world scenarios may vary. Future work entails exploring advanced techniques to further enhance our model's performance.

References

1. Aharon, N., Orfaig, R., Bobrovsky, B.Z.: BoT-SORT: robust associations multi-pedestrian tracking. arXiv preprint arXiv:2206.14651 (2022)
2. Bewley, A., Ge, Z., Ott, L., Ramos, F., Upcroft, B.: Simple online and realtime tracking. In: ICIP, pp. 3464–3468. IEEE (2016)
3. Cai, J., et al.: MeMOT: multi-object tracking with memory. In: CVPR, pp. 8090–8100 (2022)

4. Cao, J., Pang, J., Weng, X., Khirodkar, R., Kitani, K., et al.: Observation-centric sort: rethinking sort for robust multi-object tracking. arXiv preprint arXiv:2203.14360 (2022)
5. Chu, P., Fan, H., Tan, C.C., Ling, H.: Online multi-object tracking with instance-aware tracker and dynamic model refreshment. In: WACV, pp. 161–170. IEEE (2019)
6. Dendorfer, P., et al.: MOT20: a benchmark for multi object tracking in crowded scenes. arXiv preprint arXiv:2003.09003 (2020)
7. Dosovitskiy, A., et al.: An image is worth 16x16 words: transformers for image recognition at scale. arXiv preprint arXiv:2010.11929 (2020)
8. Liang, C., Zhang, Z., Zhou, X., Li, B., Zhu, S., Hu, W.: Rethinking the competition between detection and ReID in multiobject tracking. TIP, 3182–3196 (2022)
9. Milan, A., Leal-Taixé, L., Reid, I., Roth, S., Schindler, K.: MOT16: a benchmark for multi-object tracking. arXiv preprint arXiv:1603.00831 (2016)
10. Pang, J., et al.: Quasi-dense similarity learning for multiple object tracking. In: CVPR, pp. 164–173 (2021)
11. Peng, J., et al.: Chained-tracker: chaining paired attentive regression results for end-to-end joint multiple-object detection and tracking. In: Vedaldi, A., Bischof, H., Brox, T., Frahm, J.-M. (eds.) ECCV 2020. LNCS, vol. 12349, pp. 145–161. Springer, Cham (2020). https://doi.org/10.1007/978-3-030-58548-8_9
12. Peng, J., et al.: TPM: multiple object tracking with tracklet-plane matching. Pattern Recogn. **107**, 107480 (2020)
13. Shan, C., et al.: Tracklets predicting based adaptive graph tracking. arXiv preprint arXiv:2010.09015 (2020)
14. Sun, P., et al.: Transtrack: multiple object tracking with transformer. arXiv preprint arXiv:2012.15460 (2020)
15. Vaswani, A., et al.: Attention is all you need. In: NeurIPS, vol. 30, pp. 6000–6010 (2017)
16. Wang, T., et al.: Spatio-temporal point process for multiple object tracking. IEEE Trans. Neural Netw. Learn. Syst. **34**(4), 1777–1788 (2023)
17. Xu, Y., Ban, Y., Delorme, G., Gan, C., Rus, D., Alameda-Pineda, X.: TransCenter: transformers with dense queries for multiple-object tracking. arXiv e-prints, arXiv-2103 (2021)
18. Yang, F., Odashima, S., Masui, S., Jiang, S.: Hard to track objects with irregular motions and similar appearances? Make it easier by buffering the matching space. In: WACV, pp. 4799–4808 (2023)
19. Yu, F., Li, W., Li, Q., Liu, Yu., Shi, X., Yan, J.: POI: multiple object tracking with high performance detection and appearance feature. In: Hua, G., Jégou, H. (eds.) ECCV 2016. LNCS, vol. 9914, pp. 36–42. Springer, Cham (2016). https://doi.org/10.1007/978-3-319-48881-3_3
20. Yu, E., Li, Z., Han, S., Wang, H.: RelationTrack: relation-aware multiple object tracking with decoupled representation. IEEE Trans. Multimedia, 2686–2697 (2021)
21. Zeng, F., Dong, B., Zhang, Y., Wang, T., Zhang, X., Wei, Y.: MOTR: end-to-end multiple-object tracking with transformer. In: Avidan, S., Brostow, G., Cissé, M., Farinella, G.M., Hassner, T. (eds.) ECCV, vol. 13687, pp. 659–675. Springer, Cham (2022). https://doi.org/10.1007/978-3-031-19812-0_38
22. Zhang, H., et al.: ResNeSt: split-attention networks. arXiv preprint arXiv:2004.08955 (2020)

23. Zhang, Y., Wang, C., Wang, X., Zeng, W., Liu, W.: FairMOT: on the fairness of detection and re-identification in multiple object tracking. Int. J. Comput. Vis. **129**, 3069–3087 (2021)
24. Zhang, Y., et al.: ByteTrack: multi-object tracking by associating every detection box. In: Avidan, S., Brostow, G., Cissé, M., Farinella, G.M., Hassner, T. (eds.) ECCV, vol. 13682, pp. 1–21. Springer, Cham (2022). https://doi.org/10.1007/978-3-031-20047-2_1
25. Bergmann, P., Meinhardt, T., Leal-Taixe, L.: Tracking without bells and whistles. In: ICCV, pp. 941–951 (2019)
26. Chaabane, M., Zhang, P., Beveridge, J.R., O'Hara, S.: Deft: detection embeddings for tracking. arXiv preprint arXiv:2102.02267 (2021)
27. Chu, P., Wang, J., You, Q., Ling, H., Liu, Z.: TransMOT: spatial-temporal graph transformer for multiple object tracking. In: WACV, pp. 4870–4880 (2023)
28. Emami, P., Pardalos, P.M., Elefteriadou, L., Ranka, S.: Machine learning methods for data association in multi-object tracking. ACM Comput. Surv. (CSUR), 1–34 (2020)
29. Fang, K., Xiang, Y., Li, X., Savarese, S.: Recurrent autoregressive networks for online multi-object tracking. In: WACV, pp. 466–475. IEEE (2018)
30. Ge, Z., Liu, S., Wang, F., Li, Z., Sun, J.: YOLOx: exceeding yolo series in 2021. arXiv preprint arXiv:2107.08430 (2021)
31. Han, S., Huang, P., Wang, H., Yu, E., Liu, D., Pan, X.: MAT: motion-aware multi-object tracking. Neurocomputing **476**, 75–86 (2022)
32. He, L., Liao, X., Liu, W., Liu, X., Cheng, P., Mei, T.: FastReID: a Pytorch toolbox for general instance re-identification. arXiv preprint arXiv:2006.02631 (2020)
33. Hyun, J., Kang, M., Wee, D., Yeung, D.Y.: Detection recovery in online multi-object tracking with sparse graph tracker. In: WACV, pp. 4850–4859 (2023)
34. Li, W., Xiong, Y., Yang, S., Xu, M., Wang, Y., Xia, W.: Semi-TCL: semi-supervised track contrastive representation learning. arXiv preprint arXiv:2107.02396 (2021)
35. Liang, C., Zhang, Z., Zhou, X., Li, B., Hu, W.: One more check: making "fake background" be tracked again. In: AAAI, vol. 36, pp. 1546–1554 (2022)
36. Mahmoudi, N., Ahadi, S.M., Rahmati, M.: Multi-target tracking using CNN-based features: CNNMTT. Multimedia Tools Appl., 7077–7096 (2019)
37. Pang, B., Li, Y., Zhang, Y., Li, M., Lu, C.: TubeTK: adopting tubes to track multi-object in a one-step training model. In: CVPR, pp. 6308–6318 (2020)
38. Seidenschwarz, J., Braso, G., Elezi, I., Leal-Taixe, L.: Simple cues lead to a strong multi-object tracker. arXiv preprint arXiv:2206.04656 (2022)
39. Stadler, D., Beyerer, J.: Modelling ambiguous assignments for multi-person tracking in crowds. In: WACV, pp. 133–142 (2022)
40. Tokmakov, P., Li, J., Burgard, W., Gaidon, A.: Learning to track with object permanence. In: ICCV, pp. 10860–10869 (2021)
41. Wang, Q., Zheng, Y., Pan, P., Xu, Y.: Multiple object tracking with correlation learning. In: CVPR, pp. 3876–3886 (2021)
42. Wang, Y., Kitani, K., Weng, X.: Joint object detection and multi-object tracking with graph neural networks. In: ICRA, pp. 13708–13715 (2021)
43. Wang, Z., Zheng, L., Liu, Y., Li, Y., Wang, S.: Towards real-time multi-object tracking. In: Vedaldi, A., Bischof, H., Brox, T., Frahm, J.-M. (eds.) ECCV 2020. LNCS, vol. 12356, pp. 107–122. Springer, Cham (2020). https://doi.org/10.1007/978-3-030-58621-8_7
44. Wojke, N., Bewley, A.: Deep cosine metric learning for person re-identification. In: WACV, pp. 748–756. IEEE (2018)

45. Wojke, N., Bewley, A., Paulus, D.: Simple online and realtime tracking with a deep association metric. In: ICIP, pp. 3645–3649. IEEE (2017)
46. Wu, J., Cao, J., Song, L., Wang, Y., Yang, M., Yuan, J.: Track to detect and segment: an online multi-object tracker. In: CVPR, pp. 12352–12361 (2021)
47. Yang, F., Chang, X., Sakti, S., et al.: ReMOT: a model-agnostic refinement for multiple object tracking. Image Vis. Comput. **106**, 104091 (2021)
48. Zheng, L., Tang, M., Chen, Y., Zhu, G., Wang, J., Lu, H.: Improving multiple object tracking with single object tracking. In: CVPR, pp. 2453–2462 (2021)
49. Zhou, X., Koltun, V., Krähenbühl, P.: Tracking objects as points. In: Vedaldi, A., Bischof, H., Brox, T., Frahm, J.-M. (eds.) ECCV 2020. LNCS, vol. 12349, pp. 474–490. Springer, Cham (2020). https://doi.org/10.1007/978-3-030-58548-8_28

RPF3D: Range-Pillar Feature Deep Fusion 3D Detector for Autonomous Driving

Yihan Wang and Qiao Yan(✉)

Nanyang Technological University, Singapore 639798, Singapore
{WANG1517,QIAO003}@e.ntu.edu.sg

Abstract. In this paper, we present *RPF3D*, an innovative single-stage framework that explores the complementary nature of point clouds and range images for 3D object detection. Our method addresses the sampling region imbalance issue inherent in fixed-dilation-rate convolutional layers, allowing for a more accurate representation of the input data. To enhance the model's adaptability, we introduce several attention layers that accommodate a wide range of dilation rates necessary for processing range image scenes. To tackle the challenges of feature fusion and alignment, we propose the AttentiveFusion module and the Range Image Guided Deep Fusion (RIGDF) backbone architecture in the Range-Pillar Feature Fusion section, which effectively addresses the one-pillar-to-multiple-pixels feature alignment problem caused by the point cloud encoding strategy. These innovative components work together to provide a more robust and accurate fusion of features for improved 3D object detection. We validate the effectiveness of our RPF3D framework through extensive experiments on the KITTI and Waymo Open Datasets. The results demonstrate the superior performance of our approach compared to existing methods, particularly in the Car class detection where a significant enhancement is achieved on both datasets. This showcases the practical applicability and potential impact of our proposed framework in real-world scenarios and emphasizes its relevance in the domain of 3D object detection.

Keywords: Point Clouds · Deep Learning · 3D Object Detection

1 Introduction

As autonomous driving technology advances, the demand for precise and reliable 3D perception systems continues to rise to ensure safe and efficient driving. Despite significant progress, 2D object detection techniques grapple with providing accurate depth information, offering predictions of object locations in camera coordinates without depth data. This deficiency can result in imprecise predictions, particularly in intricate driving conditions such as urban environments with varying object distances. While studies like [45] have made strides in 3D detection using videos, the accuracy of these methods remains disputable due to

B. Luo et al. (Eds.): ICONIP 2023, LNCS 14449, pp. 125–146, 2024.
https://doi.org/10.1007/978-981-99-8067-3_10

their reliance on depth prediction, thus highlighting the persisting challenge of depth information loss. 3D object detection techniques employing point clouds hold potential in delivering precise depth information, a critical factor for applications like autonomous driving. However, these methods face difficulties due to the low resolution and unstructured nature of point clouds. Additionally, the practical deployment of these techniques is hindered by low efficiency and considerable memory requirements characteristic of current approaches. As the significance of 3D object detection escalates across diverse applications, researchers have proposed numerous methodologies, which can be primarily categorized into three main strategies: grid-based, point-based, and range image-based detection.

Grid-based methods divide the 3D coordinate space into voxels [2,8,27,30, 40,50] or pillars [13,37], which can be processed using 3D or 2D convolution for feature extraction. However, a trade-off exists between efficiency and information loss. Dense convolution, which can be used to encode voxel or pillar data, can lead to inefficiency [27], while sparse convolution, which is used to process point data, can result in information loss [25]. Therefore, many **point-based** approaches [24,29,30,38,44] are developed which can store more information than methods based on grids. However, the processing of a large number of points is inefficient, making these methods less suitable for real-time applications. Furthermore, point-based methods often require pre-processing steps such as downsampling or grouping, which can result in information loss and reduced accuracy. **Range images** provide a compact 2D representation of point clouds with minimal loss of the original 3D point cloud data [19]. This advantage has led to the development of numerous range image-based frameworks in recent years [1,21,37]. In addition, various multimodal-based approaches have been proposed that incorporate range images into their methodologies [11,12,15,23]. However, range images face two primary challenges: occlusion of objects within the scene and scale variation among objects. These challenges can result in reduced accuracy and performance of range image-based 3D object detection methods. To address these challenges, a second stage can be added to the detection pipeline to refine an initial group of proposals. This can potentially alleviate some of the quality concerns related to occlusion and scale variation, although it may result in a considerable increase in computational demands.

In order to tackle the aforementioned occlusion issue while maintaining computational efficiency, a novel anchor-free and end-to-end model is proposed. This innovative approach effectively combines the range-view representation with the point-view representation, resulting in a synergistic solution that delivers satisfactory performance. To tackle the serious occlusion problem, we introduce the original point cloud as guidance and information compensation in the parallel processing stream. To overcome the scale variation issue, we employ dilated convolution layers at the range image feature extraction stage. To more effectively fuse information from two distinct representations while balancing computational costs, we propose a novel backbone called RIGDF. This approach processes both range-view representation and point-view representation in the bird's-eye-view, ensuring efficient information integration.

Our proposed method, referred to as *RPF3D*, is an innovative multi-view-based deep-fusion algorithm. Distinct from existing approaches, *RPF3D* combines perspective view and 3D points in a deep-fusion module, whereas previous methods fuse two or more different 2D views in a single pipeline [3]. Moreover, *RPF3D* meticulously integrates features multiple times from two different dimensions through a carefully designed architecture, unlike previous methods that simply integrate diverse feature maps at early-stage or late-stage.

Contributions. Our primary contributions can be summarized as follows:

- We propose an anchor-free single-stage end-to-end pipeline, which integrates the multiple views of point clouds to extract dense feature maps for boosting the effectiveness of 3D object detection. This pipeline represents a significant improvement over existing methods that rely on anchor-free detection.
- We present a series of feature fusion methods for deep integration of multi-stream features, resulting in substantial accuracy and performance enhancements over existing approaches.
- We conduct extensive analysis on both the KITTI Dataset [5] and the Waymo Open Dataset [32], demonstrating the robustness of deep-level feature fusion and the impressive accuracy of our proposed *RPF3D*.

2 Related Work

Grid-Based 3D Object Detection Methods. Initial methods transformed 3D point cloud data into bird's-eye view (BEV) representations to handle point cloud irregularities. AVOD [12] improved region proposal quantization. Studies like [16,17] extended this technique for multi-modal 3D object detection, while [41] adapted it for LiDAR-only detection. [13,42] boosted efficiency with BEV. We avoid manual voxelization using 3D sparse convolution, as proposed in SEC-OND [2,6,8,30,40,50]. Voxelization introduces sparsity issues due to predefined voxel sizes, impacting performance for distant objects. A fully 3D mesh adds computational and memory load, necessitating sparse convolutions for scalability.

Point-Based 3D Object Detection Methods. Point-based methods face significant computational challenges in early cascade approaches compared to grid-based methods. In [24,38], the process begins with 2D object detection on images, followed by forming 3D proposals from projected 2D bounding boxes, and finally using PointNet [25,26] for 3D proposal regression. However, [31] lacks a sampling strategy, making it less efficient compared to PointNet. In contrast, [27,29], and [20] directly propose 3D bounding boxes through point-wise-supervised training. [28] enhances detection accuracy while addressing the time-consuming aspect of [30]. Additionally, [44] introduces an innovative sampling strategy based on feature distance, which preserves more interior points.

Range Image-Based 3D Object Detection Methods. Range image-based detection methods have not garnered as much attention as grid-based or point-based approaches. In the realm of seminars and research, LaserNet [21] directly

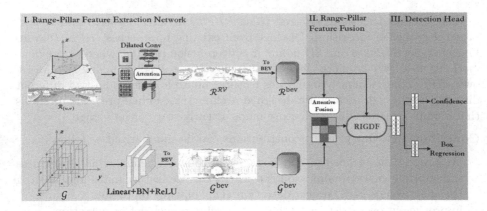

Fig. 1. *RPF3D* overview. Our architecture consists of three main modules: I. Range-Pillar Feature Extraction Network for obtaining features from the range view and point view, II. Range-Pillar Feature Fusion stage for deep integration of multi-view features, and III. Detection Head for object bounding box regression. The input includes the original 3D point cloud and its corresponding range view.

applies 2D convolutional neural networks (CNNs) to range images for bounding box regression. RangeRCNN [18] introduces the RV-PV-BEV module, which facilitates the transition from range-view (RV) to bird's-eye view (BEV) and supports continuous improvement. Building on the foundation of [18,19] proposes a multi-scale point-based module to address the boundary blurring issue of objects in pixel coordinates. RangeSparseNet [33] differentiates between foreground and background points, transforming the range image into a 3D view for bounding box regression. RCD [1] presents a novel convolution layer featuring a scale-invariant dilation rate that can be dynamically adjusted.

In this study, we propose a deep fusion method utilizing exclusively multi-view features as opposed to features from other modalities. In contrast to traditional point-based methods, *RPF3D* demonstrates superior efficiency. Furthermore, compared to other range image-based methods, *RPF3D* innovatively integrates elevation information from 3D context, resulting in enhanced performance.

3   RPF3D for Point Cloud 3D Object Detection

In this section, we present a new framework, named *RPF3D*, which is specifically designed for 3D object detection applications with an emphasis on achieving higher accuracy. We provide an overview of the architecture and subsequently introduce each section in detail.

3.1   RPF3D

Figure 1 illustrates the construction of our *RPF3D* network infrastructure, which takes the raw point cloud and the corresponding range-view image as input in

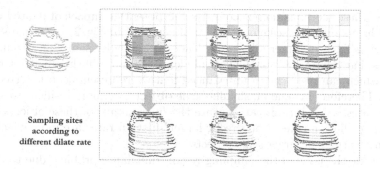

Fig. 2. Illustration of the sampling region imbalance problem mentioned in Sect. 3.2. Specifically, for a fixed-scale object, different dilate rates can abstract different features depending on the sampling site, resulting in potential variations in the resulting feature representations.

parallel. The proposed *RPF3D* framework is composed of three main components: (1) a multi-view feature extractor that captures several input features derived from the range view and the 3D point view; (2) Range-Pillar Feature Fusion module which includes the AttentiveFusion module and the RIGDF module for deeply fusing features from two different branches; and (3) a detection head for regressing 3D boxes.

In our *RPF3D* framework, we process the point cloud data through two parallel streams: a range image and a corresponding set of pillars. The raw point cloud is projected to a range image and partitioned into pillars, forming the first and second input streams, respectively. The RV representation branch, equipped with three dilated convolution layers and two attention layers, hierarchically extracts features from these streams, transforming the feature map into a dense, bird's-eye view feature map, denoted as $\mathcal{R}^{bev}$. Concurrently, the pillar-formed pseudo images are projected onto the BEV, generating $\mathcal{G}^{bev}$. These multi-view features are synchronously integrated via the AttentiveFusion module, yielding a comprehensive feature map. Subsequently, this output, in conjunction with $\mathcal{R}^{bev}$, feeds into the Range Image Guided Deep Fusion (RIGDF) backbone. Employing max-pooling, the RIGDF backbone leverages $\mathcal{R}^{bev}$ at various stages of the top-down subnetwork, guiding the formation of deeper features. The deconvolution subnetwork resizes and amalgamates the generated feature maps at each stage before routing the output to the continuous detection head. This strategy facilitates efficient and precise 3D object detection within point clouds, capitalizing on multi-view representation and sophisticated feature extraction methods.

3.2   Range-Pillar Feature Extraction Network

Range Image Feature Extracting Branch. Prior works have proposed various range image encoding methods to address the scale variation problem caused by the range image representation. For instance, RangeIoUDet [19] utilizes several fixed dilated convolutional layers to adapt to objects of different scales.

However, these methods do not consider the potential impact of dilated convolutional layers on fixed-scale objects. As illustrated in Fig. 2, when the scale of an object is fixed, a larger dilate rate leads to a smaller sampling region. This indicates that the convolutional layer only learns features in the smaller sampling region, which can result in an unbalanced feature representation for fixed-scale objects. Therefore, in general, smaller-scale objects are more resilient to smaller dilation rates, as such rates can capture the finer details of these objects better than larger dilation rates. Conversely, larger dilation rates are more effective in capturing features of larger-scale objects.

To tackle the issue of the sampling region imbalance problem due to the use of different fixed-dilation-rate layers, we introduce utilizing two attention modules in parallel to enhance the representation power of the feature extraction layers, as shown in Fig. 3. Specifically, dilate convolution layers with dilation rates $d_r = \{i|i = 1, 2, 3\}$ are first conducted to extract three different feature maps. Afterward, two attention modules are introduced hierarchically: the spatial attention module and the channel attention module, to better integrate the different feature maps generated by the different dilation rate layers. To optimize the representation of these feature maps, a channel attention layer is introduced to reweight the feature maps. The channel attention layer considers the importance of each feature map, based on the dilation rate of the corresponding convolution layer. This is important because smaller dilation rates are more effective in capturing fine details of smaller-scale objects, while larger dilation rates are more effective for larger-scale objects [19]. Following the channel attention module's decision on which feature map to prioritize based on the corresponding dilation rate, a spatial attention module is applied for in-channel feature map enhancement. In essence, the channel attention layer helps to find a more suitable solution for accommodating the range of dilation rates needed for the scene of the range image, while the spatial attention module refines the representation of the range image features by enabling the network to focus on the most salient regions of the feature maps. By considering the relative importance of each feature map, these attention layers help to refine the overall representation of the range image features, leading to improved performance in downstream tasks.

The channel attention module employs a 1D channel weight matrix $M_c \in \mathbb{R}^{1 \times 1 \times C}$ to adaptively focus on the distinct features from the three dilate convolution layers. This process facilitates the extraction of informative object features and can be formulated as:

$$M_c(F_{d_r}) = \text{sigmoid}(\text{MLP}(A.A.(F_{d_r})) + \text{MLP}(A.M.(F_{d_r}))) \tag{1}$$

where MLP is the shared multilayer perception layer, $A.A.$ and $A.M.$ indicates the adaptive average pooling and the adaptive max pooling respectively.

The spatial attention module generates a 2D spatial weight matrix $M_s \in \mathbb{R}^{H \times W \times 1}$ that highlights spatial content effectively, which can be expressed as:

$$M_s(F_{d_r}) = \text{sigmoid}(\text{Conv2d}(\text{Concat}[A.A.(F_{d_r}), A.M.(F_{d_r})]) \tag{2}$$

where F_{d_r} denotes the feature map generated by dilated convolution layers. Both Eq. (1) and Eq. (2) are activated by the sigmoid activation function.

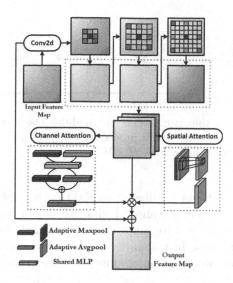

Fig. 3. Elaboration of the architecture for range image feature extraction module. This module consists of three 2D dilated convolutional layers that generate three feature maps, which are then concatenated and fed through Channel Attention and Spatial Attention mechanisms to improve feature representation.

Pillar Feature Extracting Branch. PointPillars [13] is a widely used method in state-of-the-art 3D detectors for efficiently encoding point clouds into sparse 3D feature volumes. Due to its high efficiency and accuracy, we adopt it in our pillar feature extraction branch. Specifically, we use the PointPillars method to convert the raw point cloud data into a pillar representation, where each pillar represents a local neighborhood of points within a small 3D region. We then extract the features from each pillar and generate a pillar-feature matrix, which is passed to the subsequent layers for further processing. The use of PointPillars in our framework enables efficient processing of large-scale point cloud data with high accuracy, making it a suitable choice for 3D object detection applications.

If we define the pillar feature map $\mathcal{G}$ of point cloud $\mathcal{P}$ as $\mathcal{G} = \{g_i\}_{i=0}^{\mathcal{M}-1} \subseteq \mathbb{R}^{\mathcal{M}}$, where $\mathcal{M}$ is the dimensions of $\mathcal{G}$, the pillar features can be aggregated as:

$$g_i = \mathbb{P}\{p_i | \forall p_i \in \mathrm{F_w}(\mathrm{w_j})\} \tag{3}$$

where $\mathbb{P}$ indicates the PointNet-block [25] for feature abstraction, $p_j \in w_j$ gives the relation of point p_i and pillar w_j, $\mathrm{F_w}(\mathrm{w_j})$ is a set of points in w_j.

3.3   Range-Pillar Feature Fusion

The conventional approach of concatenating intermediate features from multi-streams fails to fully explore the relationships between different feature maps, merging them only superficially without considering inherent correlations and dependencies. This could lead to a fused feature map that doesn't adequately

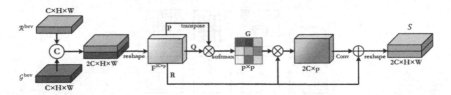

Fig. 4. An illustration of the AttentiveFusion module. The feature maps generated at the first stage are concatenated and processed using the AttentiveFusion module, resulting in an output with shape $2C \times H \times W$.

harness the synergistic effects of different modalities, possibly introducing noise or redundancy. Hence, a more advanced fusion method is crucial to model complex relationships between feature maps and extract more informative and discriminative representations. In response, we present a novel backbone for our proposed *RPF3D* framework, enabling more accurate extraction of higher-level multi-view features.

AttentiveFusion. We extract multi-view features $\mathcal{R}^{\mathcal{RV}}$ from the range view and $\mathcal{G}^{bev}$ from the bird's-eye view and pass them through the fusion module and backbone for further processing. Notably, methods like PointPainting [34] and PointAugmenting [35] employ input-level decoration, also known as early-stage fusion, wherein a one-to-one correspondence between a single camera pixel and a 3D LiDAR point is established. Additionally, work like [51] enhances the one-to-one corresponding features through attention head. Contrarily, in our proposed pipeline, each LiDAR feature is associated with a pillar containing a multitude of points, while the corresponding range view features constitute an area. Consequently, this leads to a one-pillar-to-multiple-pixels alignment issue.

Addressing this, the AttentiveFusion module is specifically designed to cope with this challenge. Leveraging a novel geometric attention strategy, it endows our model with the ability to discern and incorporate a broader range of contextual information into local features, thus enhancing its representational capability. It innovatively resolves the one-pillar-to-multiple-pixels alignment issue by adaptively learning and applying weights (**G** in Fig. 4) to the area of the range view corresponding to a specific pillar. This strategy allows the model to handle varying densities and discrepancies between the two representations effectively, ensuring a harmonized feature representation.

Figure 4 illustrates the AttentiveFusion module of feature maps from range view representation and point view representation in our proposed framework. The two sets of features are first concatenated and reshaped to form a new tensor $\mathbf{F}^{C \times p}$, where $p = H \times W$. Next, two separate convolutional layers are applied to generate two new feature maps, denoted as **P** and **Q**, with dimensions $2C \times p$. To calculate the geometrical attention map, we perform a matrix multiplication between the transpose of **P** and **Q**, followed by a softmax layer to normalize the output **G** with size $p \times p$. Once the geometrical attention map **G** is obtained, it is used to weight the origin concatenated feature $\mathbf{F}$ to produce a new feature map $\mathbf{R}$ with dimensions $2C \times p$. This feature map is then convolved and added with the origin concatenated feature $\mathbf{F}$, before being reshaped to $2C \times H \times W$. The final output feature map $\mathcal{S}$ can be defined as:

$$\mathcal{S}_k = \lambda \Sigma_{j=1}^p (\mathbf{R}_j \frac{e^{\mathbf{P}_k \cdot \mathbf{Q}_j}}{\Sigma_{j=1}^p e^{\mathbf{P}_k \cdot \mathbf{Q}_j}}) + \mathbf{F}_k \tag{4}$$

Here, j and k denote the jth position and the kth positions of $\mathbf{P}$ and $\mathbf{Q}$. The weight parameter λ is initialized to 0, progressively learns to assign more weight.

Range Image Guided Deep Fusion. The proposed Range Image Guided Deep Fusion (RIGDF), depicted in Fig. 5, consists of two subnetworks: a top-down subnetwork and a deconvolution subnetwork. Initially the attentively fused feature map $\mathcal{S}$ and features from RV $\mathcal{R}^{\mathcal{RV}}$ are concatenated and fed into the top-down subnetwork's first convolutional layer. To extract the most representative range image features, a max-pooling operation is performed, the results of which are incorporated into the ensuing convolutional layers. This selective approach to feature extraction offers valuable guidance for later processing steps. By integrating these features into the top-down subnetwork, we can effectively combine the range image information with the bird's-eye view features, leading to a more comprehensive and informative feature representation. This procedure is expressed as follows:

$$\begin{cases} ConvMD(C_{in}^1, \mathcal{F}_{df}^1, \mathcal{C}, \mathcal{H}, \mathcal{D}) \\ Conv2MD(C_{in}^2, \mathcal{F}_{df}^2, \mathcal{C}, \mathcal{H}, \mathcal{D}) \\ Conv4MD(C_{in}^3, \mathcal{F}_{df}^3, \mathcal{C}, \mathcal{H}, \mathcal{D}) \end{cases} \tag{5}$$

where $ConvMD(\cdot)$ indicates the M-Dimensional convolution, with a BatchNorm and a ReLU followed. $\mathcal{F}_{df}^1, \mathcal{F}_{df}^2, \mathcal{F}_{df}^3$ denote the output features of the first, middle and the last convolutional layer in this top-down subnetwork respectively. $\mathcal{C}, \mathcal{H}, \mathcal{D}$ indicate the kernel size, stride size and the padding size of $ConvMD(\cdot)$. Particularly, $C_{in}^1, C_{in}^2, C_{in}^3$ are the input of the corresponding convolutional layer, which can be formalized as:

$$C_{in}^1 = concat[\mathcal{S}, \mathcal{R}^{bev}]$$
$$C_{in}^2 = concat[\mathcal{F}_{df}^1, MP(\mathcal{R}^{bev})] \tag{6}$$
$$C_{in}^3 = concat[\mathcal{F}_{df}^2, MP(MP(\mathcal{R}^{bev}))]$$

where $concat[\cdot]$ denotes the concatenation of multiple features at feature channel dimension, $MP(\cdot)$ means max-pooling operation of the range feature map $\mathcal{R}^{bev}$.

Finally, each output from the top-down subnetwork layers is fed into the deconvolutional module to generate the final output. The deconvolutional module upsamples the feature maps from the top-down subnetwork, ensuring that the final output contains all relevant features needed for downstream tasks.

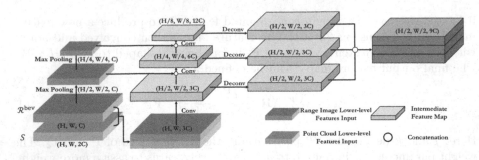

Fig. 5. Architecture of the range image-guided deep fusion (RIGDF) backbone. The output of the AttentiveFusion module, together with the range-view feature maps, serves as the input for RIGDF. RIGDF consists of a top-down subnetwork and a sequential deconvolution subnetwork, which are used to generate an output that is fed into a detection head for generating predictions.

Fig. 6. Description of the Attention Map. (a) RGB images. (b) range images. (c) feature maps generated without Channel Attention and Spatial Attention. (d) feature maps generated with Channel Attention and Spatial Attention.

Our proposed deep fusion module introduces a novel fusion strategy that distinguishes it from other multi-view fusion methods, such as [16]. Unlike these approaches, our method fuses different views of features to increasingly deeper convolutional layers, thereby enhancing the representation power of the feature maps. This approach allows for a more effective fusion of features from different views, resulting in improved 3D object detection performance. By fusing the features from different views at deeper layers of the network, our method is able to capture more high-level semantic information and spatial relationships between the features. This leads to a more comprehensive and informative representation of the input data, which is better suited to the complex and dynamic nature of real-world 3D environments.

Detection Head. Upon extracting the fusion features, the detect head of our model generates an IoU-related confidence score and performs additional box coordinate regression for each region proposal. To achieve this, our detect head employs an architecture similar to that used in prior works such as [4], which involves several fully connected layers (Fig. 6).

3.4   Loss Function

We introduce the identical loss functions as those illustrated in [13,40], MVF [49] and Pillar-OD [37] which divided the overall loss function into two parts: the regression loss and the classification loss. The regression loss defines the difference between the predicted bounding box $(x_e, y_e, z_e, l_e, w_e, h_e, \mathbf{A}_e)$ and the ground truth $(x_{gt}, y_{gt}, z_{gt}, l_{gt}, w_{gt}, h_{gt}, \mathbf{A}_{gt})$. Specifically, the angle regression loss function to effectively solve the adversarial orientations problem can be computed by:

$$\mathcal{L}_{angle} = \text{Smooth L1}[sin(\mathbf{A}_e - \mathbf{A}_{gt})] \tag{7}$$

where $\mathbf{A}$ is the angle and subscript e is the corresponding prediction.

For the direction regressing part, softmax loss function $\mathcal{L}_{dir} = -logC_k$ is applied to assess whether yaw is a non-negative angle or not as the angle localization loss cannot distinguish flipped boxes.

Loss of classification is characterized as a focal loss:

$$\mathcal{L}_{class} = -\alpha_p(1 - \mathbf{c}_p)^\gamma log\mathbf{c}_p \tag{8}$$

where $\mathbf{c}_p$ is the confidence of model prediction, α, γ are two parameters of the model and they are set as $\alpha = 0.25, \gamma = 2$ as in [40] respectively. Thus the total loss can be written as:

$$\mathcal{L}_{all} = \frac{1}{N_+}(\beta_1\mathcal{L}_{other} + \sum_{i=1}^{3}\beta_i\mathcal{L}_x) \tag{9}$$

where N_+ is the number of positive boxes, $x \in (angle, dir, class)$ sequentially, $\beta_1 = 2, \beta_2 = 0.2, \beta_3 = 1$. $other$ subscript refers to the other two characteristics: location $(\Delta x, \Delta y, \Delta z)$ and dimension $(\Delta l, \Delta w, \Delta h)$. Both $\mathcal{L}_{loc}$ and $\mathcal{L}_{dim}$ can be calculated by:

$$\mathcal{L}_{other} = \sum_{\Delta v \in \{\Delta x, \Delta y, \Delta z, \Delta l, \Delta w, \Delta h\}} \text{Smooth L1}(\Delta v) \tag{10}$$

where $\Delta x = \frac{x_{gt} - x_e}{\sqrt{(l_e)^2 + (w_e)^2}}$, $\Delta y = \frac{y_{gt} - y_e}{\sqrt{(l_e)^2 + (w_e)^2}}$, $\Delta z = \frac{z_{gt} - z_e}{h_e}$. $\Delta l = \log(l_{gt} - l_e)$, $\Delta w = \log(w_{gt} - w_e)$, $\Delta h = \log(h_{gt} - h_e)$.

4   Experiment

4.1   Dataset

To highlight the effectiveness of the present scheme, we leverage two state-of-the-art open datasets, KITTI and Waymo Open Dataset, for executing our model.

KITTI. Our experiments are conducted on the KITTI Dataset, which comprises samples of both RGB images and point clouds in urban driving scenes. However, we only utilize the point cloud data in our experiments. The dataset contains annotated objects from three common categories in the street, namely *Car, Pedestrian,* and *Cyclist.* The LiDAR point cloud data is officially divided into 7481 training samples and 7518 testing samples, the training samples are further split into a *train* set comprising 3712 samples and a *val* set comprising 3769 samples. During our analytical experiments, we train our model on the train set and evaluate it on the validation set. For the online test server, we randomly select 80% of the whole training samples to train our model and leave the rest 20% for model validation. We report the average precision of our method on both the official testing server and commonly used *val* set, thereby enabling comparison with other state-of-the-art methods.

Waymo Open Dataset. The Waymo Open Dataset is a comprehensive dataset that encompasses annotations for objects across the entire 360° view. This dataset is unique in that it provides range images as one of the input modalities and comprises a total of 1000 sequences, which are split into a training set of 798 sequences containing 15, 8361 samples and a validation set of 202 sequences containing 4, 0077 frames. In light of the vast amount of data, we only use 20% ($\sim$ 32K frames) of the training data for training purposes, while utilizing all the validation data for evaluation.

4.2   Implementation Details

Network Details. In order to generate range images for the KITTI dataset, we set the Field of View (FOV) to 90° within the range view. These range images have dimensions $C \times W \times H = 5 \times 48 \times 512$. The channel C includes $x, y, z, range, intensity$. On the other hand, for the Waymo Open Dataset, the provided range images are of size 64×2650 and can be directly used as input information. When dividing the pillars for the KITTI dataset, we stack point clouds within the range {0 m $< x <$ 69.12 m, -39.68 m $< y <$ 39.68 m, -3 m $< z <$ 1 m}. These point clouds are transformed into pillars with a spatial resolution of 0.16 m, 0.16 m, 4 m in the x, y, z axes, respectively. On the other hand, for the Waymo Open Dataset, which provides 64 layers of denser point clouds compared to KITTI, we define the range of point clouds as {-74.88 m $< x <$ 74.88 m, -74.88 m $< y <$ 74.88 m, -2 m $< z <$ 4 m}, with a spatial resolution of 0.32 m, 0.32 m, 6 m. In addition, the KITTI dataset allows a maximum of 10 points per pillar, while the Waymo Open Dataset permits up to 20 points per pillar. For the RIGDF, we utilize the PointPillars backbone as a reference. To process the BEV representation $\mathcal{R}^{bev}$, we apply a max-pooling operation, resulting in a pillar feature size of $[K, N]$, where $K = 64$ represents the number of channels, and N denotes the maximum number of pillars. Other settings such as strides align with the approach outlined in [13].

Table 1. Comparison of performance on the KITTI *test* set. For evaluation, AP with an IoU of 0.7 for the *Car* class and 0.5 for the *Pedestrian* class is calculated across 40 recall positions.

Method	Reference	Car AP$_{3D}$(%)			Ped. AP$_{3D}$(%)			Car AP$_{BEV}$(%)			Ped. AP$_{BEV}$(%)		
		Easy	Mod.	Hard	Easy	Mod.	Hard	Easy	Mod.	Hard	Easy	Mod.	Hard
PointPillars [13]	CVPR2019	82.58	74.31	68.99	51.45	41.92	38.89	90.07	86.56	82.81	57.60	48.64	45.78
Part-A^2 [30]	TPAMI2020	87.81	78.49	73.51	53.10	43.35	40.06	91.70	87.79	84.61	59.04	49.81	45.92
Point-GNN [31]	CVPR2020	88.33	79.47	72.29	51.92	43.77	40.14	93.11	89.17	83.90	55.36	47.07	44.61
PV-RCNN [27]	CVPR2020	90.25	81.43	76.82	52.17	43.29	40.29	94.98	90.65	86.14	59.86	50.57	46.74
CIA-SSD [47]	AAAI2021	89.59	80.28	72.87	-	-	-	93.74	89.84	82.39	-	-	-
RangeDet [4]	ICCV2021	85.41	77.36	72.60	-	-	-	90.93	87.67	82.92	-	-	-
S-AT GCN [36]	CoRR2021	83.20	76.04	71.17	44.63	37.37	34.92	90.85	87.68	84.20	50.63	43.43	41.58
VoxSeT [7]	CVPR2022	88.53	82.06	77.46	-	-	-	92.70	89.07	86.29	-	-	-
HMFI [14]	ECCV2022	88.90	81.93	77.30	50.88	42.65	39.78	93.04	89.17	86.37	50.88	42.65	39.78
PDV [10]	CVPR2022	90.43	81.86	77.36	47.80	40.56	38.46	94.56	90.48	86.23	51.95	45.45	43.33
SVGA-Net [9]	AAAI2022	87.33	80.47	75.91	48.48	40.39	37.92	92.07	89.88	85.59	53.09	45.68	43.30
IA-SSD [46]	CVPR2022	88.87	80.32	75.10	47.90	41.03	37.98	93.14	89.48	84.42	52.73	45.07	42.75
GD-MAE [43]	CVPR2023	88.14	79.03	73.55	-	-	-	94.22	88.82	83.54	-	-	-
RPF3D(ours)	-	88.81	82.16	77.48	51.60	43.27	40.84	92.94	91.18	86.58	55.50	48.24	45.80

Training Details. Our *RPF3D* with an end-to-end fashion is trained from scratch on NVIDIA GeForce RTX 3090 Ti with ADAM optimizer. We carry out the same procedures for augmenting data as [30], which include randomly flipping the $x - axis$, rotating the $z - axis$ globally with random angles $\mathcal{A} \in [-\frac{\pi}{4}, \frac{\pi}{4}]$, and performing a global scaling operation with random values $\int \in [0.95, 1.05]$. We predefined the learning rate for the KITTI dataset to be 0.003 for 80 epochs with a batch size of 10, and it eventually reduced to 3×10^{-8} after around 7 h of training. For the Waymo Open Dataset, we initially set the learning rate to 0.003 as well for 30 epochs with batch size 2 and it cost 60 h for the whole training process. Both of these two datasets apply a consine learning rate decay strategy.

4.3 Performance on the KITTI Dataset

Evaluation Metrics. On the KITTI dataset, the mean average precision with 40 recall positions (**mAP_R40**) is leveraged to evaluate our model's testing performance for circumstances under the 3D frame and under the BEV frame accordingly. Specifically, the intersection of union (IoU) is predefined as 0.7, 0.5 and 0.5 for three classes *Car*, *Cyclist* and *Pedestrian* respectively with three variant difficulty levels, *Easy*, *Moderate* and *Hard*.

Performance. The performance of *RPF3D* is thoroughly analyzed through comparisons with other state-of-the-art frameworks on both the KITTI *test* set and *val* set. The evaluation encompasses the 3D Average Precision (AP$_{3D}$) and Bird's Eye View Average Precision (AP$_{BEV}$) for the *Car* and the *Pedestrian* classes.

Table 1 highlights the effectiveness of *RPF3D* when compared to other state-of-the-art frameworks. The results of our work are shown in bold, while the best

Table 2. Comparison of performance on the KITTI *val* set. AP is calculated across 11 recall positions on the *Car* class.

Feature Source	Method	Reference	Car AP$_{3D}$(%)		
			Easy	Mod.	Hard
Multi-modal	AVOD-FPN [12]	IROS2018	84.41	74.44	68.65
	F-PointNet [38]	CVPR2018	83.76	70.92	63.65
	EPNet [11]	ECCV2020	<u>92.28</u>	82.59	80.14
3D	PointRCNN [29]	CVPR2019	88.88	78.63	77.38
	PV-RCNN [27]	CVPR2020	-	83.90	-
	PointGNN [31]	CVPR2020	87.89	78.34	77.38
	3DSSD [44]	CVPR2020	89.71	79.45	78.67
	PVGNet [22]	CVPR2021	89.40	85.05	79.00
	OcTr [48]	CVPR2023	88.43	78.57	77.16
BEV	PIXOR [42]	CVPR2018	86.79	80.75	76.60
	PointPillars [13]	CVPR2019	87.29	76.99	70.84
Range Image	RangeRCNN [18]	arXiv2020	91.41	82.77	80.39
	RangeDet [4]	CVPR2021	89.87	80.72	77.37
Multi-view	MVF [49]	CoRL2019	90.23	79.12	76.43
	H^{2}3D R-CNN [3]	TCSVT2021	89.63	85.20	79.08
	X-View [39]	TIP2023	88.54	80.93	77.97
	RPF3D(ours)	-	**92.10**	**85.40**	**82.44**

Table 3. Comparison of performance on the Waymo Open Dataset with 20% ($\sim 32K$ frames) training samples. $L1$ and $L2$ denote the difficulty levels of LEVEL_1 and the LEVEL_2 respectively. mAP/mAPH are calculated on the *validation* set.

Method	Reference	Vehicle(L1)		Vehicle(L2)		Pedestrian(L1)		Pedestrian(L2)		Cyclist(L1)		Cyclist(L2)	
		mAP	mAPH	mAP	mAPH	mAP	mAPH	mAP	mAPH	mAP	mAPH	mAP	mAPH
PointPillars [13]	CVPR2019	60.67	59.79	52.78	52.01	43.49	23.51	37.32	20.17	35.94	28.34	34.60	27.29
Part-A^2 [30]	TPAMI2020	71.82	71.29	64.33	63.82	63.15	54.96	54.24	47.11	65.23	63.92	62.61	61.35
VoxseT [7]	CVPR2022	74.50	74.03	65.99	65.56	80.03	72.42	72.45	65.39	71.56	70.29	68.95	67.73
HMFI [14]	ECCV2022	68.34	65.66	66.84	64.57	66.62	64.91	59.76	57.24	64.25	61.71	61.23	59.21
IA-SSD [46]	CVPR2022	70.53	69.67	61.55	60.80	69.38	58.47	60.30	50.73	67.67	65.30	64.98	62.71
PDV [10]	CVPR2022	76.85	76.33	69.30	68.81	74.19	65.96	65.85	58.28	68.71	67.55	66.49	65.36
OcTr [48]	CVPR2023	78.12	77.63	69.79	69.34	80.76	74.39	72.68	66.52	72.58	**71.50**	**69.93**	68.90
GD-MAE$_{0.2}$ [43]	CVPR2023	76.24	75.24	67.67	67.22	80.50	72.29	**73.18**	65.50	**72.63**	71.42	69.87	68.71
RPF3D(ours)	-	**78.49**	**77.98**	**70.54**	**70.02**	80.82	**77.87**	72.31	**68.76**	72.42	71.10	69.79	**69.00**

results are underlined. In general, the proposed *RPF3D*, a one-stage pipeline that fuses multi-view features through a 2D CNN module, achieves state-of-the-art performance. Specifically, the average precision (AP) for the Car class reaches 88.81%, 82.16%, and 77.48% on the Easy, Moderate, and Hard difficulty levels, respectively, in terms of 3D object detection performance. In the Bird's Eye View object detection, *RPF3D* maintains satisfactory performance for the

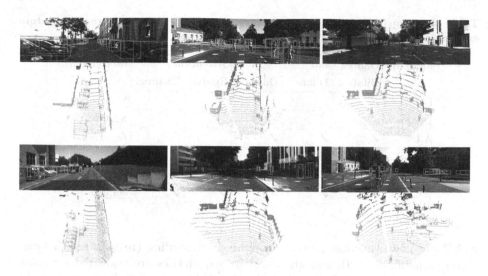

Fig. 7. The qualitative evaluations of the KITTI *test* Dataset. We provide three sets of complicated scenarios according to three categories of objects, *Cars*, *Pedestrians* and *Cyclists* (Color figure online)

Fig. 8. The qualitative evaluations on the *validation* set of the Waymo Open Dataset. The ground truths are shown in green and the predictions are shown in red. (Color figure online)

Car class, with AP scores 92.94%, 91.18%, and 86.58% accordingly on the Easy, Moderate, and Hard difficulty levels.

For further validating the effectiveness of *RPF3D*, more experiences are conducted on the KITTI validation set, as presented in Table 2. In this comparison, the AP is calculated across 11 recall positions to ensure fair comparisons. The analysis focuses on the most competitive *Car* class across three difficulty levels. As a multi-view-based framework, *RPF3D* significantly surpasses all 3D-based methods. Notably, there is a +2.30% increase in performance on the Hard difficulty level compared to the second-best method, EPNet [11].

4.4   Performance on the Waymo Open Dataset

Evaluation Metrics. On the Waymo Open Dataset, three evaluation metrics are employed to test the performance of our trained model: mean average precision (**mAP**) with Level 1 (boxes minimum with 5 points) in both BEV and 3D,

Table 4. Comparison of different dilate rates and attention in range-view feature extraction.

Method	Dilated Conv			Attention		Mod. AP$_{3D}$
	Dilate 1	Dilate 2	Dilate 3	Spatial	Channel	
D-1	✓					81.68%
D-2	✓	✓				82.45%
A.D	✓	✓	✓			83.42%
A.D.-S.	✓	✓	✓	✓		84.05%
A.D.- C.	✓	✓	✓		✓	84.38%
RPF3D	✓	✓	✓	✓	✓	**85.40**%

mAP and mean average precision weighted by heading (**mAPH**) with Level 2 (boxes minimum with a single point) in 3D. Metrics are evaluated for reference to two categories of annotated objects: *Vehicles* and *Pedestrians*, and the intersection over union thresholds are set at 0.7 and 0.5 for these classes.

Performance. As the Waymo Open Dataset solely provides range images as one of the potential inputs, we compare *RPF3D* performance on the challenging Waymo Open Dataset *validation* split to several state-of-the-art top-performing approaches. Table 3 elaborates the 3D AP across two difficulty levels, LEVEL_1 and LEVEL_2, for three object categories: Vehicle, Pedestrian, and Cyclist. The Mean Average Precision (mAP) and Mean Average Precision Height (mAPH) are reported for each method. The best results are shown in bold. Generally, it is evident that the *RPF3D* outperforms all other listed state-of-the-art methods in most categories and difficulty levels. Specifically, *RPF3D* achieves the highest mAP and mAPH scores for the Vehicle category in both LEVEL_1 and LEVEL_2 with an mAP/mAPH of 78.49%/77.98% and 70.54%/70.02% respectively. Moreover, RPF3D attains the highest mAP and mAPH scores for the Pedestrian category in LEVEL_1 and the highest mAPH score in LEVEL_2.

In conclusion, our proposed *RPF3D*, a one-stage framework that integrates a fusion strategy into a 2D CNN network, showcases notable effectiveness compared to leading frameworks that rely on 3D or multi-modal features. Despite its comparatively simple architecture, *RPF3D* exhibits competitive performance, indicating its potential as a robust alternative for object detection tasks. This underlines the power of the fusion strategy utilized and affirms *RPF3D*'s ability to deliver impressive results without dependence on intricate 3D or multi-modal feature representations.

4.5 Ablation Study

In this section, we set out the merits of our presented *RPF3D* by conducting extensive ablation experiments to substantiate the validity of multi-view features capturing and feature deep fusion modules. All experiments described in this

Table 5. Comparison of different Range-Pillar Feature Fusion strategies.

Method	Fusion Method			Mod. AP$_{3D}$
	Concat	Sum	AttentiveFusion	
Concat.-Fu.	✓			82.59%
Sum.-Fu.		✓		81.62%
RPF3D			✓	**85.40**%

Table 6. Comparison of max-pooling stage in the RIGDF.

Method	max-pooling stage in the RIGDF			Mod. AP$_{3D}$
	$i = 0$	$i = 1$	$i = 2$	
S0	✓			84.04%
S1	✓	✓		84.36%
S2	✓	✓	✓	**85.40**%

section are performed on the *val* set of the KITTI dataset and evaluated on the *Moderate* difficulty level of *Car* class.

Effects of the Range-View Feature Extraction. In our analysis of the range-view feature extraction module, we examine the effects of varying dilation rates and attention mechanisms. As indicated in Table 4, dilated convolution successfully tackles scale variance in objects of changing scales, but introduces an unbalanced sampling region issue for fixed-scale objects. This issue arises due to the shift in sampling region dependent on the dilation rate (as covered in Sect. 3.2). Employing dilation rates of 2 and 3 alongside 1 improves performance by 0.77% and 1.74% respectively. Incorporating spatial (A.D.- S.) and channel attention (A.D.- C.) further boosts performance by 0.63% and 0.96% respectively. The most effective method, our *RPF3D*, utilizes all these techniques, achieving a modified average precision of 85.40%.

Effects of the AttentiveFusion. In this section, we evaluate our AttentiveFusion module's performance, developed for Range-Pillar Feature Fusion, against common methods like concatenation and summation. As demonstrated in Table 5, AttentiveFusion successfully overcomes the one-pillar-to-multiple-pixels feature alignment challenge (discussed in Sect. 3.3), outshining the other methods. Specifically, AttentiveFusion attains a *Moderate* AP$_{3D}$ of 85.40%, superior to concatenation's 82.59% and summation's 81.62%.

Concatenation and summation falter due to their inability to handle the alignment issue between range and bird's-eye view representations. On the other hand, the AttentiveFusion module adeptly reweights multi-view features prior to their integration, establishing a correlation between these representations and boosting performance.

Effects of the RIGDF. We delve deeper into the effect of the max-pooling operation at different stages within the RIGDF, used for fusing multi-scale features across various concatenation stages. As demonstrated in Table 6, the max-pooling operation notably enhances RIGDF's performance. With an increase in the number of max-pooling stages from one to three, there's a significant performance boost, reaching a moderate AP_{3D} score of 85.40%. This underscores the value of implementing efficient multi-scale feature fusion methods to enhance 3D object detection in point clouds.

Model Complexity Analysis. Table 7 compares the complexity of various detectors, including both two-stage and one-stage models, in terms of frames per second (FPS) across different hardware devices. These detectors utilize diverse feature sources for detection. Notably, our *RPF3D* model demonstrates competitive inference speed compared to detectors of both types.

4.6   Qualitative Evaluations

We illustrate our evaluation results from both the KITTI test set and the Waymo Open Dataset's validation set in Fig. 7 and Fig. 8. Visualizations from both datasets show that *RPF3D* can effectively detect objects in complex scenarios.

Table 7. Efficiency comparison between one-stage and two-stage models. ‡: RTX 2080Ti, †: TITAN, others: RTX 3090.

Detector	Method	Feature Source	FPS(Hz)
Two-stage	F-PointNet [24]	Multi-modal	10†
	PV-RCNN [27]	3D	10
	Voxel R-CNN [2]	3D	24.5
	H²3D R-CNN [3]	Multi-view	37.1‡
	OcTr [48]	3D	15.63‡
One-stage	3DSSD [44]	3D	26.31†
	RangeDet [4]	Range Image	12‡
	X-view [39]	Multi-view	12.5†
	RPF3D(Ours)	Multi-view	23.25

5   Conclusion

In this work, we have investigated the complementary representations of point clouds and their corresponding range images and proposed an innovative single-stage framework called *RPF3D*. We have addressed the sampling region imbalance problem caused by the fixed-dilation-rate convolutional layers and introduced several attention layers to better accommodate the range of dilation rates needed for the range image scene. To further improve the feature fusion process, we have proposed an AttentiveFusion module and a novel backbone architecture called Range Image Guided Deep Fusion (RIGDF). These components are designed to better fuse features from multiple viewpoints and solve the one-pillar-to-multiple-pixels feature alignment challenge. Our experimental results on the KITTI and Waymo Open Datasets demonstrate the superior performance of our *RPF3D* compared to existing approaches. Specifically, our model achieves a significant enhancement in the performance of the *Car* class detection on both datasets, indicating the effectiveness of our approach in real-world applications.

References

1. Bewley, A., Sun, P., Mensink, T., Anguelov, D., Sminchisescu, C.: Range conditioned dilated convolutions for scale invariant 3D object detection. arXiv preprint arXiv:2005.09927 (2020)
2. Deng, J., Shi, S., Li, P., Zhou, W., Zhang, Y., Li, H.: Voxel R-CNN: towards high performance voxel-based 3D object detection. In: Proceedings of the AAAI Conference on Artificial Intelligence, vol. 35, pp. 1201–1209 (2021)
3. Deng, J., Zhou, W., Zhang, Y., Li, H.: From multi-view to hollow-3D: hallucinated hollow-3D R-CNN for 3D object detection. IEEE Trans. Circuits Syst. Video Technol. **31**(12), 4722–4734 (2021)
4. Fan, L., Xiong, X., Wang, F., Wang, N., Zhang, Z.: RangeDet: in defense of range view for lidar-based 3D object detection. In: Proceedings of the IEEE/CVF International Conference on Computer Vision, pp. 2918–2927 (2021)
5. Geiger, A., Lenz, P., Urtasun, R.: Are we ready for autonomous driving? The KITTI vision benchmark suite. In: 2012 IEEE Conference on Computer Vision and Pattern Recognition, pp. 3354–3361 (2012). https://doi.org/10.1109/CVPR.2012.6248074
6. Graham, B., Engelcke, M., Van Der Maaten, L.: 3D semantic segmentation with submanifold sparse convolutional networks. In: Proceedings of the IEEE Conference on Computer Vision and Pattern Recognition, pp. 9224–9232 (2018)
7. He, C., Li, R., Li, S., Zhang, L.: Voxel set transformer: a set-to-set approach to 3D object detection from point clouds. In: Proceedings of the IEEE/CVF Conference on Computer Vision and Pattern Recognition, pp. 8417–8427 (2022)
8. He, C., Zeng, H., Huang, J., Hua, X.S., Zhang, L.: Structure aware single-stage 3D object detection from point cloud. In: Proceedings of the IEEE/CVF Conference on Computer Vision and Pattern Recognition, pp. 11873–11882 (2020)
9. He, Q., Wang, Z., Zeng, H., Zeng, Y., Liu, Y.: SVGA-Net: sparse voxel-graph attention network for 3D object detection from point clouds. In: Proceedings of the AAAI Conference on Artificial Intelligence, vol. 36, pp. 870–878 (2022)
10. Hu, J.S., Kuai, T., Waslander, S.L.: Point density-aware voxels for lidar 3D object detection. In: Proceedings of the IEEE/CVF Conference on Computer Vision and Pattern Recognition, pp. 8469–8478 (2022)
11. Huang, T., Liu, Z., Chen, X., Bai, X.: EPNet: enhancing point features with image semantics for 3D object detection. In: Vedaldi, A., Bischof, H., Brox, T., Frahm, J.-M. (eds.) ECCV 2020. LNCS, vol. 12360, pp. 35–52. Springer, Cham (2020). https://doi.org/10.1007/978-3-030-58555-6_3

12. Ku, J., Mozifian, M., Lee, J., Harakeh, A., Waslander, S.L.: Joint 3D proposal generation and object detection from view aggregation. In: 2018 IEEE/RSJ International Conference on Intelligent Robots and Systems (IROS), pp. 1–8. IEEE (2018)

13. Lang, A.H., Vora, S., Caesar, H., Zhou, L., Yang, J., Beijbom, O.: PointPillars: fast encoders for object detection from point clouds. In: Proceedings of the IEEE/CVF Conference on Computer Vision and Pattern Recognition, pp. 12697–12705 (2019)

14. Li, X., et al.: Homogeneous multi-modal feature fusion and interaction for 3D object detection. In: Avidan, S., Brostow, G., Cissé, M., Farinella, G.M., Hassner, T. (eds.) Computer Vision-ECCV 2022: 17th European Conference, Tel Aviv, Israel, 23–27 October 2022, Proceedings, Part XXXVIII, pp. 691–707. Springer, Cham (2022). https://doi.org/10.1007/978-3-031-19839-7_40

15. Li, Y., et al.: DeepFusion: lidar-camera deep fusion for multi-modal 3D object detection. In: Proceedings of the IEEE/CVF Conference on Computer Vision and Pattern Recognition, pp. 17182–17191 (2022)

16. Liang, M., Yang, B., Chen, Y., Hu, R., Urtasun, R.: Multi-task multi-sensor fusion for 3D object detection. In: Proceedings of the IEEE/CVF Conference on Computer Vision and Pattern Recognition, pp. 7345–7353 (2019)

17. Liang, M., Yang, B., Wang, S., Urtasun, R.: Deep continuous fusion for multi-sensor 3D object detection. In: Ferrari, V., Hebert, M., Sminchisescu, C., Weiss, Y. (eds.) ECCV 2018. LNCS, vol. 11220, pp. 663–678. Springer, Cham (2018). https://doi.org/10.1007/978-3-030-01270-0_39

18. Liang, Z., Zhang, M., Zhang, Z., Zhao, X., Pu, S.: RangeRCNN: towards fast and accurate 3D object detection with range image representation. arXiv preprint arXiv:2009.00206 (2020)

19. Liang, Z., Zhang, Z., Zhang, M., Zhao, X., Pu, S.: RangeioUDet: range image based real-time 3D object detector optimized by intersection over union. In: Proceedings of the IEEE/CVF Conference on Computer Vision and Pattern Recognition (CVPR), pp. 7140–7149, June 2021

20. Ma, F., Karaman, S.: Sparse-to-dense: depth prediction from sparse depth samples and a single image. In: 2018 IEEE International Conference on Robotics and Automation (ICRA), pp. 4796–4803. IEEE (2018)

21. Meyer, G.P., Laddha, A., Kee, E., Vallespi-Gonzalez, C., Wellington, C.K.: LaserNet: an efficient probabilistic 3D object detector for autonomous driving. In: Proceedings of the IEEE/CVF Conference on Computer Vision and Pattern Recognition, pp. 12677–12686 (2019)

22. Miao, Z., et al.: PVGNet: a bottom-up one-stage 3D object detector with integrated multi-level features. In: 2021 IEEE/CVF Conference on Computer Vision and Pattern Recognition (CVPR), pp. 3278–3287 (2021). https://doi.org/10.1109/CVPR46437.2021.00329

23. Piergiovanni, A., Casser, V., Ryoo, M.S., Angelova, A.: 4D-net for learned multi-modal alignment. In: Proceedings of the IEEE/CVF International Conference on Computer Vision, pp. 15435–15445 (2021)

24. Qi, C.R., Liu, W., Wu, C., Su, H., Guibas, L.J.: Frustum PointNets for 3D object detection from RGB-D data. In: Proceedings of the IEEE Conference on Computer Vision and Pattern Recognition, pp. 918–927 (2018)

25. Qi, C.R., Su, H., Mo, K., Guibas, L.J.: PointNet: deep learning on point sets for 3D classification and segmentation. In: Proceedings of the IEEE Conference on Computer Vision and Pattern Recognition, pp. 652–660 (2017)

26. Qi, C.R., Yi, L., Su, H., Guibas, L.J.: PointNet++: deep hierarchical feature learning on point sets in a metric space. In: Advances in Neural Information Processing Systems, vol. 30 (2017)
27. Shi, S., et al.: PV-RCNN: point-voxel feature set abstraction for 3D object detection. In: Proceedings of the IEEE/CVF Conference on Computer Vision and Pattern Recognition, pp. 10529–10538 (2020)
28. Shi, S., et al.: PV-RCNN++: point-voxel feature set abstraction with local vector representation for 3D object detection. arXiv preprint arXiv:2102.00463 (2021)
29. Shi, S., Wang, X., Li, H.: PointRCNN: 3D object proposal generation and detection from point cloud. In: Proceedings of the IEEE/CVF Conference on Computer Vision and Pattern Recognition, pp. 770–779 (2019)
30. Shi, S., Wang, Z., Shi, J., Wang, X., Li, H.: From points to parts: 3D object detection from point cloud with part-aware and part-aggregation network. IEEE Trans. Pattern Anal. Mach. Intell. **43**(8), 2647–2664 (2020)
31. Shi, W., Rajkumar, R.: Point-GNN: graph neural network for 3D object detection in a point cloud. In: Proceedings of the IEEE/CVF Conference on Computer Vision and Pattern Recognition, pp. 1711–1719 (2020)
32. Sun, P., et al.: Scalability in perception for autonomous driving: waymo open dataset. CoRR abs/1912.04838 (2019). http://arxiv.org/abs/1912.04838
33. Sun, P., et al.: RSN: range sparse net for efficient, accurate lidar 3D object detection. In: Proceedings of the IEEE/CVF Conference on Computer Vision and Pattern Recognition, pp. 5725–5734 (2021)
34. Vora, S., Lang, A.H., Helou, B., Beijbom, O.: PointPainting: sequential fusion for 3D object detection. In: 2020 IEEE/CVF Conference on Computer Vision and Pattern Recognition (CVPR), pp. 4603–4611 (2019)
35. Wang, C., Ma, C., Zhu, M., Yang, X.: PointAugmenting: cross-modal augmentation for 3D object detection. In: 2021 IEEE/CVF Conference on Computer Vision and Pattern Recognition (CVPR), pp. 11789–11798 (2021). https://doi.org/10.1109/CVPR46437.2021.01162
36. Wang, L., Wang, C., Zhang, X., Lan, T., Li, J.: S-AT GCN: spatial-attention graph convolution network based feature enhancement for 3D object detection. arXiv preprint arXiv:2103.08439 (2021)
37. Wang, Y., et al.: Pillar-based object detection for autonomous driving. In: Vedaldi, A., Bischof, H., Brox, T., Frahm, J.-M. (eds.) ECCV 2020. LNCS, vol. 12367, pp. 18–34. Springer, Cham (2020). https://doi.org/10.1007/978-3-030-58542-6_2
38. Wang, Z., Jia, K.: Frustum ConvNet: sliding frustums to aggregate local pointwise features for amodal 3D object detection. In: 2019 IEEE/RSJ International Conference on Intelligent Robots and Systems (IROS), pp. 1742–1749. IEEE (2019)
39. Xie, L., Xu, G., Cai, D., He, X.: X-view: non-egocentric multi-view 3D object detector. IEEE Trans. Image Process. **32**, 1488–1497 (2023). https://doi.org/10.1109/TIP.2023.3245337
40. Yan, Y., Mao, Y., Li, B.: Second: sparsely embedded convolutional detection. Sensors **18**(10), 3337 (2018)
41. Yang, B., Liang, M., Urtasun, R.: HDNET: exploiting HD maps for 3D object detection. In: Conference on Robot Learning, pp. 146–155. PMLR (2018)
42. Yang, B., Luo, W., Urtasun, R.: PIXOR: real-time 3D object detection from point clouds. In: Proceedings of the IEEE Conference on Computer Vision and Pattern Recognition, pp. 7652–7660 (2018)
43. Yang, H., et al.: GD-MAE: generative decoder for MAE pre-training on lidar point clouds. arXiv preprint arXiv:2212.03010 (2022)

44. Yang, Z., Sun, Y., Liu, S., Jia, J.: 3DSSD: point-based 3D single stage object detector. In: Proceedings of the IEEE/CVF Conference on Computer Vision and Pattern Recognition, pp. 11040–11048 (2020)
45. Yuan, Z., Song, X., Bai, L., Wang, Z., Ouyang, W.: Temporal-channel transformer for 3D lidar-based video object detection for autonomous driving. IEEE Trans. Circuits Syst. Video Technol. **32**(4), 2068–2078 (2022). https://doi.org/10.1109/TCSVT.2021.3082763
46. Zhang, Y., Hu, Q., Xu, G., Ma, Y., Wan, J., Guo, Y.: Not all points are equal: learning highly efficient point-based detectors for 3D lidar point clouds. In: Proceedings of the IEEE/CVF Conference on Computer Vision and Pattern Recognition, pp. 18953–18962 (2022)
47. Zheng, W., Tang, W., Chen, S., Jiang, L., Fu, C.W.: CIA-SSD: confident IoU-aware single-stage object detector from point cloud. In: Proceedings of the AAAI Conference on Artificial Intelligence, vol. 35, pp. 3555–3562 (2021)
48. Zhou, C., Zhang, Y., Chen, J., Huang, D.: OcTr: octree-based transformer for 3D object detection. arXiv preprint arXiv:2303.12621 (2023)
49. Zhou, Y., et al.: End-to-end multi-view fusion for 3D object detection in lidar point clouds. In: Conference on Robot Learning, pp. 923–932. PMLR (2020)
50. Zhou, Y., Tuzel, O.: VoxelNet: end-to-end learning for point cloud based 3D object detection. In: Proceedings of the IEEE Conference on Computer Vision and Pattern Recognition, pp. 4490–4499 (2018)
51. Zou, Y., Cheng, L., Li, Z.: A multimodal fusion model for estimating human hand force: comparing surface electromyography and ultrasound signals. IEEE Rob. Autom. Mag. **29**(4), 10–24 (2022)

Traffic Signal Control Optimization Based on Deep Reinforcement Learning with Attention Mechanisms

Wenlong Ni[1,2(✉)], Peng Wang[2], Zehong Li[2], and Chuanzhuang Li[2]

[1] School of Computer Information Engineering, JiangXi Normal University, Nanchang, China
[2] School of Digital Industry, JiangXi Normal University, Shangrao, China
{wni,peng.wang,Zehong.Li,lichuanzhuang}@jxnu.edu.cn

Abstract. Deep reinforcement learning (DRL) methodology with traffic control systems plays a vital role in adaptive traffic signal controls. However, previous studies have frequently disregarded the significance of vehicles near intersections, which typically involve higher decision-making requirements and safety considerations. To overcome this challenge, this paper presents a novel DRL-based method for traffic signal control, which incorporates an attention mechanism into the Dueling Double Deep Q Network (D3QN) framework. This approach emphasizes the priority of vehicles near intersections by assigning them higher weights and more attention. Moreover, the state design incorporates signal light statuses to facilitate a more comprehensive understanding of the current traffic environment. Furthermore, the model's performance is enhanced through the utilization of Double DQN and Dueling DQN techniques. The experimental findings demonstrate the superior efficacy of the proposed method in critical metrics such as vehicle waiting time, queue length, and the number of halted vehicles when compared to D3QN, traditional DQN, and fixed timing strategies.

Keywords: Attention Mechanism · Deep Reinforcement Learning · Signal Control · Deep Q-Network

1 Introduction

How to optimize traffic flow has become an urgent problem, and traffic signal control is one of the crucial means to achieve this optimization. Traditional traffic signal control method is mainly based on fixed timing control [1]. This method is not sensitive to the change in traffic flow, and it is difficult to adapt to the change in road sections and traffic flows. In recent years, reinforcement learning [2] in traffic signal control has received more and more attention. However,

This work was supported in part by JiangXi Education Department under Grant No. GJJ191688.

the traditional reinforcement learning method is relatively difficult to deal with high-dimensional state space. Deep learning (DL) [3] has emerged and developed rapidly with improved computing power and large-scale data generation. DL is an advanced artificial intelligence technology. Therefore, the combination of deep learning and reinforcement learning, namely Deep reinforcement learning (DRL). The traffic environment state information is input into the neural network structure, and the Q value estimation of each action is output. Traditional neural networks do perform very well in terms of feature extraction. However, it has some limitations in automatically focusing on the vital state components. The contributions of this work can be summarized as follows:

- In this paper the Dueling Deep Q Network (Dueling DQN) network [4] is used to approximate the Q function, and the idea of the Double Deep Q Network (Double DQN) algorithm is used in model training to decouple the two steps of target Q value action selection and target Q value calculation, which can help to solve the overestimation problem.
- An algorithm named D3QN_AM is designed, by incorporating Attention Mechanism (AM) into the Dueling Double Deep Q Network (D3QN) model the proximity to the intersection is treated with higher priority and more weights, which can have different state components to focus on the more essential parts.
- A multi-indicator factor weighting scheme was used to design the reward function to avoid over-focusing on a single indicator at the expense of other important factors.

2   Related Work

In the field of traffic signal control, there have been many studies exploring different methods and algorithms to improve traffic flow and reduce traffic congestion. Abdulhai et al. [5] proposed a Q-learning-based acyclic signal control system including three different state representation models for testing appropriate state models under different traffic conditions. Jin et al. [6] used the SARSA algorithm to construct a signal group-based traffic signal control system.

Many types of research are now conducted based on DRL. Li et al. [7] proposed a DRL model for signal timing scheme design, which uses Stacked Autoencoder (SAE) to approximate the Q function. Liang et al. [8] proposed a DRL model for traffic light control, using the cumulative waiting time between two decision points as the reward function. Gender et al. [9] combined reinforcement learning and convolutional neural network, and the experimental results show that it can better adapt to the traffic flow. Zhang et al. [10] Dynamically Adjusting Traffic Signal Durations Using DRL to Solve Real-Life Complex Problems. Gao et al. [11] adopted experience replay and target network mechanisms to improve the algorithm's stability. However, this research often considered only a single performance metric in the design of the reward function. Mao et al. [12] evaluate seven prevailing DRL algorithms, and testing results indicate that the soft actor-critic (SAC) outperforms other DRL algorithms. However, These studies do not fully consider the impact of traffic light status on agent strategy selection.

3   System Design

In this section, we convert the problem of intersection signal control into a Markov decision process (MDP) and define the state space, action space, and reward function.

3.1   State Definition

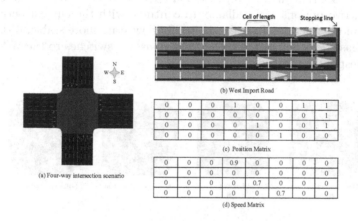

Fig. 1. Traffic status diagram.

Figure 1 (a) is a four-way intersection with four directions: east, west, south, and north. In each direction there are 4 incoming roads, where the rightmost lane is allowed to turn right only, the middle two lanes are allowed to go straight, and the leftmost lane is allowed to turn left only.

In this study, Discrete Traffic State Encoding (DTSE) is used to represent the state of the environment. As shown in Fig. 1 (b) west import lane, The lane where the E2 detector, with a length of L, is located is divided into cells of equal size to represent the vehicle information, aiming to describe the state information of the environment more accurately. The size dimension of the state space is (N, C, 2), where N denotes the number of incoming lanes, C denotes the number of cells divided uniformly from the stop line, and 2 denotes the vehicle position matrix and vehicle velocity matrix. The vehicle position matrix information and the velocity normalization matrix information are shown in Fig. 1 (c), Fig. 1 (d).

In addition to the intersection vehicle position and vehicle speed as the traffic state, this study also introduces the phase state of the traffic signal. Due to the significant impact of the previous traffic signal state on the efficiency of traffic flow at the intersection, this study utilizes one-hot encoding to represent the signal state information. For instance, if the signal state is represented as $[1, 0, 0, 0]$, the signal is in the first phase.

3.2 Action Space

The phase of signal control in this study indicates actions for the agents in reinforcement learning (RL). For the four-phase scenario of an intersection as shown in Fig. 2, there are a total of four non-conflicting phases set, which are 1) the first phase (north-south direction straight ahead); 2) the second phase (south-north direction left turn); 3) the third phase (east-west direction straight ahead); and 4) the fourth phase (east-west left turn), and the right turn is always allowed. The set that makes up the action space is A = $[0, 1, 2, 3]$. In choosing the action, two options are available: to continue with the current action or to switch to another action. To make the traffic flow state more stable, if the newly selected phase does not match the current phase, it switches to the yellow light phase corresponding to the current phase.

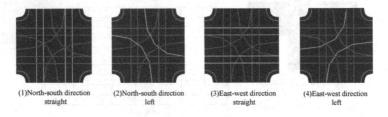

(1)North-south direction straight   (2)North-south direction left   (3)East-west direction straight   (4)East-west direction left

Fig. 2. Intersection four-phase.

3.3 Reward Function

This study adopts a multi-index coefficient weighting scheme to design the reward function. The reward function is comprehensively considered with different factors and goals, avoiding paying too much attention to a single index and ignoring other important factors, to guide the agent better to learn. Therefore, three metrics are set in this study to judge the intersection traffic efficiency, which is the length of vehicle queuing on the incoming lane at the intersection, the cumulative waiting time of vehicles, and the number of halting vehicles. Let W_t denote the sum of the cumulative waiting time of all vehicles on the incoming lane after the execution of the t-th action. Let L_t denote the sum of the queuing lengths of all vehicles on the incoming lane after the execution of the t-th action, and let H_t denote the sum of the number of halting vehicle on the incoming lane after the execution of the t-th action. The reward equation of the agent after the execution of the t-th action is shown in Eq. (1) where w_1 denotes the weight coefficient of vehicle waiting time, w_2 denotes the weight coefficient of vehicle queue length, and w_3 denotes the weight coefficient of the number of halting vehicles.

$$r_t = w_1 * (W_{t-1} - W_t) + w_2 * L_t + w_3 * H_t \tag{1}$$

After taking action, if the queue length, waiting time, and the number of halted vehicles have decreased, it indicates that some vehicles have successfully passed the intersection. In such cases, the agent receives positive feedback from the environment and continues to develop by maximizing rewards. On the other hand, if the mentioned metrics do not decrease, the agent receives negative feedback. As the agent continues to learn, it will strive to avoid this behavior (action) in the future.

4 D3QN_AM Algorithm

To enhance the neural network's ability to perceive the important state components of the traffic environment, we introduce an attention mechanism. This mechanism enables the network to automatically pay attention and assign higher weights to vehicles approaching intersections, thus providing a better understanding of the current traffic environment.

4.1 Q Network Structure

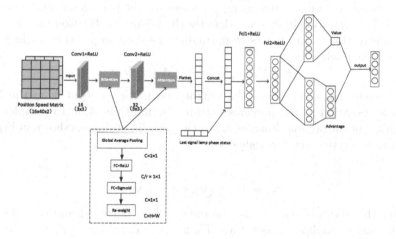

Fig. 3. Q-network structure for value functions approximation.

Inspired by the SENet (Squeeze-and-Excitation Network) [13], we design a novel algorithm called D3QN_AM with the Q-network structure shown in Fig. 3. The attention mechanism enhances the feature representation capability, and by giving higher weights to features close to the intersection, the model can better capture crucial details in that region. At the same time, the attention mechanism helps the model suppress noisy or irrelevant features far from the intersection. It allows the network to focus its resources on essential information, reducing the risk of overfitting and enhancing generalization.

The training state data used in this study include the vehicle position, speed information, and the phase information of traffic lights. The vehicle position and speed information are first used as the input of the convolution layer. An attention layer follows each convolutional layer, and the squeeze excitation block is a computational unit mainly divided into three operations: Squeeze, Excitation, and Scale.

- Squeeze operation refers to the feature compression of each channel in the global pooling layer to transform the feature map of each channel into a scalar value, as shown in Eq. (2). Where $\mathbf{F}_{sq}$ denotes the squeeze operation and $\mathbf{v}_c$ denotes that it is a two-dimensional space with dimension sizes H and W, z_c denotes a single value after global average pooling.

$$z_c = \mathbf{F}_{sq}\left(\mathbf{v}_c\right) = \frac{1}{H \times W} \sum_{i=1}^{H} \sum_{j=1}^{W} v_c(i,j) \tag{2}$$

- Excitation operation obtains the weight coefficients of each channel by first descending and then ascending the fully connected layer learning, as shown in Eq. (3) where $\mathbf{F}_{ex}$ denotes the excitation operation, $\mathbf{W}_1 \in \mathbb{R}^{\frac{C}{r} \times C}$ denotes the reduced-dimensional fully connected layer weight parameter, $\mathbf{W}_2 \in \mathbb{R}^{C \times \frac{C}{r}}$ denotes the ascending fully connected layer weight parameter, and δ denotes the ReLU activation function. σ denotes the Sigmoid activation function. The final activation function acts to normalize the data to get the weight values of each channel.

$$\mathbf{s} = \mathbf{F}_{ex}(\mathbf{z}, \mathbf{W}) = \sigma\left(\mathbf{W}_2 \delta\left(\mathbf{W}_1 \mathbf{z}\right)\right) \tag{3}$$

- Scale operation is to weight the weight information output from the Excitation operation to the previous features by multiplication to complete the rescaling of the original features in the channel dimension, as shown in Eq. (4). Where $\mathbf{F}_{scale}$ denotes the scale operation.

$$\widetilde{\mathbf{x}}_c = \mathbf{F}_{scale}\left(\mathbf{v}_c, s_c\right) = s_c \mathbf{v}_c \tag{4}$$

After this series of operations, the data will be flattened and concatenated with the last semaphore phase state. Then, these data are input into the fully connected layers fc1 and fc2. Next, the state information feature extracted from fc2 will be divided into two branches, the first branch is used to calculate the state value V(s) of the state, and the second branch is used to calculate the advantage value A(s, a) of each action. Finally, the data of the two branches are aggregated to obtain the Q value corresponding to each action. Q value calculation equation is shown in the Eq. (5), which shows that the action value function $Q(s, a; \mu, \alpha, \beta)$ is equal to the sum of the value $V(s; \mu, \alpha)$ of state s and the advantage value of zero-mean processing can make the learning process more stable and efficient. Among them, μ represents the parameters of the public part, α represents the parameters of the unique part of the value network, and β represents the parameters of the unique part of the advantage network.

$$Q(s, a; \mu, \alpha, \beta) = V(s; \mu, \alpha) + \left[A(s, a; \mu, \beta) - \frac{1}{|A|} \sum_a A(s, a; \mu, \beta) \right] \quad (5)$$

4.2 Model Structure

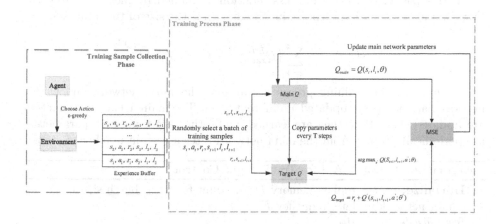

Fig. 4. Model Structure.

The model proposed in this paper is shown in Fig. 4. The full algorithm, which we call D3QN_AM, is presented in Algorithm 1. The agent uses an epsilon-greedy strategy to choose actions, and the idea behind it is to select the best action with a probability of $1 - \varepsilon$ and to select the action randomly with a probability of ε. At the beginning of training, the value of ε can be set larger, when the agent is more inclined to choose the action exploratory. As the training continues, the value of ε will become smaller and smaller, when the agent prefers to select actions exploitatively, the action corresponding to the largest Q value is selected. Once the agent takes action in the environment, the environment provides feedback by giving the reward value and next state information to the agent. The collected sample data is then stored in the experience replay buffer. In the training phase, the experience replay technique is used to train the value function network of the agents by randomly sampling a batch of experience information, which reduces the sample correlation and thus speeds up the algorithm's convergence.

In this paper, the idea of Double DQN [14] is used to train the model parameters. Both the main network Q and the target network Q' use the Q-network structure as shown in Fig. 3, and the main network is trained to choose actions, while the target network is used to compute the target Q values.

Double DQN computes the target Q-value by choosing the action of the next state S_{t+1} from the main network Q using the one with the largest Q-value, and then this action is used to compute Q_{target} through the target network Q', which can effectively avoid the overestimation problem. The equation is shown in Eq. (6), where R_t represents the immediate reward value at moment t, S_{t+1} is the next state, and Q, Q' are the main network and target network respectively. Here θ_t, θ'_t are the main network parameters and target network parameters at moment t respectively, and γ is the discount factor.

$$Q_{\text{target}}^{DDQN} = R_t + \gamma Q' \left(S_{t+1}, \text{argmax}_a Q\left(S_{t+1}, a; \theta_t\right); \theta'_t\right) \tag{6}$$

In this paper, the algorithm loss function is defined by mean square error (MSE), and the equation is shown in (7) where n is the size of the input sample.

$$MSE(\theta) = \frac{1}{n} \sum_{t=1}^{n} \left(Q_{\text{target}}^{DDQN} - Q\left(S_t, A_t; \theta_t\right)\right)^2 \tag{7}$$

To enhance the training process's stability, the main network and target network parameters are updated at different steps. The main network parameters are updated every time they are learned, while the target network parameters are periodically copied from the main network after a certain number of steps.

Algorithm 1: D3QN_AM for Traffic Light Control

---

1 **Initialize:** Q, Q', replay memory D, discount factor γ, batch_size B, target network update frequency F

2 **for** *episode* $= 1, E$ **do**

3   Initialize observation s_t, l_t

4   **for** $t = 1, T$ **do**

5   With probability ε select a random action a_t

6   Otherwise select $a_t = \arg\max_a Q(s_t, l_t, a; \theta)$

7   Execute action a_t and observe reward r_t and next state s_{t+1}, l_{t+1}

8   Store transition $(s_t, a_t, r_t, s_{t+1}, l_t, l_{t+1})$ in D

9   Set sample counter $C = C + 1$

10   Set $s_t = s_{t+1}, l_t = l_{t+1}$

11   **if** $C \geq B$ **then**

12   Sample random minibatch of transitions from D

13             $Q_{\text{eval}} = Q(s_j, l_j, a_j; \theta)$

14             $Q_{\text{target}} = r_j + \gamma Q'(s_{j+1}, l_{j+1}, \arg\max_{a'} Q(s_{j+1}, l_{j+1}, a'; \theta); \theta')$

15   Calculate $Loss = \text{MSE}(Q_{\text{target}}, Q_{\text{eval}})$

16   Update the parameters θ using the Adam optimizer

17   Set $L \leftarrow L + 1$

18   **if** $L \% F == 0$ **then**

19   update $\theta' = \theta$

20   **end**

21   **end**

22   **end**

23 **end**

---

5   Simulation Study

This study conducts experiments within the Simulation of Urban Mobility (SUMO) simulation environment, encompassing varying traffic flow scenarios, including low, medium, and high peak periods. The yellow light time designed in this study is 3 s. To make the training more stable, the maximum green light time is set to 50 s, beyond which it is forced to switch to the next phase.

The total duration of each round of operation in this study is 5200 s, where 0–1000 s is the middle peak, 1000 s–2500 s is the peak, and the low peak occurs at 2500 s–5200 s. The specific vehicle parameters are set as follows: the vehicle length is 5 m, the acceleration is $0.8 \, \mathrm{m/s^2}$, deceleration of $4.5 \, \mathrm{m/s^2}$ and minimum spacing of 2.5 m; vehicles enter the road network from any entrance, and the ratios of straight ahead, right turn and left turn are 0.6, 0.2 and 0.2 respectively. To simulate the low peak, medium peak, and peak situations in real traffic, the vehicle arrival time distribution follows the Weibull distribution [15].

The experimental parameters are set as follows: the learning rate α is set to 0.001, the maximum exploration rate $\varepsilon_{\max}$ is set to 1, the minimum exploration rate $\varepsilon_{\min}$ is set to 0.01, the exploration decay rate $\varepsilon_{\mathrm{decay}}$ is set to 0.96, the training batch size is 64, the buffer size for storing experience is 2000, the discount rate γ is 0.75, the target network update frequency is 100, the number of training rounds is 140, and the training duration per round is 5200, and the optimizer is Adam.

To verify the effectiveness of the algorithm, in this experiment, D3QN_AM is compared with the following three benchmark algorithms:

- Fixed Timing Control (FTC), a pre-defined set of timing schemes based on Webster's Timing Method [16].
- The traffic signal control based on D3QN, state, action, reward, and training algorithms is consistent with this study's approach. The difference is that the attention mechanism is not incorporated.
- Traffic signal control based on traditional DQN [17] with state, action, and reward definitions are consistent with this study's approach.

After 140 training rounds, each round of simulation time is 5200 s. In each round, the average reward value, the average loss value, the average waiting time of vehicles, the average queue length, and the average halting vehicle number are recorded. Finally, the following experimental results are obtained by drawing. From Fig. 5 (a), it can be seen that the average reward value tends to rise more obviously with the increase of training times, which means that the performance of deep reinforcement learning gradually improves. Similarly, D3QN_AM achieves better reward values and convergence than the two benchmark algorithms, D3QN and DQN. Also, from Fig. 5 (c) (d) (e), we observe that the average vehicle queue length and the average vehicle waiting time, as well as the average vehicle queue length, decrease significantly, which indicates that the algorithm achieves better results in controlling traffic flow and optimizing vehicle waiting time. Besides, D3QN_AM also performs well on these traffic metrics of average waiting time, average queue length, and the number of halting vehicles, outperforming the benchmark algorithms D3QN and DQN. From Fig. 5 (b), it can be observed that the loss value gradually decreases and eventually tends to zero. In the beginning stage, the loss value of the initial model is high. However, with the increase in training times, the loss value gradually decreases, indicating that the model optimizes during the learning process, making its prediction results closer to the actual value.

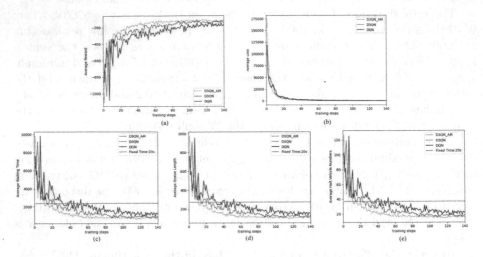

Fig. 5. The performance metrics for the 140 rounds of algorithmic training are as follows: (a), (b), (c), (d), and (e) indicate the average cumulative reward value, average loss value, average vehicle waiting time, average vehicle queue length, and the average number of halted vehicles, respectively.

In the early stage of training, due to the agent's high exploration rate of action selection, it will randomly select actions and act on the environment, so the data fluctuates wildly. However, as the number of training increases, the exploration rate gradually decreases. The agent gradually learns how to choose the optimal action in the current state and gradually improves the action selection strategy by updating the weight parameters of the Q function. The action with a large Q value is selected with a high probability of gradually stabilizing the data.

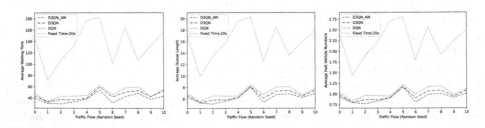

Fig. 6. Comparison of the metrics of different algorithms

In the test phase, the traffic flow data generated by eleven groups of random seeds uses to verify the effectiveness of the D3QN_AM algorithm. By recording the average waiting time, the average queue length of all vehicles on the lane every 52 s, and the average halting vehicle number with an experiment of 5200 s, the curves are plotted as Fig. 6. The experimental results show that the D3QN_AM algorithm performs best in terms of average waiting time, average queue length, and average halting vehicle number, which are all the better than the D3QN algorithm, the traditional DQN algorithm, and the fixed timing scheme of the 20 s.

Table 1 compares the metrics under different methods, which shows the experimental results more intuitively. Compared with the D3QN algorithm, the algorithm in this paper reduces the average waiting time by 12.9%, the average queue length by 5.7%, and the average halting vehicle number by 5.15%; compared with the traditional DQN algorithm, the algorithm in this paper reduces the average waiting time by 20.1%, the average queue length by 11.8%, and the average halting vehicle number by 10.7%; compared with the fixed duration the 20 s, the algorithm in this paper reduces 70.9% of the average waiting time, 60.6% of the average queue length, and 58.6% of the average halting vehicle number.

Table 1. Comparison of the indicators under different methods

Method	Average waiting time	Average queue length	Average halting number
Fixed Time: 20 s	134.08	15.88	2.22
DQN	48.91	7.09	1.03
D3QN	44.84	6.63	0.97
D3QN_AM	**39.07**	**6.25**	**0.92**

These results show that the algorithm in this paper can better learn traffic signal timing strategies and thus significantly reduce traffic congestion and waiting time. In summary, by introducing this attention mechanism, our model can adaptively focus on more essential state components, thus improving perception and achieving better performance in traffic signal control tasks.

6   Conclusion

In this paper, the authors proposed an innovative algorithm called D3QN_AM, which incorporates an attention mechanism to address the traffic signal control problem. The core idea of this algorithm is to enable the intelligent agent to focus more on critical information in the traffic environment and assign it higher weights and more attention. It enhances the agent's perception of vehicles near intersections and improves the effectiveness of traffic signal control. The experimental results demonstrate that compared to D3QN, traditional DQN methods, and fixed timing strategies, the proposed D3QN_AM algorithm outperforms critical metrics such as vehicle waiting time, queue length, and the number of halted vehicles. In conclusion, the introduction of the D3QN_AM algorithm in this research holds significant innovative significance for addressing traffic signal control problems. By incorporating the attention mechanism, the intelligent agent becomes more attentive to critical information, thereby effectively enhancing the performance of traffic signal control. The algorithm exhibits remarkable advantages in our experiments, providing valuable references and insights for future research in traffic signal control.

References

1. Papageorgiou, M., Diakaki, C., Dinopoulou, V., Kotsialos, A., Wang, Y.: Review of road traffic control strategies. Proc. IEEE **91**(12), 2043–2067 (2003). https://doi.org/10.1109/JPROC.2003.819610
2. Sutton, R.S., Barto, A.G., et al.: Introduction to Reinforcement Learning, vol. 135. MIT Press Cambridge (1998)
3. LeCun, Y., Bengio, Y., Hinton, G.: Deep learning. Nature **521**(7553), 436–444 (2015)
4. Wang, Z., Schaul, T., Hessel, M., Hasselt, H., Lanctot, M., Freitas, N.: Dueling network architectures for deep reinforcement learning. In: International Conference on Machine Learning, pp. 1995–2003. PMLR (2016)
5. El-Tantawy, S., Abdulhai, B.: An agent-based learning towards decentralized and coordinated traffic signal control. In: 13th International IEEE Conference on Intelligent Transportation Systems, pp. 665–670. IEEE (2010)
6. Jin, J., Ma, X.: A group-based traffic signal control with adaptive learning ability. Eng. Appl. Artif. Intell. **65**, 282–293 (2017)
7. Li, L., Lv, Y., Wang, F.Y.: Traffic signal timing via deep reinforcement learning. IEEE/CAA J. Automatica Sinica **3**(3), 247–254 (2016)
8. Liang, X., Du, X., Wang, G., Han, Z.: A deep q learning network for traffic lights' cycle control in vehicular networks. IEEE Trans. Veh. Technol. **68**(2), 1243–1253 (2019)
9. Genders, W., Razavi, S.: Using a deep reinforcement learning agent for traffic signal control (2016)
10. Zhang, L., et al.: DynamicLight: dynamically tuning traffic signal duration with DRL (2022)
11. Gao, J., Shen, Y., Liu, J., Ito, M., Shiratori, N.: Adaptive traffic signal control: deep reinforcement learning algorithm with experience replay and target network. arXiv preprint arXiv:1705.02755 (2017)
12. Mao, F., Li, Z., Li, L.: A comparison of deep reinforcement learning models for isolated traffic signal control. IEEE Intell. Transp. Syst. Mag. **15**(1), 160–180 (2023). https://doi.org/10.1109/MITS.2022.3144797
13. Hu, J., Shen, L., Albanie, S., Sun, G., Wu, E.: Squeeze-and-excitation networks (2019)
14. Van Hasselt, H., Guez, A., Silver, D.: Deep reinforcement learning with double Q-learning. In: Proceedings of the AAAI Conference on Artificial Intelligence, vol. 30 (2016)
15. Hallinan, A.J. Jr.: A review of the Weibull distribution. J. Qual. Technol. **25**(2), 85–93 (1993)
16. Webster, F.V.: Traffic signal settings. Road Research Technical Paper 39 (1958)
17. Mnih, V., et al.: Playing Atari with deep reinforcement learning (2013)

CMCI: A Robust Multimodal Fusion Method for Spiking Neural Networks

Runhao Jiang[1], Jianing Han[1], Yingying Xue[2], Ping Wang[2], and Huajin Tang[1,3(✉)]

[1] College of Computer Science and Technology, Zhejiang University, Hangzhou, China
{RhJiang,jnhan}@zju.edu.cn
[2] Biosensor National Special Laboratory Key Laboratory for Biomedical Engineering of Education Ministry, Department of Biomedical Engineering, Zhejiang University, Hangzhou, China
{12115038,cnpwang}@zju.edu.cn
[3] Zhejiang Lab, Hangzhou, China
htang@zju.edu.cn

Abstract. Human understand the external world through a variety of perceptual processes such as sight, sound, touch and smell. Simulating such biological multi-sensory fusion decisions using a computational model is important for both computer and neuroscience research. Spiking Neural Networks (SNNs) mimic the neural dynamics of the brain, which are expected to reveal the biological multimodal perception mechanism. However, existing works of multimodal SNNs are still limited, and most of them only focus on audiovisual fusion and lack systematic comparison of the performance and robustness of the models. In this paper, we propose a novel fusion module called Cross-modality Current Integration (CMCI) for multimodal SNNs and systematically compare it with other fusion methods on visual, auditory and olfactory fusion recognition tasks. Besides, a regularization technique called Modality-wise Dropout (ModDrop) is introduced to further improve the robustness of multimodal SNNs in missing modalities. Experimental results show that our method exhibits superiority in both modality-complete and missing conditions without any additional networks or parameters.

Keywords: spiking neural networks · multimodal learning · missing modality

1 Introduction

Humans leverage multiple perceptual processes, including sight, sound, touch, and smell, to understand the world comprehensively through multi-perspective

Supported by the National Key Research and Development Program of China under Grant 2020AAA0105900 and the National Natural Science Foundation of China under Grant 62236007.

observations. These multimodal signals can be efficiently integrated at multiple spatial and temporal scales in the brain, but the underlying neural mechanisms are not yet clear [1]. While multimodal fusion has been extensively studied in artificial neural networks (ANNs), the focus has primarily been on multimedia data consisting of images, audio, and text, or domain-specific multimodal data acquired by multiple sensors (e.g., medical imaging) [2]. Both the data and the model are far from the neural processing of the brain, making it challenging to simulate or reveal the real multimodal perception mechanism.

Recently, spiking neural networks (SNNs) [3], which mimic the neural dynamics and spiking communication processes of the brain, have gained much attention due to their better biologically plausible and low power consumption and several spike-based multimodal fusion methods have been proposed. Chen et al. [4] focused on the SNN-based auditory and olfactory encoding process, and temporally fused the encoded spike trains of the two modalities. Zhang et al. [5] proposed a cross-modal coupling module that promotes synchronization and integration of different modalities through excitatory and inhibitory lateral connections. However, these fusion methods require pre-trained unimodal encoding or recognition networks and lack collaborative learning of multiple modalities, thus making it difficult to capture inter-modal associations. Rathi et al. [6] introduced unsupervised spike-timing dependent plasticity (STDP) to simultaneously learn unimodal networks and cross-modal connections for bimodal digital recognition of images and speech. Liu et al. [7] proposed an event-based cross-modal attention method for end-to-end multimodal deep SNNs and achieved event-based audiovisual joint digit classification. In short, existing multimodal SNNs have produced satisfactory results and confirmed the advantages of multimodal fusion recognition, but they have three major limitations:

* Most current works focus on two modal fusion recognition tasks (e.g., toy audiovisual joint digit classification), while the applicability and performance for more than two modalities remain unknown.
* Existing spike-based multimodal fusion methods lack systematic comparison. Due to many differences in datasets (static or event-based), network architectures (shallow or deep network), action levels (input, feature or decision) and learning algorithms (unsupervised or supervised), it is difficult to judge which method is more effective.
* Most established multimodal SNNs are based on the modal integrity assumption, i.e., all modality inputs are available and clean in training and inference. However, in the real world, there is inevitably some degree of loss or noise in the modal signal due to the sensor itself and some external factors (e.g., transmission failure [8]). The robustness of existing models in this situation has not been discussed.

To tackle the above issues, we focus on the fusion recognition task for the visual, auditory and olfactory modalities in end-to-end deep SNNs trained with Surrogate Gradient Learning (SGL) [9–14] and propose a simple and effective fusion module called Cross-modality Current Integration (CMCI). The module measures the importance of different modalities and performs adaptive fusion through the internal dynamics of the spiking neurons. Experiments demonstrate that our proposed CMCI is suitable for any stage of multimodal SNN (e.g., feature or decision) and exhibits superiority in both modality-complete and missing conditions without any additional networks or parameters. Besides, a regularization technique called Modality-wise Dropout (ModDrop) is introduced for the first time to improve the robustness of multimodal SNN in missing modalities. We validate the effectiveness of ModDrop and further investigate how it affects the fusion process of different fusion methods.

2   Methodology

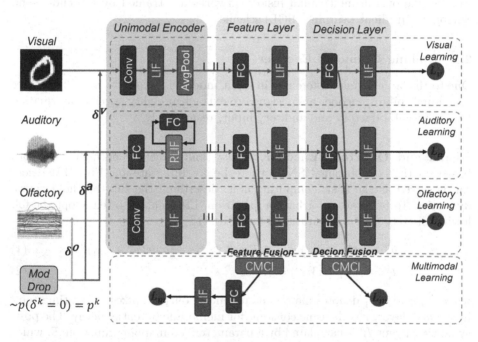

Fig. 1. The spike-based multimodal fusion framework based on the Cross-modality Current Integration (CMCI) module and ModDrop training strategy. LIF and RLIF correspond to validated (Eq. 1–2) and recurrent (Eq. 3–4) LIF, where the input currents are derived from either fully connected (FC) or convolutional (Conv) operations.

2.1   Architecture Overview

To fully capture inter-modal correlations while retaining modality-specific components, the proposed spike-based multimodal fusion framework consists of unimodal and multimodal learning (Fig. 1). For unimodal learning, the pipeline of each unimodal spiking neural network is unified into three parts to facilitate multi-level modal fusion: spike-based unimodal encoding, feature representation and decision output. The unimodal encoder is designed specificly to each modal, allowing full extraction of modal information. Followed by the unimodal encoder are two spike-based fully connected (FC) layers, where the feature layer transforms the modal feature of different dimensions to the same scale and the decision layer outputs the final prediction for the given modalities. For multimodal learning, a simple and effective Cross-modality Current Integration (CMCI) module is proposed to conduct feature-level or decision-level fusion of spiking neuron network from different modalities. Considering the inevitably modal data miss or corruption, we introduce a modality-wise dropout mechanism (ModDrop) [15] to encourage the CMCI to adaptively select and fuse according to the modal condition, and further mitigating the network sensitivity to missing modalities. Finally, the overall multimodal fusion networks are trained by the end-to-end Surrogate Gradient Learning (SGL) scheme.

2.2 Spiking Unimodal Encoder

Due to the large format differences in input modalities (refer to Sect. 3.1), three individual spiking encoder are designed to extract modality-specific information for the visual, olfactory and auditory input, respectively.

Visual and Olfactory Encoder. Visual and olfactory encoders are both based on the convolutional SNN, where Leaky Integrate-and-Fire (LIF) neuron is served as the basic computing unit, describing the sub-threshold neural dynamics. To facilitate error backpropagation, LIF neurons often adopt the following discrete-time computational form [10,11]:

$$u_t^n = \tau(1 - o_{t-1}^n)u_{t-1}^n + I_t^n; o_t^n = Heaviside(u_t^n - v_{th}); \quad (1)$$
$$I_t^n = \mathcal{F}(o_t^{n-1}; W^n) = W^n o_t^n \quad (2)$$

where u_t^n and o_{t-1}^n denotes membrane potential, output spikes at the timestep t in the n-th layer. τ is the time constant for membrane potential decay. The post-synaptic current I_t^n is calculated by a parameterized mapping function $\mathcal{F}$, which involves fully connected (FC) or convolutional operation, where W^n represents the weight matrix. Heaviside step function simulates the spike firing process, i.e., a spike will be emitted when the membrane potential reaches the threshold v_{th}.

Due to the spatiotemporal redundancy of visual stimuli, the visual encoder adopts additional average pooling operation to reduce the feature size of convolutional output spikes, while olfactory encoder maintains the original input dimension.

Auditory Encoder. Auditory stimuli contain richer temporal features, so the recurrent LIF (RLIF) neurons [7] with stronger dynamic representation capabilities are used in the auditory encoder. Unlike LIF neurons that only receive spikes from the previous layer, RLIF neurons also receive their membrane potentials as inputs, whose neural dynamic can be written as:

$$u_t^n = \tau(1 - o_{t-1}^n)u_{t-1}^n + I_t^n; o_t^n = Heaviside(u_t^n - v_{th}); \qquad (3)$$
$$I_t^n = \mathcal{F}(o_t^{n-1}; W^n) + \mathcal{F}(u_{t-1}^n; W_{rec}^n) \qquad (4)$$

where W_{rec}^n is the recurrent weight matrix in layer n.

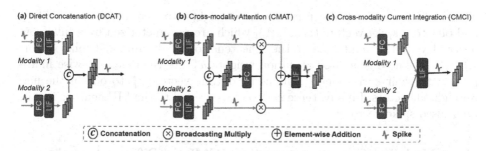

Fig. 2. Fusion module comparison: Fusion via (a) Direct Concatenation (DCAT), (b) Cross-modality Attention (CMAT), (c) Cross-modality Current Integration (CMCI)

2.3 Multimodal Fusion Module

In this section, we focus on general multimodal fusion modules that can fuse information of unimodal networks at different levels (feature or decision-level) to provide a more comprehensive representation and reliable prediction. To explore which fusion module is simpler and more effective in deep SNNs, we introduce two typical fusion modules, Direct Concatenation (DCAT) and Cross-modality Attention (CMAT) [7], as baselines for comparison with the proposed Cross-modality Current Integration (CMCI). For the convenience of description, we assume that the input spike vector of the feature layer or decision layer is represented as $H = \{h^v, h^a, h^o\}$, where superscript v, a, o denotes the visual, auditory and olfactory modality, respectively.

Direct Concatenation (DCAT). A straightforward approach for implementing multimodal fusion is to directly concatenate the spike vector z_t^v from visual modality, z_t^a from auditory modality and z_t^o from olfactory modality. Hence, DCAT fusion is modeled as:

$$z_t^k = \mathcal{LIF}^k(\mathcal{FC}^k(h_t^k; W^k)); k \in \{v, a, o\} \qquad (5)$$
$$F_t = [z_t^v, z_t^a, z_t^o]; \qquad (6)$$

where $\mathcal{FC}^k$ and $\mathcal{LIF}^k$ represent the fully connection (refer to Eq. 2) and LIF (refer to Eq. 1) operation of the feature or decision layer for the k-th modality, respectively. F_t represents the final fused spike vector.

Cross-Modality Attention (CMAT). The CMAT method [7] automatically focus on more reliable modal information through attention mechanism, i.e., reweighting the spike vector z_t^k. This process can be described as:

$$z_t^k = \mathcal{LIF}^k(\mathcal{FC}^k(h_t^k; W^k)); k \in \{v, a, o\} \tag{7}$$

$$[w_t^v, w_t^a, w_t^o] = \mathcal{FC}([z_t^v, z_t^a, z_t^o]) \tag{8}$$

$$F_t = \mathcal{LIF}(w_t^v \odot z_t^v + w_t^a \odot z_t^a + w_t^o \odot z_t^o) \tag{9}$$

where w_t^v, w_t^a and w_t^o are the attention weights applied to visual, auditory and olfactory spike vectors (z_t^v, z_t^a, z_t^o), which are parameterized by a fully connected layer $\mathcal{FC}$, conditioned on the concatenation of all unimodal spike vectors $[z_t^v, z_t^a, z_t^o]$. These generated attention weights w_t^k are then element-wise multiplied with their corresponding unimodal output vectors z_t^k to obtain the new reweighted vectors. Finally, these new vectors are fed to the LIF neurons for the final fused spike vector F_t.

Cross-Modality Current Integration (CMCI). Different from the concatenation or reweighting on the spike vector z_t^v, our proposed CMCI method perform the adaptive fusion of modal information by the internal dynamics of LIF neurons. Specifically, the input spike vector h_t^k of features or decision layer is converted into the postsynaptic currents I_t^k from different modalities by the $\mathcal{FC}^k$ operation. Subsequently, LIF neurons receive I_t^k and adaptively generate multimodal spike representations F_t. The formula is expressed as:

$$I_t^k = \mathcal{FC}^k(h_t^k; W^k); k \in \{v, a, o\} \tag{10}$$

$$F_t = \mathcal{LIF}(I_t^v + I_t^a + I_t^o) \tag{11}$$

As illustrated in Fig. 2, the proposed CMCI is more lightweight compared to other fusion approaches, which not only avoids the multiplying the dimensions of the fusion representation like DCAT, but also does not require any extra attention generation and reweighting process like CMAT. Besides, stronger postsynaptic current can indicate more confident modal information, and further produce a greater effect on the fused spike representations. Thus, competition for postsynaptic current from different modalities may induce potential attention allocation in LIF neurons, prompting better integration and spike representations.

2.4 ModDrop: Modality-Wise Dropout

To address the issue of missing modality, we introduce a regularization technique called modality-wise dropout (ModDrop) [15] to improve the robustness of SNN under arbitrary modal missing conditions. Unlike the original dropout which randomly drops a portion of neurons, ModDrop drops the entire input of a modality with a certain probability during training. Specifically, let $\{x^k\}, k \in$

$\{1, ..., K\}$ be the multi-modal training samples. The input data after ModDrop can be expressed as:

$$\tilde{x}^k = \delta^k x^k, \tag{12}$$

where $\delta^k \in \{0, 1\}$ is a binary mask indicated by Bernoulli selector $p(\delta^k = 0) = p^k$ for each modality. Here, we set $p^k = 0.1$ for all ModDrop operations throughout this research.

In this way, each input modality x^k is set to 0 with a certain probability p^k and there are $2^K - 1$ possible missing modality configurations. Furthermore, the learning objective of ModDrop is:

$$\underset{\theta_s}{\text{argmin}} L(y, F(\tilde{x} \mid \theta_s)) \tag{13}$$

where F is the fused network parameterized by θ_s and $\mathcal{L}$ is the loss function for specific task. Therefore, ModDrop simulates multiple missing modality configurations and force the network to create robust representations that can adapt to arbitrary missing conditions.

2.5   Loss Function

The final losses $\mathcal{L}_{total}$ consist of all unimodal losses and multimodal fusion losses:

$$\mathcal{L}_{total} = \lambda_v \mathcal{L}_v + \lambda_a \mathcal{L}_a + \lambda_o \mathcal{L}_o + \mathcal{L}_m \tag{14}$$

where $\mathcal{L}_v$, $\mathcal{L}_a$, $\mathcal{L}_o$ and $\mathcal{L}_m$ denotes the loss of visual, auditory, olfactory and fused network. The coefficient λ_v, λ_a and λ_o controls the proportion of visual, auditory and olfactory loss. When ModDrop is not applied, the coefficients for each modality are the same, i.e., $\lambda_v = \lambda_a = \lambda_o = 1$. When applying ModDrop, we use the binary mask δ^k of each modality as the coefficient to avoid the negative effects of training with zero-valued inputs, i.e., $\lambda_k = \delta^k, k \in \{v, a, o\}$.

All the losses are calculated by the mean square error (MSE) loss function that measures the distance between the label vector y and the firing rate of the output neuron:

$$\mathcal{L} = \left\| y - \frac{1}{T} \sum_{t=1}^{T} o_t \right\|_2^2 \tag{15}$$

where T is the given time window and o_t is the final output spike vector of the last layer.

3   Experiments

3.1   Datasets and Preprocess

Due to the lack of benchmark event-based multimodal (more than 2 modalities) dataset for SNN, we constructed a new multimodal dataset with visual, auditory and olfactory modalities by combining MNIST [16], Speech Commend [17] and Gas sensor arrays [18] dataset.

MNIST Dataset. The MNIST dataset has ten classes (0-9) of handwritten digits with a total of 70,000 grayscale images (28 × 28 pixel). The image can be fed directly to the SNN at each time step without any other preprocessing [19].

Speech Commend Dataset. To maintain category consistency, we construct a ten-category subset of digital speech (0-9) from the Speech Command dataset. There are a total of 38908 speech samples with a recording length of 1 s, which are divided into training and test sets with a ratio of 9:1. The original speech signal needs to be converted into a mel-spectrogram for further processing by the SNN, where the frame number of the spectrogram is equal to the given time windows ($T = 20$) and the number of Mel bands is set to 40. Finally, the delta and deltas-deltas feature extractions are performed on the mel-spectrogram respectively, and combined with the original spectrogram to form a $120 \times T$ input vector.

Gas Sensor Arrays Dataset. This gas dataset contains 18000 sensor measurement recordings obtained from 10 chemical sources (e.g., acetone, butanol). There are nine sensor arrays and each sensor array has eight different metal-oxide gas sensors. The response of the sensor to the gas is represented by the electrical conduction through the metal oxide film of the sensors. Each gas sample has 72 time series of 260 s with a sampling rate of 100 Hz. The collected data contains various configurations, including three wind speeds, five different temperatures, and six different locations.

We performed a series of preprocessing on this data. First, the sensor data are downsampled at $260/T$ Hz to ensure a consistent time window. Since the values of the sensors do not vary much in short periods, smaller sampling rate is feasible and do not affect the overall trend of the time series. Then, considering the data drift of the gas sensor [20,21], the sensor data is further normalized to [0.0, 1.0]. Finally, 72 gas waveforms are converted into images sequences of size $8 \times 9 \times T$ for the convolutional SNN. The data obtained from the first, second, third, and fifth locations are used for training and the fourth location for test[1].

For constructing the multimodal dataset, we randomly select 30000 (6000) groups of samples from the training (test) set of the above three data sets for training (test), where each group contains randomly selected three samples with the same label from three modalities.

Table 1. The network architectures for different modality

Modal branches	Unimodal encoder	Feature layer	Decision layer
Visual	16C3-AP2	128F	10F
Auditory	240R	128F	10F
Olfactory	16C3	128F	10F

xCy represents x convolution filters (y × y kernel size) with zero padding and APy represents average pooling layer with y × y pooling kernel size.; R represents recurrent connection and F represents the fully connection.

[1] The sixth location is excluded due to the missing data.

3.2 Network Architectures

The detailed architecture of the different modal networks is shown in Table 1, where all unimodal encoders use a single-layer structure and the dimension of the feature layer is unified to 128. We also compare additional networks and parameters required for different fusion methods in Table 2. It can be seen that our proposed CMCI method does not introduce additional networks and parameters, except for the necessary output layer for feature fusion. CMAT method requires an additional network to generate attention score, leading to extra parameters and calculations. Although the DCAT method does not introduce any additional network, it increases the size of the fusion output layer due to the concatenation operation and the parameters grow linearly with the number of modalities.

Table 2. The comparison of networks and parameters of different fusion methods

Fusion methods	Feature fusion (#Params)	Decision fusion (#Params)
DCAT	384-10 (3840)	30-10 (300)
CMAT	128-10;384-3* (2432)	30-3* (30)
CMCI	128-10 (1280)	- (0)

The attention network architectures.

3.3 Multimodal Fusion Comparison Without Missing Modalities

Without considering the missing modalities, we compare the performance of different multimodal fusion methods conducted at the feature or decision level. As shown in Table 3, all fusion methods significantly improve the recognition accuracy compared to unimodal recognition, except for the DCAT method with decision fusion. This indicates that the DCAT method is not suitable for decision-level fusion. We can also notice that the proposed CMCI method achieves state-of-the-art fusion performance both at the feature and decision levels without introducing extra parameters and computations. In addition, even if there are modalities with poor performance (e.g., the auditory modality with 80.50%), the CMCI method still achieves 99.95% recognition accuracy, which reveals that the CMCI method can effectively utilize the complementarity of modal data to achieve effective multimodal integration.

Table 3. Accuracy comparison of multimodal methods without missing modalities

Unimodal		Multimodal	Feature fusion (%)	Decision fusion (%)
Visual Modality	97.87	DCAT	99.93	93.45
Auditory Modality	80.50	CMAT [7]	99.72	99.61
Olfactory Modality	97.78	CMCI	**99.95**	**99.95**

Table 4. Comparison of multimodal fusion methods with missing modalities

Modal*			Feature fusion (%)			Decision fusion (%)		
V	O	A	DCAT	CMAT	CMCI	DCAT	CMAT	CMCI
●	o	o	31.32—97.48	10.10—97.97	69.83—97.58	86.92—97.67	94.80—95.38	86.38—96.95
o	●	o	91.10—95.37	90.45—97.32	70.00—97.87	64.63—97.18	76.78—92.92	95.50—96.23
o	o	●	10.88—78.18	12.58—77.95	20.82—81.60	34.60—81.97	10.07—80.07	39.93—81.98
●	●	o	99.82—99.45	99.65—99.38	99.68—99.83	94.95—99.70	99.52—98.98	99.70—99.70
●	o	●	41.48—98.63	51.88—98.15	77.48—98.90	88.88—98.93	95.62—98.13	96.10—98.60
o	●	●	96.97—97.37	82.88—98.57	96.68—98.82	82.92—98.17	94.75—97.92	98.17—98.25
●	●	●	99.93—99.73	99.72—99.55	99.95—99.95	93.45—99.88	99.61—99.70	99.95—99.88
Average			67.36—95.17	63.89—95.56	**76.35—96.36**	78.08—**96.21**	81.59—94.73	**87.96**—95.94

The modal column gives all possible missing conditions of V (Vsual), O (Olfactory), A (Auditory) modality, where ● and o represent the corresponding modality is present and absent, respectively. Average means the average accuracy under all possible missing conditions. The left and right sides of the symbol | represent the fusion accuracy without and with ModDrop, respectively

3.4 Multimodal Fusion Comparison with Missing Modalities

Table 4 shows the performance of different multimodal fusion methods for all possible missing conditions. First, we observe that the performance of all fusion methods drops significantly without ModDrop, especially when two modalities are missing. The multimodal fusion accuracy under many missing conditions is even much lower than the single-modal results (Table 3), demonstrating that multimodal fusion is not robust, and missing modality branches may cause large disturbances to the fused representation. It is worth noting that the overall performance of feature fusion is worse than decision fusion when the modality is missing. A possible reason is that the missing modal component has a greater impact on the feature distribution, and this effect will be further amplified by the subsequent network. However, our proposed CMCI still outperforms other fusion methods, with 8.99% and 12.64% (9.88% and 6.37%) higher accuracy than DCAT and CMAT under feature (decision) fusion, demonstrating the inherent robustness of CMCI.

By adding the ModDrop training, the multimodal fusion performances of all missing conditions are improved consistently and most multimodal fusion accuracy is higher than the unimodal results. This indicates that the ModDrop training can effectively mitigate the network sensitivity to missing modalities. With the ModDrop training, our proposed CMCI achieve better recognition accuracy, which is still higher than CMAT and DCAT methods.

3.5 The Effect of ModDrop

We further investigate how ModDrop affects the fusion process by visualizing the modal components and fusion representations (Fig. 3). Here, we take the decision fusion process of CMAT and CMCI as an example and assume that both

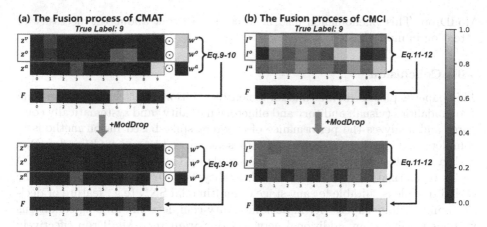

Fig. 3. The decision fusion process of (a) CMAT and (b) CMCI. z^k and I^k $k \in \{v, o, a\}$ represents the components generated by visual, olfactory and auditory modality, where the missing modal components are highlighted by the red rectangle. F represents the fusion results (Color figure online)

the visual and olfactory modalities are missing. The CMAT method achieves multimodal fusion by weighting the generated attention score with the spike output from different modalities, while our proposed CMCI achieves multimodal fusion by integrating the postsynaptic currents from different modalities. We define the firing rate $z^k, k \in \{v, o, a\}$ and the average postsynaptic current $I^k, k \in \{v, o, a\}$ as the modal components of the two fusion methods, respectively. The influence of different modalities on the final fusion results can be inferred from the distribution of the modal components. Note that the missing modal branch also produces a non-zero component due to the spontaneous response of the unimodal network (the bias of the convolution and linear layers).

Without ModDrop training, the components of missing modalities, i.e., z^v and z^o for (Fig. 3(a)), I^v and I^o for (Fig. 3(b)), produces a strong incorrect decision distribution, which may severely interfer with the fused decisions. Besides, the fusion representations of the CMAT method also rely on the attention score, but it incorrectly suppresses the correct modal component, instead of focusing on reliable modal inputs as expected. The worse performance of CMAT with missing modalities is most likely due to the dual interference of the attention scores and the missing modal components.

By adding the ModDrop training, the activation values of the missing modal components are all significantly reduced, thereby alleviating the interference with the complete modality. An interesting phenomenon is that while ModDrop can effectively drive the model to focus on reliable modal inputs, it does not seem to be achieved through the modulation of attention scores in CMAT (Fig. 3(a) bottom), i.e., ModDrop has little effect on the attention distribution. The attention still has a detrimental effect on modal fusion, which may be the reason why our CMCI method still performs better than the CMAT method after adding

ModDrop. This phenomenon also inspires us to think about the necessity of attention in modal fusion.

3.6 Conclusion

This paper explores the multimodal fusion recognition of SNNs with more than two modalities (visual, auditory and olfactory modality) and systematically compares and analyzes the performance of different spike-based fusion methods in complete and missing modalities. We present an end-to-end multimodal SNN with a novel fusion module called Cross-modality Current Integration (CMCI) and a regularization technique called Modality-wise Dropout (ModDrop). The CMCI module is suitable for any stage of multimodal SNN (e.g., feature or decision) and exhibits superiority in both modality-complete and missing conditions without requiring any additional networks or parameters. ModDrop effectively improves the robustness of multimodal SNN in missing modalities. By visualizing the fusion process of different fusion methods, we further find that the incorrect distribution of the missing modal components is the main reason for the drastic degradation of the model performance under the modality-missing condition. ModDrop reduces the activation values of the missing modal components, thereby effectively driving the model to focus on reliable modal inputs. The experiments also reveal that the effect of cross-modal attention on multimodal fusion is not always positive, which inspired us to re-examine the need for attention in multimodal fusion and design attention mechanisms that are more suitable for multimodal SNNs.

Acknowledgments. This work was supported by the National Key Research and Development Program of China under Grant 2020AAA0105900 and the National Natural Science Foundation of China under Grant 62236007.

References

1. Tan, H., Zhou, Y., Tao, Q., Rosen, J., van Dijken, S.: Bioinspired multisensory neural network with crossmodal integration and recognition. Nat. Commun. **12**(1), 1120 (2021)
2. Baltrušaitis, T., Ahuja, C., Morency, L.P.: Multimodal machine learning: a survey and taxonomy. IEEE Trans. Pattern Anal. Mach. Intell. **41**(2), 423–443 (2018)
3. Roy, K., Jaiswal, A., Panda, P.: Towards spike-based machine intelligence with neuromorphic computing. Nature **575**(7784), 607–617 (2019)
4. Chen, C., Xue, Y., Xiong, Y., Liu, M., Zhuang, L., Wang, P.: An auditory and olfactory data fusion algorithm based on spiking neural network for mobile robot. In: 2022 IEEE International Symposium on Olfaction and Electronic Nose (ISOEN), pp. 1–4. IEEE (2022)
5. Zhang, M., et al.: An efficient threshold-driven aggregate-label learning algorithm for multimodal information processing. IEEE J. Sel. Top. Signal Process. **14**(3), 592–602 (2020)
6. Rathi, N., Roy, K.: STDP based unsupervised multimodal learning with cross-modal processing in spiking neural networks. IEEE Trans. Emerg. Top. Comput. Intell. **5**(1), 143–153 (2018)

7. Liu, Q., Xing, D., Feng, L., Tang, H., Pan, G.: Event-based multimodal spiking neural network with attention mechanism. In: ICASSP 2022-2022 IEEE International Conference on Acoustics, Speech and Signal Processing (ICASSP), pp. 8922–8926. IEEE (2022)
8. Chavarriaga, R., et al.: The opportunity challenge: a benchmark database for on-body sensor-based activity recognition. Pattern Recognit. Lett. **34**(15), 2033–2042 (2013)
9. Gu, P., Xiao, R., Pan, G., Tang, H.: STCA: spatio-temporal credit assignment with delayed feedback in deep spiking neural networks. In: Twenty-Eighth International Joint Conference on Artificial Intelligence IJCAI 2019, pp. 1366–1372 (2019)
10. Wu, Y., Deng, L., Li, G., Zhu, J., Shi, L.: Spatio-temporal backpropagation for training high-performance spiking neural networks. Front. Neurosci. **12**, 331 (2018)
11. Wu, Y., Deng, L., Li, G., Zhu, J., Xie, Y., Shi, L.: Direct training for spiking neural networks: faster, larger, better. In: Proceedings of the AAAI Conference on Artificial Intelligence, vol. 33, pp. 1311–1318 (2019)
12. Li, Y., Guo, Y., Zhang, S., Deng, S., Hai, Y., Gu, S.: Differentiable spike: rethinking gradient-descent for training spiking neural networks. Adv. Neural. Inf. Process. Syst. **34**, 23426–23439 (2021)
13. Guo, Y., et al.: IM-loss: information maximization loss for spiking neural networks. Adv. Neural. Inf. Process. Syst. **35**, 156–166 (2022)
14. Ma, G., Yan, R., Tang, H.: Exploiting noise as a resource for computation and learning in spiking neural networks. arXiv preprint arXiv:2305.16044 (2023)
15. Neverova, N., Wolf, C., Taylor, G., Nebout, F.: Moddrop: adaptive multi-modal gesture recognition. IEEE Trans. Pattern Anal. Mach. Intell. **38**(8), 1692–1706 (2015)
16. LeCun, Y.: The MNIST database of handwritten digits (1998). http://yann.lecun.com/exdb/mnist/
17. Warden, P.: Speech commands: a dataset for limited-vocabulary speech recognition. arXiv preprint arXiv:1804.03209 (2018)
18. Vergara, A., Fonollosa, J., Mahiques, J., Trincavelli, M., Rulkov, N., Huerta, R.: On the performance of gas sensor arrays in open sampling systems using inhibitory support vector machines. Sens. Actuators B Chem. **185**, 462–477 (2013)
19. Rathi, N., Roy, K.: DIET-SNN: a low-latency spiking neural network with direct input encoding and leakage and threshold optimization. IEEE Trans. Neural Netw. Learn. Syst. (2021)
20. Choi, J.H., Lee, J.S.: Embracenet: a robust deep learning architecture for multimodal classification. Inf. Fusion **51**, 259–270 (2019)
21. Wang, S.H., Chou, T.I., Chiu, S.W., Tang, K.T.: Using a hybrid deep neural network for gas classification. IEEE Sens. J. **21**(5), 6401–6407 (2020)

A Weakly Supervised Deep Learning Model for Alzheimer's Disease Prognosis Using MRI and Incomplete Labels

Zhi Chen, Yongguo Liu[✉], Yun Zhang, Jiajing Zhu, and Qiaoqin Li

Knowledge and Data Engineering Laboratory of Chinese Medicine, School of
Information and Software Engineering, University of Electronic Science and
Technology of China, Chengdu 610054, China
ygliu_uestc@163.com

Abstract. Predicting cognitive scores using magnetic resonance imaging (MRI) can aid in the early recognition of Alzheimer's disease (AD) and provide insights into future disease progression. Existing methods typically ignore the temporal consistency of cognitive scores and discard the subjects with incomplete cognitive scores. In this paper, we propose a Weakly supervised Alzheimer's Disease Prognosis (WADP) model that incorporates an image embedding network and a label embedding network to predict cognitive scores using baseline MRI and incomplete cognitive scores. The image embedding network is an attention consistency regularized network to project MRI into the image embedding space and output the cognitive scores at multiple time-points. The attention consistency regularization captures the correlations among time-points by encouraging the attention maps at different time-points to be similar. The label embedding network employs a denoising autoencoder to embed cognitive scores into the label embedding space and impute missing cognitive scores. This enables the utilization of subjects with incomplete cognitive scores in the training process. Moreover, a relation alignment module is incorporated to make the relationships between samples in the image embedding space consistent with those in the label embedding space. The experimental results on two ADNI datasets show that WADP outperforms the state-of-the-art methods.

Keywords: Alzheimer's disease · Cognitive score prediction · Deep learning · Disease progression · Weakly supervised learning

1 Introduction

Alzheimer's disease (AD), characterized by memory loss and behavioral issues, is an irreversible, progressive disease and the leading cause of dementia in elderly subjects [13]. It is reported that the population of Americans age 65 and older with AD is expected to surpass 13.8 million in 2050 [4]. To aid in the diagnosis of AD and the assessment of treatment effects, several cognitive assessments have been designed [15], such as Mini-Mental state examination (MMSE) [9],

AD assessment scale-cognitive (ADAS-Cog) [24], and clinical dementia rating-sum of boxes (CDR-SB) [11]. However, conducting assessments by clinicians is a highly time-consuming task relying on their expertise and experience. As a result, there has been an increasing interest in developing machine learning models capable of automatically estimating cognitive scores from neuroimaging measurements (such as structural magnetic resonance imaging (MRI) data) [14, 19]. These models not only provide estimates of current cognitive scores but also have the ability to predict future cognitive scores based on baseline MRI data, enabling the prognosis of AD progression [22].

Inspired by the development of deep learning in visual analysis tasks, recent studies have aimed to learn discriminative features automatically from MRI images based on deep neural networks [5,10,20,22,28]. These methods can be divided into four categories based on the input type of networks [28]: 2D slice-based methods [23,27], ROI-based methods [2,18], patch-based methods [21,22], and subject-based methods [5,16]. 2D slice-based methods take 2D slices of 3D MRI images as input, without considering the 3D spatial information. ROI-based and patch-based methods require preselecting regions (e.g., AD-sensitive regions [32] or patches). However, as region selection and subsequent prediction are treated as independent stages, the preselected regions may not be optimal for prognosis models, leading to performance degradation [17]. To tackle these limitations, subject-based methods that use the whole-brain MRI images as the input are proposed. However, existing subject-based methods still have the following limitations:

1) Neglecting the temporal consistency of cognitive scores. Existing studies have primarily focused on identifying informative brain regions with structural changes related to pathological features using attention mechanisms [16,19,31]. However, these approaches have mainly considered attention maps at the baseline, overlooking the relationships between attention maps at different time-points. From the biological standpoint, brain structural changes are thought to be associated with cognitive decline in aging [1] and AD [29]. These changes not only impact cognitive performance at the baseline [7,26] but also have implications for cognitive performance at future time-points [8,29]. Therefore, it is desirable to focus on the same abnormal regions when predicting the cognitive performance at different time-points so as to extract features related to the structural changes in the baseline MRI data associated with cognitive performance. However, previous methods rarely exploit the consistency between different time-points.

2) Discarding the subjects with incomplete cognitive scores. The ground-truth cognitive scores may be missing during long-term follow-up. For instance, due to time conflicts, patients may be unable to undergo assessments at the designated time, resulting in missing cognitive scores. Many AD prognosis models assume complete training labels and only utilize subjects with complete cognitive scores for training [22]. Therefore, the number of training subjects is reduced. Additionally, there are often correlations between assessments and time-points. For example, it is reported that different assessments evaluate some overlapping cognitive abilities [6]. These correlations can contribute to

improved prediction performance, but they are lost when missing cognitive scores are discarded.

In this paper, we propose a Weakly supervised AD Prognosis (WADP) model with two main network structures: an image embedding network and a label embedding network, to predict cognitive scores using baseline MRI and incomplete labels. For image embedding, an attention consistency-guided convolutional network is developed to locate the informative brain regions and learn the feature embedding of the whole MRI image simultaneously. An attention consistency regularization, which makes the attention maps at different time-points similar, is introduced to cover the similar atrophic regions and capture the correlations among time-points. For label embedding, a denoising autoencoder is developed to explore inter-score correlations and learn the feature embedding of cognitive scores. We adopt dropout to mask the input to simulate the presence of missing cognitive scores and train the autoencoder to impute the missing cognitive scores so as to increase the number of available training samples. Moreover, considering that both image embedding and label embedding should contain the semantic information about cognitive performance, we design a relation alignment module to capture robust and discriminative features by explicitly enhancing the consistency of the relationships among samples in different embedding spaces. The key contributions of this work can be summarized as follows:

- An attention consistency-guided image embedding network that enhances the consistency between the attention maps at different time-points is designed to make the discriminative regions for different time-points similar.
- A denoising autoencoder-based label embedding network is designed to explore the correlations between cognitive scores and make full use of the subjects with incomplete cognitive scores.
- A relation alignment regularization term that captures the correlations between MRI images and cognitive scores is presented to make the embedding spaces discriminative.
- We evaluate the proposed model on two ADNI datasets in comparison with the state-of-the-art methods in AD prognosis.

2   Method

Let $\mathcal{D} = (\boldsymbol{X}_i, \mathbf{s}_i)_{i=1}^{N}$ denotes the set of training data, where $\boldsymbol{X}_i$ and $\mathbf{s}_i = (s_{i1}, \cdots, s_{iC}) \in \mathbb{R}^{C}$ are the MRI image at baseline and cognitive scores of the ith subject, respectively. s_{ir} is the rth cognitive score of the ith subject. Here, $s_{ir} = -1$ means that this cognitive score is missing. N denotes the number of training subjects and C denotes the number of cognitive scores. Our objective is to predict three assessments (MMSE, CDR-SB, and ADAS-Cog) at four time-points (BL, M06, M12, and M24). The baseline (BL) defined in the ADNI database is the date when the patient performs the screening in the hospital for the first time. The other three time-points are determined based on the duration from the baseline, including 6th month after BL (M06), 12th month after BL

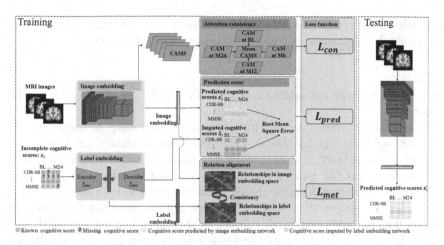

Fig. 1. Illustration of WADP model, which contains an image embedding network and a label embedding network. The image embedding network learns the image embeddings and extracts the CAMs. The attention consistency module makes the CAMs at different time-points similar. The label imputing module imputes missing cognitive scores and extracts the label embeddings. The relation alignment module makes the relationships in the image embedding space consistent with those in the label embedding space.

(M12), and 24th month after BL (M24). Therefore, s_i contains the scores of three assessments at four time-points and $C = 12$ in this paper.

The overall framework of WADP is depicted in Fig. 1, which consists of an image embedding network and a label embedding network. The image embedding network utilizes a convolutional network to embed MRI images into the image embedding space and generate attention maps. The label embedding network employs a denoising autoencoder to embed cognitive scores into the label embedding space and impute missing scores. Additionally, three modules are incorporated into the framework: an attention consistency module to ensure the similarity of attention maps across different time-points, a relation alignment module to establish consistency between the relationships in the image embedding space and those in the label embedding space, and a prediction loss module to quantify the prediction error.

2.1   Image Embedding Network

We design a 3D convolutional network to automatically extract the deep features of whole-brain MRI images. Specifically, the architecture owns 11 convolutional blocks and a global average pooling layer. Each convolutional block consists of a 3D convolutional layer, a 3D batch-normalization layer, and a ReLU activation layer. The numbers of feature maps are set as 16, 16, 16, 32, 32, 32, 64, 64, 64, 128, and m for each block, respectively, where m is the number of feature maps at the last convolutional layer. Since a global average pooling layer is added on the last convolutional layer, m also denotes the image embedding size. The convolutional network can be regarded as a nonlinear mapping, which projects MRI image $\boldsymbol{X}_i$ of the ith subject to the image embedding space:

$$\boldsymbol{x}_i = f_{cnn}\left(\boldsymbol{X}_i, \boldsymbol{\theta}_{cnn}\right), \tag{1}$$

where $\boldsymbol{x}_i \in \mathbb{R}^m$ is the embedding vector of $\boldsymbol{X}_i$ and $\boldsymbol{\theta}_{cnn}$ is the weight matrix of the network.

In addition to predicting cognitive scores, we also want to locate discriminative regions for prognosis. We describe the class activation mapping (CAM) [30], which can extract attention maps produced in convolutional neural network and highlight the regions semantically relevant to labels. The key idea is that the discriminative features should cooccur with the regions related to the labels/cognitive scores [30]. To facilitate description, we omit the subscript i representing the index of the ith subject in this part. For a given MRI image, let $\boldsymbol{F} \in \mathbb{R}^{m \times L \times W \times H}$ represents the feature maps from the last convolutional layer, where m, L, W, H are the number of feature maps, length, width, and height of the feature maps, respectively, and $f_k(x, y, z) = F[k, x, y, z]$ represents the activation of the kth feature map at spatial location (x, y, z). A global average pooling layer is applied on $\boldsymbol{F}$ to produce the image embedding vector $\boldsymbol{x} = (f_1, \ldots, f_m)^T$ and capture the semantic information of the MRI image, where $f_k = \sum_{x,y,z} f_k(x, y, z)/(L \times W \times H)$. Then, the elements in $\boldsymbol{x}$ are plugged into the predicted score s'_c via a fully connected layer:

$$s'_c = \boldsymbol{w}^{c^T} \boldsymbol{x} = \sum_k w^c_k f_k, \tag{2}$$

where $\boldsymbol{w}^c = (w^c_1, w^c_2, \ldots, w^c_m)^T$ is the weight for score s'_c and w^c_k represents the weight corresponding to score s'_c for the kth feature map. The predicted cognitive score s'_c can be rewritten as

$$s'_c = \sum_k w^c_k f_k = \frac{\sum_{x,y,z} \sum_k w^c_k f_k(x, y, z)}{L \times W \times H} = \frac{\sum_{x,y,z} A_c(x, y, z)}{L \times W \times H}, \tag{3}$$

where $A_c(x, y, z) = \sum_k w^c_k f_k(x, y, z)$ and $\boldsymbol{M}_c = \text{softmax}(\boldsymbol{A}_c)$ is the attention map for the cth cognitive score. The softmax function is to constrain each element in $\boldsymbol{M}_c$ within the range of $[0, 1]$, which can indicate the contribution of the activation at the spatial location (x, y, z) to the prediction of the cth cognitive score. The intensity of each pixel in the attention map $\boldsymbol{M}_c$ is proportional to its discriminative power for the cth cognitive score.

By using CAM, we can find label-relevant regions in MRI images. The labels in this paper are the cognitive scores that indicate the cognitive function measured at baseline (BL) and future time-points (M06, M12, and M24). Existing studies have found that there are different atrophy patterns in baseline MRI scans of AD patients [8]. These patterns differ in the location of atrophic regions and the amount of pathology in each region, associating with different cognitive trajectories. It means that the cognitive performance at baseline and future timepoints can be inferred based on the baseline brain atrophy patterns. Therefore, it is desirable to focus on the same atrophic regions when predicting cognitive performance at different time-points. In other words, the regions that the model pays attention to at different time-points for an assessment should be consistent over time to cover the same pathological regions in the MRI image at baseline. Therefore, we develop a regularization term to make the attention maps at different time-points similar:

$$\mathcal{L}_{\text{con}} = \frac{1}{3} \sum_{c=1}^{3} \mathcal{L}_{\text{con},c} = \frac{1}{3} \times \frac{1}{4|M|} \sum_{t=1}^{4} \left\| M_{ct} - \bar{M}_c \right\|_F, \quad (4)$$

where $\|\cdot\|_F$ is the Frobenius norm of a matrix, $\mathcal{L}_{\text{con},c}$ is the attention consistency loss for the cth assessment, $\bar{M}_c = 1/4 \sum_{t=1}^{4} M_{ct}$ is the mean attention map of the cth assessment, M_{ct} is the attention map of the cth assessment at the tth time-point, and $|M|$ denotes the size of the attention map.

2.2   Label Embedding Network

It is worth noting that there exist some subjects with incomplete cognitive scores. The number of training subjects significantly decreases if these subjects are discarded at the training stage, leading to a low prognosis performance [25]. We aim to leverage the correlations among cognitive scores to impute the missing cognitive scores so as to deal with the problem of incomplete cognitive scores. A denoising autoencoder-based label embedding network is designed so as to learn the label embedding vector and impute missing cognitive scores simultaneously. For dataset $\mathcal{D} = (X_i, s_i)_{i=1}^{N}$ with N training samples, we denote the subdataset containing the subjects with complete cognitive scores as $\widetilde{\mathcal{D}} = \{X_i, s_i\}_{i=1}^{\widetilde{N}}$, where $\widetilde{N}$ is the number of the subjects with complete cognitive scores. The following two steps are performed for imputing the missing cognitive scores:

Step 1: Denoising autoencoder training. We train the denoising autoencoder on $\widetilde{\mathcal{D}}$. To simulate the presence of missing cognitive scores, the cognitive score vector s_i is corrupted, in which a fraction of cognitive scores in s_i chosen at random are masked. The corrupted cognitive score vector is denoted as $\bar{s}_i$. Then, $\bar{s}_i$ is fed into the encoder to obtain label embedding vector:

$$l_i = f_{\text{enc}}\left(\bar{s}_i, \theta_{\text{enc}}\right), \quad (5)$$

where θ_{enc} is the weight of the encoder and $l_i \in \mathbb{R}^m$ is the label embedding vector. Next, l_i is mapped back with a decoder to reconstruct the original cognitive scores:

$$\tilde{s}_i = f_{\text{dec}}\left(l_i, \theta_{\text{dec}}\right), \quad (6)$$

where θ_{dec} is the weight of the decoder and $\tilde{s}_i$ is the reconstructed cognitive score vector. The mean square error is used to train the autoencoder:

$$\mathcal{L}_{rec} = \frac{1}{\widetilde{N}} \sum_{i=1}^{\widetilde{N}} \left\| s_i - \tilde{s}_i \right\|_F^2. \quad (7)$$

After training the denoising autoencoder, label embedding vector l_i captures the correlations among cognitive scores and we can fill the artificially introduced blanks with the reconstructed cognitive score vector $\tilde{s}_i$.

Step 2: Missing cognitive score imputing. Since that the trained denoising autoencoder has the ability to impute missing elements, here we feed the incomplete cognitive scores in $\mathcal{D}$ into the denoising autoencoder and fill the missing elements with the reconstructed cognitive scores. Specifically, we adopt the elements in $\tilde{s}_i$ to fill the corresponding missing values in s_i:

$$\hat{s}_i = \tilde{s}_i \odot v + (s_i \odot (1 - v)), \tag{8}$$

where $\odot$ denotes element-wise multiplication, s_i is the imputed cognitive score vector, and v is a mask vector with element $v_c = 1$ if the cth cognitive score is missing and $v_c = 0$ otherwise. Thus, we can train the WADP model with $\hat{s}_i$ as the target and all data can be fully utilized in the following training without discarding any subject or cognitive score.

2.3   Relation Alignment

Based on the image embedding network, we can learn the embedding for MRI image. Since we aim to predict the AD progression by estimating the cognitive scores based on MRI data, it is excepted that the image embedding contains the semantic information about AD progression. Moreover, we also project the cognitive scores into the label embedding space by the label embedding network and the label embedding captures the high-level semantic features of cognitive scores. As a result, both image embedding space and label embedding space preserve the semantic information about AD progression, which is of vital importance for cognitive score prediction. Therefore, the relationships between samples in image embedding space and label embedding space should be consistent if the semantic information about AD has been captured. In light of this, a relation alignment module is designed to make the relationship between two image embedding vectors consistent with that between their corresponding label embedding vectors. Specifically, we calculate the distance between two label embedding vectors as

$$lab_{i,j} = (l_i - l_j)^T (l_i - l_j). \tag{9}$$

For convenience, we construct the similarity matrix of the label embedding vectors B whose element $b_{i,j} = e^{-lab_{i,j}}$. Then, we define the image embedding matrix $E = [x_1, x_2, \ldots, x_N] \in \mathbb{R}^{m \times N}$. We consider if two subjects have similar label embedding vectors, the same relationships should be preserved in the corresponding image embedding space. Therefore, we define the relation alignment regularization term:

$$\mathcal{L}_{\text{met}} = \frac{1}{2} \sum_{i=1}^{N} \sum_{j=1}^{N} b_{i,j} \|x_i - x_j\|_2^2 = \text{tr}(E P_G E^T), \tag{10}$$

where $P_G = D_G - B$ is the related Laplacian matrix and D_G is a diagonal matrix with the ith diagonal element $D_G(i,i) = \sum_{j=1}^{N} b_{i,j}$.

Table 1. Number of subjects with three cognitive assessments at four time-points

Dataset		TOT	MMSE				CDR-SB				ADAS-Cog			
			BL	M06	M12	M24	BL	M06	M12	M24	BL	M06	M12	M24
ADNI1	AD	192	192	180	162	136	192	179	160	135	191	180	161	135
	MCI	397	397	378	356	302	397	378	356	300	397	377	355	301
	NC	229	229	221	211	203	229	217	207	197	229	221	210	202
	TOT	818	818	779	729	641	818	774	723	632	817	778	726	638
ADNI2	AD	146	146	131	105	31	146	131	110	31	145	129	105	31
	MCI	339	339	315	313	270	339	312	307	278	338	315	312	270
	NC	291	291	272	206	245	291	272	201	244	291	273	206	244
	TOT	776	776	719	624	546	776	715	618	553	774	717	623	545

2.4 Overall Loss Function

Since the missing cognitive scores are imputed by the label embedding network, the image embedding network can be trained on weakly labeled data only after the label embedding network training is finished. Therefore, we first train the label embedding network on subdataset $\widetilde{\mathcal{D}}$ with loss defined in (7). Then, the missing cognitive scores in $\mathcal{D}$ are imputed by the corresponding reconstructed values according to (8). Next, we jointly minimize the following loss function for training the whole framework on $\mathcal{D}$.

$$\mathcal{L} = \mathcal{L}_{\text{pred}} + \lambda_1 \mathcal{L}_{\text{met}} + \lambda_2 \mathcal{L}_{\text{con}}, \tag{11}$$

where λ_1 and λ_2 are trade-off hyperparameters. Here, $\mathcal{L}_{\text{met}}$ defined in (10) and $\mathcal{L}_{\text{con}}$ defined in (4) are the relation alignment loss and the attention consistency loss, respectively. The prognosis loss L_{pred} is the mean square error between the predicted cognitive scores and the target cognitive scores.

3 Experiments

3.1 Datasets and Experimental Settings

Datasets. We conduct experiments on two datasets (i.e., ADNI1 and ADNI2) from the ADNI database (adni.loni.usc.edu) [12] to perform experimental analysis. The datasets comprise three categories of subjects, namely Alzheimer's disease (AD) patients, mild cognitive impairment (MCI) patients, and normal controls (NC). All subjects have the BL MRI scans and some subjects miss cognitive scores at certain time-points. We show the detailed information of the datasets in Table 1. MRI preprocessing is carried out using the CAT12 in SPM12 running on MATLAB 2020a. First, the MRI data are denoised, interpolated, bias field inhomogeneity corrected and affine registered. Then the skull and other non-brain elements are removed. Finally, the images are registered into the standard MNI space using the deformable registration algorithm DARTEL [3].

Table 2. Regression results on ADNI2 obtained by models trained on ADNI1.

		CDR-SB				ADAS-Cog				MMSE				Avg.
		BL	M06	M12	M24	BL	M06	M12	M24	BL	M06	M12	M24	
CC	RRFS	0.4588	0.4683	0.4484	0.4137	0.3814	0.4323	0.4383	0.3789	0.4478	0.4699	0.4464	0.4398	0.4353
	DM2L	0.5241	0.5560	0.5261	0.4713	0.5902	0.5940	0.5596	0.4988	0.5402	0.5728	0.5309	0.5043	0.5390
	wiseDNN	0.5788	0.6014	0.5538	0.5095	**0.6352**	0.6299	0.5933	0.5548	0.5761	**0.6200**	0.5644	0.5362	0.5795
	MWAN	0.5744	0.5994	0.5599	0.5106	0.5887	0.6129	0.5871	0.5276	0.5527	0.5948	0.5610	0.5414	0.5675
	ICAM	0.5672	0.5980	0.5673	0.5160	0.5979	0.6147	0.5847	0.5315	0.5403	0.5963	0.5636	0.5497	0.5689
	WADP	**0.5955**	**0.6255**	**0.6022**	**0.5677**	0.6202	**0.6313**	**0.6299**	**0.5684**	**0.5781**	0.6176	**0.5998**	**0.5714**	**0.6006**
nRMSE	RRFS	0.1560	0.1831	0.2183	0.2270	0.2366	0.2458	0.2600	0.2672	0.1523	0.1804	0.2060	0.2135	0.2122
	DM2L	0.0916	0.1086	0.1331	0.1253	0.0890	0.0951	0.1112	0.1057	0.0989	0.1130	0.1094		0.1048
	wiseDNN	0.0903	0.1086	0.1347	0.1171	0.0854	0.0888	0.1035	0.0965	0.0766	0.0928	0.1096	0.0985	0.1002
	MWAN	**0.0886**	0.1069	0.1358	0.1182	0.0850	0.0914	0.1054	0.0998	0.0769	0.0945	0.1090	0.0994	0.1009
	ICAM	0.0905	0.1071	**0.1328**	0.1190	0.0866	0.0936	0.1093	0.0990	0.0798	0.0962	0.1119	0.1007	0.1022
	WADP	0.0892	**0.1005**	0.1338	**0.1118**	**0.0810**	**0.0880**	**0.1024**	**0.0937**	**0.0750**	**0.0909**	**0.1074**	**0.0965**	**0.0975**

Table 3. Regression results on ADNI1 obtained by models trained on ADNI2.

		CDR-SB				ADAS-Cog				MMSE				Avg.
		BL	M06	M12	M24	BL	M06	M12	M24	BL	M06	M12	M24	
CC	RRFS	0.4125	0.4359	0.4440	0.3841	0.3118	0.3625	0.4149	0.1054	0.4028	0.4027	0.4506	0.2561	0.3653
	DM2L	0.4073	0.4232	0.4239	0.4806	0.3645	0.3445	0.3585	0.4185	0.3771	0.3804	0.3966	0.4307	0.4005
	wiseDNN	**0.5123**	0.5156	0.4829	0.5014	0.5218	**0.5357**	0.5348	0.5267	0.4460	0.4823	0.4793	0.4961	0.5029
	MWAN	0.4524	0.4709	0.4703	0.4897	0.4595	0.4947	0.4941	0.4879	0.4094	0.4436	0.4561	0.4749	0.4670
	ICAM	0.4907	0.4952	0.5059	0.5169	0.4887	0.4938	0.4898	0.4913	0.4636	**0.4843**	0.4773	0.4949	0.4910
	WADP	0.5021	**0.5160**	**0.5308**	**0.5723**	**0.5262**	0.5325	**0.5528**	**0.5659**	**0.4808**	0.4810	**0.5051**	**0.5421**	**0.5256**
nRMSE	RRFS	0.1617	0.1818	0.2025	0.2698	0.2540	0.2578	0.2592	0.2937	0.1655	0.1913	0.2100	0.2536	0.2251
	DM2L	0.1018	0.1240	0.1391	0.1841	0.1011	0.1217	0.1344	0.1608	0.0934	0.1186	0.1344	0.1710	0.1320
	wiseDNN	0.1020	0.1168	0.1277	0.1724	0.0938	**0.1013**	**0.1077**	0.1482	0.0854	0.1113	0.1282	0.1704	0.1221
	MWAN	0.0932	0.1164	0.1334	0.1870	**0.0820**	0.0995	0.1156	0.1431	0.0834	0.1106	0.1299	0.1650	0.1216
	ICAM	0.1062	0.1286	0.1465	0.1923	0.0939	0.1099	0.1273	0.1513	0.0856	0.1129	0.1314	0.1803	0.1305
	WADP	**0.0921**	**0.1135**	**0.1261**	0.1670	0.0897	0.1016	**0.1077**	**0.1338**	**0.0811**	**0.1076**	**0.1213**	**0.1499**	**0.1160**

Experimental Settings. We use the PyTorch package for implementation and Nvidia RTX 3090 GPU for training. The Adam optimizer with a learning rate of 0.001 is used for training. The batch size is set as 8. Hyperparameters m, λ_1, λ_2 are set as 64, 0.05, and 0.005, respectively. Following the experimental setup described in previous studies [16,22], we initially employ ADNI1 as the training set and test the performance using ADNI2. In the second group, we switch the roles, employing ADNI2 as the training set and testing with ADNI1. Moreover, we select 20% samples from the training set as the validation set. All experiments are repeated 10 times to offer the average performance. Normalized root mean square error (nRMSE) and correlation coefficient (CC) are used to evaluate experimental methods.

3.2   Performance Comparison

Here, we compare WADP with existing state-of-the-art methods, including RRFS [33], DM2L [21], wiseDNN [22], MWAN [16], and ICAM [5]. The prediction results achieved by different methods are reported in Tables 2 and 3. Some observations can be summarized as follows: 1) Deep learning-based methods that learn task-oriented features (i.e., DM2L, wiseDNN, MWAN, ICAM, and WADP)

Table 4. Ablation results on ADNI2 obtained by models trained on ADNI1.

		CDR-SB				ADAS-Cog				MMSE				Avg.
		BL	M06	M12	M24	BL	M06	M12	M24	BL	M06	M12	M24	
CC	WADP-M	0.5721	0.5923	0.5613	0.5078	**0.6366**	**0.6415**	0.6133	0.5446	0.5653	0.6086	0.5773	0.5465	0.5806
	WADP-C	0.5693	0.5952	0.4209	0.5067	0.6117	0.6186	0.5554	0.5475	0.5532	0.5958	0.5304	0.5395	0.5537
	WADP-L	0.5674	0.5868	0.5539	0.5105	0.6020	0.6257	0.5941	0.5474	0.5411	0.5950	0.5543	0.5393	0.5681
	WADP	**0.5955**	**0.6255**	**0.6022**	**0.5677**	0.6202	0.6313	**0.6299**	**0.5684**	**0.5781**	**0.6176**	**0.5998**	**0.5714**	**0.6006**
nRMSE	WADP-M	0.0919	0.1039	0.1291	0.1208	0.0818	0.0894	0.1039	0.1009	0.0782	0.0944	0.1088	0.1060	0.1008
	WADP-C	0.0915	0.1014	0.1422	0.1209	0.0825	0.0901	0.1278	0.1044	0.0783	0.0937	0.1231	0.1075	0.1053
	WADP-L	0.0902	0.1054	**0.1303**	0.1239	0.0846	0.0911	0.1052	0.1004	0.0781	0.0948	0.1092	0.1051	0.1015
	WADP	**0.0892**	**0.1005**	0.1338	**0.1118**	**0.0810**	**0.0880**	**0.1024**	**0.0937**	**0.0750**	**0.0909**	**0.1074**	**0.0965**	**0.0975**

Table 5. Ablation results on ADNI1 obtained by models trained on ADNI2.

		CDR-SB				ADAS-Cog				MMSE				Avg.
		BL	M06	M12	M24	BL	M06	M12	M24	BL	M06	M12	M24	
CC	WADP-M	0.4857	0.5012	**0.5471**	**0.5761**	0.4997	0.4997	**0.5734**	0.5481	0.4569	0.4746	0.4943	0.5370	0.5161
	WADP-C	0.4793	0.4667	0.4910	0.5390	0.4754	0.5004	0.5186	0.5331	0.4484	0.4620	0.4758	0.5142	0.4920
	WADP-L	0.4776	0.4786	0.4608	0.5109	0.4529	0.4837	0.4852	0.4757	0.4061	0.4276	0.4285	0.4586	0.4622
	WADP	**0.5021**	**0.5160**	0.5308	0.5723	**0.5262**	**0.5325**	0.5528	**0.5659**	**0.4808**	**0.4810**	**0.5051**	**0.5421**	**0.5256**
nRMSE	WADP-M	0.0916	0.1173	0.1281	0.1731	0.0922	0.1032	0.1037	0.1368	0.0847	0.1109	0.1179	0.1509	0.1175
	WADP-C	0.0923	0.1158	0.1301	**0.1667**	0.0914	0.1027	0.1105	0.1377	0.0823	0.1086	0.1237	0.1553	0.1181
	WADP-L	0.0918	0.1140	0.1303	0.1747	0.0911	0.1034	0.1103	0.1390	0.0822	0.1085	0.1256	0.1606	0.1193
	WADP	**0.0921**	**0.1135**	**0.1261**	0.1670	**0.0897**	**0.1016**	**0.1077**	**0.1338**	**0.0811**	**0.1076**	**0.1213**	**0.1499**	**0.1160**

usually yield better results than the method using handcrafted features (i.e., RRFS), which verifies that combining feature extraction and prediction model construction does help to improve the prognosis performance. 2) Compared with the methods (i.e., DM2L and wiseDNN) that extract deep features from predefined regions, WADP that jointly extracts informative dementia-related regions from the whole brain MRI images and predicts multiple scores yields lower nRMSEs and higher CCs. It suggests that our joint learning strategy can boost the learning performance by implicitly exploiting the locations related to cognitive scores. 3) WADP outperforms other attention mechanism-based methods (i.e., MWAN and ICAM) in terms of nRMSE and CC. All subjects in the datasets can be included in the training process of WADP, while the subjects with incomplete cognitive scores are discarded in MWAN and ICAM. WADP effectively exploits the subjects with incomplete cognitive scores and significantly increases the number of training subjects. As the original data distribution can be described more accurately with more samples, the prognosis performance is boosted. 4) WADP yields better prognosis performance than another weakly supervised method (i.e., wiseDNN). The reason is that, equipped with the relation alignment, WADP can learn discriminative feature representations for MRI images and cognitive scores. Besides, wiseDNN discards missing cognitive scores and ignores the correlations among cognitive scores. However, WADP uses a label imputing strategy to fill missing cognitive scores to exploit the correlations among cognitive scores, leading to the improvement of the prognosis performance.

3.3 Ablation Study

Effect of Label Imputing. To show the effectiveness of label imputing, we implement a variant of WADP called WADP-L. The difference between WADP-L and WADP is that WADP-L is trained on subdataset $\widetilde{\mathcal{D}}$ instead of dataset $\mathcal{D}$. Tables 4 and 5 show the results of WADP and WADP-L. For the testing results on ADNI2, WADP achieves better nRMSE than WADP-L in 11 cognitive scores and better CC than WADP-L in 12 cognitive scores. The average nRMSE and CC values of WADP are 0.0975 and 0.6006, respectively, which are better than those of WADP-L (i.e., 0.5681 and 0.1015, respectively), which show the effectiveness of the label imputing for exploiting weakly labeled data.

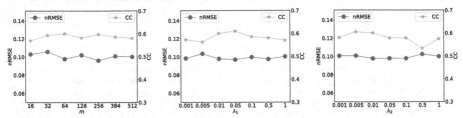

Fig. 2. The influence of different hyperparameters on the performance of WADP

Effect of Attention Consistency. We implement a variant of WADP named WADP-C without attention consistency regularization to analyze the impact of the attention consistency regularization, as shown in Tables 4 and 5. It is found that the attention consistency regularization can further boost prognosis performance, which suggests the potential of restraining the similarity between the attention maps at different time-points in AD prognosis. The reason is that the attention consistency regularization allows the model to focus on the same regions when predicting the scores of an assessment at different time-points.

Effect of Relation Alignment. Here, we design a variant of WADP without relation alignment named WADP-M to evaluate the effectiveness of relation alignment as shown in Tables 4 and 5. We observe that WADP yields better prognosis results than WADP-M for all of the 12 cognitive scores. Taking the testing results on ADNI2 as an example, WADP achieves a CC value of 0.5955, which is better than that of WADP-M (i.e., 0.5721) in predicting CDR-SB at BL. The average nRMSE and CC of WADP-M are 0.1008 and 0.5806, respectively, while WADP obtains an average nRMSE of 0.0975 and an average CC of 0.6006, demonstrating that the relation alignment module can improve the prognosis performance. The main reason behind this is that the relation alignment encourages the subjects with similar label embeddings to have similar image embeddings, making the embedding spaces more discriminative.

3.4 Influence of Hyperparameters

Influence of Embedding Size. We train WADP with different embedding sizes $m = \{16, 32, 64, 128, 256, 384, 512\}$ on ADNI1, and the testing results on

ADNI2 are shown in Fig. 2. As can be seen, the performance decreases if m is too small, and it improves and relatively keeps stable when $64 \leq m \leq 256$. If m is too small, the capacity of embeddings may be restricted. A larger m increases the risk of overfitting since more parameters have to be learned. Therefore, the final embedding size is set as 64 in the experiments.

Influence of Trade-Off Parameters. In order to clearly know how trade-off parameters λ_1 and λ_2 influence experimental results, we make parameter sensitivity analysis as shown in Fig. 2. The WADP model is trained on ADNI1 and tested on ADNI2. We can notice that WADP can achieve the best performance when λ_1 is around 0.05 and λ_2 is around 0.005. Besides, WADP shows relatively poor performance when λ_1 or λ_2 is overly small or large. Too small values of λ_1 eliminate the influence of the relation alignment module, while too large values overemphasize. When λ_2 is very small, limited knowledge is borrowed from attention maps, and the model is degraded to a non-attention consistency model. When λ_2 is very large, the impact of prediction loss decreases and the prediction error gradually increases. According to Fig. 2, the optimal values for λ_1 and λ_2 are 0.05 and 0.005, respectively.

4 Conclusion

In this paper, we present a novel deep learning model named WADP for AD prognosis to jointly predict cognitive scores at four time-points using weakly labeled data. The attention consistency regularized convolutional network is presented to project MRI images into the embedding space and guide the attention maps at different time-points to be similar. Instead of discarding subjects with incomplete cognitive scores, the proposed autoencoder-based label embedding network can effectively impute missing cognitive scores and make full use of all available samples. To make the embedding spaces discriminative, we integrate a relation alignment regularization term into the objective function so as to align the sample relationships in image embedding space and those in label embedding space. Experimental results on public datasets demonstrate the promising performance of WADP on AD prognosis task.

Acknowledgments. This research was supported in part by the National Key R&D Program of China under grant 2019YFC1710300, Yibin Science and Technology Plan Project under grant 2022ZYD10, Key Laboratory of State Administration of Traditional Chinese Medicine for Scientific Research & Industrial Development of Traditional Chinese Medicine Regimen and Health under grant GZ2022009, Key Laboratory of Sichuan Province for Traditional Chinese Medicine Regimen and Health under grant GZ2022009 and the Sichuan Science and Technology Program under grants 2020YFS0283, 2021YJ0184, 2021YFS0152, and 2019YFS0019.

Data collection and sharing for this project was funded by the Alzheimer's Disease Neuroimaging Initiative (ADNI) (National Institutes of Health Grant U01 AG024904) and DOD ADNI (Department of Defense award number W81XWH-12-2-0012). As such, the investigators within the ADNI contributed to the design and implementation of ADNI and/or provided data but did not participate in analysis or writing of this paper.

References

1. Adak, S., et al.: Predicting the rate of cognitive decline in aging and early Alzheimer disease. Neurology **63**(1), 108–114 (2014)
2. Aderghal, K., Khvostikov, A., Krylov, A., Benois-Pineau, J., Afdel, K., Catheline, G.: Classification of Alzheimer disease on imaging modalities with deep CNNs using cross-modal transfer learning. In: IEEE International Symposium on Computer-Based Medical Systems (CBMS), pp. 345–354. IEEE (2018)
3. Ashburner, J.: A fast diffeomorphic image registration algorithm. Neuroimage **38**, 95–113 (2007)
4. Association, A.: 2020 Alzheimer's disease facts and figures. Alzheimers Dement. **16**(3), 391–460 (2020)
5. Bass, C., et al.: ICAM-Reg: interpretable classification and regression with feature attribution for mapping neurological phenotypes in individual scans. IEEE Trans. Med. Imaging **42**(4), 959–970 (2023)
6. Bobholz, J.H., Brandt, J.: Assessment of cognitive impairment: relationship of the dementia rating scale to the mini-mental state examination. J. Geriatr. Psychiatry Neurol. **6**(4), 210–213 (1993)
7. Dinomais, M., Celle, S., Duval, G.T., Roche, F., Bartha, R., Beauchet, O.: Anatomic correlation of the mini-mental state examination: a voxel-based morphometric study in older adults. PLoS ONE **11**(10), e0162889 (2016)
8. Dong, A., Toledo, J.B., Honnorat, N., Doshi, J., Varol, E., Sotiras, A., et al.: Heterogeneity of neuroanatomical patterns in prodromal Alzheimer's disease: links to cognition, progression and biomarkers. Brain **140**(3), 735–747 (2017)
9. Folstein, M.F., Folstein, S.E., McHugh, P.R.: Mini-mental state. A practical method for grading the cognitive state of patients for the clinician. J. Psychiatr. Res. **12**(3), 189–198 (1975)
10. Gu, P., Xu, X., Luo, Y., Wang, P., Lu, J.: BCN-GCN: a novel brain connectivity network classification method via graph convolution neural network for Alzheimer's disease. In: Mantoro, T., Lee, M., Ayu, M.A., Wong, K.W., Hidayanto, A.N. (eds.) ICONIP 2021. LNCS, vol. 13108, pp. 657–668. Springer, Cham (2021). https://doi.org/10.1007/978-3-030-92185-9_54
11. Hughes, C.P., Berg, L., Danziger, W.L., Coben, L.A., Martin, R.L.: A new clinical scale for the staging of dementia. Br. J. Psychiatry **140**, 566–572 (1982)
12. Jack, C.R., Jr., et al.: The Alzheimer's disease neuroimaging initiative (ADNI): MRI methods. J. Magn. Reson. Imaging **27**(4), 685–691 (2008)
13. Jia, J., Wei, C., Chen, S., Li, F., Tang, Y., Liu, Z., et al.: The cost of Alzheimer's disease in china and re-estimation of costs worldwide. Alzheimers Dement. **14**(4), 483–491 (2018)
14. Jin, H., Chien, S.P., Meijer, E., Khobragade, P., Lee, J.: Learning from clinical consensus diagnosis in India to facilitate automatic classification of dementia: machine learning study. JMIR Ment. Health **8**(5), e27113 (2021)
15. Landau, S.M., et al.: Associations between cognitive, functional, and FDG-PET measures of decline in AD and MCI. Neurobiol. Aging **32**(7), 1207–1218 (2011)
16. Lian, C., Liu, M., Wang, L., Shen, D.: Multi-task weakly-supervised attention network for dementia status estimation with structural MRI. IEEE Trans. Neural Networks Learn. Syst. **33**(8), 4056–4068 (2022)

17. Lian, C., Liu, M., Zhang, J., Shen, D.: Hierarchical fully convolutional network for joint atrophy localization and Alzheimer's disease diagnosis using structural MRI. IEEE Trans. Pattern Anal. Mach. Intell. **42**(4), 880–893 (2020)
18. Lin, W., et al.: Convolutional neural networks-based MRI image analysis for the Alzheimer's disease prediction from mild cognitive impairment. Front. Neurosci. **12**, 777 (2018)
19. Liu, M., Tang, J., Yu, W., Jiang, N.: Attention-based 3D ResNet for detection of Alzheimer's disease process. In: Mantoro, T., Lee, M., Ayu, M.A., Wong, K.W., Hidayanto, A.N. (eds.) ICONIP 2021. LNCS, vol. 13108, pp. 342–353. Springer, Cham (2021). https://doi.org/10.1007/978-3-030-92185-9_28
20. Liu, M., Zhang, J., Adeli, E., Shen, D.: Landmark-based deep multi-instance learning for brain disease diagnosis. Med. Image Anal. **43**, 157–168 (2018)
21. Liu, M., Zhang, J., Adeli, E., Shen, D.: Joint classification and regression via deep multi-task multi-channel learning for Alzheimer's disease diagnosis. IEEE Trans. Biomed. Eng. **66**(5), 1195–1206 (2019)
22. Liu, M., Zhang, J., Lian, C., Shen, D.: Weakly supervised deep learning for brain disease prognosis using MRI and incomplete clinical scores. IEEE Trans. Cybern. **50**(7), 3381–3392 (2020)
23. Raghu, M., Zhang, C., Kleinberg, J., Bengio, S.: Transfusion: understanding transfer learning for medical imaging. In: Advances in Neural Information Processing Systems (NIPS), pp. 3342–3352 (2019)
24. Rosen, W.G., Mohs, R.C., Davis, K.L.: A new rating scale for Alzheimer's disease. Am. J. Psychiat. **141**(11), 1356–1364 (1984)
25. Shi, Y.X., Wang, D.B., Zhang, M.L.: Partial label learning with gradually induced error-correction output codes. In: Tanveer, M., Agarwal, S., Ozawa, S., Ekbal, A., Jatowt, A. (eds.) ICONIP 2022. LNCS, vol. 13623, pp. 200–211. Springer, Cham (2022). https://doi.org/10.1007/978-3-031-30105-6_17
26. Sluimer, J.D., Vrenken, H., Blankenstein, M.A., Bouwman, F.H., Barkhof, F., van der Flier, W.M.: Whole-brain atrophy rate and CSF biomarker levels in mci and ad: a longitudinal study. Neurobiol. Aging **31**(5), 758–764 (2010)
27. Wang, S., Shen, Y., Chen, W., Xiao, T., Hu, J.: Automatic recognition of mild cognitive impairment from MRI images using expedited convolutional neural networks. In: Lintas, A., Rovetta, S., Verschure, P.F.M.J., Villa, A.E.P. (eds.) ICANN 2017. LNCS, vol. 10613, pp. 373–380. Springer, Cham (2017). https://doi.org/10.1007/978-3-319-68600-4_43
28. Wen, J., et al.: Convolutional neural networks for classification of Alzheimer's disease: overview and reproducible evaluation. Med. Image Anal. **63**, 1–20 (2020)
29. Young, A.L., et al.: Uncovering the heterogeneity and temporal complexity of neurodegenerative diseases with subtype and stage inference. Nat. Commun. **9**(1), 4273 (2018)
30. Zhou, B., Khosla, A., Lapedriza, A., Oliva, A., Torralba, A.: Learning deep features for discriminative localization. In: IEEE Conference on Computer Vision and Pattern Recognition (CVPR), pp. 2921–2929. IEEE (2016)
31. Zhu, W., Sun, L., Huang, J., Han, L., Zhang, D.: Dual attention multi-instance deep learning for Alzheimer's disease diagnosis with structural MRI. IEEE Trans. Med. Imaging **40**(9), 2354–2366 (2021)
32. Zhu, X., Suk, H.I., Shen, D.: A novel matrix-similarity based loss function for joint regression and classification in ad diagnosis. Neuroimage **100**, 91–105 (2014)
33. Zhu, X., Suk, H., Wang, L., Lee, S., Shen, D.: A novel relational regularization feature selection method for joint regression and classification in AD diagnosis. Medical Image Anal. **38**, 205–214 (2017)

Two-Stream Spectral-Temporal Denoising Network for End-to-End Robust EEG-Based Emotion Recognition

Xuan-Hao Liu[1], Wei-Bang Jiang[1], Wei-Long Zheng[1], and Bao-Liang Lu[1,2,3(✉)]

[1] Department of Computer Science and Engineering, Shanghai Jiao Tong University,
800 Dongchuan Road, Shanghai 200240, China
{haogram_sjtu,935963004,weilong,bllu}@sjtu.edu.cn

[2] RuiJin-Mihoyo Laboratory, RuiJin Hospital, Shanghai Jiao Tong University School
of Medicine, 197 Ruijin 2nd Road, Shanghai 200020, China

[3] Key Laboratory of Shanghai Commission for Intelligent Interaction and Cognitive
Engineering, Shanghai Jiao Tong University, 800 Dongchuan Road, Shanghai 200240,
China

Abstract. Emotion recognition based on electroencephalography (EEG) is attracting more and more interest in affective computing. Previous studies have predominantly relied on manually extracted features from EEG signals. It remains largely unexplored in the utilization of raw EEG signals, which contain more temporal information but present a significant challenge due to their abundance of redundant data and susceptibility to contamination from other physiological signals, such as electrooculography (EOG) and electromyography (EMG). To cope with the high dimensionality and noise interference in end-to-end EEG-based emotion recognition tasks, we introduce a Two-Stream Spectral-Temporal Denoising Network (TS-STDN) which takes into account the spectral and temporal aspects of EEG signals. Moreover, two U-net modules are adopted to reconstruct clean EEG signals in both spectral and temporal domains while extracting discriminative features from noisy data for classifying emotions. Extensive experiments are conducted on two public datasets, SEED and SEED-IV, with the original EEG signals and the noisy EEG signals contaminated by EMG signals. Compared to the baselines, our TS-STDN model exhibits a notable improvement in accuracy, demonstrating an increase of 6% and 8% on the clean data and 11% and 10% on the noisy data, which shows the robustness of the model.

Keywords: EEG · EMG · Emotion Recognition · End-to-end · Denoising · Robust Classification

1 Introduction

The rapid development of deep learning techniques opens up new possibilities for brain-computer interfaces (BCI). Different from the BCI applications in helping

B. Luo et al. (Eds.): ICONIP 2023, LNCS 14449, pp. 186–197, 2024.
https://doi.org/10.1007/978-981-99-8067-3_14

a paralytic patient walk again [1], affective brain-computer interface (aBCI) is aiming to detect, analyze, and respond to human emotions. Emotions are essential in our daily lives and influence our behaviors and mental states consciously or unconsciously.

In recent decades, EEG-based emotion recognition has obtained great interest due to the reason that EEG signals are inherently correlated to brain activity [2]. The previous EEG-based emotion recognition studies are almost based on manually extracted features, e.g., power spectral density (PSD) [10] and differential entropy (DE) features [14]. Significant progress has been achieved by using these handcrafted features [10,17]. However, they could be biased in specific domains and ignore rich information in the temporal domain. To fully excavate emotion-related information in raw EEG signals and eliminate the complicated process of handcrafted feature extraction, end-to-end models are promising approaches.

Basically, there are two types of BCIs for EEG recordings: the invasive BCIs [1] and the non-invasive BCIs [10]. The non-invasive BCIs are used more widely in research and treatments due to their hurtlessness and safety. However, EEG signals acquired by the non-invasive BCI are more easily contaminated by other physiological signals such as EMG caused by facial muscle movements, especially when some patients are unable to control muscle movements because of their illnesses [3]. Figure 1 shows the influence of an EMG signal on clean EEG data. It can be seen that the clean EEG signal is almost destroyed by the EMG signal. The presence of noise interference presents formidable challenges in the realm of end-to-end EEG-based emotion recognition. Thus, it is important to design better end-to-end denoising neural networks.

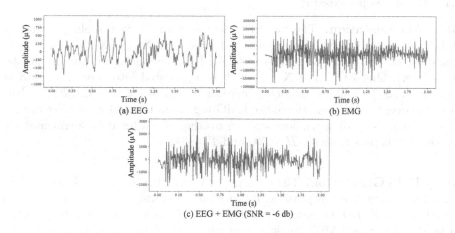

Fig. 1. Examples of a raw EEG segment and EMG interference

For end-to-end EEG-based emotion recognition, lots of endeavors have been made in the past few years. EEGnet [9] is a compact convolutional neural network (CNN) designed for EEG-based BCI to extract features from raw EEG signals.

To detect the valance and arousal levels, an end-to-end regional-asymmetric CNN was proposed and achieved an accuracy of over 95%, which was better than other methods using handcrafted features [16]. After that, Tao *et. al.* improved the accuracies to over 97% by an attention-based convolutional recurrent neural network (ACRNN) network on the same tasks [15]. These achievements show the superiority of end-to-end emotion recognition.

Deep learning methods have been proven to be effective in the denoising of EEG signals, which can learn the neural oscillations in EEG for eliminating noise from other artifacts. A benchmark dataset called EEGdenoiseNet was proposed for the research of EEG denoising [8]. However, it focuses on the EEG denoising task and not considering other EEG-based tasks.

To the best of our knowledge, addressing the processing of noisy data in the domain of end-to-end EEG-based emotion recognition remains largely unexplored. In this paper, we pioneerly introduce a novel Two-Stream Spectral-Temporal Denoising Network to achieve robust classification against EMG interference. An EEG noise-adding approach is proposed to simulate real-world muscle artifacts. Comprehensive experiments are conducted to test the proposed TS-STDN model in the cases of using clean data and EMG-contaminated data, respectively. Experimental results demonstrate the outperforming ability of our TS-STDN model in robust recognition. The code of our model and the noise-adding approach is published in https://github.com/XuanhaoLiu/TS-STDN.

2   Methodology

2.1   Data Preprocessing

Data Augmentation. The raw EEG signals with H Hz sampling frequency and duration T_{all} of a subject are denoted as $\mathbf{X}_{all} = [\mathbf{X}_{train}, \mathbf{X}_{test}] \in \mathbb{R}^{M \times C \times L}$, where M, C, and L is the number of trials, EEG channels, and sample points, respectively. One EEG trial $\mathbf{X}_s \in \mathbb{R}^{C \times L}$ is segmented into several slices $S = \{S_1, S_2, \ldots, S_n\}$ by sliding window, where slices $S_i(i = 1, 2, \ldots, n) \in \mathbb{R}^{C \times T}$. Due to the reason that the lengths of the EMG segments in EEGdenoiseNet are 2 s, we employ a 2-s sliding window with an overlap of 1-s for data augmentation. Hence, in this paper, we set $H = 200$ and $T = 400$.

Noisy Data Generation. The strategy for generating noisy data is to simulate real situations as much as possible. By carefully scanning a large amount of EEG records in the SEED dataset, we conclude that the brain areas which are prone to be disturbed by EMG signals are mainly distributed in the temporal areas on both sides of the scalp, while the frontal area and occipital area are less likely to be impacted. The influence of EMG signals is often asymmetric on the cortex as human muscle movements are not always symmetrical. Moreover, it cannot be ignored that individual differences are significant among different subjects. Based on the observation, forty electrodes that are easily affected by EMG signals are selected. The ten EMG groups are shown in Fig. 2, each group

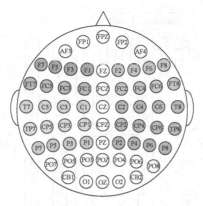

Fig. 2. The ten EMG noise-adding groups of electrodes, electrodes with the same color are in a group.

has four electrodes. We divided the forty electrodes into ten groups and for each particular EEG slice S_i, six out of ten groups are chosen randomly to be added the same EMG signals with a constant signal-noise ratio (SNR) calculated by Eq. (1), while there is no noise added to the remaining four groups. Notably, the EMG signals added to each EEG slice are selected randomly as well. The randomness creates asymmetry and individual differences in an easy way.

Let $x \in \mathbb{R}^T$ denotes a single channel of EEG signals, and $z \in \mathbb{R}^T$ denotes EMG signals, we generate noisy data by linearly combining x and λ times z with a constant SNR.

$$SNR = 10 \log \frac{RMS(x)}{RMS(\lambda \cdot z)},\tag{1}$$

in which the RMS stands for root mean squared (RMS). Let $\mathbf{S} = \{\mathbf{S}_1, \mathbf{S}_2, \ldots, \mathbf{S}_n\}$, where slices $\mathbf{S}_i (i = 1, 2, \ldots, n) \in \mathbb{R}^{C \times T}$, denotes the noisy slices generated by the strategy we propose in this paper.

2.2 Two-Stream Spectral-Temporal Denoising Network

Inspired by the two-stream network for action recognition in videos [6], we propose a novel two-stream spectral-temporal denoising network with self-attention to fully excavate the emotional information from spectral and temporal aspects of EEG signals. Figure 3 illustrates the overall architecture of the two-stream spectral-temporal denoising network. The input noisy data are passed to both of the streams, which extract the temporal and spectral features, respectively. Meanwhile, two U-net networks [5] are employed to reconstruct the clean EEG signals in the temporal and spectral domains. After that, the features extracted by the CNN are concatenated and fed into an LSTM network [12]. Finally, all features are multiplied by the weight calculated with the self-attention module and classified by a linear layer to predict emotions.

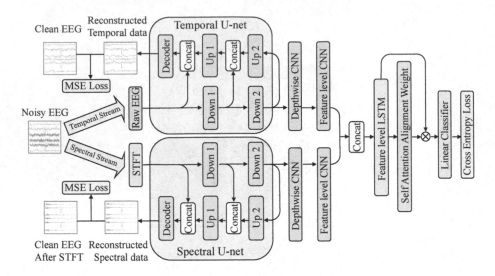

Fig. 3. Two-Stream Spectral-Temporal Denoising Network

Short Time Fourier Transform. To fully consider both the spectral and temporal information of EEG signals, short time Fourier transform (STFT) is utilized for extracting spectral information:

$$STFT(x,t) = \int_{-\infty}^{\infty} x(\tau)h(\tau - t)e^{-j2\pi f\tau} d\tau, \tag{2}$$

in which h is the window function. Let the $\tilde{S}' = \{\tilde{S}'_1, \tilde{S}'_2, \ldots, \tilde{S}'_n\}$ denotes the slices after STFT. For the purpose of balancing the size of two streams in our TS-STDN model, we elaborately choose the parameter of STFT to make the size of $\tilde{S}'$ approximately equal to raw EEG slices S.

Spectral and Temporal U-Net. By carefully setting the parameter of the STFT function, the slices $\tilde{S}$ are similar in size to S. Consequently, the spectral U-net and temporal U-net are designed to have the same structure symmetrically for balancing the stream scale and improving parallel performance. A batch of noisy EEG slices $\mathbf{S}_b \in \mathbb{R}^{B \times C \times T}$ is fed in the TS-STDN, where B is the size of each batch. During the training stage, the clean EEG batches S_b corresponding to the input noisy data $\mathbf{S}_b$ are used for training the denoising ability of TS-STDN. While at the testing stage, no clean data are available to our model.

The U-nets [5] can be regarded as encoder-decoder structures using several down or up units with shortcut concatenation, where each unit consists of a 1-D convolutional layer and an average pooling layer or an upsampling layer. The convolutional layers keep the length of the input and output tensors consistent by padding but change the channel numbers simultaneously. The lengths of tensors are doubled up by the upsampling layer, or reduced by half by the average pooling layer with a kernel of size (1,2).

We only introduce the temporal U-net due to the symmetry. Firstly, the input batch S_b is transformed to $S_b' \in \mathbb{R}^{B \times 1 \times C \times T}$ for subsequent calculating. Let $\sigma()$ denote the exponential linear unit (ELU) activation function $ELU()$, and $BN()$ stands for the batch normalization. Consequently, the outputs of down1 and down2 are:

$$X_{d1} = AvgPool(\sigma(BN(Conv_{d1}(S_b')))) \in \mathbb{R}^{B \times 2 \times C \times (T/2)}, \tag{3}$$

$$X_{d2} = AvgPool(\sigma(BN(Conv_{d2}(X_{d1})))) \in \mathbb{R}^{B \times 4 \times C \times (T/4)}. \tag{4}$$

The up1 and up2 units are designed to reduce half of the channel number but double the length of the input tensor by replacing the average pooling layer with an upsampling layer in the down units. Consequently, the outputs of up1 and up2 are calculated as:

$$X_{u2} = \sigma(BN(Conv_{u2}(Upsample(X_{d2})))) \in \mathbb{R}^{B \times 2 \times C \times (T/2)}, \tag{5}$$

$$X_{u1} = \sigma(BN(Conv_{u1}(Upsample(Concat(X_{u2}, X_{d1}))))) \in \mathbb{R}^{B \times 2 \times C \times T}. \tag{6}$$

Afterward, the output of up1 X_{u1} concatenated with the input noisy data S_b' is fed into the decoder, whose output has the same shape of tensor S_b'. The decoder has two convolutional layers to better reconstruct the clean EEG signals without EMG interference $\hat{S}_b$. The reconstructed data $\hat{S}_b$ is used to compute the Mean-Squared Loss (MSE) loss with the clean data S_b:

$$\mathcal{L}_{temporal} = MSELoss(\hat{S}_b, S_b). \tag{7}$$

Same to the reconstruction loss in the temporal domain, the reconstruction loss of spectral signals is calculated by:

$$\mathcal{L}_{spectral} = MSELoss(\hat{\tilde{S}}_b, \tilde{S}_b). \tag{8}$$

Depthwise CNN and Feature Level CNN & RNN. Inspired by the Xception [7], a depthwise CNN with a kernel size of $(C, 1)$ is employed for aggregating the spatial information between EEG channels. Let D denotes the depth multiplier number, the output of the depthwise CNN is:

$$X_D = AvgPool(\sigma(BN(Conv_D(X_{d2})))) \in \mathbb{R}^{B \times 4D \times 1 \times (T/8)}. \tag{9}$$

The feature level CNN is utilized to compress temporal information. Let F_t and F_s denote the filter numbers of the feature level CNN in the temporal and spectral streams, the output of the feature level CNN is:

$$X_{F_{s/t}} = AvgPool(\sigma(BN(Conv_F(\tilde{X}_D/X_D)))) \in \mathbb{R}^{B \times F_{s/t} \times 1 \times (T/32)}. \tag{10}$$

An LSTM network is adopted for the spectral and temporal information fusion. The output of the feature level CNN of both streams X_{F_s} and X_{F_t} are reshaped, concatenated and linearly embedded to $X_F \in \mathbb{R}^{B \times (F_s + F_t) \times d}$, where d is the embedding dimension. As the LSTM is a seq2seq model, the X_F is transformed to $X_L \in \mathbb{R}^{B \times (F_s + F_t) \times d}$ by regarding each feature as a token of d dimension.

Self-Attention Alignment Weight. After extracting the spatial, spectral, and temporal information through two-stream U-net, CNN, and LSTM, the output tensor X_L is highly semantic and discriminative. However, some of the $F_a = F_s + F_t$ features are not closely related to human emotions. Hence, a self-attention module [4] is utilized for concentrating on the features mostly related to emotions. The alignment weight is calculated from the average of each feature with two linear layers activated by the $tanh()$ function. Let $X_a \in \mathbb{R}^{B \times F_a}$ denotes the average tensor of X_L. The first dimensionality-reduction layer has a parameter $W_1 \in \mathbb{R}^{F_a \times d_r}$ and a bias $b_1 \in \mathbb{R}^{d_r}$, while the second dimensionality increasing layer has a parameter $W_2 \in \mathbb{R}^{d_r \times F_a}$ and a bias $b_2 \in \mathbb{R}^{F_a}$, where the d_r is the reduction dimension number. The alignment weight w is calculated by:

$$w = softmax(W_2 \cdot (tanh(W_1 \cdot X_a + b_1)) + b_2) \in \mathbb{R}^{B \times F_a}. \tag{11}$$

Finally, multiply each feature of X_L by the corresponding coefficient in the attention weights w to produce X_{sa}. X_{sa} is $\{w_1 x_1, w_2 x_2, w_3 x_3, \ldots, w_{F_a} x_{F_a}\}$. Then, X_{sa} is flattened and fed into a linear classifier to get the final prediction $\hat{y}$. The cross-entropy loss is applied to compute the classification loss:

$$\mathcal{L}_{cls} = -\sum_{i=1}^{N} y_i \log \hat{y}_i , \tag{12}$$

where y stands for the ground truth emotion label. Consequently, the entire objective function is given as minimizing the linear combination of the classification loss and reconstruction loss:

$$\arg\min_{\Theta} \mathcal{L}_{all} = \arg\min_{\Theta}(\mathcal{L}_{cls} + \mathcal{L}_{spectral} + \mathcal{L}_{temporal}). \tag{13}$$

3 Experiment

3.1 Dataset

SEED and SEED-IV. The SJTU Emotion EEG datasets are a series of datasets that record the EEG signals of subjects while they are watching emotion videos. The original SEED dataset [10] chooses fifteen Chinese film clips in order to induce three target emotions: positive, neutral, and negative. The SEED-IV dataset [11] contains four categories of emotion including happy, sad, neutral, and fear. Seventy-two film clips are chosen as stimuli. The 62-channel ESI NeuroScan System was employed to capture EEG signals in both datasets, using the EEG cap with 62 channels positioned according to the international 10–20 system at 1000 Hz. All EEG signals are then downsampled to 200 Hz.

Specifically, for the SEED dataset, each recording contains fifteen trials of EEG signals, we use the samples from the first nine clips as the training set and the samples from the remaining six clips as the testing set. However, for the SEED-IV dataset, the last eight clips are unequal in the number of emotion types. Hence, we choose the two video clips that appear at the end of each session for each emotion as the testing set, and the remaining sixteen clips compose the training set.

EEGdenoiseNet. EEGdenoiseNet [8] is a benchmark dataset designed for the purpose of training and evaluating deep learning denoising models. The dataset consists of single-channel EEG, EOG, and EMG signals collected from diverse publicly available datasets. To ensure the quality and reliability of the dataset, rigorous preprocessing procedures are conducted on all physiological signals. Notably, the signals are segmented into 2-second intervals and meticulously examined by an expert to confirm their cleanliness and suitability for analysis. There are 5598 pure EMG segments in EEGdenoiseNet, and as the original EMG signals are 512 Hz, we downsampled them to 200 Hz to match the EEG signals from SEED datasets.

3.2 Implementation Details

We compare our TS-STDN model with other end-to-end approaches including LSTM [12], EEGnet [9], and ACRNN [15] under subject-dependent conditions. In this paper, the number of EEG channels is $C = 62$, the sampling frequency is $H = 200$ Hz, and the length of slices is $T = 400$. For the TS-STDN model, the depth multiplier number $D = 4$, the filter number of the feature level CNN of both streams are $F_s = F_t = 16$, and the embedding dimension $d = 16$. The LSTM model regards the input EEG data as a sequence that contains T tokens, each token is obtained by embedding a sample point in a 32-dim tensor through two linear layers connected by a ReLU activation function. The hyperparameters of the EEGnet model and ACRNN remain the same as those in the original papers. All models are implemented by PyTorch [13] deep learning framework and trained with Adam optimizer with a learning rate of $\eta = 0.001$ and a batch size of $B = 64$.

3.3 Ablation Experiments

Ablation experiments are conducted by removing some modules to evaluate the effectiveness of each module in our TS-STDN model. We design several variant models as follows:

- Spectral DN: This model only contains the spectral stream of the TS-STDN.
- Temporal DN: This model only contains the temporal stream of the TS-STDN.
- TS-STDN w/o SA: This model only removes the self-attention alignment weight module.
- TS-STDN w/o MSE: This model is trained without using the reconstruction MSE Loss computed with clean data in both spectral and temporal domains.

3.4 Results on Clean Data

We use the raw EEG signals without adding noise to evaluate the performance of each model. Subject-dependent experiments on two public datasets SEED

Table 1. The accuracies (Avg./Std.) of different methods on SEED and SEED-IV

Method	SEED		SEED-IV	
	Avg. (%)	Std. (%)	Avg. (%)	Std. (%)
LSTM [12]	67.50	12.06	43.95	10.30
ACRNN [15]	57.37	10.93	40.19	07.27
EEGnet [9]	72.49	12.52	54.26	12.19
Spectral DN	76.88	09.89	54.90	11.01
Temporal DN	73.69	12.70	55.50	12.17
TS-STDN w/o SA	75.73	09.96	56.64	10.12
TS-STDN w/o MSE	76.93	10.98	58.84	12.11
TS-STDN	**78.45**	10.49	**62.13**	12.18

[10] and SEED-IV [11] demonstrates the outperforming performance of our TS-STDN model, the classification accuracies are presented in Table 1. Remarkably, our TS-STDN achieves the highest accuracies of 78.45% and 62.13% on the SEED and SEED-IV datasets, respectively. It is worth noting that relying solely on a single stream results in varying decreases in accuracy, which shows the complementary properties of the temporal and spectral information. The Spectral DN demonstrates better performance on the SEED dataset, whereas the Temporal DN exhibits higher accuracy on the SEED-IV dataset. As a result, it remains uncertain which stream holds greater importance over the other. Without employing the self-attention module, TS-STDN fails to extract the key features closely related to emotions and has 3% and 6% reductions in accuracy. It can be seen that the denoising U-net is still effective when processing clean data, which can be regarded as an autoencoder.

The confusion matrices illustrating the classification performance of our TS-STDN model on the SEED and SEED-IV datasets are presented in Fig. 4(a) and (b). Regarding the SEED dataset, the positive emotion exhibits the highest accuracy, whereas classifying the negative emotion proves to be notably challenging. In contrast, for the SEED-IV dataset, the accuracies for all emotions are relatively similar to each other.

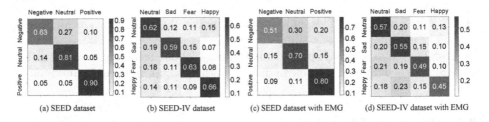

(a) SEED dataset   (b) SEED-IV dataset   (c) SEED dataset with EMG   (d) SEED-IV dataset with EMG

Fig. 4. Confusion matrices of our TS-STDN model

Table 2. The average accuracies (%) of different methods on the SEED dataset under varying levels of EMG interference.

Method	The SNR of contaminated EEG						Avg
	−6 db	−7 db	−8 db	−9 db	−10 db	−11 db	
LSTM [12]	51.95	55.22	55.02	48.57	48.74	48.57	51.34
ACRNN [15]	48.30	49.68	49.30	47.05	47.43	46.90	48.11
EEGnet [9]	57.32	59.63	59.08	54.52	53.67	54.19	56.40
Spectral DN	65.51	68.26	68.10	63.23	62.58	63.83	65.25
Temporal DN	64.79	67.89	66.77	61.29	61.24	63.05	64.17
TS-STDN w/o SA	64.62	68.58	68.94	64.08	63.64	64.05	65.65
TS-STDN w/o MSE	65.50	68.36	68.72	61.26	61.00	61.03	64.31
TS-STDN	**67.81**	**70.38**	**71.03**	**65.27**	**64.79**	**64.67**	**67.33**

3.5 Results on Noisy Data

To fully investigate the robustness of our TS-STDN model against the interference of EMG signals, we tested our model under conditions where the raw EEG signals are contaminated by EMG signals with varying signal-to-noise ratios. The clean EEG data are from the SEED dataset [10] and the SEED-IV dataset [11], while the EMG data are from the EEGdenoiseNet dataset [8]. Specifically, we conducted our experiment on noisy data with SNR from −6 dB to −11 dB, simulating the disturbance from relatively moderate to extremely intense. The SNR is calculated by the Eq. (1). For fairness, all methods are tested on the same noisy data which had been generated before evaluation.

Table 2 and Table 3 display the results of each model under various noise intensity conditions on the noisy SEED and SEED-IV datasets, respectively. It can be seen that our TS-STDN model acquires the best classification accuracies under all conditions, exhibiting an increase of about 11% to the EEGnet on the SEED dataset, and 10% on the SEED-IV dataset to the EEGnet. All baseline models, which possess no denoising modules, perform worse when processing contaminated EEG. The Spectral DN model shows better performance than the Temporal DN model when facing noisy data on both datasets, suggesting that spectral information is more robust to EMG interference. Through the observation of reduced accuracy on both datasets of the TS-STDN w/o SA model, it is evident that the self-attention module remains effective when facing noisy EEG data. It is worth noting that by employing reconstruction loss, our TS-STDN model exhibits a notable improvement in accuracy, demonstrating an increase of 3% on the SEED dataset and 7% on the SEED-IV dataset to the TS-STDN w/o MSE model, which proves the effectiveness of the reconstruction module. Moreover, compared to the improvement of the reconstruction module on the clean data, which are 2% and 4% on two datasets, we find the reconstruction module is more suitable for denoising. The average confusion matrices of our

TS-STDN model on noisy datasets are depicted in Fig. 4(c) and (d). All emotion classification accuracies are influenced by EMG disturbance.

Table 3. The average accuracies (%) of different methods on the SEED-IV dataset under varying levels of EMG interference.

Method	The SNR of contaminated EEG						Avg
	−6 db	−7 db	−8 db	−9 db	−10 db	−11 db	
LSTM [12]	33.64	33.33	33.63	33.61	33.48	33.22	33.48
ACRNN [15]	37.40	37.67	37.36	37.72	37.35	37.00	37.42
EEGnet [9]	42.57	42.86	42.41	42.83	42.07	41.98	42.46
Spectral DN	48.93	49.11	48.21	49.03	49.51	48.73	48.92
Temporal DN	46.21	45.13	45.22	45.34	46.02	45.72	45.61
TS-STDN w/o SA	50.12	49.15	49.15	49.34	50.31	49.66	49.62
TS-STDN w/o MSE	45.09	44.80	45.73	44.85	45.79	45.07	45.22
TS-STDN	**52.17**	**52.40**	**52.84**	**52.25**	**52.33**	51.92	**52.32**

4   Conclusion

In this paper, we introduce a novel Two-Stream Spectral-Temporal Denoising Network to thoroughly exploit emotion-related features from both spectral and temporal views in contaminated EEG signals while learning the ability to eliminate noise interference. The TS-STDN model acquires the denoising capability by reconstructing clean EEG signals using U-net architectures, which can be seen as a denoising autoencoder. To simulate the real-world EMG disturbance on EEG signals, we propose a random algorithm for EMG noise adding in the raw EEG recordings. The experimental results demonstrate that the TS-STDN model performs the best on both clean and contaminated data, and is robust when facing extremely intense noise interference. The source code of the noisy data production and TS-STDN model implementation is shared in https://github.com/XuanhaoLiu/TS-STDN.

Acknowledgments. This work was supported in part by grants from National Natural Science Foundation of China (Grant No. 61976135), STI 2030-Major Projects+2022ZD0208500, Shanghai Municipal Science and Technology Major Project (Grant No. 2021SHZDZX), Shanghai Pujiang Program (Grant No. 22PJ1408600), Medical-Engineering Interdisciplinary Research Foundation of Shanghai Jiao Tong University "Jiao Tong Star" Program (YG2023ZD25), and GuangCi Professorship Program of RuiJin Hospital Shanghai Jiao Tong University School of Medicine.

References

1. Lorach, H., Galvez, A., Spagnolo V., Martel, F., et al.: Walking naturally after spinal cord injury using a brain-spine interface. Nature 1–8 (2023)
2. Alarcao, S.M., Fonseca, M.J.: Emotions recognition using EEG signals: a survey. IEEE Trans. Affect. Comput. **10**(3), 374–393 (2017)
3. Supriya, S., Siuly, S., Wang, H., Zhang, Y.: Epilepsy detection from EEG using complex network techniques: a review. IEEE Rev. Biomed. Eng. **16**, 292–306 (2021)
4. Vaswani, A., Shazeer, N., Parmar, N.: Attention is all you need. In: Advances in Neural Information Processing Systems (NeurIPS), pp. 5998–6008 (2017)
5. Ronneberger, O., Fischer, P., Brox, T.: U-Net: convolutional networks for biomedical image segmentation. In: Navab, N., Hornegger, J., Wells, W.M., Frangi, A.F. (eds.) MICCAI 2015. LNCS, vol. 9351, pp. 234–241. Springer, Cham (2015). https://doi.org/10.1007/978-3-319-24574-4_28
6. Simonyan, K., Zisserman, A.: Two-stream convolutional networks for action recognition in videos. In: Advances in Neural Information Processing Systems (NeurIPS), pp. 568–576 (2014)
7. Chollet, F.: Xception: deep learning with depthwise separable convolutions. In: Proceedings of the IEEE Conference on Computer Vision and Pattern Recognition (CVPR), pp. 1251–1258 (2017)
8. Zhang, H., Zhao, M., Wei, C., Mantini, D., Li, Z., Liu, Q.: EEGdenoiseNet: a benchmark dataset for deep learning solutions of EEG denoising. J. Neural Eng. **18**(5), 056057 (2021)
9. Lawhern, V.J., Solon, A.J., Waytowich, N.R., Gordon, S.M., Hung, C.P., Lance, B.J.: EEGNet: a compact convolutional neural network for EEG-based brain-computer interfaces. J. Neural Eng. **15**(5), 056013 (2018)
10. Zheng, W.L., Lu, B.L.: Investigating critical frequency bands and channels for EEG-based emotion recognition with deep neural networks. IEEE Trans. Auton. Ment. Dev. **7**(3), 162–175 (2015)
11. Zheng, W.L., Liu, W., Lu, Y., Lu, B.L., Cichocki, A.: Emotionmeter: a multimodal framework for recognizing human emotions. IEEE Trans. Cybern. **49**(3), 1110–1122 (2018)
12. Hochreiter, S., Schmidhuber, J.: Long short-term memory. Neural Comput. **9**(8), 1735–1780 (1997)
13. Paszke, A., et al.: Automatic differentiation in pytorch (2017)
14. Duan, R.N., Zhu, J.Y., Lu, B.L.: Differential entropy feature for EEG-based emotion classification. In: 2013 6th International IEEE/EMBS Conference on Neural Engineering (NER), pp. 81–84. IEEE (2013)
15. Tao, W., et al.: EEG-based emotion recognition via channel-wise attention and self attention. IEEE Trans. Affect. Comput. **14**(1), 382–393 (2023)
16. Cui, H., Liu, A., Zhang, X., Chen, X., Wang, K., Chen, X.: EEG-based emotion recognition using an end-to-end regional-asymmetric convolutional neural network. Knowl.-Based Syst. **205**, 106243 (2020)
17. Li, R., Wang, Y., Zheng, W.L., Lu, B.L.: A multi-view spectral-spatial-temporal masked autoencoder for decoding emotions with self-supervised learning. In: Proceedings of the 30th ACM International Conference on Multimedia, pp. 6–14 (2022)

Brain-Inspired Binaural Sound Source Localization Method Based on Liquid State Machine

Yuan Li[1], Jingyue Zhao[2], Xun Xiao[1], Renzhi Chen[2], and Lei Wang[2]([✉])

[1] College of Computer, National University of Defence Technology,
Changsha 410071, Hunan, China
{liyuan22,xiaoxun520}@nudt.edu.cn
[2] Defense Innovation Institute, Academy of Military Sciences, Beijing, China
leiwang@nudt.edu.cn

Abstract. Binaural Sound Source Localization (BSSL) is a remarkable topic in robot design and human hearing aid. A great number of algorithms flourished due to a leap in machine learning. However, prior approaches lack the ability to make a trade-off between parameter size and accuracy, which is a primary obstacle to their further implementation on resource-constrained devices. Spiking Neural Network (SNN)-based models have also emerged due to their inherent computing superiority over sparse event processing. Liquid State Machine (LSM) is a classic Spiking Recurrent Neural Network (SRNN) which has the natural potential of processing spatiotemporal information. LSM has been proved advantageous on numerous tasks once proposed. Yet, to our best knowledge, it is the first proposed BSSL model based on LSM, and we name it BSSL-LSM. BSSL-LSM is lightweight with only 1.04M parameters, which is a considerable reduction compared to CNN (10.1M) and D-BPNN (2.23M) while maintaining comparable or even superior accuracy. Compared to SNN-IID, there is a 10% accuracy improvement for 10° interval localization. To achieve better performance, we introduce Bayesian Optimization (BO) for hyperparameters searching and a novel soft label technique for better differentiating adjacent angles, which can be easily mirrored on related works. **Project page:** https://github.com/BSSL-LSM.

Keywords: Binaural sound source Localization · Spiking neural network · Liquid state machine · Bayesian optimization

1 Introduction

Sound Source Localization (SSL) is a key issue in human-computer interaction, virtual reality, and digital hearing aid design. As an important branch of SSL approaches, binaural SSL (BSSL, with two microphones) is prevalent as it gets rid of deliberately designed multi-microphone array configuration [6,17,32], making it more suitable to be employed in resource-restricted scenarios such as edge

B. Luo et al. (Eds.): ICONIP 2023, LNCS 14449, pp. 198–213, 2024.
https://doi.org/10.1007/978-981-99-8067-3_15

computing devices. Nevertheless, previous BSSL solutions based on CNN and DNN still face the problems of enormous memory consumption and computational complexity problems due to their artistically crafted pre-processing module and network structure. We endeavor to make a trade-off between a lightweight model and high localization accuracy.

In BSSL, interaural acoustic cues mainly consist of two primary features that determine the source angle in the horizontal plane [26], Interaural Time Difference (ITD) and Interaural Level Difference (ILD). Many BSSL models that rely on ITD and ILD cues have emerged. The bionic ones inspired by the transmission and processing mechanism of sound in the mammal auditory pathway were favored earlier. Unfortunately, they are limited by developments in physiology and psychology and can only show promising results in ideal anechoic room [5].

As neural networks showed powerfulness in various problems, attempts to use machine learning (ML) based algorithms for BSSL tasks exploded. These methods showed satisfying results in challenging acoustic environments because they usually involved multifarious information filtering and multiple hybrid feature extraction. But this is generally with the cost of huge parameters, complex processing, and high computational overload. For instance, multitask learning of TF-CNN [20] can accurately classify the azimuth and elevation of a sound source simultaneously, even in low signal-to-noise ratio environments. However, it sequentially employed SIFT preprocess, TF-CNN extraction module, and a multitask neural network module consisting of a series of fully connected layers for angle classification. It entails a laborious process and time-consuming training.

Spiking neural networks (SNN), which is described as the third-generation neural network, is well-known for low power consumption and excellent parallelism due to sparse event processing mechanism [12,28,29,35]. One striking merit of SNN is its inherent spatiotemporal information processing nature, which is in tune with sound signal processing. Researchers have tried to use SNN for BSSL tasks and achieved passable results. For example, Goodman et al. [11] proposed a system including 1 million neurons that manually carried out delay lines and gains but did not report its performance in noisy acoustic environments. Jorge Dvila-Chac et al. [4] realized MSO based on Jeffress coincidence detector [13] and LSO module to extract ITD cues and ILD cues from spiking binaural signals, respectively. The system is effective but too complex to apply to small terminal equipment. Furthermore, they only provide the classification results at 15° spatial resolution. The above two are based on the delay lines configuration, whose mechanism is still unclarified. To sum up, SNN-based BSSL approaches remain potential but challenging.

A particular while biologically plausible form of SNN, LSM, or reservoir, has been proven effective on various tasks such as robot control and image recognition subsequently once proposed [3,24,36]. Structurally, LSM contains a liquid layer of randomly connected spike neurons in recurrent structure, which can perform dynamic responses to input spike trains with decaying transient memories. Reasonably, it intrinsically holds greater promise in dealing with temporal patterns such as sound signals. LSM has showed comparable or superior performance to state-of-the-art algorithms on speech recognition problems [34]. But to the best of our knowledge, none has attempted to demonstrate the efficiency of LSM on BSSL tasks, and this is also our work in this paper. It is worth noting that LSM is a data-dependent model whose structure must readapt to new input data when the dataset changes. To this end, proposing a fast and efficient hyperparameters optimization method is indispensable.

In light of the analysis and challenges mentioned above, we investigate the availability of bio-inspired LSM on BSSL task and specifically propose BSSL-LSM in this paper. Our main contributions can be summarized as follows:

- BSSL-LSM is the first BSSL model based on LSM, which involves a LSM module for binaural cues extraction and a simple MLP classifier for angle prediction.
- BSSL-LSM outperforms previous approaches regarding localization accuracy and parameter size. We enhance the performance of adjacent-angle classification by introducing a novel soft label technique, which can be easily applied in other SSL works.
- We further adopt BO to support fast hyperparameters searching on the top of the LSM and MLP classifier jointly. It enables BSSL-LSM a faster and more flexible adaptation to new data distribution.

Testing the efficiency of LSM on BSSL tasks and reducing parameter size while maintaining high localization accuracy are the primary two problems we expect to address in this paper. We conducted all the experiments based on the AISHELL speech and CIPIC HRTF databases. Cochlea toolbox [22] with Zilany model [37] are used for encoding sound signals to spike trains. Experiments show that our proposed BSSL-LSM can achieve the localization accuracy of $94.29 \pm 0.5\%$ at $15°$ spatial resolution with only almost 1.04M parameters. There is a commendable improvement compared to one of our baseline models, CNN with 10.1M parameters. BSSL-LSM can improve the accuracy by 10% compared to another baseline model SNN-IID. More details can be seen in Sect. 5.

The remainder of the paper goes as follows. In Sect. 2, the background is described. Related works are introduced in Sect. 3. Our method is explained in Sect. 4. We discussed the experiments and results in Sect. 5. At last, a short conclusion is made in Sect. 6.

2   Background

Liquid State Machine: A representative model of the SRNN with randomly cyclic synaptic connections. LSM mainly consists of three parts, as shown in Fig. 1: 1) Input layer, where input spike trains randomly connect to the reservoir neurons. 2) The Liquid layer, consisting of excitatory and inhibitory neurons, can filter and map input information to state vectors. 3) Readout layer, usually serves as the classifier of neuron state vectors in the reservoir layer and allows for different forms such as support vector machines (SVMs), linear regression, multiple layer perceptron (MLP), and so on. The success of LSM is highly dependent on two properties: input separability and fading memory. A randomly generated liquid structure nearly impossiblely performs well on a new dataset as it is input-dependent. Hyperparameters like the neuron dynamics (e.g., threshold, membrane time constant, refractory period, and so on) and network topology (e.g., neuron interconnection probability, synaptic density, and so on) play a vital role in the final results. More explanations can be seen in Table 1.

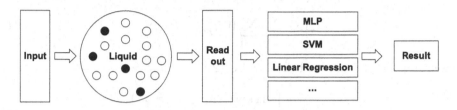

Fig. 1. Diagram of the LSM.

We select MLP and treat BSSL as a classification work in our work. For each sample, each element in the state vector is equal to the number of spikes a neuron outputs throughout the simulation. Mathematically, it can be described as $O_i = \sum_t^T Spike_{i,t}$, T is the time steps for each simulation, i represents the $i-th$ neuron in the liquid. $Spike_{i,t}$ is 1 when the neuron fires and otherwise is 0 at each step. The state vector transformed from each sample can be expressed as $[O_0, O_1, \ldots, O_i, \ldots, O_{N-1}]$, where "N" is the number of neurons in the liquid.

Bayesian Optimization: As we state above, there are numerous configurable hyperparameters, and so is the MLP classifier. Hyperparameter optimization (HPO) is usually a laborious and computationally expensive process. BO is a sequential HPO method and befits any black-box problems. Compared with the Particle Swarm Optimization (PSO) algorithm and grid-search algorithm, BO is more effective when exploring high-dimensional hyperparameter space and converges faster to the global optimal [16]. The key idea behind BO is to use prior belief about the function and the observations to update a posterior distribution over the function. At each iteration, BO selects the set of hyperparameters combination x_i from the parameter space χ based on its prior belief and follows $x_i \in \arg\max_{x \in \chi} a(x; D)$, $a(x)$ is the acquisition function, D is the set of previously evaluated points. The true black-box function $f(x)$ evaluates $f(x_i) = y_i$. Then, $D \cup (x_i, y_i)$ is used to update the surrogate model and select the next point. This process repeats over and over until the search time runs out or it reaches the maximum number of iterations.

STDP Learning Rule: The Spike Timing Dependent Plasticity (STDP) learning rule is a biologically plausible unsupervised learning rule and is an extension of the Hebbian learning rule [7,8]. The Hebbian learning rule can be described as the correlation between pre and post-synaptic neural activities affecting the strengthening or weakening of the synapse between them. The precise timing information of pre-synaptic and post-synaptic spiking plays a crucial role in determining the direction and weight changes in the synapse. The general form of the STDP learning rule can be expressed as follows: if a post-synaptic neuron fires after a pre-synaptic neuron fires during a given time window, then the synaptic weight between them will be increased; otherwise, it will be decreased, which is known as preemption phenomenon. The general formula for the timing characteristics between pre-synaptic and post-synaptic spikes on synaptic weight changes can be expressed as:

$$\Delta\omega = \sum_{i=1}^{N}\sum_{j=1}^{N} W(t_{pre}^i - t_{post}^j) \tag{1}$$

among Eq. (1), W can be defined as Eq. (2):

$$W(x) = \begin{cases} A_+ exp(-x/\tau_+), & x > 0 \\ -A_- exp(x/\tau_-), & x < 0 \end{cases} \tag{2}$$

where A_+ and A_- are determined by current synapse state, τ_+ and τ_- are both time constant.

3   Related Work

We will provide a short and systemic review from three aspects: CNN or DNN-based BSSL models; SNN-based BSSL models; LSM applications and its structure optimization.

CNN or DNN Based BSSL Models: [23] proposed D-BPNN and CNN with complete binaural sound signals as input, which can classify an azimuthal interval of 15°, 30° and 45° in horizontal plane. WaveLoc [26] is an end-to-end BSSL model that can work in both anechoic and reverberant environments. It estimates azimuth from −90° to 90° with an error rate of 1.5% to 3%. DeepEar [32] is a well-trained BSSL model that can substantially outperform other BSSL models in multi-source scenarios. Ying et al. [31] proposed a DNN architecture BSSL model and performed azimuth prediction with a root mean square localization error of 3.68° in the −90° to 90° range. [20] proposed a binaural SSL method based on a time-frequency convolutional neural network (TF-CNN) with multi-task learning to simultaneously localize azimuth and elevation under unknown acoustic conditions. [14] presented an end-to-end Binaural Audio Spectrogram Transformer (BAST) model to predict the sound azimuth in both anechoic and reverberation environments, which achieved an angular distance of 1.29° and a mean square error of 1e−3 at all azimuths. Youssef et al. [33] proposed a BSSL model in a humanoid robot that used a multi-layer perceptron network. It performs both azimuth and elevation estimation and has error rates from 1.2% to 2% and 2.6% respectively. CNN or DNN-based approaches usually profit from complex network architectures and possess large amounts of training parameters. Unlike the above methods, the proposed BSSL-LSM can make a trade-off between localization accuracy and training complexity.

SNN Based BSSL Models: The Delay line and coincidence detection model is still the design foundation of many SNN-based sound source localization algorithms. Robert Luke and David McAlpine [18] presented a binaural lateralization system of 100% accuracy, which is free of delay lines and generalized to noisy environments. Yet it can just tell left or right rudely and lacks fine-grained localization. But as they said in their work, it was the first BSSL model exploiting spatiotemporal properties based on SNN without pre-defined delay lines. Inspired by the mammalian auditory pathway system, Wall et al. [27] introduced a biologically inspired BSSL model that can process raw sound signals directly based on time encoding and backpropagation techniques. It is a conspicuous step forward in natural sound localization modeling. Glackin et al. [10] proposed a BSSL approach based on the medial superior olive (MSO) and measured over a azimuthal range of $[-180°, 170°]$. To our knowledge, none has ever tested the efficiency of LSM on BSSL task; it is a meaningful try to do it.

LSM Applications and Structure Optimization: The first LSM structure with randomly interconnected spiking neurons was proposed by Maass et al. in 2002 [19]. Numerous researchers have introduced LSM to their work [15,25]. Xuhu Yu et al. [34] proposed a multimodal learning of audio-visual speech recognition with LSM and achieved recognition accuracy of 86.8% on the LRW dataset. The gesture recognition approach designed by Xun et al. [30] achieved 98.42% accuracy on the DVS128 Gesture Dataset, while the parameters were reduced by 90% compared to baseline. Proper hyperparameter configuration determines the feature extraction ability of LSM, while manual tuning cannot always find the optimal solution. Xuhu et al. and Xun et al. used the PSO algorithm to search for a liquid structure. Reynolds [21] introduced genetic algorithms to generate reservoirs and showed powerful performance over manually tuning ones on classification tasks. Unlike the above works, we first introduce LSM in the sound source localization domain and apply BO for hyperparameters optimization.

4   Method

The overall structure of the BSSL-LSM model can be divided into three parts: Encoding unit, LSM feature extraction unit, and MLP classifier unit. BO is on the top of LSM and MLP classifier. To put it in another way, BO collaboratively optimizes hyperparameters of LSM and MLP classification module. The entire model is shown in the Fig. 2.

4.1   Encoding Unit

The first step to input to the LSM is preprocessing and encoding sound signal to spiking trains. A large amount of encoding methods exist, and a summary can be found in [9]. Referring to [18], we select the cochlear toolbox [22], which implements the Zilany model [37], a biologically plausible approach. The Zilany model effectively approximates the process by which the mammalian cochlea

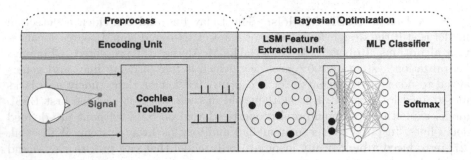

Fig. 2. The overall structure of the BSSL-LSM model.

converts mechanical signals into electrical signals and the action of the basilar membrane in frequency selection. We draw on the work of Robert Luke et al. [18] for parameter settings of the Zilany model. 100 IF neurons are used to simulate the pulse signals of a single ear, and the left and right ear signals are concatenated to obtain an input signal of 200 neurons. The encoding process is shown in the Preprocess part in Fig. 2.

4.2   LSM Feature Extraction Unit

We select the LIF neuron for the liquid layer, and its mathematical expression can be represented as:

$$\tau_m \frac{du}{dt} = -[u - u_{rest}] + RI(t) \tag{3}$$

There are 800 LIF neurons in the reservoir, 80% of which are excitatory neurons and 20% are inhibitory ones. This distribution has been proved reasonable early. To reduce the searching space, we manually set the dynamic parameters of excitatory and inhibitory neurons in the reservoir based on our experience. We adopt a two-dimensional liquid structure and apply Euclidean Distance to measure the distance of neuron m and neuron n in the liquid layer, which is mathematically expressed as:

$$D_{m,n} = \sqrt{\sum_{i=1}^{2}(m_i - n_i)^2} \tag{4}$$

A bio-inspired rule is that when two neurons are closer, they are more likely to interconnect and enhance synaptic density. Thus, the connection probability can be further formulated as:

$$P_{m,n} = W * e^{-(D_{m,n}/\lambda)} \tag{5}$$

P is the configurable connection probability consisting of 5 forms. P_{input} means the connection probability between input and excitatory neurons in the reservoir. P also include P_{ee}, P_{ei}, P_{ie}, P_{ii} and their explanations can be seen in Table 1. W and λ can be seen as scale factors related to P.

4.3   MLP Classifier

In our work, we take BSSL as a classification task. Each output feature vector O of the reservoir is normalized first and then connected to an MLP classifier. MLP classifier consists of a hidden layer, a relu activation layer, a batch normalization layer, and an output layer. The number of neurons in the hidden layer is a hyperparameter we will decide through BO. The number of neurons in the output layer is up to the granularity of classification. We conduct all the experiments from the azimuthal range of $[-45°, 45°]$. Therefore, for 10-degree resolution, it is a 9-classification problem ($[-40° : 10° : 40°]$), for 20-degree resolution, it is a 5-classification task ($[-40° : 20° : 40°]$), and for 15-degree resolution, it is a 7-classification task ($[-45° : 15° : 45°]$).

It is more difficult to distinguish two adjacent positions in BSSL tasks. For example, $10°$ is more easily misclassified to $0°$ and $20°$ than $40°$. We propose a soft label technique to make our model pay more attention to adjacent angle classification and get higher accuracy. For each true label θ^*, we use $cos\Delta\theta^* = [cos(\theta_0 - \theta^*), cos(\theta_2 - \theta^*), \ldots, cos(\theta_{k-1} - \theta^*)]$ as its ground-truth soft label, k is the class number. Therefore, the softmax output of the last layer can be expressed as:

$$T = softmax(t \cdot cos\Delta\theta) = [T_0, T_1, \ldots, T_{k-1}], \tag{6}$$

$$T_i = \frac{e^{cos(\theta_i - \theta^*) \cdot t}}{\sum_j e^{cos(\theta_j - \theta^*) \cdot t}} \tag{7}$$

t is a flexible parameter that decides how "hard" or "soft" the label is. Therefore, the CrossEntropy loss used in this work can be expressed as Eq. (8), T^* is the ground-truth soft label, while T is the predicted value.

$$Loss_{CE} = -\sum T^* \cdot logT \tag{8}$$

4.4   Bayesian Optimization

As we stated before, the efficiency of LSM is mainly decided by the topology of the reservoir. It is usually data-independent. Thus, a pool must be reconstructed when given a new task. Apart from the LSM, there are also a few configurable hyperparameters in the MLP classifier. We introduce BO for simultaneously optimizing the LSM structure and MLP classifier. As the liquid structure generation is random, each iteration costs different durations. We fixed the max iteration to 900 iterations and set no early stop mechanism. We use Tree-structured Parzer Estimator (TPE) for the surrogate model and Expected Improvement (EI) for the acquisition function. TPE defines two probability density functions for $P(x|y)$:

$$P(x|y) = \begin{cases} l(x), & y < y^* \\ g(x), & y \le y^* \end{cases} \tag{9}$$

According to Bayes' theorem, $p(y|\boldsymbol{x}) \cdot p(\boldsymbol{x}) = p(\boldsymbol{x}|y) \cdot p(y)$,

$$p(\boldsymbol{x}) = \int_{\Theta} p(\boldsymbol{x}|y)p(y)dy \qquad (10)$$
$$= \gamma l(\boldsymbol{x}) + (1 - \gamma)g(\boldsymbol{x})$$

$\gamma = p(y < y^*)$ takes the default value 0.25 in the hypterOpt package. With TPE, the EI function can be further expressed as:

$$EI_{y^*}(\boldsymbol{x}) = \frac{\int_{-\infty}^{y^*}(y^* - y)p(y)dy}{\gamma + (1 - \gamma)g(\boldsymbol{x})/l(\boldsymbol{x})} \qquad (11)$$

It can be derived that $EI \propto (\gamma + (1 - \gamma)g(\boldsymbol{x})/l(\boldsymbol{x}))^{-1}$. To maximize expected improvement, at point $\boldsymbol{x}$, $l(\boldsymbol{x})$ needs to be higher while $g(\boldsymbol{x})$ needs to be lower.

The detailed explanations and searching ranges of hyperparameters in LSM and MLP classifier are shown in Table 1. It is a 10-dimensional optimization problem. Different hyperparameter configurations greatly impact the final result, as seen in Table 2.

Table 1. Hyper parameters explanation

Name	Hyper Parameter	Search Range	Best
P_{input}	Connection probability of **input** to **excitatory** neurons	$[0.1, 0.9]$	0.1397
P_{ee}	Connection probability of **excitatory** to **excitatory** neurons	$[0.1, 0.9]$	0.1004
P_{ei}	Connection probability of **excitatory** to **inhibitory** neurons	$[0.1, 0.9]$	0.2052
P_{ie}	Connection probability of **inhibitory** to **excitatory** neurons	$[0.1, 0.9]$	0.7961
P_{ii}	Connection probability of **inhibitory** to **inhibitory** neurons	$[0.1, 0.9]$	0.5867
hidden_size	Number of neurons in the hidden layer	$[32, 640]$	516
w_decay	L2 regularization coefficient	$[0.0001, 0.01]$	0.0077
step_size	Learning rate change step	$[10, 35]$	31
gamma	Decay factor of learning rate	$[0.05, 0.5]$	0.4049
t	Factor of soft label	$[10, 1000]$	625.99

5   Experiment and Evaluation

5.1   Experiment Platform

All experiments are conducted using Python 3.6, the SNN simulation framework Brian 2.4, and PyTorch 1.0. However, the cochlear toolbox, which converts sound signals to spike trains, must be installed based on Python 2.7. All experiments are performed on the CPU.

Table 2. Results of different hyperparameter configuration

P_{input}	P_{ee}	P_{ei}	P_{ie}	P_{ii}	hidden size	w_decay	step size	gamma	t	result (acc)
0.3010	0.6181	0.2242	0.8621	0.4091	592	0.0049	21	0.4785	708.29	**0.1211**
0.3090	0.2632	0.1661	0.4403	0.4027	542	0.0060	21	0.2054	693.54	0.32
0.1004	0.2677	0.2882	0.3344	0.5055	509	0.0088	25	0.4282	645.93	0.58
0.2494	0.2134	0.1391	0.8271	0.4282	640	0.0056	14	0.4842	770.03	0.7622
0.1397	0.1004	0.2052	0.7961	0.5867	516	0.0077	31	0.4049	625.99	**0.8633**

5.2   Dataset

The dataset consists of the CIPIC HRTF database [1] and the speech signal dataset. The original single-channel sound signal is firstly convolved with position-specific HRTF to obtain binaural signals, which simulate the reflection and diffraction effects of the human head, trunk, and outer ear on the incoming sound.

HRTF Dataset: We select the spatial HRTF collected from the KEMAR with a large pinnae subject. In the CIPIC HRTF database, the distance from the sound source to the microphone is a fixed value of $1m$. In our work, we just select the region from $[-45° : 45°]$ with a step $5°$ for the azimuth range and fix the elevation angle at $0°$. Azimuth $0°$ means the signal source is in the front of the head. To put it directly, we only conduct localization of the horizontal plane from $-45°$ to $45°$.

Speech Signal Dataset: AISHELL-ASR0009-OS1 database [2] is selected to simulate source sound signal. The original audio signal is sampled at a frequency of 16kHz. Each signal is segmented into 3 s. There are 600 pieces of original audio data divided into training and testing sets in a ratio of 500:100. Each piece of data is convolved with HRTF at different azimuth angles, resulting in different training and testing dataset sizes for different classification tasks, as shown in Table 3. Attention, our BSSL-LSM model is not affected by the content of the speakers words but only relies on binaural cues.

Table 3. Dataset configuration

Classification task	Interval	Training set size	Testing set size
9-classification	10°	4500	900
7-classification	15°	3500	700
5-classification	20°	2500	500

5.3   Experiments and Results

We conduct experiments aiming at answering the following questions:

- **RQ1:** Is LSM an effective solution for BSSL tasks? How is the performance of BSSL-LSM compared with baseline models?
- **RQ2:** Is our proposed soft label technique necessary?
- **RQ3:** Is there an improvement when introducing STDP learning rule?
- **RQ4:** Is BO more competitive than PSO?

We apply six-fold cross-validation and take the average test accuracy of each fold as the final test accuracy. We set hyperparameters to the searched best values to answer RQ1, RQ2, and RQ3.

A. Test Efficiency and Comparison with Baseline Models

We select D-BPNN [23], CNN [23], and SNN-IID [27] as baseline models. We compare their accuracy and parameter size. These methods are tested across different azimuthal ranges and angular resolutions, as it is hard to find ideal baseline models that tackle ultimately the same localization tasks. However, it is still clear that BSSL-LSM can trade between localization accuracy and parameter size. Compared with SNN-IID, localization accuracy of BSSL-LSM at 10° interval improve by approximately 10.5%. Compared with D-BPNN, there is an improvement in both accuracy (7.41% at 15° interval) and parameter size (less than 1/2 of D-BPNN). Compared with CNN, localization accuracy of BSSL-LSM at 15° interval decreases a little (3.5%) with a massive reduction of parameters (almost only 1/10 of CNN). Meanwhile, BSSL-LSM performs better at 15° resolution localization than CNN at 20° resolution. More details can be found in Table 4.

Table 4. Comparison results between BSSL-LSM and other methods.

Method	Azimuth range	Angular resolution	Accuracy(%)	Parameter size
D-BPNN [23]	[0°, 360°]	15°, 30°, 45°	86.88, 89.27, 93.29	2.23 M
CNN [23]	[0°, 360°]	15°, 30°, 45°	97.75, <u>98.89</u>, 99.51	10.1 M
SNN-IID [27]	[−60°, 60°]	10°	<u>75.76</u>	-
BSSL-LSM (Ours)	[−45°, 45°]	10°, 15°, 20°	**86.33**, 94.29, **99.00**	**1.04 M**

Answer to RQ1: LSM is an effective solution to BSSL task, BSSL-LSM can reduce a large number of parameters while maintaining a satisfying accuracy.

B. Test Efficiency of Proposed Soft Label Technique

Intuitively speaking, adjacent locations are more difficult to distinguish. Soft label outperforms when the truth label is amphibious, or two labels are more likely to be misclassified than others, as stated in Sect. 4.3. We make experiments based on 9-class classification (10° interval) to test the effect of soft level on final results by selecting different values of t. The t-value determines how smooth the label is: the lower, the more smooth.

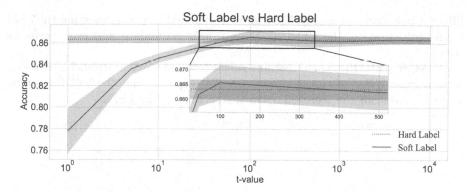

Fig. 3. Accuracy under different t-value and hard label.

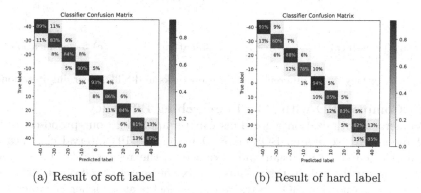

(a) Result of soft label   (b) Result of hard label

Fig. 4. Comparison of hard label and soft label.

We did five groups of experiments using different seeds to mitigate randomness. Figure 3 presents the mean and variance of the test results under different seeds. It is not hard to find that when setting t at a proper value (e.g., $t = 100$), it outperforms the classifier that uses a hard label. It improves the result by about 0.53% on average. We also find that when applying cross-validation, the accuracy of using soft labels is more stable than that of hard labels. Furthermore, we present the confusion matrix of classification under different labeling techniques. From Fig. 4, we can clearly find that the soft label technique alleviates the misclassification phenomenon of adjacent angles to a certain extent.

Answer to RQ2: Our proposed soft label technique is necessary and can improve the accuracy by about 0.53% compared with the one-hot label.

C. Test Impact of STDP Learning

STDP learning is a well-known training method for SNN networks. We are curious whether the performance can improve when applying the STDP learning rule on a searched best LSM structure. Therefore, we made a series of experiments under $10°, 15°, 20°$ resolution localization and tested the influence under different iterations. From Fig. 5, it can be concluded that when setting the LSM at an optimal state, the STDP learning rule is redundant and even performs worse.

Answer to RQ3: STDP learning rule is not necessary in BSSL-LSM model, or even bring a decreasement in accuracy. We suppose that the STDP learning process breaks out the searched optimal structure of the LSM.

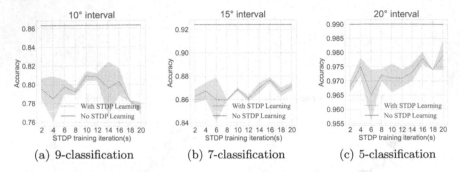

Fig. 5. Accuracy under different classification tasks and STDP learning iterations.

D. Compare BO with PSO In Searching Efficiency

Numerous optimization algorithms can be applied to our proposed BSSL-LSM model. We compared BO and PSO under the 10° interval classification task and calculated the mean and variance of the results. We aim to get a satisfying result in a given time. Thus, we fix the search iteration at 900. From Fig. 6, we can find that BO always converges faster at a higher accuracy than PSO. BO constantly seeks to explore better solutions, while PSO always stops searching at the local optimal.

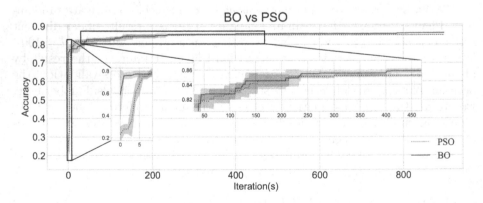

Fig. 6. Comparison between BO and PSO.

Answer to RQ4: BO is a more competitive hyperparameter searching algorithm than PSO for our BSSL-LSM model.

6   Conclusion and Future Work

We propose a lightweight LSM based binaural sound source localization model (BSSL-LSM) in this study. This is the first work that tries to apply LSM on sound source localization tasks. BSSL-LSM can receive binaural signals as input and automatically extract binaural acoustic cues to do angle classification without manual intervention. We introduce Bayesian Optimization for hyperparameters searching to make BSSL-LSM a fast adaptation to new datasets without needing STDP learning. The soft label technique is another striking point that alleviates classification confusion between adjacent angles, which can be easily migrated to a broader series of related problems. Experiments have certified that BSSL-SSL outperforms baseline models in terms of parameter size, localization resolution, and accuracy.

For future work, we intend to demonstrate further whether LSM is a desirable approach for noisy environments or multi-sound source localization scenarios. Another line of study is better and faster hyperparameters optimization as it unquestionably affects the performance of LSM a lot.

Acknowledgement. This work was supported in part by the National Natural Science Foundation of China under Grants 62372461, 62032001 and 62203457, and in part by the Key Laboratory of Advanced Microprocessor Chips and Systems.

References

1. Algazi, V.R., Duda, R.O., Thompson, D., Avendaño, C.: The CIPIC HRTF database. In: Proceedings of the 2001 IEEE Workshop on the Applications of Signal Processing to Audio and Acoustics (Cat. No. 01TH8575), pp. 99–102 (2001)
2. Bu, H., Du, J., Na, X., Wu, B., Zheng, H.: AISHELL-1: an open-source mandarin speech corpus and a speech recognition baseline. In: Oriental COCOSDA 2017 (2017, submitted)
3. Das, D., Bhattacharya, S., Pal, U., Chanda, S.: PLSM: a parallelized liquid state machine for unintentional action detection. ArXiv abs/2105.09909 (2021)
4. Dávila-Chacón, J., Liu, J., Wermter, S.: Enhanced robot speech recognition using biomimetic binaural sound source localization. IEEE Trans. Neural Netw. Learn. Syst. **30**(1), 138–150 (2018)
5. Desai, D., Mehendale, N.: A review on sound source localization systems. Arch. Comput. Methods Eng. **29**(7), 4631–4642 (2022)
6. Faraji, M.M., Shouraki, S.B., Iranmehr, E.: Spiking neural network for sound localization using microphone array. In: 2015 23rd Iranian Conference on Electrical Engineering, pp. 1260–1265 (2015)
7. Gerstner, W., Kempter, R., van Hemmen, J.L., Wagner, H.: A neuronal learning rule for sub-millisecond temporal coding. Nature **383**, 76–78 (1996)
8. Gerstner, W., Ritz, R., van Hemmen, J.L.: Why spikes? Hebbian learning and retrieval of time-resolved excitation patterns. Biol. Cybern. **69**, 503–515 (1993)
9. Ghani, A., McGinnity, T.M., Maguire, L.P., McDaid, L.J., Belatreche, A.: Neuro-inspired speech recognition based on reservoir computing (2010)

10. Glackin, B.P., Wall, J.A., Mcginnity, T.M., Maguire, L.P., McDaid, L.J.: A spiking neural network model of the medial superior olive using spike timing dependent plasticity for sound localization. Frontiers Comput. Neurosci. **4** (2010)
11. Goodman, D.F.M., Pressnitzer, D., Brette, R.: Sound localization with spiking neural networks. BMC Neurosci. **10**, 1 (2009)
12. Guo, S., et al.: A systolic SNN inference accelerator and its co-optimized software framework. In: Proceedings of the 2019 on Great Lakes Symposium on VLSI (2019)
13. Jeffress, L.A.: A place theory of sound localization. J. Comp. Physiol. Psychol. **41**(1), 35–9 (1948)
14. Kuang, S., van der Heijden, K., Mehrkanoon, S.: BAST: binaural audio spectrogram transformer for binaural sound localization. ArXiv abs/2207.03927 (2022)
15. Li, S., Wang, L., Wang, S., Xu, W.: Liquid state machine applications mapping for NoC-based neuromorphic platforms. In: Dong, D., Gong, X., Li, C., Li, D., Wu, J. (eds.) ACA 2020. CCIS, vol. 1256, pp. 277–289. Springer, Singapore (2020). https://doi.org/10.1007/978-981-15-8135-9_20
16. Li, Y., Zhang, Y., Zhou, G., Gong, Y.: Bayesian optimization with particle swarm. In: 2021 International Joint Conference on Neural Networks (IJCNN), pp. 1–6 (2021)
17. Liaquat, M.U., Munawar, H.S., Rahman, A., Qadir, Z., Kouzani, A.Z., Mahmud, M.A.P.: Sound localization for ad-hoc microphone arrays. Energies (2021)
18. Luke, R., McAlpine, D.: A spiking neural network approach to auditory source lateralisation. In: ICASSP 2019–2019 IEEE International Conference on Acoustics, Speech and Signal Processing (ICASSP), pp. 1488–1492. IEEE (2019)
19. Maass, W., Natschläger, T., Markram, H.: Real-time computing without stable states: a new framework for neural computation based on perturbations. Neural Comput. **14**, 2531–2560 (2002)
20. Pang, C., Liu, H., Li, X.: Multitask learning of time-frequency CNN for sound source localization. IEEE Access **7**, 40725–40737 (2019)
21. Reynolds, J.J.M., Plank, J.S., Schuman, C.D.: Intelligent reservoir generation for liquid state machines using evolutionary optimization. In: 2019 International Joint Conference on Neural Networks (IJCNN), pp. 1–8 (2019)
22. Rudnicki, M., Schoppe, O., Isik, M., Völk, F., Hemmert, W.: Modeling auditory coding: from sound to spikes. Cell Tissue Res. **361**, 159–175 (2015)
23. Song, H., Liu, X., Yu, S.: Binaural localization algorithm based on deep learning. Technical Acoust. **41** (2022)
24. Tang, C., Ji, J., Lin, Q., Zhou, Y.: Evolutionary neural architecture design of liquid state machine for image classification. In: ICASSP 2022–2022 IEEE International Conference on Acoustics, Speech and Signal Processing (ICASSP), pp. 91–95 (2022)
25. Tian, S., Qu, L., Wang, L., Hu, K., Li, N., Xu, W.: A neural architecture search based framework for liquid state machine design. Neurocomputing **443**, 174–182 (2021)
26. Vecchiotti, P., Ma, N., Squartini, S., Brown, G.J.: End-to-end binaural sound localisation from the raw waveform. In: ICASSP 2019–2019 IEEE International Conference on Acoustics, Speech and Signal Processing (ICASSP), pp. 451–455. IEEE (2019)
27. Wall, J.A., McDaid, L.J., Maguire, L.P., McGinnity, T.M.: Spiking neural network model of sound localization using the interaural intensity difference. IEEE Trans. Neural Netw. Learn. Syst. **23**(4), 574–586 (2012)

28. Wang, S., et al.: A power efficient hardware implementation of the if neuron model. In: Conference on Advanced Computer Architecture (2018)
29. Wu, J., Chua, Y., Zhang, M., Li, H., Tan, K.C.: A spiking neural network framework for robust sound classification. Frontiers Neurosci. **12** (2018)
30. Xiao, X., et al.: Dynamic vision sensor based gesture recognition using liquid state machine. In: International Conference on Artificial Neural Networks (2022)
31. Xu, Y., Afshar, S., Singh, R.K., Wang, R., van Schaik, A., Hamilton, T.J.: A binaural sound localization system using deep convolutional neural networks. In: 2019 IEEE International Symposium on Circuits and Systems (ISCAS), pp. 1–5. IEEE (2019)
32. Yang, Q., Zheng, Y.: DeepEar: sound localization with binaural microphones. IEEE Trans. Mob. Comput. (2022)
33. Youssef, K., Argentieri, S., Zarader, J.L.: A binaural sound source localization method using auditive cues and vision. In: 2012 IEEE International Conference on Acoustics, Speech and Signal Processing (ICASSP), pp. 217–220 (2012)
34. Yu, X., Wang, L., Chen, C., Tie, J., Guo, S.: Multimodal learning of audio-visual speech recognition with liquid state machine. In: International Conference on Neural Information Processing (2022)
35. Zheng, H., Wu, Y., Deng, L., Hu, Y., Li, G.: Going deeper with directly-trained larger spiking neural networks. In: AAAI Conference on Artificial Intelligence (2020)
36. Zhu, J., et al.: An event based gesture recognition system using a liquid state machine accelerator. In: Proceedings of the Great Lakes Symposium on VLSI 2022 (2022)
37. Zilany, M.S.A., Bruce, I.C., Carney, L.H.: Updated parameters and expanded simulation options for a model of the auditory periphery. J. Acoust. Soc. Am. **135**(1), 283–6 (2014)

A Causality-Based Interpretable Cognitive Diagnosis Model

Jinwei Zhou[1], Zhengyang Wu[2,3(✉)], Changzhe Yuan[4], and Lizhang Zeng[5]

[1] School of Artificial Intelligence, South China Normal University,
Foshan 528225, China
[2] School of Computer Science, South China Normal University,
Guangzhou 510631, China
wuzhengyang@m.scnu.edu.cn
[3] Pazhou Lab, Guangzhou 510330, China
[4] School of Electronics and Information, Guangdong Polytechnic Normal University,
Guangzhou 510665, China
[5] Institute for Brain Research and Rehabilitation, South China Normal University,
Guangzhou 510631, China

Abstract. Cognitive diagnosis model (abbr.CDM) aims to assess students' cognitive processes during learning, enabling personalized support based on their needs. Nevertheless, deep learning-based CDMs are inherently opaque, posing challenges in providing psychological insights into the reasoning behind predicted outcomes. We address this by creating three interpretable parameters: skill mastery, exercise difficulty, and exercise discrimination. Inspired by Bayesian networks and neural networks, we use feature engineering for extraction of interpretable parameters and tree-enhanced naive Bayes classifiers for prediction. Our method balances interpretability and accuracy. Experimentally, we compare our approach to traditional and advanced models on four datasets, analyzing each feature's impact. We conduct ablation studies on each feature to examine their contribution to student performance prediction. Thus, causality-based interpretable cognitive diagnosis model (CBICDM) has great potential for providing adaptive and personalized instructions with causal reasoning in real-world educational systems.

Keywords: Cognitive diagnosis · Knowledge tracing ·
Interpretability · Bayesian network · Causal relations

1 Introduction

With the increasing demand for personalized learning, it is becoming more important to accurately and effectively track student's knowledge states based on their learning process, and then provide personalized teaching services for them. Therefore, Cognitive Diagnosis and Knowledge Tracing (KT) are proposed for modelling student's knowledge state. CD focuses on analyzing the learning process of student's to improve learning efficiency, while KT emphasizes predicting student's knowledge status in the next time step. Specifically, CD predicts

B. Luo et al. (Eds.): ICONIP 2023, LNCS 14449, pp. 214–226, 2024.
https://doi.org/10.1007/978-981-99-8067-3_16

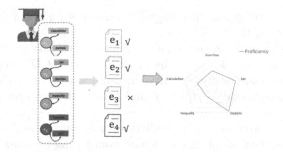

Fig. 1. An example of cognitive diagnosis.

the student's knowledge status by analyzing their historical interaction data. It reflects the student's knowledge mastery level in different dimensions, rather than a single score of exercise answers. As a simple example, a teaching environment found that most of the educated students were generally poor at answering exercises on inequalities. The CD model is used to identify that the difficulty information of the knowledge concept is higher than the average level of students at this stage. Figure 1 shows the process of CD for a student.

Many CDMs have been proposed. Such as DINA (Deterministic Inputs, Noisy "And" gate model) [4], IRT (the item response theory) [5], FuzzyCDF (the fuzzy cognitive diagnosis framework) [10,21], and DeepCDM (the deep cognitive diagnosis framework) [8] etc. Unfortunately, none of these models propose specific solutions for interpretability [2] and computational complexity. However, these two issues have also caught the attention of researchers. Wang et al. proposed the Neural Cognitive Diagnosis (NeuralCDM) [17] in 2020. This model combines neural networks to analyze the complex interactions of students' exercises and modeling. Yang et al. proposed a novel quantitative relationship neural network for explainable cognitive diagnosis model (QRCDM) [22] in 2022.

Summarizing the advantages and disadvantages of the above models, we propose a causality-based interpretable cognitive diagnosis model namely CBICDM. The model mainly includes two parts. In the first part, the feature parameters with psychological significance are extracted and quantified by MLP combined with feature engineering to reduce the computational complexity of the model. In the second part, the extracted feature parameters are put into a Tree-Augmented Naive Bayesian Classifier (TAN) [11], and the causal relationship between exercises, knowledge concepts and feature relationships is used to predict the student's score. The application of TAN greatly simplifies the structure of the model while meeting the requirements of intrinsic interpretability [3] In this paper we summarize the respective advantages of CDs and KTs fields and propose CBICDM. The main contributions of this study are summarized as follows:

1. This study employs three psychologically meaningful interpretability parameters for post-hoc analysis [20]. However, most existing CDMs rely on only two interpretability parameters, i.e., guessing and skipping. The inclusion of

additional psychologically meaningful parameters significantly enhances the interpretability of the model's outcomes.

2. This study helps to improve the intrinsic interpretability of the model, which is limited by the complexity of the model. Existing CDMs do not meet the criterion of intrinsic interpretability. In our model, the TAN module effectively constrains model complexity and fulfills the requirement of intrinsic interpretability.

3. This study performs accuracy prediction experiments and ablation experiments on three datasets of varying scales, comprising two small-scale datasets and one large-scale dataset. The results of experiments demonstrate that the CBICDM outperforms the traditional CDs in terms of predictive performance.

2   Related Work

2.1   Cognitive Diagnosis

Current research on CDs is based on theories of cognitive psychology, educational measurement, and data analysis. The two most influential studies are DINA [4] and IRT [5]. IRT theory describes student's probability of answering exercises through a probability model. The original IRT model is a two-parameters model (abbr.2PL), which assumes that a student's probability of answering an exercise correctly is only influenced by two parameters: the difficulty of the exercise, and the student's ability. In addition, there are some variations of IRT, such as the Three-parameters IRT (TIRT) and the Multidimensional IRT (MIRT) [16]. The principle of the DINA model is to establish a binary response model. The model assumes that students' mastery of a certain knowledge concept is only affected by the test exercises related to the knowledge concept, and is not affected by other factors. The student's cognitive state is simply classified as 0 or 1, but this is inconsistent with actual situations. In addition, DINA's theory regards knowledge concepts as independent of each other, ignoring the correlation between them. Liu et al. improve the DINA model to propose the FuzzyCDF, which have multi-level scoring capabilities. DeepCDM [8] uses the student test score matrix (X-matrix) and knowledge concept inspection matrix (Q-matrix) to obtain the internal relationship between knowledge concepts-knowledge concepts, knowledge concepts-exercises, and exercises-exercises. Wang et al. propose that NeuralCDM added the relationship between students and exercises based on DeepCDM. Yang et al. proposed the QRCDM [22] to quantify the implicit relationship as a number from 0 to 1, and finally use the contribution matrix as a prediction, and retain the respondent's errors and guesses as explanations parameters.

2.2   Knowledge Tracing

Knowledge tracing is an important technology in the field of personalized learning. Bayesian knowledge tracing (BKT) [1] is a popular modeling method in the current industry, which treats the student's potential knowledge state as a set of binary variables. However, as a standard Hidden Markov Model (HMM) [15], it cannot observe short-term state change values and requires discrete time points. In 2015, Piech et al. proposed a deep learning knowledge tracing model(DKT) [14] using Long Short-Term Memory (LSTM) to apply recurrent neural networks to knowledge tracing tasks for the first time. The emergence of DKT has brought knowledge tracing to a new height in predicting students' future learning outcomes. In 2017, Zhang et al. proposed a DKVMN model [23] with a static key matrix and a dynamic value matrix, which addressed two issues: BKT's inability to observe short-term state changes and DKT's insufficient modeling of interactions between different knowledge concepts in the system. The Deep Knowledge Tracing and Dynamic Student Classification (DKT-DSC) [13] aims to use k-means clustering to predict students' mastery levels. Although these methods can achieve good results in predicting results, they are opaque and requiring more resources and time for large sample datasets. To address this issue, the Interpretable Knowledge Tracing Model (IKT) [12] uses TAN to reduce the number of parameters and a tree structure to simplify the complexity of the model. Compared to KT, CD does not have such stringent requirements for predicting accuracy. The primary focus of the model is to provide more interpretable implicit and explicit parameters that have psychological significance and can be effectively employed in practical teaching activities. It is imperative that researchers put more effort into enhancing the credibility of the model. Based on this, the experiments in the paper only compare the relevant algorithms in CD, such as DINA, FuzzyCDF, NeuralCDM, and QRCDM.

3   Methodology

3.1   Problem Definition

Table 1. Some important notations.

Notation	Description
X	Matrix of student scores on exercises, $X \in \mathrm{R}^{N \times M}$
X_{ij}	The score of student ith on exercise jth
Q	Matrix of relationships between exercise and knowledge, $Q \in \mathrm{R}^{M \times K}$
q_{jk}	The correlation between exercise jth and knowledge kth
a_{ik}	The mastery level of knowledge kth for student ith
η_{ij}	The potential answering situation of student ith in exercise jth
s_j	The probability of a student slipping in the jth exercise
g_j	The probability of a student guessing in the jth exercise

Assuming that there are N Students, M Exercises and K Knowledge concepts in the online teaching system, we express them as $S = \{s_1, s_2, \ldots, s_N\}$,

$E = \{e_1, e_2, \ldots, e_M\}$ and $K_n = \{k_1, k_2, \ldots, k_K\}$ respectively. Each learning interaction is denoted as a set of triplet (s, e, r), where $r \in \{1, 0\}$ is a binary that equals 1 if correct and 0 if incorrect. In addition, we have Q-matrix (usually labeled by experts) $\mathbf{Q} = \{Q_{ij}\}_{M \times K}$, where $Q_{ij} = 1$ if exercise e_i relates to knowledge concept k_j and $Q_{ij} = 0$ otherwise. Some important notations of this paper are listed in Table 1.

3.2 Model Framework

It is important to note that the mastery value of a skill is the probability of having learned that skill, rather than the probability of correctly applying it to the next exercise. The CBICDM consists of two main stages, i.e., Feature Extraction and Classification, as illustrated in Fig. 2.

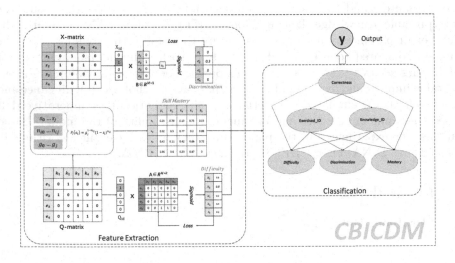

Fig. 2. The architecture of CBICDM.

During the feature extraction stage, we transform the student's answering status and the relationships between knowledge concepts and exercises into Q and X matrices. Then, we employ convolutional neural networks and statistical method to mine the implicit relationships between knowledge concepts and exercise exercises, and form three feature parameters i.e., *skill mastery*, *exercise difficulty*, and *exercise discrimination*. It should be noted that we do not feed all the data into the neural network. It is evident that using a neural network to extract parameters influenced by multiple factors, such as *skill mastery* and *exercise discrimination*, is the most suitable approach. However, for the parameter of exercise difficulty, a statistical method is more appropriate. This setting ensures that the extracted interpretable parameters closely align with the actual situation, while also reducing the complexity of the model. Finally, the three feature parameters are input into the TAN classifier to make predictions using the dependencies between nodes.

3.3   Feature Extraction

In general, when it comes to CDMs, three elements need to be taken into consideration: student factors, exercise factors, and their interaction. Drawing upon the theories of DINA and IRT, we have designed student factors as interpretable vectors similar to DINA. Exercise factors typically prioritize problem difficulty as a crucial condition. Typically, an exercise encompasses multiple knowledge concepts. Here, we treat the exercise's difficulty as the highest difficulty value among the knowledge concepts it encompasses. Apart from these two conditions, accurately answering a question also requires students to correctly identify the knowledge concepts within the exercise. Therefore, we have chosen three feature values with psychological significance Skill mastery, Exercise Difficulty, and Exercise Discrimination to serve as interpretable parameters. To better align with a genuine teaching environment and more effectively assist students in self-diagnosing, we have processed all three of these feature values as continuous variables.

Skill Mastery. It is the student's mastery of the skill, and it comes from DINA. As a typical discrete CDM, DINA describes students as a multidimensional vector of knowledge mastery, diagnosing based on actual student responses. The DINA model is simple, with good interpretability of its parameters, and its complexity is not affected by the number of attributes. The skill mastery is denoted by h^s here. The probability of answering jth exercise correctly when that ith student's knowledge concept mastering condition a_i is given. Table 1 shows some important symbols for this paper. The potential response of student ith to exercise jth:

$$\eta_{ij} = \prod_{k=1}^{k} a_{ik}^{q}, \tag{1}$$

$$h^s = P_j(a_i) = P\left(X_{ij} = 1 \mid a_i\right) = g_j^{1-\eta_{ij}}\left(1 - s_j\right)^{\eta_{ij}}, \tag{2}$$

Exercise Difficulty. Knowledge difficulty indicates the difficulty of each knowledge concept examined by the exercise, and exercise difficulty depends on the most difficult knowledge concept in the exercise. In the following formula we express it as e^{diff}.

Exercise Discrimination. Exercise discrimination refers to the ability of students to correctly identify the knowledge concept being tested in the exercise. Both them are range of 0 to 1, and are derived by using IRT and MIRT models.We use e^{disc} to represent exercise discrimination. They can be obtained by:

$$e^{disc} = \text{sigmoid}\left(x^e \times A\right), \quad A \in R^{M \times K}, \tag{3}$$

$$e^{diff} = \text{sigmoid}\left(x^e \times B\right), \quad B \in R^{M \times 1}, \tag{4}$$

Where $x^e \in \{0,1\}^{1 \times M}$ is the one-hot representation of the exercise, A and B are two trainable matrices It is important to note that a single exercise may encompass multiple knowledge concepts, and the difficulty of the exercise is determined by the most difficult one. Therefore, in the subsequent processing, we define the exercise difficulty e^{diff} as the highest difficulty among the knowledge concepts it contains.

3.4   Classification Module

In CBICDM, the Bayesian network paradigm is being adopted to develop explanations with diagnosis and predictive capabilities. The TAN structure, which is a simple extension of the naive Bayesian network, plays a crucial role in the classification module. The root node in the TAN structure represents the correctness of the exercise, while the evidence nodes (including knowledge concept ID, knowledge concept mastery, exercise discrimination, and exercise difficulty) have causal relationships with it. The TAN model enhances the independence assumption presented in the Naive Bayesian Network and significantly improves prediction accuracy. Moreover, it has shown excellent scalability for large datasets. To further investigate the effectiveness of this approach, we conducted a study on the Minimum Weight Spanning Tree (MWST) [7] method. Specifically, our algorithm selects a vertex from the graph as the root node through the greedy strategy and then adds new edges in order of increasing weight until a spanning tree containing all vertices is constructed. In the experiments of this paper, a more convenient method was used to directly use the data mining tool Weka [9] to detect the model's performance.

CBICDM performs inference by using three meaningful extracted features f_t: skill mastery, exercise discrimination, and exercise difficulty as evidence at the current timestamp t.

$$P(correctness_t = y \mid f_t) = \frac{P(y)P(f_t \mid y)}{\sum_{y'} P(y')P(f_t \mid y')}, \tag{5}$$

$$P(f_t \mid y) = P(s_t \mid y)P(h_t^s \mid y)P(e_j^{disc} \mid y)P(e_j^{diff} \mid y) \tag{6}$$

Those are assessed h_t^s of h^s at time t, the e_j^{disc} and e_j^{diff} are e^{disc} and e^{diff} of problem j. The inference is estimated in the context of discretized values, i.e. conditional probability tables. It doesn't handle continuous variables. Discretization algorithm bin all features into sets for best discrimination among classes. The class node (correctness) represents the predicted probability that the student would answer the problem with the associated skill correctly. Thus the prediction of problem associated with skill can be retrieved from $correctness_t$ as described in Fig. 2.

Interpretation can be achieved by utilizing the conditional probability tables associated with each node and their causal connections. By examining the evidence available at each timestamp, we can determine whether the failure in students' problem-solving stems from a lack of skill practice or the individual students finding the problem excessively challenging.

4   Experiments

4.1   Dataset

To validate the proposed model, we are testing it on three publicly available datasets from two different tutoring scenarios. All of these datasets were derived from actual interactions between students and online educational systems. This is shown in the following Table 2.

- **ASSISTments**[1]: An online tutoring system used in this experiment was originally created in 2004 and is designed for middle and high school students to complete math exercises using a step-by-step prompting approach. If a student correctly answers a exercise with the aid of the system's assistance, they are presented with a new exercise. However, if they answer incorrectly, the system provides them with brief instructional videos to guide them. We utilized the ASSISTments 2009–2010 (skill builder) dataset for this experiment.
- **Math1&Math2**[2]: Math is from 'NeurIPS 2020 Education Challenge', collected by the online education website Eedi,and contains answer records from September 2018 to May 2020. Math1 contains 84,180 data points, 4,209 students, 15 objective practice exercises, 5 subjective practice exercises, and 11 knowledge concepts. Math2 contains 78,200 data points, 3,911 students, 16 objective practice exercises, 4 subjective practice exercises, and 16 knowledge concepts.
- **JunYi**[3]: JunYi is taken from the online learning platform Junyi Academy, and the dataset includes answer records from October 2012 to January 2015. Each exercise contains only one concept, and one concept is contained by only one exercise. It provides the interaction among concepts marked by experts. Specifically, JunYi marks the dependency or similar relations between concepts as natural numbers from 1 to 9.

Table 2. The specifics of the three datasets

Dataset	Learners	Objective	Subjective	Concepts	Interactions
Math1	4209	15	5	11	84180
Math2	3911	16	4	16	78220
ASSIST0910	4163	17746	0	123	324572
JunYi	36591	721	0	721	1550016

---

[1] https://sites.google.com/site/assistmentsdata/home/2009-2010-assistment-data/skill-builder-data-2009-2010.

[2] https://eedi.com/projects/neurips-education-challenge.

[3] https://pslcdatashop.web.cmu.edu/DatasetInfo?datasetId=1198.

4.2 Baselines and Experimental Settings

In order to establish the superiority of CBICDM, we compare several existing
CDs with it. The details of these baseline models are as follows:

- **DINA** [4] takes into account the certainty and noise of the input, meaning that students may provide incorrect answers due to guessing or being
  unfamiliar with the exercise.
- **FuzzyCDF** [10] uses fuzzy set theory to represent students' ability levels in
  different knowledge areas, in order to adapt to various assessment needs and
  objectives.
- **NeuralCDM** [17] combines traditional CDMs with deep learning techniques
  to achieve more accurate predictions.
- **QRCDM** [22] can quantitatively express the relationship between exercises
  and knowledge concepts and express it as a weight of a real number between
  0 and 1.

4.3 Implementation Details

We compared our model with four other models: DINA, Fuzzy, NCDM, and
QRCDM. These models are primarily based on the IRT theory for DINA and
deep learning for Fuzzy, NCDM, and QRCDM. We did not compare our model
with other variants because they did not show significant performance differences. We implemented all neural network models using the PyTorch framework
and used mini-batch stochastic gradient descent to minimize the loss function
to speed up the training process. The dimensions of the full connection layers
are 512, 256, 1 respectively, and Sigmoid is used as activation function for all
of the layers. The batch size was set to 32, and we trained the model with a
learning rate of 0.01 while using dropout to avoid overfitting. For skill mastery
feature extraction, we set the epoch value to 40, exercise difficulty and exercise
discrimination to 60 to ensure optimal feature values. It should be noted that
the feature engineering mining of the three parameters is carried out in steps,
so the epoch values are not the same. They are determined based on the loss
value during training. In our experiments, we divided each dataset into 80% for
training and 20% for testing. We analyzed and compared the results using the
area under the curve (AUC), prediction accuracy (ACC), and root mean square
error (RMSE). Additionally, for CBICDM, we extracted all features into the
WEKA data format (arff), which automatically divided the dataset into test
and training sets in WEKA.The specific results are shown in Table 3.

4.4 Ablation Experiment

This section will further analyze this issue. We will compare the CBICDM
with various features using an ablation study. This study contributes to our
understanding of the contribution of each feature in predicting student performance. CBICDM-1 only considers skill ID and skill mastery for student performance prediction, and it achieves higher performance than the original DINA.

Table 3. Performance comparison of all CD methods.

Dataset		Math1		Math2		ASSIST0910		JunYi	
		AUC ↑	ACC ↑	AUC ↑	ACC ↑	AUC ↑	ACC ↑	AUC ↑	ACC ↑
Baseline	DINA	0.4986	0.4871	0.5957	0.5591	0.6704	0.6691	0.6238	0.5015
	FuzzyCDF	0.6642	0.6181	0.6740	0.6273	–	–	–	–
	NCDM	0.6928	0.6531	0.6726	0.6367	0.7486	0.7278	0.6953	0.7370
	QRCDM	0.7679	0.6958	**0.7964**	0.7172	**0.7840**	**0.7383**	0.6327	0.8069
	CBICDM	**0.804**	**0.7163**	0.793	**0.7220**	0.764	0.7323	**0.785**	**0.8073**

When CBICDM-2 takes into account the student's skill ID, skill mastery, and knowledge concept difficulty, it shows similar (or slightly lower) performance to QRCDM and higher performance than NCDM on two Assistments datasets. Our proposed model, CBICDM-3, outperforms any CDM in this experiment. The difficulty of knowledge concepts leads to a slight improvement in AUC between CBICDM-1 and CBICDM-2 of less than 4%. The results of each model with different combinations of features show that our features provide more information in predicting student performance on different datasets. When we apply item discriminability to CBICDM-3, its AUC increases by approximately 1% to 5% (for example, CBICDM-1 and CBICDM-3 increase from 0.713 to 0.764 in the Assist09 dataset). The specific results are shown in Table 4.

- CBICDM-1: item ID, skill mastery.
- CBICDM-2: CBICDM-1+exercise difficulty.
- CBICDM-3: CBICDM-2+exercise discrimination.

Table 4. Results of ablation experiment.

Dataset		Math1		Math2		ASSIST0910		Junyi	
		AUC↑	RMSE↓	AUC↑	RMSE↓	AUC↑	RMSE↓	AUC↑	RMSE↓
Model	CBICDM-1	0.779	0.4252	0.789	0.4318	0.713	0.4618	0.773	0.4206
	CBICDM-2	0.781	0.4252	0.789	0.4318	0.757	0.4451	0.784	0.3805
	CBICDM-3	**0.793**	**0.4220**	**0.804**	**0.4269**	**0.764**	**0.4255**	**0.797**	**0.3801**

4.5   Model Interpretability Analysis

This study adopts two metrics to evaluate and analyze the interpretability of CDMs: DOA [6]. DOA was proposed by the NeuralCDM [18]. This metric is based on the intuition that if student i has a better mastery on knowledge concept k than student v, then i is more likely to answer exercises related to k

correctly than v [19]. For knowledge concept $k, DOA(K)$ is formulated as:

$$DOA_k = \frac{1}{Z} \sum_{i=1}^{N} \sum_{v=1}^{N} I(a_{ik} > a_{vk}) \frac{\sum_{j=1}^{M} I(Q_{jk} = 1) \wedge J(j,i,v) \wedge I(r_{ij} > r_{vj})}{Z_0},$$

(7)

$$Z_0 = \sum_{j=1}^{M} I(Q_{jk} = 1) \wedge J(j,i,v) \wedge I(r_{ij} \neq r_{vj}),$$

(8)

$$Z_1 = \sum_{i=1}^{N} \sum_{v=1}^{N} I(a_{ij} > a_{vj}) I(Z_0 > 0),$$

(9)

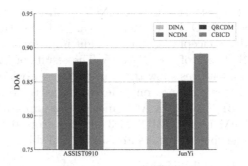

Fig. 3. DOA of CDMs.

Where a_{ik} is the proficiency of student i on knowledge concept $k. I(Event)$ = 1 if $Event$ is true and $I(Event) = 0$ otherwise. $J(j,i,v) = 1$ if both student i and v did exercise j and $J(j,i,v) = 0$ otherwise. If $Z_0 = 0$, the corresponding (i,v,k) triplet is excluded from the calculation of DOA. It should be noted that the DOA experiment overlooked the scenario where a single exercise encompasses multiple knowledge concepts. Therefore, in our experiment, we only utilized the ASSIST0910 and JunYi datasets as sources of experimental data. Among traditional models, we only compare with DINA, since for IRT, MIRT and PMF, there are no clear correspondence between their latent features and knowledge concepts. Furthermore, in our experiment based on the deep learning-based CDM model, we selected NCDM and QRCDM as control groups for comparison. As depicted in Fig. 3, the performance of the four model groups in terms of DOA on the ASSIST0910 dataset is relatively similar, with results fluctuating around 0.87. In contrast, on the JunYi dataset, it is noticeably evident that the DOA performance of CBICDM surpasses that of the other models. The Math dataset contains some subjective questions, and the knowledge concepts of the questions are not rich enough, so we did not conduct DOA experiments on the Math dataset. This once again highlights the superior interpretability of causal-based CDMs compared to statistical-based CDMs and neural network-based CDMs.

5   Conclusion

By utilizing data mining techniques in feature engineering, we were able to extract three meaningful potential features from student behavior data. The causally related TAN structure based on these features leads to improved performance of CDMs in predicting student performance, without requiring a large number of parameters or complex structures. Compared to using deep learning models alone, this approach saves significant computational resources and provides causal explanations for meaningful features. The proposed approach extracts three meaningful student characteristics: skill mastery, exercise difficulty, and exercise discrimination. By utilizing features extracted from skills, knowledge, and student behavior data, CBICDM is able to predict student scores while providing meaningful causal explanations. Our experiments using four public datasets demonstrate that CBICDM outperforms known CDMs, including machine learning-based models, while requiring less computational power. Feature engineering enables us to obtain more variance in the data, while TAN provides crucial causal explanations, allowing for more accurate and personalized performance predictions. In the field of CDs research, we believe that the focus should not solely be on the predictive performance of models but also on simplifying the model structure to meet the requirements of intrinsic and post-hoc interpretability. This is important in adapting to personalized learning systems. Finding a balance between student performance prediction and interpretability, designing and conducting more experiments to understand more knowledge acquisition and learning behavior, and optimizing student learning in future educational environments are essential.

Acknowledgements. This research was supported by the National Natural Science Foundation of China (NSFC) under the Grant No. 62377015.

References

1. Abdelrahman, G., Wang, Q., Nunes, B.P.: Knowledge tracing: a survey. ACM Comput. Surv. **55**(11), 224:1–224:37 (2023)
2. Agarwal, N., Das, S.: Interpretable machine learning tools: a survey. In: 2020 IEEE Symposium Series on Computational Intelligence, SSCI 2020, Canberra, Australia, 1–4 December 2020, pp. 1528–1534. IEEE (2020)
3. Chen, V., Li, J., Kim, J.S., Plumb, G., Talwalkar, A.: Interpretable machine learning: moving from mythos to diagnostics. Commun. ACM **65**(8), 43–50 (2022)
4. De La Torre, J.: Dina model and parameter estimation: a didactic. J. Educ. Behav. Stat. **34**(1), 115–130 (2009)
5. Embretson, S.E., Reise, S.P.: Item Response Theory. Psychology Press, London (2013)
6. Fouss, F., Pirotte, A., Renders, J.M., Saerens, M.: Random-walk computation of similarities between nodes of a graph with application to collaborative recommendation. IEEE Trans. Knowl. Data Eng. **19**(3), 355–369 (2007)
7. Gallager, R.G., Humblet, P.A., Spira, P.M.: A distributed algorithm for minimum-weight spanning trees. ACM Trans. Program. Lang. Syst. (TOPLAS) **5**(1), 66–77 (1983)

8. Gao, L., Zhao, Z., Li, C., Zhao, J., Zeng, Q.: Deep cognitive diagnosis model for predicting students' performance. Futur. Gener. Comput. Syst. **126**, 252–262 (2022)
9. Kotak, P., Modi, H.: Enhancing the data mining tool WEKA. In: 5th International Conference on Computing, Communication and Security, ICCCS 2020, Patna, India, 14–16 October 2020, pp. 1–6. IEEE (2020)
10. Liu, Q., et al.: Fuzzy cognitive diagnosis for modelling examinee performance. ACM Trans. Intell. Syst. Technol. (TIST) **9**(4), 1–26 (2018)
11. Mack, D.L., Biswas, G., Koutsoukos, X.D., Mylaraswamy, D.: Using tree augmented Naive Bayes classifiers to improve engine fault models. In: Uncertainty in Artificial Intelligence: Bayesian Modeling Applications Workshop. Citeseer (2011)
12. Minn, S., Vie, J.J., Takeuchi, K., Kashima, H., Zhu, F.: Interpretable knowledge tracing: simple and efficient student modeling with causal relations. In: Proceedings of the AAAI Conference on Artificial Intelligence, pp. 12810–12818 (2022)
13. Minn, S., Yu, Y., Desmarais, M.C., Zhu, F., Vie, J.J.: Deep knowledge tracing and dynamic student classification for knowledge tracing. In: 2018 IEEE International Conference on Data Mining (ICDM), pp. 1182–1187. IEEE (2018)
14. Piech, C., et al.: Deep knowledge tracing. In: Advances in Neural Information Processing Systems, vol. 28 (2015)
15. Rabiner, L., Juang, B.: An introduction to hidden Markov models. IEEE ASSP Mag. **3**(1), 4–16 (1986)
16. Su, Y., et al.: Time-and-concept enhanced deep multidimensional item response theory for interpretable knowledge tracing. Knowl. Based Syst. **218**, 106819 (2021)
17. Wang, F., et al.: Neural cognitive diagnosis for intelligent education systems. In: Proceedings of the AAAI Conference on Artificial Intelligence, pp. 6153–6161 (2020)
18. Wang, F., et al.: Neuralcd: a general framework for cognitive diagnosis. IEEE Trans. Knowl. Data Eng. (2022)
19. Wang, W., Ma, H., Zhao, Y., Li, Z., He, X.: Tracking knowledge proficiency of students with calibrated Q-matrix. Expert Syst. Appl. **192**, 116454 (2022)
20. Wojtas, M., Chen, K.: Feature importance ranking for deep learning. Adv. Neural. Inf. Process. Syst. **33**, 5105–5114 (2020)
21. Wu, R., et al.: Cognitive modelling for predicting examinee performance. In: Twenty-Fourth International Joint Conference on Artificial Intelligence (2015)
22. Yang, H., et al.: A novel quantitative relationship neural network for explainable cognitive diagnosis model. Knowl.-Based Syst. **250**, 109156 (2022)
23. Zhang, J., Shi, X., King, I., Yeung, D.Y.: Dynamic key-value memory networks for knowledge tracing. In: Proceedings of the 26th International Conference on World Wide Web, pp. 765–774 (2017)

RoBrain: Towards Robust Brain-to-Image Reconstruction via Cross-Domain Contrastive Learning

Che Liu[1,2], Changde Du[1], and Huiguang He[1,2(✉)]

[1] Laboratory of Brain Atlas and Brain-Inspired Intelligence, State Key Laboratory of Multimodal Artificial Intelligence Systems, Institute of Automation, Chinese Academy of Science, Beijing, China
{liuche2022,changde.du,huiguang.he}@ia.ac.cn
[2] School of Future Technology, University of Chinese Academy of Sciences, Beijing, China

Abstract. With the development of neuroimaging technology and deep learning methods, neural decoding with functional Magnetic Resonance Imaging (fMRI) of human brain has attracted more and more attention. Neural reconstruction task, which intends to reconstruct stimulus images from fMRI, is one of the most challenging tasks in neural decoding. Due to the instability of neural signals, trials of fMRI collected under the same stimulus prove to be very different, which leads to the poor robustness and generalization ability of the existing models. In this work, we propose a robust brain-to-image model based on cross-domain contrastive learning. With deep neural network (DNN) features as paradigms, our model can extract features of stimulus stably and generate reconstructed images via DCGAN. Experiments on the benchmark *Deep Image Reconstruction* dataset show that our method can enhance the robustness of reconstruction significantly.

Keywords: Neural decoding · Brain-to-image reconstruction · fMRI · Contrastive learning

1 Introduction

Brain-to-image reconstruction is one of the neural decoding tasks [11,16], which aims to reconstruct and recover stimulus images using neural signals. It requires accurate measurement of brain activities, involving many neuroimaging techniques and methods, of which functional Magnetic Resonance Imaging (fMRI) is used commonly [12,13]. fMRI records brain activities by measuring the contrast of blood oxygen level dependent (BOLD) of human brain [1], which reflects the activation of neurons. With its high spatial resolution and no side effects, fMRI has become the main tool for studying brain-to-image reconstruction in non-invasive brain imaging techniques. As a result, we focus on the task of brain-to-image reconstruction based on fMRI in this paper.

In general, researchers use the same image to stimulate one subject multiple times and then collect multiple trials of fMRI for each stimulus. Theoretically, fMRI signals under the same stimulus store similar stimulus-related information and the results of brain-to-image reconstruction should be close. However, it is not the case. We reconstruct a few images separately from fMRI signals of different trials under the same stimulus. As shown in Fig. 1, there are large differences in the color and shape of those images indicating the instability of brain-to-image reconstruction.

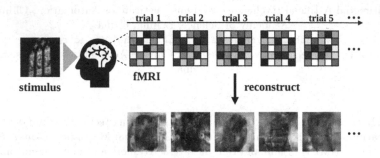

Fig. 1. Reconstruction from fMRI of different trials under the same stimulus. Since the neural activities of the same subject are quite different, cross-trial decoding is unstable, thus becoming a challenge.

The challenge is known as the lack of robustness in cross-trial neural decoding. When processing multiple trials of the same stimulus, previous works generally 1) take the average across all trials [5,17–19] or 2) consider each trial as an individual sample [17,18]. Models trained in this way perform poorly when confronted with extreme data or when applied to different subjects, which shows the poor robustness and generalization ability of existing models.

The primary challenge in cross-trial decoding arises from the presence of both external stimulus information and individual subject characteristics, along with experimental noise within fMRI signals. During the decoding process, it becomes essential to filter out irrelevant information while retaining stimulus-related information.

Inspired by the text-to-image task [22], we adopt the strategy for contrastive learning in this paper. Considering that there is a one-to-many relationship between stimulus images and fMRI signals, the information of stimulus stored in fMRI becomes the shared information for all trials of fMRI under the same stimulus. By employing a contrastive loss to encourage the proximity of decoded neural features from the same stimulus, the fMRI decoder can prioritize the extraction of shared features within the fMRI data, specifically those relevant to the stimulus. This approach diminishes the impact of noise, resulting in improved robustness and generalization capabilities of the brain-to-image model.

The robustness of neural decoding is of great significance to the development of brain-like intelligent algorithms [14], brain-computer interfaces (BCI) [6], diagnosis and treatment of brain diseases [2] and other fields. Researchers can explore the anatomical and functional connectivity of brain regions through neural decoding, thereby constructing the visual processing pathway of the human brain. In addition, if the corresponding technologies and methods are applied to BCI, mechanical equipment can be directly controlled by brain signals and advanced artificial intelligence of human-computer integration can be realized.

Our main contributions in this paper are listed as follows: 1) We build a complete brain-to-image reconstruction framework and verify the feasibility of our method. 2) We validate the effectiveness of our method for improving the robustness of brain-to-image reconstruction through qualitative and quantitative assessment. Codes are available at https://github.com/cheee2000/RoBrain.

2   Method

Our reconstruction method is divided into three steps: Firstly, train a brain decoder which decodes fMRI signals into neural features in the form of DNN features; Secondly, train GANs to recover images from DNN features; Thirdly, generate images from neural features via the trained GANs.

2.1   Neural Decoding

The goal of this stage is to train a brain decoder based on contrastive learning that can robustly extract stimulus-related features from fMRI signals. The training architecture is shown in Fig. 2, including a pre-trained image encoder and a brain decoder to be trained. Stimulus images and the corresponding fMRI data of all trials are fed into the network to extract image and neural features respectively. We establish a linear mapping from neural features to image features, enabling the brain decoder to learn the information of stimulus images from fMRI.

Image Encoder and Image Feature. A pre-trained ResNet50 [8] network is used in our paper as an image encoder. Features of its 7 main parts (5 convolution parts, 1 pooling layer and 1 linear layer) are extracted in turn and flattened into one dimension. In order to improve the regression accuracy, we perform PCA on each feature to reduce the dimension. Finally, all features are concatenated into a 7115-dimensional image feature. The DNN feature acts as a bridge in the following two stages, which can not only establish the mapping with the neural features but also recover stimulus images, which is of great significance to our brain-to-image method.

Brain Decoder. The role of the brain decoder is to decode neural features from the fMRI signals, which receives the preprocessed fMRI data and outputs the decoded features. The brain decoder used in this paper is a L2-regularized linear regression model. The dimension of the output layer is set to the dimension of image features. In this way, neural features are extracted and projected to the space of image features at the same time.

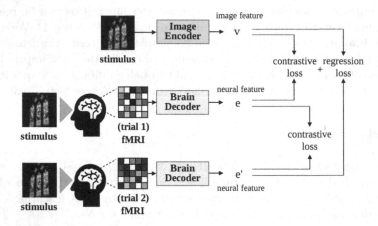

Fig. 2. Training architecture of brain decoder. *Cross-domain* and *fMRI-domain* contrastive losses are utilized during the regression from neural features to image features.

Loss Function. Given that e_{ij} is the neural feature decoded from the jth trial of the ith stimulus and v_i is the feature of the ith stimulus image, we define a regression loss with mean squared error as

$$L_{reg}(i,j) = \frac{1}{d} \|e_{ij} - v_i\|_2^2 \tag{1}$$

where $d = 7115$ is the dimension of image feature and neural feature. Moreover, we utilize two contrastive losses. The first one is computed across the domains of image and fMRI in order to make e_{ij} closer to v_i than other image features. Let $sim(a,b) = a^T b/\|a\|\|b\|$ denote cosine similarity between a and b. Then the loss function is defined as

$$L_1(i,j) = -log \frac{exp(sim(e_{ij}, v_i)/\tau_1)}{\sum_{p=1}^N exp(sim(e_{ij}, v_p)/\tau_1)} \tag{2}$$

where N is the batch size. The other is an intra-modal contrastive loss in the domain of fMRI, which is defined as

$$L_2(i,j) = -log \frac{\sum_{q=1}^T exp(sim(e_{ij}, e_{iq})/\tau_2)}{\sum_{p=1}^N \sum_{q=1}^T exp(sim(e_{ij}, e_{pq})/\tau_2)} \tag{3}$$

where T is the number of trials of a stimulus. In this way, we push neural features of different trials under the same stimulus together and neural features under different stimuli away. The training loss of brain decoder is the weighted addition of $L_{reg}(i,j)$, $L_1(i,j)$, and $L_2(i,j)$, which is defined as

$$L(i,j) = L_{reg}(i,j) + \lambda_1 L_1(i,j) + \lambda_2 L_2(i,j) \tag{4}$$

and the overall loss is the average across all trials of all stimuli, which is defined as

$$L = \frac{1}{NT} \sum_{i=1}^{N} \sum_{j=1}^{T} L(i,j). \tag{5}$$

2.2   Image Reconstruction

In this stage, we train a DCGAN [15] (Deep Convolutional Generative Adversarial Network) to recover the original images from the DNN features extracted above. The generator G contains 5 deconvolution layers, which only receives image features as input and outputs reconstructed images of size 64×64. Given that the brain-to-image task in this paper does not require the diversity of generated images, we do not take Gaussian noise as input. Correspondingly, the discriminator D with 5 convolution layers receives images as input and outputs probability values.

The discriminator D is optimized by an adversarial loss to improve the output probability of real images and reduce the probability of reconstructed images. For the generator G, on the one hand, generated images require to be similar to the original images in pixels. On the other hand, the generator is optimized to fool the output of the discriminator. Overall, we define the loss of G as a weighted sum of a MSE loss and an adversarial loss.

2.3   Brain-to-Image Reconstruction

In the first two stages, we train a brain decoder for neural feature extraction from fMRI, and a DCGAN for image reconstruction. We integrate them into a complete brain-to-image model to reconstruct stimulus images using fMRI signals. The reconstruction process can be simply described as: Firstly, fMRI features are extracted via the brain decoder, and then the neural features are input into the generator of DCGAN, reconstructing the stimulus images through the information stored in the features.

To further improve the quality of reconstructed images and the robustness of models, we fine-tune the brain decoder using paired data of stimulus images and fMRI with the generator fixed. The training architecture is shown in Fig. 3 where a pre-trained image encoder is adopted to extract features of reconstructed images. Except for the MSE loss between original and reconstructed images, an extra intra-modal contrastive loss in the domain of images is defined as

$$L_c(i,j) = -log \frac{\sum_{q=1}^{T} exp(sim(u_{ij}, u_{iq})/\tau)}{\sum_{p=1}^{N} \sum_{q=1}^{T} exp(sim(u_{ij}, u_{pq})/\tau)} \tag{6}$$

where u_{ij} denotes the feature of the image reconstructed from the jth trial of the ith stimulus. In this way, we push together the semantic information of images reconstructed from the same stimulus.

To simplify the computation of the gradient, we adopt a pre-trained Inception-V3 [20] model as the image encoder in this stage instead of ResNet50 to avoid PCA.

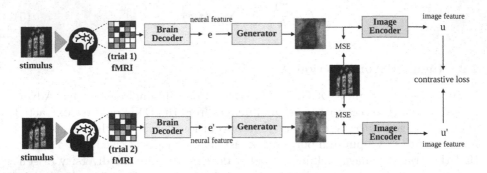

Fig. 3. Training architecture of brain-to-image model with paired data of stimulus images and fMRI. With generator and image encoder fixed, brain decoder is optimized by MSE loss and *image-domain* contrastive loss.

3   Experiments

3.1   Datasets

We evaluate our method on the dataset *Deep Image Reconstruction*[1] [19] or *DIR* for short. The *DIR* dataset was proposed by Kamitani Lab, including 1250 natural images and corresponding fMRI data. These images are totally selected from the *ImageNet* dataset [4]. Researchers used a 3T MRI scanner to collect fMRI signals of three healthy subjects (S1: age 33, male; S2: age 23, male; S3: age 23, female), covering brain regions of lower visual cortex V1-V4 and higher visual cortex LOC, FFA and PPA. For the training set of 1200 images, each image was presented 5 times, and for the test set of 50 images, each image was presented 24 times. As a result, each subject contains 6000 training trials and 1200 test trials of fMRI.

In addition to the *DIR* dataset, this paper collects an extra image set called *ImageNet20* dataset or *IN20* for short. The dataset is selected from the *ImageNet-1k* dataset where 20 images are chosen randomly from each of $1k$ categories, for a total of 20,000 images. The *IN20* dataset is superior to the *DIR* dataset in both categories and quantities of images, which can improve the quality of the images reconstructed by DCGAN.

---

[1] https://github.com/KamitaniLab/DeepImageReconstruction.

3.2   Experimental Settings

Hyperparameters. In our experiments, the hyperparameters in the loss functions are set as $\tau_1 = 4.0, \tau_2 = 0.2, \lambda_1 = 0.01, \lambda_2 = 0.1, \tau = 1.0$. We trained the brain decoder for 200 epochs using Adam optimizer with an initial learning rate of 8e−5, and 1% learning rate drop every epoch until 8e−6. We trained the DCGAN on the two datasets until it converged using Adam optimizer with a learning rate of 2e−4. Besides, we fine-tuned the brain decoder for only 10 epochs with a learning rate of 8e−6.

Baseline. Since we are the first to discuss the differences between trials of fMRI, making a direct comparison in robustness with SOTA methods is difficult. As a result, we set two baselines as follows, both of which share the same model architectures as ours, but use naive processing approaches mentioned in the introduction section.

- **Base1**: taking the average across all trials under the same stimulus as fMRI samples without any contrastive loss (CL) or fine-tuning.
- **Base2**: considering different trials as independent samples like ours, but no use of CL or fine-tuning.

3.3   Evaluation Metric

We choose MSE (mean squared error), SSIM (structural similarity index measure) and PSM (perceptual similarity metric) as metrics to evaluate the quality of reconstructed images. MSE measures the differences between pixels of two images while SSIM captures the global information in three aspects of luminance, contrast and structure [21], both of which evaluate the lower-level information of images. Studies [7,10] find out that higher-level perceptual metric is more similar to humans' judgments. Following Zhang et al. [23], we define PSM with a pre-trained AlexNet [9] as

$$PSM(x,\hat{x}) = \sum_l \frac{1}{H_l W_l} \sum_{h,w} \|w_l \odot (f_x^l - f_{\hat{x}}^l)\|_2^2 \tag{7}$$

where $x, \hat{x}$ denote the original image and the reconstructed image, $f_x^l, f_{\hat{x}}^l \in \mathbb{R}^{H_l \times W_l \times C_l}$ denote features of layer l and $w_l \in \mathbb{R}^{C_l}$ denotes channelwise activations.

Table 1. Reconstruction results of DCGAN on the two datasets. We consider the results of image reconstruction as the upper bound for brain-to-image reconstruction.

Dataset	MSE↓	SSIM↑	PSM↓
DIR	0.0315	0.434	0.478
IN20	0.0255	0.449	0.303

3.4   Reconstruction Results

DCGAN converges after 400 epochs with the training of *DIR* dataset, which we call *DIR-DCGAN*. As a comparison, DCGAN converges after 200 epochs with the training of *IN20* dataset, which is called *IN20-DCGAN*. We then evaluate both models by calculating MSE, SSIM and PSM on the test set of *DIR* dataset. As shown in Table 1, *IN20-DCGAN* behaves better in all three metrics than *DIR-DCGAN*. Therefore, we consider reconstruction results of *IN20-DCGAN* as an upper bound for brain-to-image reconstruction. It can be seen that both low-level information and semantic information is presented better using the larger dataset.

Since 24 trials of fMRI data are collected for each stimulus in the *DIR* dataset, each original image corresponds to 24 reconstructed images. Examples of reconstructed images are shown in Fig. 4.

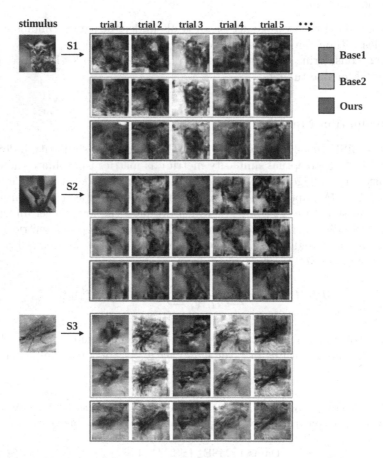

Fig. 4. Reconstruction results for three subjects. The first and second rows for each subject show the reconstructed images in **Base1** and **Base2**. The last row shows the reconstructed images in our method which are lower in noise and more similar.

We calculate MSE, SSIM and PSM of each pair of reconstructed image and original stimulus image, and then count the mean and standard deviation of 24 values under the same stimulus. After that, we calculate the average of the means and standard deviations of 50 stimuli respectively. Results are shown in Table 2, where metrics are displayed in the form of mean ± standard deviation.

Table 2. Average results across all stimuli in brain-to-image reconstruction. The standard deviations of all reconstructed images under the same stimulus are reduced by 20.5%, 21.3% and 11.6% at least in MSE, SSIM and PSM compared with Base1 and Base2, while the means are also optimized. The robustness of the brain-to-image reconstruction is enhanced.

	MSE↓	SSIM↑	PSM↓
Base1	0.0970 ± .0331	0.2363 ± .0560	0.4990 ± .0602
+fine-tune	0.0862 ± .0272	0.2519 ± .0464	0.5107 ± .0533
Base2	0.0969 ± .0340	0.2434 ± .0488	0.4845 ± .0611
+CL	0.0972 ± .0313	0.2411 ± .0440	**0.4768 ± .0563**
+fine-tune	0.0847 ± .0271	0.2588 ± .0412	0.5025 ± .0545
Ours	**0.0840 ± .0263**	**0.2594 ± .0384**	0.4912 ± .0532

Qualitative and quantitative results indicate that fine-tuning can reduce the noise of reconstructed images, and contrastive learning can enhance the robustness of reconstruction significantly with the quality of images remaining the same or slightly optimized, all of which demonstrates the effectiveness of our method.

3.5 n-way Identification Task

In order to further illustrate that our method maintains the quality of reconstructed images while enhancing the robustness, we perform an n-way identification task on the *DIR* dataset and compare the results with several state of the art methods. In the task, each reconstructed image is required to identify the ground truth from a set of n randomly selected candidate images. We choose SSIM and PSM as metrics and calculate the average identification accuracy across three subjects.

State of the art methods [16] on the *DIR* dataset include: 1) encoder-decoder based method by Beliy et al. [3], DNN-based methods 2) *ShenDNN* and 3) *ShenDNN+DGN* by Shen et al. [19], 4) GAN-based method by Shen et al. [18] and 5) shape-semantic GAN based method by Fang et al. [5]. We set $n = 2, 5, 10, 50$ (full) and compute the n-way identification accuracy.

As shown in Fig. 5, accuracy based on SSIM performs well, second only to *ShenDNN* while accuracy based on PSM is close to other methods as well, which indicates that our method increases the robustness without losing the quality of reconstruction.

3.6   Further Discussion

In our method, contrastive learning is used to make the fMRI decoder more inclined to extract the common features of fMRI. In the process of decoding, the stimulus-related information will be retained, with the irrelevant information discarded.

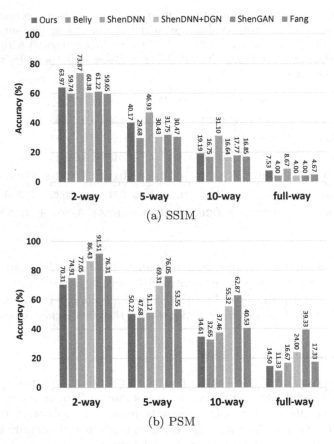

Fig. 5. n-way identification accuracy based on (a) SSIM and (b) PSM on the *DIR* dataset.

To verify this, we perform experiments on decoded neural features. We choose MSE and PCC (Pearson correlation coefficient) to measure the similarity between neural and image features. Moreover, we propose a new metric **SIM** to measure the similarity between neural features of different trials under the same stimulus, which is defined as

$$SIM = \frac{1}{NT^2} \sum_{k=1}^{N} \sum_{i=1}^{T} \sum_{j=1}^{T} sim(e_{ki}, e_{kj}) \tag{8}$$

where $sim(a, b) = a^T b / \|a\| \|b\|$ denotes cosine similarity.

As shown in Table 3, MSE and PCC remain almost unchanged while SIM is significantly improved, which shows that the robustness of decoding is enhanced.

Table 3. Results on the decoded neural features. The similarity of the decoded features under the same stimulus is increased significantly, which indicates that the robustness of cross-trial decoding is enhanced.

	MSE↓	PCC↑	SIM↑
Base1	92.41	0.1886	0.2439
Base2	60.14	0.2443	0.3600
Ours	54.62	0.2162	**0.5196**

4   Conclusion and Future Work

In this paper, we propose a 3-stage brain-to-image method and successfully reconstruct images from fMRI signals. To improve the robustness of the reconstruction, we adopt three strategies for contrastive learning during the training. We evaluate our model on *Deep Image Reconstruction* dataset. The results show that our method can improve the robustness of neural decoding and make the reconstructed images under the same stimulus more similar.

The future work includes several aspects: 1) introducing the contrastive learning of image categories, so that the brain decoder can decode more high-level semantic features; 2) applying our method to more subjects and more DNNs; 3) modularizing our method and applying it to other neural decoding tasks.

Acknowledgements. This work was supported in part by the National Key R&D Program of China 2022ZD0116500; in part by the National Natural Science Foundation of China under Grant 62206284, Grant 62020106015 and Grant 82272072; and in part by the Beijing Natural Science Foundation under Grant J210010.

References

1. Bandettini, P.A.: Twenty years of functional MRI: the science and the stories. Neuroimage **62**(2), 575–588 (2012)
2. Beauchamp, M.S., et al.: Dynamic stimulation of visual cortex produces form vision in sighted and blind humans. Cell **181**(4), 774–783 (2020)
3. Beliy, R., Gaziv, G., Hoogi, A., Strappini, F., Golan, T., Irani, M.: From voxels to pixels and back: self-supervision in natural-image reconstruction from fMRI. In: Advances in Neural Information Processing Systems, vol. 32 (2019)
4. Deng, J., Dong, W., Socher, R., Li, L.J., Li, K., Fei-Fei, L.: Imagenet: a large-scale hierarchical image database. In: Proceedings of the IEEE Conference on Computer Vision and Pattern Recognition, pp. 248–255 (2009)

5. Fang, T., Qi, Y., Pan, G.: Reconstructing perceptive images from brain activity by shape-semantic GAN. Adv. Neural. Inf. Process. Syst. **33**, 13038–13048 (2020)
6. Fernández, E., et al.: Visual percepts evoked with an intracortical 96-channel microelectrode array inserted in human occipital cortex. J. Clin. Invest. **131**(23) (2021)
7. Güçlütürk, Y., Güçlü, U., Seeliger, K., Bosch, S., van Lier, R., van Gerven, M.A.: Reconstructing perceived faces from brain activations with deep adversarial neural decoding. In: Advances in Neural Information Processing Systems, vol. 30 (2017)
8. He, K., Zhang, X., Ren, S., Sun, J.: Deep residual learning for image recognition. In: Proceedings of the IEEE Conference on Computer Vision and Pattern Recognition, pp. 770–778 (2016)
9. Krizhevsky, A., Sutskever, I., Hinton, G.E.: Imagenet classification with deep convolutional neural networks. In: Advances in Neural Information Processing Systems, vol. 25 (2012)
10. Mozafari, M., Reddy, L., VanRullen, R.: Reconstructing natural scenes from fMRI patterns using BigBiGAN. In: 2020 International Joint Conference on Neural Networks (IJCNN), pp. 1–8. IEEE (2020)
11. Naselaris, T., Kay, K.N., Nishimoto, S., Gallant, J.L.: Encoding and decoding in fMRI. Neuroimage **56**(2), 400–410 (2011)
12. Nestor, A., Lee, A.C., Plaut, D.C., Behrmann, M.: The face of image reconstruction: progress, pitfalls, prospects. Trends Cogn. Sci. **24**(9), 747–759 (2020)
13. Poldrack, R.A., Farah, M.J.: Progress and challenges in probing the human brain. Nature **526**(7573), 371–379 (2015)
14. Pulvermüller, F., Tomasello, R., Henningsen-Schomers, M.R., Wennekers, T.: Biological constraints on neural network models of cognitive function. Nat. Rev. Neurosci. **22**(8), 488–502 (2021)
15. Radford, A., Metz, L., Chintala, S.: Unsupervised representation learning with deep convolutional generative adversarial networks. arXiv preprint arXiv:1511.06434 (2015)
16. Rakhimberdina, Z., Jodelet, Q., Liu, X., Murata, T.: Natural image reconstruction from fMRI using deep learning: a survey. Front. Neurosci. **15**, 795488 (2021)
17. Seeliger, K., Güçlü, U., Ambrogioni, L., Güçlütürk, Y., van Gerven, M.A.: Generative adversarial networks for reconstructing natural images from brain activity. Neuroimage **181**, 775–785 (2018)
18. Shen, G., Dwivedi, K., Majima, K., Horikawa, T., Kamitani, Y.: End-to-end deep image reconstruction from human brain activity. Frontiers in Computational Neuroscience **13**, 21 (2019)
19. Shen, G., Horikawa, T., Majima, K., Kamitani, Y.: Deep image reconstruction from human brain activity. PLoS Comput. Biol. **15**(1), e1006633 (2019)
20. Szegedy, C., Vanhoucke, V., Ioffe, S., Shlens, J., Wojna, Z.: Rethinking the inception architecture for computer vision. In: Proceedings of the IEEE Conference on Computer Vision and Pattern Recognition, pp. 2818–2826 (2016)
21. Wang, Z., Bovik, A.C., Sheikh, H.R., Simoncelli, E.P.: Image quality assessment: from error visibility to structural similarity. IEEE Trans. Image Process. **13**(4), 600–612 (2004)
22. Ye, H., Yang, X., Takac, M., Sunderraman, R., Ji, S.: Improving text-to-image synthesis using contrastive learning. arXiv preprint arXiv:2107.02423 (2021)
23. Zhang, R., Isola, P., Efros, A.A., Shechtman, E., Wang, O.: The unreasonable effectiveness of deep features as a perceptual metric. In: Proceedings of the IEEE Conference on Computer Vision and Pattern Recognition, pp. 586–595 (2018)

High-Dimensional Multi-objective PSO Based on Radial Projection

Dekun Tan, Ruchun Zhou[✉], Xuhui Liu, Meimei Lu, Xuefeng Fu, and Zhenzhen Li

School of Information Engineering, Nanchang Institute of Technology, Nanchang 330099, Jiangxi, China
rczhou_nit@163.com

Abstract. When solving multi-objective problems, traditional methods face increased complexity and convergence difficulties because of the increasing number of objectives. This paper proposes a high-dimensional multi-objective particle swarm algorithm that utilizes radial projection to reduce the dimensionality of high-dimensional particles. Firstly, the solution vector space coordinates undergo normalization. Subsequently, the high-dimensional solution space is projected onto 2-dimensional radial space, aiming to reduce computational complexity. Following this, grid partitioning is employed to enhance the efficiency and effectiveness of optimization algorithms. Lastly, the iterative solution is achieved by utilizing the particle swarm optimization algorithm. In the process of iteratively updating particle solutions, the offspring reuse-based parents selection strategy and the maximum fitness-based elimination selection strategy are used to strengthen the diversity of the population, thereby enhancing the search ability of the particles. The computational expense is significantly diminished by projecting the solution onto 2-dimensional radial space that exhibits comparable characteristics to the high-dimensional solution, while simultaneously maintaining the distribution and crowding conditions of the complete point set. In addition, the offspring reuse-based parents selection strategy is used to update the external archive set, further avoiding premature convergence to local optimal solution. The experimental results verify the effectiveness of the method in this paper. Compared with four state-of-the-art algorithms, the algorithm proposed in this paper has high search efficiency and fast convergence in solving high-dimensional multi-objective optimization problems, and can also obtain higher quality solutions.

Keywords: High-dimensional multi-objective particle swarm · Radial projection · Parent selection strategy · Elimination selection strategy

1 Introduction

High-dimensional multi-objective optimization problems have extensive applications in various disciplines and industries, but facing big challenge to solve it. To address this challenge, swarm intelligence algorithms have emerged as effective approaches. Among them, the particle swarm algorithm, proposed by Kennedy and Eberhart [1], is a population-based global optimization algorithm. It utilizes individual particles and

B. Luo et al. (Eds.): ICONIP 2023, LNCS 14449, pp. 239–252, 2024.
https://doi.org/10.1007/978-981-99-8067-3_18

global optimal positions to determine search directions and movement speeds, iteratively generating new particle swarms to solve the problem. The algorithm excels in both local and global search capabilities.

Many scholars have conducted research on the high-dimensional multi-objective optimization problem, employing various strategies to enhance algorithmic performance. Lin proposed the NMPSO [2] algorithm, which compares the merits of the generated variant particles with the current particles and selects the better ones, but the processing of the multi-peak function is poor. Zhang [3] proposed a decomposition-based multi-objective evolutionary algorithm, in the case of high-dimensional complexity, it is difficult to set a reasonable weight vector. NSGAIII proposed by M. Carvalho [4] is a reference point-based method for solving multi-objective optimization problems, and the solutions obtained from solving high-dimensional objectives is easily fall into local optimum. Wang proposed MaOEA with two archives [5], which achieved good results in terms of convergence, complexity, and diversity, but did not work well on a large number of targets. Most improved high-dimensional multi-objective particle swarm algorithms fail to achieve a satisfactory balance between convergence and diversity when addressing complex and challenging high-dimensional multi-objective optimization problems [6], often resulting in being trapped in local optima. The algorithm proposed in this paper preserves particle diversity while reducing algorithm complexity.

To address the issue of selecting optimal solutions from a large pool of non-dominated solutions in high-dimensional multi-objective problems, this study incorporates a radial projection mechanism for particle dimensionality reduction. By considering the fitness values of the particles, the algorithm performs parent selection strategy and elimination strategy, aiming to enhance local search capability while preserving population diversity and avoiding premature convergence.

2 High-Dimensional Multi-objective PSO Based on Radial Space Projection (RHMPSO)

2.1 High-Dimensional Multi-objective Problem

When the number of objectives in a multi-objective optimization problem exceeds four, the difficulty of solving the problem significantly increases. Such problems are referred to as high-dimensional multi-objective problems [7]. Without loss of generality, considering the minimization case, the general form of a high-dimensional multi-objective problem is defined in Eq. (1).

$$\begin{cases} MinF(X) = (f_1(X), f_2(X), \cdots, f_m(X))^T \\ s.t. g_i(x) \geq 0, i = 1, 2, \cdots, p \\ h_j(x) = 0, j = 1, 2, \cdots, q \end{cases} \tag{1}$$

where $m \geq 4$, Ω is the decision space, $X = (x_1, \cdots x_n) \in \Omega$ is the decision variable, $g_i(x) \geq 0$ is the inequality constraint condition, $h_j(x) = 0$ is the equality constraint condition.

2.2 Radial Projection

For high-dimensional multi-objective problems, traditional multi-objective optimization algorithms based on Pareto dominance sorting is difficult to find the approximate set of Pareto optimal solutions. Therefore, the use of evolutionary algorithm to solve high-dimensional multi-objective optimization problems has become a current research hotspot. Dimensionality reduction is an effective approach to reduce algorithm complexity. In this regard, the algorithm introduces a radial projection mechanism to achieve dimensionality reduction of high-dimensional particles.

Radial projection was initially applied in the field of bioinformatics and later introduced to the field of multi-objective optimization [8–11]. Radial projection enables the visualization of high-dimensional individuals' neighborhood relationships in a two-dimensional space. Let's consider two particles, F_1 and F_2, belongs to a m-dimensional objective space, their corresponding radial space coordinates are Y_1 and Y_2, and the Euclidean distance between them in the objective space is denoted as $E_F = \|F_1 - F_2\|$. Similarly, the Euclidean distance between Y_1 and Y_2 in the projected space is denoted as $E_Y = \|Y_1 - Y_2\|$. Due to E_Y linearly approximates the objective space distance E_F, and hence the distribution of projected individuals can effectively reflect the spatial distribution of the original objective individuals [12].

The representation of particle projection from a m-dimensional objective space onto 2-dimensional plane is shown in Eq. (2).

$$f(\theta_k) = \sum_{k=0}^{k-1} p(\gamma_k \cos\theta, \gamma_k \sin\theta) \tag{2}$$

When the projection center is $(0, 0)$, the projection radius is γ, the coordinates of point p can be expressed as $p(m, n) = (\gamma\cos\theta, \gamma\cos\theta)$, θ is the direction of the projection. The projection coordinates are $(\gamma_k \cos(2\pi/N), \gamma_k \sin(2\pi/N))$. Here, N denotes all non-projection directions, and k refers to the multi-scale parameter, which indicates the different projection radius. Normalizing data features can eliminate the effects of different feature scales and make the features comparable. The normalized numerical features are more stable, and thus the model converges more easily. Therefore, we first normalize the particle initialization position [13], mapping the particles to the range of [0,1] to achieve proportional scaling of the original data, the normalized process is shown in Eq. (3), where the initial velocities of the particles are all set as 0.

$$x_i' = \frac{x_i - x_{\min}}{x_{\max} - x_{\min}} \tag{3}$$

where x_i, $x_{\min}$, and $x_{\max}$ represent the ith individual in the population, the minimum and maximum vectors formed by the objective function of the population individuals, respectively.

The target vector is projected from the high-dimensional target space to the 2-dimensional radial space by projection and normalization. Then, the distribution of each individual in the projected 2-dimensional radial space is evaluated by Eq. (4) to calculate the crowding distance between particles, and the larger the crowding distance is, the better the distribution is.

$$Crow = Crowd \sum_{i=1}^{N} |x_i' - x_{i-1}'| \tag{4}$$

At this point, each non-inferior solution can be mapped to 2-dimensional radial space by convergence and distribution.

The 2-dimensional radial space is divided into equal-sized grids [14], and the projected particles are assigned to these grids. To ensure neighborhood coverage, the grid division is adjusted by determining the area of the rectangle occupied by each individual in the radial space based on the upper and lower bounds of the projected solutions. This allows for the projection plane to be divided into equidistant grids. The label G of each solution in the radial grid is calculated by Eq. (5).

$$G = \lfloor S(Y - B_l)/S(B_u - B_l) \rfloor \tag{5}$$

where, B_l and B_u are the lower and upper bounds of the radial coordinates. S is the spacing, determined by the population size N.

To calculate the fitness evaluation value F of the particle, as shown in Eq. (6).

$$F = e^{1/Crowd} \tag{6}$$

According to Eq. (2)–Eq. (6), the normalized target space individual can be projected onto 2-dimensional radial space and the projected area can be equally divided into different regions. Figure 1 gives a schematic diagram of 3-dimensional radial projection and the final projection result.

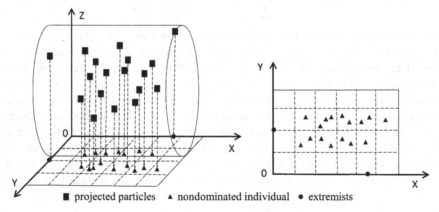

■ projected particles   ▲ nondominated individual   • extremists

Fig. 1. 3-Dimensional radial projection and Radial space particle position distribution.

By radial projection, the high-dimensional problem is downscaled to 2-dimensional radial space. After the particles undergo a projection transformation, they are displayed as two-dimensional coordinates, which can be readily utilized in the particle updating formula. This simplifies the particle update process and facilitates the iterative solution of the problem by using the particle swarm algorithm. In this paper, the generative adversarial network (GAN) [15] is used to reconstruct the data after dimensionality reduction.

2.3 Global Extreme Value Selection Strategy

After projecting high-dimensional objective space data onto the 2-dimensional radial space, the velocity of each particle can be set to the projected 2-dimensional velocity vector. This allows the particles to navigate within the projected space in search of the global optimal solution. In the standard PSO algorithm, the global optimal solution g_{best} guides the particles towards the Pareto front. However, traditional PSO algorithms encounter challenges such as low accuracy and slow search speed when solving high-dimensional multi-objective problems. To address these challenges and enhance the convergence and diversity of the algorithm, this approach incorporates the offspring reuse-based parents selection strategy [16] and the maximum fitness-based elimination selection strategy [17] to adjust the global extremes. By employing these two strategies simultaneously, the algorithm accelerates the convergence rate and increases the robustness, thereby improving its performance in high-dimensional multi-objective search. In particle swarm optimization algorithm, each particle i represents a potential solution to a multi-objective problem. It moves at a certain speed within the search space and dynamically adjusts its flight speed based on collective flight experience. The particle i which undergoes parent selection strategy and elimination selection strategy then updated by the following velocity and position update Eq. (7) and Eq. (8).

$$v_{i,d}^{(K+1)} = \omega v_{i,d}^{(K)} + c_1 r_1 (p_{i,d} - v_{i,d}^{(K)}) + c_2 r_2 (p_{g,d} - v_{i,d}^{(K)}) \tag{7}$$

According to Eq. (7), it can be observed that the movement of particle i is influenced by both its own historical best position and the global best position. This influence occurs during the selection of the global extremes, which updates the historical best p_{best} and the global best g_{best} for each individual.

$$x_{i,d}^{(K+1)} = x_{i,d}^{(K)} + v_{i,d}^{(K+1)} \tag{8}$$

where, ω represents the inertia weight, $v_{i,d}^{(K)}$ and $x_{i,d}^{(K)}$ denote the velocity and position of the d-dimensional of particle i at the Kth iteration respectively. In this paper, $d = 2.p_i$ and p_g represent the historical best position of particle i and the global best guide position. c_1 and c_2 are learning factors, while r_1 and r_2 are random values uniformly distributed within the range [0, 1].

The implementation process of global extremum selection strategy: calculate the fitness of all particles and find the particle with the highest fitness as the current particle optimal solution. For the particles with relatively high fitness, the parent selection strategy of offspring reuse is used for further optimization. For the remaining particles, the particles with fitness values exceeding a preset threshold are eliminated through elimination selection strategy. To better control the probability of the ith individual being selected in parent selection, a sorting selection mechanism is used. After each iteration, particles are sorted from big to small according to their fitness values, and the probability of the ith particle is denoted as $L(i)$, as shown in Eq. (9).

$$L(i) = h(1 - h)^{i-1} \tag{9}$$

where h is a random number between 0 and 1.

If the two parents in the process of solving Z_1 and Z_2 are locally optimal solutions, and the parent individual Z_1 has a better fitness, then the optimal individuals is found in the region with the better fitness value. In this case, neighborhood crossover can be used to generate offspring individual Z_3, and specifically, Eq. (10) can be used to generate the offspring individual Z_3.

$$Z_3 = \lambda(Z_1 - Z_2) + Z_2 \tag{10}$$

where λ is the random number between $(0, 1)$.

In order to enhance the algorithm's ability to search for non-inferior solutions, the non-inferior solution individuals in the radial space are explored using the offspring reuse-based parents selection strategy. At first, the parent population is initialized. Then, the offspring population is generated by crossover and mutation. Offspring individuals' fitness is evaluated, and a subset of parents is selected. Each offspring individual's fitness is compared to the selected parents. If it is non-dominated, it joins the parent population; otherwise, it is discarded. This process is repeated until the desired number of offspring individuals is reached. Then the maximum fitness-based elimination selection strategy is added to retain the good individuals and eliminate the particles with fitness greater than a threshold, followed by an initial method to supplement the number of eliminations, keeping the number of individuals in the population constant and enhancing the diversity of the population. Global extreme selection is shown in Fig. 2.

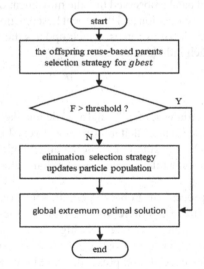

Fig. 2. Flow chart for global extreme selection.

In our improved algorithm, the offspring reuse-based parents selection strategy enhances the algorithm's capability to explore and search for solutions, thereby preventing premature convergence of the search individuals. The maximum fitness-based elimination selection strategy helps control the size of the particle swarm, reduces computational costs, and improves convergence speed. This elimination selection strategy preserves excellent individuals while promptly supplementing missing individuals, thereby increasing population diversity to some extent.

2.4   Archive Maintenance Strategy

The external archive set stores global optimal solutions and is shared among all particles. The process of updating the external archive set using the parent selection strategy involves several steps. Firstly, the external archive set is initialized, denoted as *Rep*. Then, the fitness of the solutions in the external archive set is evaluated. Next, the parent population is initialized. Offspring individuals are generated through crossover and mutation operations, and their fitness is evaluated. A subset of parent individuals is selected. For each offspring individual, its fitness is compared to the fitness of the selected parents. If the offspring individual is non-dominated by any of the selected parents, it is added to the parent population. If the offspring individual is dominated by any of the selected parents, it is discarded. The external archive set is then updated by comparing the offspring individuals with the solutions in the external archive set, performing non-dominated sorting, and adding non-dominated offspring individuals to the external archive set while removing dominated solutions. This process is repeated until the desired number of offspring individuals is selected, typically in each iteration or generation of the algorithm. The archive maintenance strategy is shown in Fig. 3.

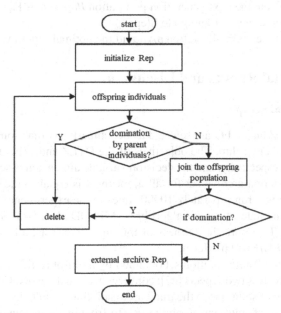

Fig. 3. The flowchart of archive maintenance strategy.

3   The Workflow of RHMPSO Algorithm

Based on the above ideas, the main steps of RHMPSO algorithm are described as follows:

Step1: Initialize relevant parameters, including setting the population size N; initializing the population Pop; current iteration number gen; maximum iteration number gen_{max}.

Step2: Project the high-dimensional solution vector space coordinates radially onto 2-dimensional radial space.

Step3: Calculate the fitness value F of the individuals in Pop after projection.

Step4: Set $gen = 1$.

Step5: If $gen = gen_{max}$, go to Step6, otherwise go to Step14.

Step6: Remove the individuals with the lowest fitness in Pop_{sorted} to obtain the offspring population $Pop_{offspring}$.

Step7: Merge selected Pop_{parent} and offspring population $Pop_{offspring}$ to obtain the new population Pop_{new}.

Step8: Update the velocity and position of the individuals in Pop_{new} using Eq. (7) and Eq. (8).

Step9: Update the fitness value F of the current individuals.

Step10: Select two groups with the highest fitness values for crossover and mutation.

Step11: Merge the mutated group with the PSO particles to form the new individuals Pop_{new}.

Step12: Maintain the external archive of the merged population, delete redundant solutions, and generate the next generation population $Pop(gen + 1)$.

Step13: Set $gen = gen + 1$ and go to Step5.

Step14: Output the optimal solution p_{best} and the optimal fitness value F_{best}.

4   Experimental Results and Discussion

4.1   Experimental Setup

In this paper, DTLZ and WFG are selected as benchmark test functions [18] to test the effectiveness of the algorithm. The objectives of the DTLZ and WFG functions are set to 3, 4, 6, and 10, respectively, and the decision variable dimension is set to $M - 1 + 10$. The size of the external archive set for all algorithms is equal to the population size. Each test function was run separately 10000 times in the experiment. The algorithm is compared with four state-of-the-art algorithms: MPSOD [19], NMPSO, SPEA2 [20], and IDBEA [21]. To ensure the fairness of the algorithm tests, all experiments were conducted on the *PlatEMO* platform.

Evaluation of the Pareto solution set obtained by multi-objective optimization algorithms typically involves two aspects [22]: convergence and diversity. Convergence measures the proximity of solutions to the true Pareto solutions, while diversity assesses the uniform distribution of solutions along the Pareto frontier. To evaluate the overall performance of the RHMPSO algorithm, this paper uses the inverse generational distance (IGD) [23] and hypervolume (HV) [24] to evaluate the comprehensive performance of the algorithm. Among the two indicators, IGD calculates the average Euclidean distance between all solutions on the true Pareto frontier and the non-dominated solutions obtained by the optimization algorithm. A smaller IGD value indicates a closer proximity of the non-dominated solution set to the true Pareto frontier, as well as a more evenly distributed solution set, indicating better convergence and diversity. HV, measures the volume of the objective space dominated by at least one non-dominated solution in the non-dominated solution set. HV effectively captures the convergence and diversity of the algorithm. Table 1 and Table 2 present a comparative analysis of the average IGD and HV values for five algorithms on various dimensions of the DTLZ test function and WFG1 test function.

Table 1. Average IGD values obtained by five algorithms on test problems.

Problem	M	RHMPSO	MPSOD	NMPSO	SPEA2	IDBEA
DTLZ2	3	**5.4752e−2** **(5.96e−4)+**	5.7095e−2 (6.01e−4)−	7.6601e−2 (2.75e−3)−	5.7803e−2 (3.93e−4)−	5.6151e−2 (7.55e−4)
	4	1.6422e−1 (3.62e−3)−	1.4337e−1 (1.31e−3)−	1.5703e−1 (2.83e−3)−	**1.3686e−1** **(2.20e−3)+**	1.4256e−1 (1.04e−3)
	6	3.3511e−1 (7.38e−3)−	3.0188e−1 (4.14e−3)−	2.9328e−1 (2.93e−3)−	9.8874e−1 (1.33e−1)−	**2.8673e−1** **(2.88e−3)**
	10	**5.4347e−1** **(8.45e−3)+**	5.6749e−1 (1.29e−2)+	5.4489e−1 (7.70e−2)+	2.3641e+0 (2.19e−1)−	1.2417e+0 (5.10e−5)
DTLZ6	3	**4.1142e−3** **(1.03e−4)+**	2.7766e−1 (2.39e−1)+	1.3829e−2 (2.43e−3)+	5.9338e−3 (4.59e−3)+	3.1201e−1 (4.60e−1)
	4	**2.5129e−2** **(2.59e−3)+**	5.5193e−1 (3.34e−1)+	4.0174e−2 (1.83e−2)+	1.0072e+0 (6.92e−1)=	9.6669e−1 (6.80e−1)
	6	**3.9568e−2** **(1.38e−2)+**	8.2281e−1 (4.30e−1)+	9.2392e−2 (9.00e−2)+	8.9565e+0 (6.56e−1)−	2.5390e+0 (1.34e+0)
	10	**7.5612e−2** **(2.82e−2)+**	1.2210e+0 (5.56e−1)+	7.3853e−1 (1.95e−2)+	9.7346e+0 (1.23e−1)=	8.8931e+0 (1.80e+0)
WFG1	3	**5.4532e−1** **(7.63e−2)+**	1.8792e+0 (7.22e−2)−	1.0683e+0 (8.16e−2)−	6.4407e−1 (7.25e−2)+	9.0125e−1 (1.22e−1)
	4	**7.4468e−1** **(5.75e−2)+**	2.1040e+0 (6.25e−2)−	1.3004e+0 (1.03e−1)−	8.1253e−1 (1.28e−1)+	1.0390e+0 (1.56e−1)
	6	**1.2600e+0** **(1.18e−1)+**	2.4706e+0 (6.71e−2)−	1.9512e+0 (2.01e−1)−	1.8207e+0 (1.54e−1)−	1.3868e+0 (2.25e−1)
	10	**1.9291e+0** **(1.51e−1)+**	3.1410e+0 (6.21e−2)+	2.4012e+0 (3.37e−1)=	2.9639e+0 (2.20e−1)+	4.4299e+0 (4.00e+0)

From Table 1, it can be observed that RHMPSO achieves the highest number of optimal IGD values in the same experimental environment, particularly in the experiments of DTLZ6 and WFG1. The experimental data of RHMPSO significantly outperforms the results of the other four algorithms. In the experiments of DTLZ2, RHMPSO attained greater IGD values under 3 and 10 objectives. From Table 2, RHMPSO achieved 8 optimal HV values, while NMPSO, SPEA2 and IDBEA obtained 2, 1, and 1 optimal

Table 2. Average HV values obtained by five algorithms on test problems.

Problem	M	RHMPSO	MPSOD	NMPSO	SPEA2	IDBEA
DTLZ2	3	**5.5948e−1** **(1.53e−3)+**	5.4591e−1 (2.39e−3)−	5.5337e−1 (1.18e−3)+	5.5222e−1 (1.74e−3)=	5.5172e−1 (1.67e−3)
	4	**6.9465e−1** **(1.71e−3)+**	6.5583e−1 (4.58e−3)−	6.6452e−1 (4.04e−3)−	6.6041e−1 (6.63e−3)−	6.7680e−1 (4.10e−3)
	6	**8.3796e−1** **(3.15e−3)+**	6.9692e−1 (1.42e−2)−	7.7414e−1 (1.01e−2)−	1.2313e−2 (2.14e−2)−	7.9555e−1 (7.49e−3)
	10	**9.1606e−1** **(3.69e−2)+**	7.2851e−1 (3.81e−2)+	8.3585e−1 (1.45e−2)+	4.5866e−5 (2.51e−4)−	9.0852e−2 (7.47e−5)
DTLZ6	3	**2.0001e−1** **(1.76e−4)+**	9.4770e−2 (6.24e−2)−	1.9630e−1 (7.86e−4)+	1.9888e−1 (2.28e−3)+	1.1693e−1 (8.07e−2)
	4	1.2684e−1 (1.14e−2)+	6.1406e−2 (5.54e−2)+	**1.4235e−1** **(1.60e−3)+**	4.2506e−3 (1.85e−2)=	1.5145e−2 (3.41e−2)
	6	9.9183e−2 (8.70e−3)+	2.4771e−2 (4.10e−2)+	**1.0438e−1** **(4.15e−3)+**	0.0000e+0 (0.00e+0)=	2.7185e−5 (1.48e−4)
	10	**9.0909e−2** **(0.00e+0)+**	1.8352e−2 (3.73e−2)+	5.8390e−2 (1.55e−2)+	0.0000e+0 (0.00e+0)=	0.0000e+0 (0.00e+0)
WFG1	3	6.6692e−1 (3.59e−2)+	9.1880e−2 (3.19e−2)−	4.2640e−1 (3.64e−2)−	**6.9950e−1** **(4.25e−2)+**	5.2926e−1 (4.99e−2)
	4	**7.5233e−1** **(4.72e−2)+**	1.8017e−1 (1.96e−2)−	4.5178e−1 (3.74e−2)−	6.8884e−1 (6.13e−2)+	5.9501e−1 (6.60e−2)
	6	**6.1736e−1** **(1.37e−1)**	2.4817e−1 (1.85e−2)−	3.7805e−1 (3.62e−2)−	4.5357e−1 (4.31e−2)−	6.0247e−1 (5.30e−2)=
	10	5.2676e−1 (5.74e−2)−	2.2857e−1 (1.14e−2)−	4.1366e−1 (5.97e−2)−	3.3290e−1 (2.66e−2)−	**6.2443e−1** **(2.42e−1)**

HV value, respectively. In the tests of DTLZ2, all the results are noticeably better than the other four algorithms. In the tests of DTLZ6, RHMPSO attained greater HV values under 3 and 10 objectives. Therefore, the results from Table 1 and Table 2 indicate that RHMPSO has better convergence and diversity, that is, the algorithm can simultaneously obtain more higher quality solutions and better coverage range of the solution set.

To further demonstrate the advantages of our algorithm over other algorithms under the same conditions, the results of the average IGD value of RHMPSO and the five comparison algorithms are compared in Fig. 4. It can be seen from the figure that RHMPSO had the fastest IGD value convergence speed in the four-objective DTLZ2 and DTLZ4 problems.

To highlight the advantages of RHMPSO, we compare its Pareto images with those of five other advanced algorithms in Fig. 5. Specifically, we examine their performance on the DTLZ1 (multi-modal and linear) problem. Among the algorithms, SPEA2 and IDBEA exhibit poor diversity, while MPSOD and NMPSO fails to converge near the true Pareto front. In conclusion, RHMPSO stands out in the evaluation of the DTLZ1

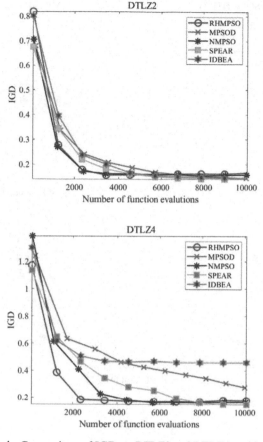

Fig. 4. Comparison of IGD on DTLZ2 and DTLZ4 problems.

problem, showcasing a uniform distribution of solutions that ensures good diversity and convergence. In contrast, other algorithms like SPEAR and IDBEA struggle with solution distribution. Therefore, RHMPSO indicating its superior global search ability and convergence characteristics.

4.2 Conclusion

This paper presents an enhanced algorithm called High-Dimensional Multi-Objective Particle Swarm Optimization Algorithm based on Radial Projection (RHMPSO) to improve solution diversity and reduce computational complexity in high-dimensional multi-objective optimization problems. The algorithm projects high-dimensional particles onto 2-dimensional radial space and divides them into equal-sized rectangles by grid division method. It selects individuals with better diversity as initial solutions based on crowding distance. The initial solutions are input into the PSO algorithm, utilizing a fusion of parent selection and elimination strategies to update the particle population and obtain the global optimal solution. Furthermore, the algorithm employs parent selection

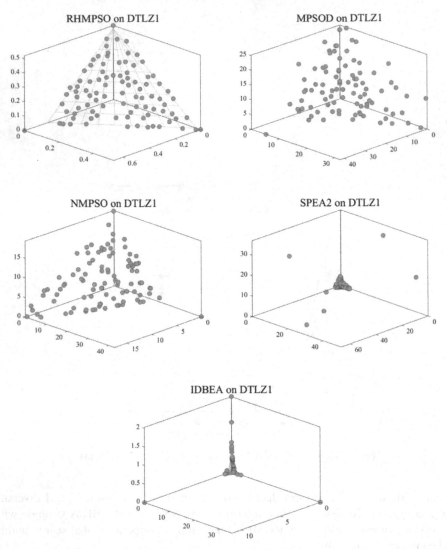

Fig. 5. Non-dominated front solutions obtained by RHMPSO and comparison algorithms on the DTLZ1 test function.

to update the external archive set, ensuring solution diversity while improving efficiency and accuracy. Experimental comparisons with four other algorithms, based on DTLZ and WFG test functions, demonstrate the superior performance of the RHMPSO algorithm. It achieves excellent results in terms of the IGD and HV indicators, as well as generating high-quality non-dominated front solutions across 3, 4, 6, and 10 objectives.

Acknowledgement. The work was supported by Science and Technology Project of Jiangxi Provincial Department of Education under grant GJJ190958.

References

1. Russell, E., James, K.: Particle swarm optimization. In: Proceedings of the IEEE International Conference on Neural Networks, vol. 4, pp. 1942–1948 (1995)
2. Qiuzhen, L.: Particle swarm optimization with a balanceable fitness estimation for many-objective optimization problems. IEEE Trans. Evolution. Comput. **22**(1), 32–46 (2018)
3. Qingfu, Z., Hui, L.: MOEA/D: a multi objective evolutionary algorithm based on decomposition. IEEE Trans. Evolution. Comput. **11**(6), 712–731 (2007)
4. Yuan, Y., Hua, X., Bo, D.: An improved NSGA-III procedure for evolutionary many-objective optimization. In: Proceedings of the 2014 Annual Conference on Genetic and Evolutionary Computation (2014)
5. Zhang, Z.: A many-objective optimization based intelligent intrusion detection algorithm for enhancing security of vehicular networks in 6G. IEEE Trans. Veh. Technol. **70**(6), 5234–5243 (2021)
6. Mohamad, Z., Mohd Z.: A multi-objective particle swarm optimization algorithm based on dynamic boundary search for constrained optimization. Appl. Soft Comput. **70**, 680–700 (2018)
7. Tianyou, C., Weijian, K., Jinliang, D.: Review of high-dimensional multi-objective evolutionary algorithms. Control Decision **4**(3), 6 (2010)
8. Castellanos-Garzón, J.A., Armando García, C.: A visual analytics framework for cluster analysis of DNA microarray data. In: Expert Systems with Applications, pp.758–774 (2013)
9. David, J., Walker, R.M., Jonathan, E.: Visualizing mutually non-dominating solution sets in many-objective optimization. IEEE Trans. Evolution. Comput. **17**(2), 165–184 (2013)
10. Ibrahim, A.: 3D-RadVis: visualization of Pareto front in many-objective optimization. In: Evolutionary Computation (2016)
11. Cheng, H.: A radial space division based evolutionary algorithm for many-objective optimization. Appl. Soft Comput. **61**, 603–621 (2017)
12. Qinmu, P.: Retinal vessel segmentation based on radial projection and semi-supervised learning. Ph.D. thesis, Huazhong University of Science and Technology (2011)
13. EngAik, L., WeiHong, T., KadriJunoh, A.: An improved radial basis function networks based on quantum evolutionary algorithm for training nonlinear datasets. IAES Int. J. Artif. Intell. 120–131 (2019)
14. Pingan, D.: Basic principles of finite element meshing. Mech. Des. Manuf. **4**, 34–36 (2000)
15. Pujia, W.: Research on Dimensionality Reduction Algorithm for scRNAseq Data Based on Generative Adversarial Networks and Autoencoders, p. 1 (2021)
16. Yuan, L.: Research on Environmental Selection Strategies for High-Dimensional Multi Objective Optimization Algorithms, p. 1 (2017)
17. Minqiang, L.: The fundamental theory and application of genetic algorithm. Artif. Intell. Robot. Res. (2002)
18. Ishibuchi, H.: Performance of decomposition-based many-objective algorithms strongly depends on Pareto front shapes. IEEE Trans. Evolution. Comput. **21**(2), 169–190 (2017)
19. Shanbhag, G.V.: "Mesoporous sodalite: a novel, stable solid catalyst for base-catalyzed organic transformations. J. Catal. **264**(1), 88–92 (2009)
20. Mifa, K.: SPEA2+: improving the performance of the strength Pareto evolutionary algorithm 2. In: Xin, Y. (ed.) Parallel Problem Solving from Nature - PPSN VIII, pp. 742–751. Springer, Heidelberg (2004). https://doi.org/10.1007/978-3-540-30217-9_75
21. Ying, Z., Rennong, Z., Jialiang, Z.: Improving decompostion based evolutionary algorithm for solving dynamic firepower allocation multi-objective optimization model. Acta Armament. **36**,1533–1540 (2015)

22. Xiaopeng, W.: Pareto genetic algorithm in multi-objective optimization design. J. Syst. Eng. Electron. **25**(12), 4 (2003)
23. Yanan, S., Gary, G.Y., Zhang, Y.: IGD indicator-based evolutionary algorithm for many-objective optimization problems. IEEE Trans. Evolution. Comput. **23**(2), 173–187 (2019)
24. Hub, S., Hingston, P.: An evolution strategy with probabilistic mutation for multi-objective optimisation. In: The 2003 Congress on Evolutionary Computation, 2003 (CEC 2003) (2004)

Link Prediction Based on the Sub-graphs Learning with Fused Features

Haoran Chen(✉) ⓘ, Jianxia Chen ⓘ, Dipai Liu ⓘ, Shuxi Zhang ⓘ, Shuhan Hu ⓘ,
Yu Cheng ⓘ, and Xinyun Wu ⓘ

Hubei University of Technology, Wuhan, Hubei, China
1810863407@qq.com

Abstract. As one of the important research methods in the area of the knowledge graph completion, link prediction aims to capture the structural information or the attribute information of nodes in the network to predict the link probability between nodes, In particular, the graph neural networks based on the sub-graphs provide a popular approach for the learning representation to the link prediction tasks. However, they cannot solve the resource consumption in large graphs, nor do they combine global structural features since they often simply stitch attribute features and embedding to predict. Therefore, this paper proposes a novel link prediction model based on the Sub-graphs Learning with the Fused Features, named SLFF in short. In particular, the proposed model utilizes random walks to extract the sub-graphs to reduce the overhead in the process. Moreover, it utilizes the Node2Vec to process the entire graph and obtain the global structure characteristics of the node. Afterward, the SLFF model utilizes the existing embedding to reconstruct the embedding according to the neighborhood defined by the graph structure and node attribute space. Finally, the SLFF model can combine the attribute characteristics of the node with the structural characteristics of the node together. The extensive experiments on datasets demonstrates that the proposed SLFF has better performance than that of the state-of-the-art approaches.

Keywords: Link Prediction · Feature Fusion · Structural Features · Property Characteristics · Graph Neural Networks · Subgraph Representation

1 Introduction

Knowledge graph (KG) is a semantic-based graph database used to describe the relationships and attributes between entities [1]. It consists of a set of nodes connected by directed edges, where each node represents an entity and each edge represents the relationship between entities. However, with the continuous changes in the real world, most knowledge graphs require constant updates and improvements, which is known as the knowledge graph completion (KGC).

As one of the important research methods for the KGC, link prediction (LP) aims to capture the structural information of the network or the attribute information of nodes in the network to predict the probability of linking between two nodes without edges

B. Luo et al. (Eds.): ICONIP 2023, LNCS 14449, pp. 253–264, 2024.
https://doi.org/10.1007/978-981-99-8067-3_19

[2–4]. LP models have a broad application scenario such as the biology, social networks, e-commerce and so on.

Many approaches for solving the LP problem have been proposed, which mainly include traditional rule-based methods, such as Common Neighbors (CN) [5], Adamic/Adar (AA) [6], Jaccard [7] and Katz [8], knowledge graph embedding (KGE) methods, and graph neural networks (GNNs) -based methods. CN approach calculates the number of common neighbors as a similar judgment basis to determine whether nodes are connected. Jaccard approach judges the similarity based on the relative number of common neighbors. CN, Jaccard, and AA can only capture neighbors within two hops, while Katz can capture multi-hop neighbors and give different weights to neighbor nodes.

Usually, KGE approaches map nodes into different vectors of the same dimension and calculate the similarity between node vectors. For example, DeepWalk [9] utilizes random walks to capture nodes and the skip-gram model [10] to get node embedding; Node2Vec [11] extends the random walk strategy via both the depth first search (DFS) and the breadth first search (BFS). LINE [12] utilizes BFS in the weighted graphs. However, KGEs-based models only consider the relationship between nodes without the characteristics of the nodes themselves.

As GNNs are evolved, variants of GNNs emerged one after another. Kipf et al. proposed a GCN [13] to extract features except that its object is graph data. Hamilton et al. introduced GraphSAGE [14] to efficiently generate node embeddings based on the textual information. DGCNN [15] proposed a sort pool layer that contains local information, which can be extracted to the global information through stacking. Velickovic et al. [16] proposed a graph attention network (GAT) model to utilize an attention mechanism to assign different weights to different domain nodes. However, it is inefficient for the hidden state of these GNNs-based models to update nodes. Also, there are some informative features on the edges that cannot be modeled by them.

From the above analysis, this paper proposes a novel Sub-graphs Learning with the Fused Features link prediction model, named SLFF in short, First, the proposed SLFF model adopts the ScaLed [17] approach to design the sampling of sub-graphs to enhance their scalability. Moreover, the random walks have been utilized to extract the sub-graphs to reduce the overhead in the process and enable it to be applied to larger graphs. Afterward, this paper utilizes the Node2Vec to process the entire graph and obtain the global structure characteristics of the node. For the fusion of the node features and its attribute features, the SLFF model adopts the graph embedding retrofitting (GEREF) approach [18], which is a composite node embedding approach that can fuse the node structure and attribute features. In particular, the SLFF model utilizes the existing embedding to reconstruct the embedding according to the neighborhood, which has been defined by the graph structure and node attribute space. Therefore, the SLFF model can combine the attribute characteristics of the node with the structural characteristics of the node together. The extensive experiments on datasets demonstrates that the proposed SLFF has better performance than that of the state-of-the-art approaches. The contributions of this paper are presented as follows:

(1) Designs the sampling of sub-graphs via, the random walk approach to extract the sub-graphs with fewer resources and less time, which can be able to be scaled to the large graphs.

(2) Utilizes the Node2Vec approach to process the entire graph and obtain the global structure features of the nodes, which are prepared for the subsequent feature fusion.

(3) Utilizes the existing embedding to reconstruct the embedding according to the neighborhood, which has been defined by the graph structure and node attribute space.

(4) Combines the attribute characteristics of the node with the structural characteristics of the node together to obtain the final node embedding.

2 Related Work

In this section, this paper briefly introduces the work related to our approaches, including GNNs-based link prediction, and feature fusion.

2.1 GNNs-Based Link Prediction

GNNs have been utilized for the LP recently with the sub-graphs but not taken into account the attribute characteristics of the node itself. For example, Zhou et al. [19] proposed a structure-enhanced neural network (SEG), which utilizes different paths to obtain the structural information of sub-graphs, and then incorporates its structure into a common GNN model. SEG integrated the topology and node characteristics to make full use of the graph information, but it may not perform well in large and complex graphs. Li et al. [20] introduced another SEAL-like node marking trick that directly utilizes the shortest distance to the target node, but it doesn't work as well as expected.

In particular, the subgraph-based learning approach is one of important approaches of GNNs, It focuses not only on the characteristics of nodes and edges, but also on the structural features of the graph. For example, the SEAL [21] utilized GNNs instead of fully connected neural networks in Weisfeiler-Lehman Neural Machine (WLNM), so as to improve the performance of the graph feature learning. Moreover, the SEAL approach first extracts a closed sub-graph around the target node, then marks its nodes according to the different positions of the nodes in the sub-graph. Finally, it utilizes a GNN framework called DGCNN [15] to predict the existence of the target link. These works simply extract closed sub-graphs with node labeling techniques, and directly connect local structural features with original node attribute features as input to the GNN. However, subgraph-based learning approaches lack the extraction ability of large-scale closed sub-graphs. And the way in which global structural features and node attribute features are fused is not considered.

The SLFF model in this paper utilizes the tagging technique introduced in SEAL. However, SEAL simply connects the structural features with the attribute features of the nodes as the input of the GNN, while the SLFF model utilizes the existing embedding to reconstruct the embedding according to the neighborhood defined by the graph structure and node attribute space, so as to fuse the attribute characteristics of the node and the global structural characteristics of the node, and then connect the local structural features with the fused embedding as the input of the GNN. This makes the information obtained by the model more complete during the training process.

2.2 Feature Fusion

Some common feature fusion approaches include Concat [21], Add [22], and Attention [23]. In particular, the concat approach stitches two features directly, in which if one feature dimension is too large, it may cause bad effects. For example, the method of feature fusion in the SEAL is the use of Concat. The Add approach is to add the two features directly, in which if the feature dimensions are different, the data needs to be supplemented to make the dimensions consistent. Residual Network (RN) [22] often uses the Add method to fuse features. Usually, Attention is suitable for most common scenarios and fuses features by learning the weight distribution of the individual features. For example, Self-attention [24] assigns different degrees of the attention to each feature through the attention mechanism to achieve the purpose of fusing features.

Inspired by the GEREF approach [18], the SLFF model combines the node attribute features with the global features of the node together. Different from the GEREF model, however, the SLFF utilizes the Concat approach to fuse with the local structural characteristics of the node in order to generate the node embedding finally. Therefore, the SLFF not only integrates the attribute characteristics of the node but also does not lose the global and local features of the node, so that it can obtain more complete information.

3 Methodology

As shown in Fig. 1, the SLFF model consists of the following three parts:

(1) Subgraph extraction sub-module: Extract the closed subgraph of the target node through the random walk approach.
(2) Feature fusion sub-module: Fuse the attribute characteristics of the node and the global structure characteristics of the node to obtain the entire input vector.
(3) Dynamic graph neural network sub-module: Obtain the final prediction result through the Multi-Layer Perceptron (MLP).

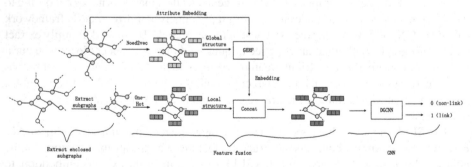

Fig. 1. The SLFF model consists of three parts, (1) sub-graphs extraction, (2) feature fusion, and (3) graph neural network.

3.1 Problem Definition

Let G represents the graph, V represents the set of nodes in the graph, $E \in V \times V$ represents the edge set that connects the node pairs, use u and v to represent two different target nodes, k to represent the number of random walks in the DRNL method, and h to represent the step size of each walk. W_u and W_v represent the set of nodes that pass through a random walk from the target node u or v k times with a step of h.

3.2 Subgraphs Extraction

Given a graph G, the target node (u, v) starts to randomly walk k times, and the number of steps is h, two node sets W_u and W_v can be obtained, $W_u \cup W_v$ can get the node set of the closed subgraph, and the edge will be get in the closed subgraph. After obtaining the closed subgraph, the DRNL approach is utilized to label the nodes to obtain the closed subgraph marked by the nodes. Figure 2(b) is the closed subgraph obtained by the random walk sampling and marking of the target node pair (u,v) in Fig. 2(a), where $k = 2, h = 2, W_u = \{u, f, g, h\}, W_v = \{v, d, e, a, b\}$.

The DRNL Approach. This paper uses the function $f_l: V \to N$ to represent the label of the node, and $f_l(i)$ to represent the label of node i. There are two criteria for the DRNL method as follows: (1) the two target nodes u,v have a unique label 1 (2) if nodes $d(i,u)$ $= d(j,u)$ and $d(i,v) = d(j,v)$, then nodes i and j have the same label. $d(i,u)$ indicates how far apart node i and target node u are apart. So, the topological position of a node i in a closed subgraph can be described by its radius relative to the two central nodes $(d(i, u), d(i, v))$.

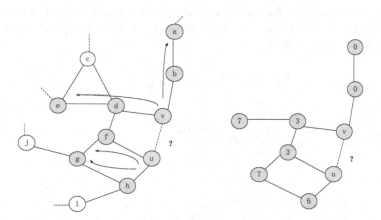

Fig. 2. (a) Target nodes u and v randomly walk; (b) Closed sub-graphs.

The label $f_l(i)$ and double radius $(d(i, u), d(i, v))$ are satisfied as follows:

- if $d(i, u) + d(i, v) \neq d(j, u) + d(j, v)$, then $d(i, u) + d(i, v) < d(j, u) + d(j, v)$, $f_l(i) < f_l(j)$;

- if $d(i, u) + d(i, v) = d(j, u) + d(j, v)$, then $d(i, u)d(i, v) < d(j, u)d(j, v), fl(i) < fl(j)$.

The Hash function is defined in the Eq. (1) as follows:

$$fl(i) = 1 + min(d_u, d_v) + (d/2)[(d/2) + (d\%2) - 1] \tag{1}$$

where $d_u := d(i, u), d_v := d(i, v), d := d_u + d_v, (d/2)$ and $(d\%2)$ are the integer quotient and remainder of d divided by 2, respectively. For nodes where $d(i, u) = \infty$ or $d(i, v) = \infty$, this paper gives them an empty label 0. As shown in the Fig. 2, nodes d, f can be represented as $(1,2)$, so their label is 3, and node h is represented as $(1,4)$, so they are labeled as 6. Because $d(a, u) = \infty$ and $d(b, u) = \infty$, they are both labeled as 0.

3.3   Feature Fusion

The purpose of the feature fusion sub-module is to reconstruct the embedding according to the neighborhood defined by the graph structure and node attribute space. In particular, it fuses the attribute characteristics of the node and the global structure characteristics of the node in order to connect the obtained local structural features and the fused embedding to obtain the input vector. As shown in the Fig. 3, after extracting the closed subgraph, a one-hot code is set for each node according to the label to obtain the local structural characteristics of the node.

Node2Vec Algorithm.   The purpose of the Node2vec algorithm is to obtain the global structure characteristics of nodes. As shown in the Table 1, the parameter p controls the probability of repeatedly visiting the vertices that have just been visited, if p is high, the probability of visiting the vertices that have just been visited will become lower and vice versa. q controls whether the walk is outward or inward, and if $q > 1$, random walks tend to visit nodes close to the target node, and if $q < 1$, they tend to visit nodes that are far away from the target node.

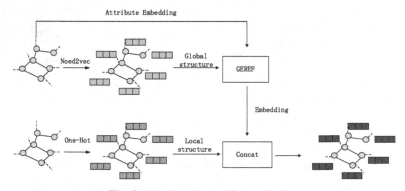

Fig. 3.   The flowchart of feature fusion.

In order to map the embedding to the low-dimensional embedding and retain the attribute and structural information of the node to the greatest extent, this paper uses the GEREF approach [18] to fuse the attribute features and structural features.

Table 1. The Node2vec Algorithm.

Algorithm 1 The node2vec algorithm.

LearnFeatures*(Graph G = (V, E, W), Dimensions d, Walks per node r, Walk length*
  l, Context size k, Return p, In-out q)
  π = *PreprocessModifiedWeights(G, p, q)*
  G' = (V, E, π)
  Initialize walks to Empty
  for *iter = 1* **to** *r* **do**
    for *all nodes u $\in$ V* **do**
      walk = node2vecWalk(G', u, l)
      Append walk to walks
  f = StochasticGradientDescent(k, d, walks)
  return *f*

node2vecWalk*(Graph G' = (V, E, π), Start node u, Length l)*
  Inititalize walk to [u]
  for *walk_iter = 1* **to** *l* **do**
    curr = walk[-1]
    V_{curr} = *GetNeighbors(curr, G')*
    s = AliasSample(V_{curr}, π)
    Append s to walk
  return *walk*

The global structure characteristics of the node and the attribute characteristics of the node can be fused by the objective function of GEREF in the following Eq. (2). Finally the local structure features of the node and the fused node features can be connected to obtain node embedding.

$$\mathcal{L}(Z^*) = (1 - \lambda_G - \lambda_X) \sum_{i=1}^{n} |z_i^* - z_i|^2 + \lambda_G \sum_{i=1}^{n} \sum_{j:v_j \in \mathcal{N}(v_i)} \frac{|z_i^* - z_j^*|^2}{|\mathcal{N}(v_i)|}$$
$$+ \lambda_X \sum_{i=1}^{n} \sum_{j:v_j \in \mathcal{N}_X(v_i)} \frac{|z_i^* - z_j^*|^2}{|\mathcal{N}_X(v_i)|} \tag{2}$$

where,

- $Z = (z_1,...,z_n)$ is the structure embedding pre-trained for each node,
- $Z^* = (z_1^*,...,z_n^*)$ is the new embedding obtained by combining structural embedding and attribute embedding,
- the hyperparameters $\lambda_G > 0$ *and* $\lambda_X > 0$ control the importance of structures and attributes,
- $N_X(v_i)$ represents the neighbor of v_i in the attribute neighborhood,
- $N(v_i)$ represents the neighbor of v_i in the graph neighborhood.

Then the Adam optimizer is utilized to optimize the function and update the embedding vector, in order to obtain the global structure feature with attribute characteristics.

3.4 Dynamic Graph CNN

Since the DGCNN learns global shape information by fusing local neighbor information and stackable multi-layer EdgeConv modules, it has excellent point cloud classification and segmentation functions and it is utilized as the GNN part to learn the sub-graph for the classification task, the input is the embedding after feature fusion, the output dimension is 1 for subgraph classification.

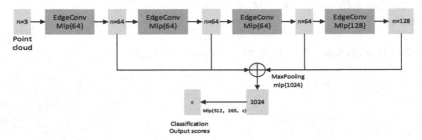

Fig. 4. Flowchart of the classification module in DGCNN.

As shown in the Fig. 4, the DGCNN contains four EdgeConv layers, and performs the pooling operations and multiple fully connected layers, in order to obtain the classification results. In particular, the first step is to use the EdgeConv module to characterize the point-wise feature step by step for the input, the output of the previous EdgeConv module is the input of the next EdgeConv module. Second, it stitches different levels of point-wise features together to obtain global features through max pooling. Finally, a few of MLPs are utilized for the classification.

4 Experiment and Analysis

In this paper, extensive experiments are carried out to evaluate the proposed SLFF model, and the experimental results show that SLFF model is an excellent LP framework with the good prediction effect in various networks.

4.1 Datasets

This paper considers some undirected graph datasets with node properties, using the following public datasets: Cora, Citeseer, Pubmed, which are commonly utilized in many LP tasks. It contains the specific information of datasets in the Table 2. The datasets is divided into the training set, verification set and test set at the ratio of 85%:5%:10%. In addition, each datasets is added unlinked negative sample and positive samples randomly at the ratio of 1:1.

4.2 Comparison of Models

The SLFF model has been compared with the baselines including four types as follows.

Models based on the heuristics: CN [5], AA [6]. Models based on graph feature extraction: GCN [13], SAGE [14], GIN [25]. Models based on the latent features: MF [26], Node2Vec [11]. Models based on the sub-graphs: SEAL [21].

Table 2. Datasets Statistics.

Datasets	Nodes	Edge	Attribute dimensions
Cora	2708	5429	1433
Citeseer	3327	4732	3705
Pubmed	19717	44338	500

4.3 Hyper-parameters of Model

Table 3 shows the following hyper-parameter settings for the SLFF model.

Table 3. Hyper-parameters of model.

Hyper-parameters	Description	Value
K	The number of random walks	20
h	The number of steps to walk randomly	3
lr	Learning rate	0.0001
epochs	Number of training rounds	50
dropout	The proportion of discarded neural units	0.5

4.4 Evaluation Metrics

AUC and average accuracy (AP) are utilized as evaluation metrics, and all experiments are run 5 times and AUC and AP results are reported.

4.5 Results and Analysis

Comparative Experiments. The results of the comparative experiments are shown in Table 4. It can be observed that SLFF outperforms most baselines, which can be achieved by reducing resource consumption and fusing global structure information and attribute information.

Table 4. Comparative experimental results.

Models	Cora Datasets		Citeseer Datasets		Pubmed Datasets	
	AUC	AP	AUC	AP	AUC	AP
CN	71.32	71.18	66.37	66.37	64.26	64.22
AA	71.40	71.58	66.37	66.37	64.26	64.28
GCN	66.79	70.30	66.02	67.90	89.29	87.11
SAGE	60.72	61.21	68.85	67.58	71.47	74.92
GIN	62.61	67.38	66.27	72.02	82.56	84.28
MF	50.02	50.66	46.91	49.70	72.74	74.45
N2V	54.65	52.55	51.43	50.74	55.54	53.14
SEAL	89.92	91.67	87.93	90.49	95.51	95.98
SLFF	**91.95**	**92.96**	**88.28**	**91.09**	**96.42**	**96.60**

Ablation Experiments. In order to verify the effectiveness of the link prediction model based on fusion feature subgraph learning, ablation experiments are carried out. As shown in Table 5, the experimental results of ablation are described according to the following model. In particular, SLFF (GEREF) model is a link prediction model based on fusion feature subgraph learning proposed in this paper. SLFF (Concat) model does not use the GEREF Feature Fusion module, but simply connects the features.

Table 5. AUC and AP using GEREF and Concat.

Models	Cora Datasets		Citeseer Datasets		Pubmed Datasets	
	AUC	AP	AUC	AP	AUC	AP
SLFF (GEREF)	91.95	92.96	88.28	91.09	96.42	96.60
SLFF (Concat)	89.01	90.81	87.36	89.32	96.89	97.00

The results are only slightly worse on the Pubmed dataset, which we believe is caused by the dataset being too large and the loss of structure or attribute features. The above results show that the GEREF module is very important for the model, it integrates the attribute characteristics of the node, and can obtain a more complete embedding of information.

5   Conclusion

Most advanced LP approach utilizes the subgraph models to learn the embedding of closed subgraphs formed by nodes to surrounding nodes. However, the existing methods cannot scale to large, real-world diagrams, and global and attribute features are not properly embedded. This paper solves these problems through the proposed SLFF model, using the sparse sub-graphs combined with global and attribute features to predict the links in the knowledge graph. The extensive experiments show that the SLFF model can achieve comparable or even better accuracy than existing state-of-the-art LP models while reducing overhead. In the future work, the paper will think about how to reasonably extract global structural features, and how to integrate subgraph embedding, global features and attribute features more reasonably to obtain a subgraph representation with the perfect information.

Acknowledgments. This work is supported by National Natural Science Foundation of China (61902116).

References

1. Zou, X.: A survey on application of knowledge graph. In: Conference 2020, CCEAI, vol. 1487, Singapore (2020)
2. Chen, Y., Ma, T., Yang, X., Wang, J., Song, B., Zeng, X.: MUFFIN: multi-scale feature fusion for drug-frug interaction prediction. Bioinformatics **37**(17), 2651–2658 (2021)
3. Chen, L., Xie, Y., Zheng, Z., Zheng, H., Xie, J.: Friend recommendation based on multi-social graph convolutional network. IEEE Access **8**, 43618–43629 (2020)
4. Oh, S., Choi, J., Ko, N., Yoon, J.: Predicting product development directions for new product planning using patent classification-based link prediciton. Scientometrics **125**(3), 1833–1876 (2020)
5. Newman, M.E.J.: Clustering and preferential attachment in growing networks. Phys. Rev. E **64**, 025102(R) (2001)
6. Adamic, L.A., Adar, E.: Friends and neighbors on the web. Soc. Netw. **25**(3), 211–230 (2003)
7. Fitz-Gerald, S.J., Wiggins, B.: Introduction to Modern Information Retrieval. McGraw-Hill, Inc., New York (1986)
8. Katz, L.: A new status index derived from sociometric analysis. Psychometrika **18**(1), 39–43 (1953)
9. Perozzi, B., AI-Rfou, R., Skiena, S.: DeepWalk: online learning of social representations. In: CONFERENCE 2014, KDD, vol. 14, pp. 701–710 (2014)
10. Mikolov, T., Chen, K., Corrado, G., Dean, J.: Efficient estimation of word representations in vector space. Computation and Language (cs.CL). arXiv preprint arXiv:1301.3781 (2013)
11. Grover, A., Leskovec, J.: node2vec: scalable feature learning for networks. Social and Information Networks. arXiv preprint arXiv:1607.00653 (2016)
12. Tang, J., Qu, M., Wang, M., Zhang, M., Yan, J., Mei, Q.: LINE: large-scale information network embedding. Machine Learning. arXiv preprint arXiv:1503.03578 (2015)
13. Kipf, T.N., Welling, M.: Semi-supervised classification with graph convolutional networks. In: Conference 2017, ICLR. arXiv preprint arXiv:1609.02907 (2017)
14. Hamilton, W.L., Ying, Z., Leskovec, J.: Inductive representation learning on large graphs. Social and Information Networks. arXiv preprint arXiv:1706.02216 (2017)
15. Zhang, M., Cui, Z., Neumann, M., Chen, Y.: An end-to-end deep learning architectire for graph classification. In: Conference 2018, AAAI, vol. 554, pp. 4438–4445 (2018)
16. Velickovic, P., Cucurull, G., Casanova, A., Romero, A., Lio, P., Bengio, Y.: Graph attention networks. Machine Learning. arXiv preprint arXiv:1710.10903 (2017)
17. Louis, P., Jacob, S.A., Salehi-Abari, A.: Sampling enclosing sub-graphs for link prediction. Machine Learning. arXiv preprint arXiv:2206.12004 (2022)

18. Bielak, P., Puchalska, D., Kajdanowicz, T.: Retrofitting structural graph embeddings with node attribute information. In: Conference 2022, ICCS, London, part 1, pp. 178–191 (2022)

19. Ai, B., Qin, Z., Shen, W., Li, Y.: Structure enhanced graph neural networks for link prediction. Machine Learning. arXiv preprint arXiv:2201.05293 (2022)

20. Li, P., Wang, Y., Wang, H., Leskovec, J.: Distance encoding: design provably more powerful neural networks for graph representation learning. Machine Learning. arXiv preprint arXiv: 2009.00142 (2020)

21. Zhang, M., Chen, Y.: Link prediction based on graph neural networks. Machine Learning. arXiv preprint arXiv:1802.09691 (2018)

22. He, K., Zhang, X., Ren, S., Sun, J.: Deep residual learning for image recognition. Computer Vision and Pattern Recognition. arXiv preprint arXiv:1512.03385 (2015)

23. Bahdanau, D., Cho, K., Bengio, Y.: Neural machine translation by jointly learning to align and translate. Computation and Language. arXiv preprint arXiv:1409.0473 (2014)

24. Vaswani, A., et al.: Attention is all you need. Computation and Language. arXiv preprint arXiv:1706.03762 (2017)

25. Xu, K., Hu, W., Leskovec, J., Jegelka, S.: How powerful are graph neural networks? Machine Learning. arXiv preprint arXiv:1810.00826 (2018)

26. Xue, H.-J., Dai, X., Zhang, J., Huang, S., Chen, J.: Deep matrix factorization models for recommender systems. In: Conference 2017, IJCAI, Melbourne, vol. 17, pp. 3203–3209 (2017)

Naturalistic Emotion Recognition Using EEG and Eye Movements

Jian-Ming Zhang[1], Jiawen Liu[1], Ziyi Li[1], Tian-Fang Ma[1], Yiting Wang[1], Wei-Long Zheng[1], and Bao-Liang Lu[1,2,3(✉)]

[1] Department of Computer Science and Engineering,
Shanghai Jiao Tong University, Shanghai 200240, China
{jmzhang98,ljw_venn,liziyi,matianfang2676,weilong,bllu}@sjtu.edu.cn
[2] RuiJin-Mihoyo Laboratory, RuiJin Hospital, Shanghai Jiao Tong University School of Medicine, Shanghai 200020, China
[3] Key Laboratory of Shanghai Commission for Intelligent Interaction and Cognitive Engineering, Shanghai Jiao Tong University, Shanghai 200240, China

Abstract. Emotion recognition in affective brain-computer interfaces (aBCI) has emerged as a prominent research area. However, existing experimental paradigms for collecting emotional data often rely on stimuli-based elicitation, which may not accurately reflect emotions experienced in everyday life. Moreover, these paradigms are limited in terms of stimulus types and lack investigation into decoding naturalistic emotional states. To address these limitations, we propose a novel experimental paradigm that enables the recording of physiological signals in a more natural way. In our approach, emotions are allowed to arise spontaneously, unrestricted by specific experimental activities. Participants have the autonomy to determine the start and end of each recording session and provide corresponding emotion label. Over a period of three months, we recruited six subjects and collected data through multiple recording sessions per subject. We utilized electroencephalogram (EEG) and eye movement signals in both subject-dependent and cross-subject settings. In the subject-dependent unimodal condition, our attentive simple graph convolutional network (ASGC) achieved the highest accuracy of 76.32% for emotion recognition based on EEG data. For the cross-subject unimodal condition, our domain adversarial neural network (DANN) outperformed other models, achieving an average accuracy of 71.90% based on EEG data. These experimental results demonstrate the feasibility of recognizing emotions in naturalistic settings. The proposed experimental paradigm holds significant potential for advancing emotion recognition in various practical applications. By allowing emotions to unfold naturally, our approach enables the future emergence of more robust and applicable emotion recognition models in the field of aBCI.

Keywords: Affective Brain-computer Interfaces · Naturalistic Emotion Recognition · EEG · Eye Movements

B. Luo et al. (Eds.): ICONIP 2023, LNCS 14449, pp. 265–276, 2024.
https://doi.org/10.1007/978-981-99-8067-3_20

1 Introduction

In the field of mental illness diagnosis, scale tests and empirical judgments by physicians are often considered the gold standard. However, patients may provide false answers and conceal their true condition for various reasons. As a result, there is a growing interest in objective methods for determining emotional and mental states based on patients' physiological signals [1]. With the rapid advancement of artificial intelligence technology, emotional intelligence (EI) has shown great potential in medical and other domains, offering the possibility of emotion recognition based on physiological signals [2].

In recent studies, EEG and eye movement signals have played a significant role in capturing human emotions [3–5]. In the field of aBCI, existing emotion recognition paradigms based on physiological signals typically rely on passive elicitation, which can place a considerable burden on subjects [6]. These paradigms involve collecting signals in controlled laboratory environments while evoking emotions through specific stimuli. Subjects are immersed in emotional stimuli such as pictures, audio, and videos to induce specific emotional states corresponding to the stimuli. The limited experimental settings are due to the lack of technology for collecting high-quality labeled emotion data in daily life, limiting the experimentation to specific tasks conducted in a laboratory. While these experiments fall under the emotion elicitation experiment paradigm and have demonstrated that EEG and eye movement signals can detect human emotions, they still have two limitations: 1) Emotions are passively evoked in unnatural states, and 2) The stimuli are limited to pictures and videos, with the paradigm reliant on stimuli.

To address these limitations, we propose a novel paradigm for emotion recognition that operates in natural states without relying on passive stimuli. Our paradigm introduces several key innovations: 1) Emotions are actively evoked rather than relying on stimuli, aiming to closely resemble daily life experiences; 2) There are no limitations on the forms of stimuli used; 3) The data collection process does not require consistency across subjects, allowing for a diversity of data; and 4) The emotion labels in our collected data are more accurate as they are provided by the subjects themselves.

To validate the feasibility of recognizing emotions in naturalistic settings using our paradigm, we collected EEG and eye movement signals from six subjects engaged in daily emotional activities. Subsequently, we conducted a series of experiments using commonly used models in aBCI. The results demonstrated the viability of recognizing emotions in a naturalistic manner.

2 Methods

2.1 Paradigm Design

The EEG signals were recorded using the DSI-24 Dry Electrode EEG Cap, while eye movement signals were recorded with the Tobii Screen-based Eye Tracker. To ensure that our experiments closely resembled the daily lives of the subjects, we selected six graduate students (three males and three females; Avg. of ages: 24.33, Std. of ages: 1.97). The experiments took place in proximity to their work-places. The start and end of the experiments were determined by the subjects themselves. The objective of the experiments was to record the subjects' natural emotional states during their daily activities, which included watching movies, reading, playing games, etc.

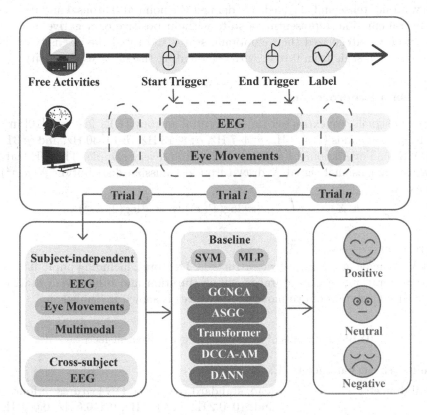

Fig. 1. The procedure of data collecting and emotion recognition from collected signals.

Each experiment consisted of multiple trials. The subjects initiated each trial by clicking the mouse to send triggers. When the subjects perceived a significant change in their emotions, they would click the mouse to send the start trigger, and they would label the trial after sending the end trigger. Several trials were recorded within a single experiment, corresponding to the subjects' emotional states. The subject would also retrace the entire experiment and refine the emo-tion labels at the end of each experiment. A brief illustration of the procedure is provided at the top of Fig. 1, while the bottom shows the models used for emotion recognition (Sect. 2.3) and our experiment settings (Sect. 3.1). After the experiment, the information of each trial included the start and end time, EEG and eye movement signals, emotion labels, and the activities performed.

We utilized the 2-D valence and arousal space [7] as a popular method for dimensional emotion representation. Based on the 2-D Emotion Wheel [8,9], we selected several common typical emotions as our emotion labels, including astonished, excited, delighted, happy, pleased, satisfied, relaxed, calm, sleepy, tired, droopy, bored, depressed, sad, miserable, frustrated, distressed, annoyed, angry, afraid, tense and alarmed. We defined the point at 0 arousal and 0 valence as neutral emotion, representing a state without positive or negative emotions. The subjects categorized their emotional states and rated the intensity of their emotions on a scale from 0 to 5, with 5 indicating the strongest intensity.

2.2 Data Preprocessing

For EEG signals, we extracted the differential entropy (DE) features [10] in the five frequency bands (δ: 1–3 Hz, θ: 4–7 Hz, α: 8–13 Hz, β: 14–30 Hz, and γ: 31–50 Hz) with non-overlapping 4-s time window from every sample. The DE feature on a one-dimensional signal X drawn from a Gaussian distribution $N(\mu, \delta^2)$ is defined as

$$h(X) = -\int_{-\infty}^{\infty} P(x) \log(P(x)) dx = \frac{1}{2} \log(2\pi e \sigma^2), \tag{1}$$

where $P(x) = \frac{1}{\sqrt{2\pi\sigma^2}} e^{-\frac{(x-\mu)^2}{2\sigma^2}}$.

For eye movement signals, we extracted 23 features, including pupil diameter, fixation duration, saccade duration, blink duration and other event statistics. Detailed information of eye movement features is shown in Table 1.

Table 1. Details of the extracted eye movement features.

Eye movement parameters	Extracted features
Pupil diameter	Mean, standard deviation, DE features in four bands (0–0.2 Hz, 0.2–0.4 Hz, 0.4–0.6 Hz, 0.6–1 Hz)
Fixation duration	Mean, standard deviation, maximum
Saccade duration	Mean, standard deviation
Blink duration	Mean, standard deviation
Event statistics	Fixation frequency, saccade frequency, saccade latency, blink frequency

We carried out the emotion classification tasks with positive (emotions on the right side of the Emotion Wheel), negative (emotions on the left side of the Emotion Wheel) [8,9] and neutral emotions. Since the emotions were generated under the natural states of subjects, statistical results have shown an obvious data imbalance. Specifically, the numbers of neutral and negative trials were significantly less than that of positive ones. The statistics of different emotions for six subjects are shown in Table 2.

Table 2. The number of trials for EEG and eye movement data over six subjects. Each row represents statistic for each subject. Each column represents the statistic for positive, neutral and negative emotions.

Subject	EEG				Eye Movement			
	Positive	Neutral	Negative	Total	Positive	Neutral	Negative	Total
01	12	4	2	18	6	1	2	9
02	13	2	8	23	13	2	8	23
03	7	3	10	20	7	3	10	20
04	7	5	5	17	7	4	5	16
05	18	5	6	29	17	5	6	28
06	13	6	1	20	5	4	1	10

2.3   Models

We carefully selected several classification models commonly employed in emotion recognition based on EEG and eye movement signals, and we created appropriate training and test data sets tailored to their characteristics.

To effectively capture the interdependencies between input channels, we employed a graph convolutional network (GCN) that utilizes an adjacency matrix as a weight representation. This approach helps prevent overfitting among the channels of features and allows us to examine the significance of each channel. We treated the EEG and eye movement signals as graphs and applied a graph convolutional network with channel attention (GCNCA) model [11]. This model achieved notable performance in classifying three emotions: anger, surprise, and neutrality.

Similarly, we utilized an attentive simple graph convolutional network (ASGC) [12], a graph neural network-based model for processing EEG signals in tasks related to measuring human decision confidence. ASGC incorporates a learnable adjacency matrix and a simple graph convolutional network (SGC) to capture the coarse-grained relationships between EEG channels. It then employs a self-attention mechanism to capture the fine-grained relationships between channels. Finally, a confidence distribution loss is used to calculate the discrepancy between the predicted class distribution and the true confidence distribution. ASGC leverages the topological structure of EEG signal channels through graph neural networks, dynamically adjusts channel weights for each sample using self-attention, and addresses the challenges of limited training samples and ambiguous labels using a confidence distribution loss, as opposed to relying solely on simple one-hot encoding. The main process of the model is as follows.

Considering each EEG channel as a graph node, a feature matrix $\mathbf{X} \in \mathbb{R}^{n \times d}$ is constructed, where n denotes the number of channels and d denotes the feature dimension of each channel. Using the learnable adjacency matrix $\mathbf{A} \in \mathbb{R}^{n \times n}$ in SGC to capture the coarse-grained relationships between EEG channels, a high-dimensional feature matrix $\mathbf{Z} \in \mathbb{R}^{n \times h}$ is obtained, where h denotes the hidden layer size. Specifically, $\mathbf{Z}$ can be expressed as follows

$$\mathbf{Z} = \mathbf{S}^K \mathbf{X} \mathbf{W} = \tilde{\mathbf{X}} \mathbf{W}, \tag{2}$$

where $\mathbf{S} = \tilde{\mathbf{D}}^{-\frac{1}{2}} \tilde{\mathbf{A}} \tilde{\mathbf{D}}^{-\frac{1}{2}}$, which is a normalized adjacency matrix. Meanwhile, $\tilde{\mathbf{A}} = \mathbf{A} + \mathbf{I}_n$, $\tilde{\mathbf{D}}_{ii} = \sum_j \tilde{\mathbf{A}}_{ij}$, $\mathbf{W} \in \mathbb{R}^{d \times h}$, and K represents the number of graph convolution layers. Then, the fine-grained relationships between EEG channels are captured by a self-attentive mechanism to obtain a weighted feature matrix $\hat{\mathbf{X}} \in \mathbb{R}^{n \times d}$. Specifically, $\hat{\mathbf{X}}$ can be denoted as

$$\hat{\mathbf{X}} = \mathrm{softmax}\left(\mathbf{Z}\mathbf{Z}^T\right) \tilde{\mathbf{X}}, \tag{3}$$

where the softmax operation is normalized along each row so that each row sums to 1. Finally, the features of all nodes in $\hat{\mathbf{X}}$ are stitched into a vector, and the final class distribution is obtained by a fully connected layer and a softmax activation function.

Transformer and attention-based fusion have been employed to extract complementary properties of EEG and eye movements, which proved the effectiveness of Transformer on interpreting temporal resolution for both modalities [13]. Besides improving the performance of emotion recognition, Transformer achieves better parallelism than sequential models like recurrent neural networks (RNN).

For multimodal approaches, by extending deep canonical correlation analysis (DCCA) model, Liu et al. [14] introduced deep canonical correlation analysis with attention mechanism (DCCA-AM), which added an attention-based fusion module assisting the representations of multiple modalities by passing them to multiple nonlinear transform layers for better emotion recognition. Through adaptive weighted-sum fusion, attention-based fusion produced no worse results than weighted-sum fusion since the weights computed by attention-based fusion can be the same as the weighted-sum fusion. In addition, DCCA-AM can handle different dimensions, different distributions, different sampling rates, etc. It is worth mentioning that the loss function of DCCA-AM consists of two parts,

$$\mathbf{L} = \alpha \mathbf{L}_{cca} + \beta \mathbf{L}_{classification}, \tag{4}$$

where $\mathbf{L}$ is the final loss, α and β are trade-offs that control the synergy of the two loss terms. Let $\mathbf{X_1} \in \mathbb{R}^{N \times d_1}$ and $\mathbf{X_2} \in \mathbb{R}^{N \times d_2}$ be the instance matrices for two modalities respectively, where d_1 and d_2 represent the dimensions of two different features and N represents the number of instances. By constructing two deep neural networks f_1, f_2 with parameters W_1, W_2 respectively, we can obtain $\mathbf{L}_{cca}$ which represents the opposite number of correlations between EEG and eye movement signals.

$$\mathbf{L}_{cca} = -\mathrm{corr}(f_1(X_1; W_1), f_2(X_2; W_2)), \tag{5}$$

where $f_1(X_1; W_1)$ and $f_2(X_2; W_2)$ represent the outputs of the neural networks and corr represents the correlation between them. $\mathbf{L}_{classification}$ represents the cross entropy generated by fusing the EEG and eye movement signals into the fully connected layer.

For cross-subject models, domain adversarial neural network (DANN) [15] can extract the shared representations between the source domain and the target domain. It utilizes an ingenious gradient reversal layer (GRL) to bridge differences between domains, resulting in learning domain-independent features. In cross-subject emotion recognition, DANN can eliminate the differences across subjects and achieve better generalization performance.

3   Experiments and Results

3.1   Experiment Settings

In the field of emotion recognition in aBCI, two paradigms are commonly used: subject-dependent and cross-subject [14]. Previous studies have demonstrated that both EEG and eye movement modalities are effective in measuring the emotional state of subjects [4]. Furthermore, combining multiple modalities can provide a more comprehensive understanding of human emotions by capturing different aspects. Additionally, research has shown that different modalities can complement each other in emotion measurement [16]. However, collecting multimodal signals can be costlier due to the various types of signals involved. To ensure fair and comprehensive results, we conducted experiments using both unimodal and multimodal approaches.

In this study, we designed four experimental settings that constructed different training and test data: subject-dependent emotion recognition (including EEG-based, eye movement-based, and multimodal) and cross-subject emotion recognition based on EEG to testify our paradigm under different conditions. We utilized stratified K-fold cross-validation technique for the subject-dependent experiment settings, while leave-one-subject-out validation was applied for the cross-subject experiment setting. Notably, Table 2 indicates that there is only one trial in eye movement signals of *Subject 01* and both EEG, eye movement signals of *Subject 06* under neutral or negative emotion. For these three cases, we performed binary classification tasks, which are marked with stars in the result tables. The average and standard deviation of accuracies reported in the results do not include the binary classification tasks. In all four settings, we used traditional support vector machine (SVM) and multilayer perceptron (MLP) as the baseline classifiers.

Table 3. Results of different models on EEG-based subject-dependent unimodal emotion recognition task.

Model	SVM	MLP	GCNCA	Transformer	ASGC
01	60.53	58.99	67.16	72.69	83.79
02	77.53	77.02	71.20	64.63	72.21
03	67.35	72.01	69.55	91.81	90.54
04	46.45	38.21	40.15	58.81	55.94
05	79.77	70.98	64.58	89.01	79.11
*06	64.49	72.00	64.70	83.64	90.55
Avg.	66.33	63.44	62.53	75.39	**76.32**
Std.	12.13	13.94	**11.41**	13.06	11.82

* Binary classification due to the number of trials

3.2 Experimental Results

Subject-Dependent Emotion Recognition. In the subject-dependent unimodal EEG-based emotion recognition, we further evaluated the performance of GCNCA [11], Transformer [13], and ASGC [12]. From the results presented in Table 3, we can observe that ASGC and Transformer achieved the highest average accuracies of 76.32% and 75.39%, respectively, across the three classification tasks. When focusing on the subject-dependent unimodal eye movement-based emotion recognition, we considered GCNCA and Transformer as well. Table 4 illustrates that Transformer outperformed the other algorithms with an accuracy rate of 75.65%. Notably, the average accuracy of eye movement signals using Transformer was higher than that of EEG signals.

Table 4. Results of different models on eye movement-based subject-dependent unimodal emotion recognition task.

Model	SVM	MLP	GCNCA	Transformer
*01	97.12	99.74	97.38	99.19
02	44.22	54.57	70.62	72.76
03	72.51	71.29	77.04	84.09
04	39.32	62.80	50.73	57.99
05	71.67	69.27	77.09	87.75
*06	82.79	70.29	92.44	87.47
Avg.	56.93	64.48	68.87	**75.65**
Std.	15.26	**6.53**	10.80	11.60

* Binary classification due to the number of trials

For the multimodal setting, we fused EEG and eye movement signals and used Transformer and DCCA-AM [14] additionally. From Table 5, an average accuracy of 74.91% is obtained using DCCA-AM which achieves the best performance.

Table 5. Results of different models under subject-dependent multimodal setting.

Method	SVM	MLP	Transformer	DCCA-AM
*01	89.96	94.19	81.41	99.74
02	56.38	52.61	57.54	68.16
03	77.92	75.22	84.24	76.99
04	52.60	41.05	45.62	60.55
05	80.28	67.90	72.69	93.93
*06	64.49	80.83	81.77	74.62
Avg.	66.80	59.20	65.02	**74.91**
Std.	**12.41**	13.28	14.67	12.43

* Binary classification due to the number of trials

Cross-Subject Emotion Recognition. For the cross-subject emotion recognition task, we adopted leave-one-subject-out validation, where signals from one subject were used as the test set, and signals from the remaining subjects served as the training set. The corresponding results are displayed in Fig. 2. In this evaluation, we solely utilized EEG data, which aligns with the prevailing trend in current emotion recognition tasks based on physiological signals [17]. We also assessed the performance of GCNCA, Transformer, and DANN [15] models. Among these models, DANN with transfer learning techniques demonstrated the best performance, achieving an average accuracy of 71.90%. The success of DANN can be attributed to the effective utilization of the Gradient Reversal Layer (GRL) to minimize subject-related differences and extract domain-independent features. This approach eliminates the negative impact caused by inter-subject variations [18]. It is worth noting that the lowest performance was observed when the test set exclusively contained data from *Subject 04*.

Based on the aforementioned results, it is evident that the performance of *Subject 04* is consistently poor across all settings. The confusion matrices depicting the performance of *Subject 04* and *Subject 05* using ASGC, Transformer, and DCCA-AM models under the subject-dependent setting can be seen in Fig. 3. We can see that the results for *Subject 05* exhibit a significant effect across all three emotions, particularly positive emotion. Conversely, *Subject 04* demonstrates proficiency in recognizing neutral emotion but struggles with differentiating between positive and negative emotions when relying on EEG signals.

Upon closer examination of the source data, we discovered that, except for *Subject 04*, the remaining subjects actively or consciously engaged in activities

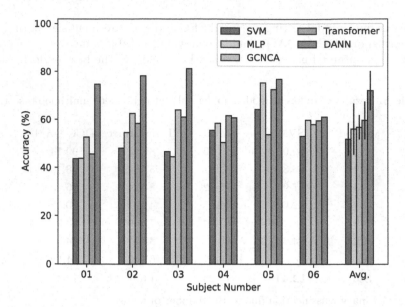

Fig. 2. The accuracies for each subject of each model and the averages under the cross-subject setting.

that evoked or elicited emotions, albeit in varying forms. Analyzing the activities among the subjects reveals that the other participants engaged in a wider range of behaviors, some of which led to noticeable changes in emotion, such as watching movies. In contrast, *Subject 04* predominantly participated in less emotionally stimulating activities, such as reading papers and performing official work. During these activities, emotions such as distress or calmness naturally

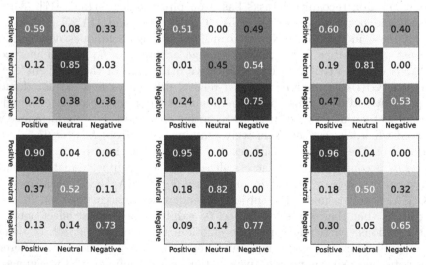

Fig. 3. Confusion matrices of *Subject 04* (top) and *Subject 05* (bottom) under subject-dependent experimental setting. Each column includes matrices under settings of EEG, eye movement and multimodal signals, respectively.

arise, but they are discontinuous and less apparent due to the subjects' primary focus on thinking or cognitive tasks.

4   Conclusions

In this paper, we have introduced a novel paradigm for emotion recognition and conducted a series of experiments to demonstrate its feasibility. Our approach involved recording various physiological signals while minimizing the influence of external stimuli, allowing us to capture naturally occurring emotions rather than relying on passively induced responses. The results obtained from our current models have shown promising performance, indicating the potential for non-passive stimuli-based emotion recognition in future applications, particularly in everyday contexts.

Moving forward, we plan to expand our dataset by collecting emotional data from a larger number of subjects under natural conditions. This will enable us to enhance the diversity and representativeness of our dataset, leading to more robust and reliable emotion recognition models. Additionally, we aim to develop efficient annotation tools that facilitate the continuous and natural collection of emotion labels, further improving the quality and granularity of our data.

By advancing our understanding of emotion recognition through non-passive stimuli and continuously refining our methodology, we believe our research will contribute to the development of more accurate and practical emotion recognition systems in various real-world applications.

Acknowledgments. This work was supported in part by grants from National Natural Science Foundation of China (Grant No. 61976135), STI 2030-Major Projects+2022ZD0208500, Shanghai Municipal Science and Technology Major Project (Grant No. 2021SHZDZX), Shanghai Pujiang Program (Grant No. 22PJ1408600), Medical-Engineering Interdisciplinary Research Foundation of Shanghai Jiao Tong University "Jiao Tong Star" Program (YG2023ZD25), and GuangCi Professorship Program of RuiJin Hospital Shanghai Jiao Tong University School of Medicine.

References

1. Faust, O., Hagiwara, Y., Hong, T.J., Lih, O.S., Acharya, U.R.: Deep learning for healthcare applications based on physiological signals: a review. Comput. Methods Programs Biomed. **161**, 1–13 (2018)
2. Brunner, C., et al.: BNCI horizon 2020: towards a roadmap for the BCI community. Brain-Comput. Interfaces **2**(1), 1–10 (2015)
3. Bijanzadeh, M., et al.: Decoding naturalistic affective behaviour from spectro-spatial features in multiday human iEEG. Nat. Hum. Behav. **6**(6), 823–836 (2022)
4. Lu, Y., Zheng, W., Li, B., Lu, B.: Combining eye movements and EEG to enhance emotion recognition. In: Proceedings of the Twenty-Fourth International Joint Conference on Artificial Intelligence (IJCAI), pp. 1170–1176 (2015)
5. Petrantonakis, P.C., Hadjileontiadis, L.J.: Emotion recognition from EEG using higher order crossings. IEEE Trans. Inf. Technol. Biomed. **14**(2), 186–197 (2009)

6. Zheng, W.L., Lu, B.L.: Investigating critical frequency bands and channels for EEG-based emotion recognition with deep neural networks. IEEE Trans. Auton. Ment. Dev. **7**(3), 162–175 (2015)
7. Kensinger, E.A.: Remembering emotional experiences: the contribution of valence and arousal. Rev. Neurosci. **15**(4), 241–252 (2004)
8. Plutchik, R.: Emotion. A Psychoevolutionary Synthesis. Harper & Row, New York (1980)
9. Kollias, D., et al.: Deep affect prediction in-the-wild: aff-wild database and challenge, deep architectures, and beyond. Int. J. Comput. Vision **127**(6), 907–929 (2019)
10. Duan, R.N., Zhu, J.Y., Lu, B.L.: Differential entropy feature for EEG-based emotion classification. In: 2013 6th International IEEE/EMBS Conference on Neural Engineering (NER), pp. 81–84. IEEE (2013)
11. Jiang, W.B., Zhao, L.M., Guo, P., Lu, B.L.: Discriminating surprise and anger from EEG and eye movements with a graph network. In: 2021 IEEE International Conference on Bioinformatics and Biomedicine (BIBM), pp. 1353–1357. IEEE (2021)
12. Liu, L.-D., Li, R., Liu, Y.-Z., Li, H.-L., Lu, B.-L.: EEG-based human decision confidence measurement using graph neural networks. In: Mantoro, T., Lee, M., Ayu, M.A., Wong, K.W., Hidayanto, A.N. (eds.) ICONIP 2021. CCIS, vol. 1517, pp. 291–298. Springer, Cham (2021). https://doi.org/10.1007/978-3-030-92310-5_34
13. Wang, Y., Jiang, W.B., Li, R., Lu, B.L.: Emotion transformer fusion: complementary representation properties of EEG and eye movements on recognizing anger and surprise. In: 2021 IEEE International Conference on Bioinformatics and Biomedicine (BIBM), pp. 1575–1578. IEEE (2021)
14. Liu, W., Zheng, W.L., Li, Z., Wu, S.Y., Gan, L., Lu, B.L.: Identifying similarities and differences in emotion recognition with EEG and eye movements among Chinese, German, and French people. J. Neural Eng. **19**(2), 026012 (2022)
15. Ganin, Y., et al.: Domain-adversarial training of neural networks. J. Mach. Learn. Res. **17**(1), 2030–2096 (2016)
16. Zhao, L.M., Li, R., Zheng, W.L., Lu, B.L.: Classification of five emotions from EEG and eye movement signals: complementary representation properties. In: 2019 9th International IEEE/EMBS Conference on Neural Engineering (NER), pp. 611–614. IEEE (2019)
17. Zhao, L.M., Yan, X., Lu, B.L.: Plug-and-play domain adaptation for cross-subject EEG-based emotion recognition. In: Proceedings of the AAAI Conference on Artificial Intelligence, pp. 863–870 (2021)
18. Zhang, J.-M., et al.: A cross-subject and cross-modal model for multimodal emotion recognition. In: Mantoro, T., Lee, M., Ayu, M.A., Wong, K.W., Hidayanto, A.N. (eds.) ICONIP 2021. CCIS, vol. 1517, pp. 203–211. Springer, Cham (2021). https://doi.org/10.1007/978-3-030-92310-5_24

Task Scheduling with Improved Particle Swarm Optimization in Cloud Data Center

Yang Bi, Wenlong Ni$^{(\boxtimes)}$, Yao Liu, Lingyue Lai, and Xinyu Zhou

School of Computer and Information Engineering, Jiangxi Normal University, Nanchang, China
{byang,wni,liuy,laily,xyzhou}@jxnu.edu.cn

Abstract. This paper proposes an improved particle swarm optimization algorithm with simulated annealing (IPSO-SA) for the task scheduling problem of cloud data center. The algorithm uses Tent chaotic mapping to make the initial population more evenly distributed. Second, a non-convex function is constructed to adaptively and decreasingly change the inertia weights to adjust the optimization-seeking ability of the particles in different iteration periods. Finally, the Metropolis criterion in SA is used to generate perturbed particles, combined with an modified equation for updating particles to avoid premature particle convergence. Comparative experimental results show that the IPSO-SA algorithm improves 13.8% in convergence accuracy over the standard PSO algorithm. The respective improvements over the other two modified PSO are 15.2% and 9.1%.

Keywords: Cloud Data Center · Task Scheduling · Particle Swarm Optimization · Simulated Annealing

1 Introduction

In cloud data centers, improving computing efficiency and reducing resource costs is always a challenging problem. Due to the heterogeneity and dynamics of cloud environments, resource allocation and task scheduling is considered as a NP hard problem, which is a non-deterministic problem of polynomial complexity. For such problems with long solution time and high complexity, it is intuitive to use simulated annealing algorithm (SA), genetic algorithm (GA), ant colony algorithm (ACA), particle swarm algorithm (PSO), etc.

PSO originated from the research on the foraging behavior of birds, and was first proposed by Dr. Eberhart and Dr. Kennedy [1]. The original purpose of this algorithm is to achieve the study of complex social behavior by simulating a simple social system [2]. After further research, it was found that PSO can be used to solve complex optimization problems. A particle in the swarm is a candidate solution. Through the cooperation and information sharing among individuals in the swarm, the optimal solution to the problem to be optimized is found in the search space of a given dimension. PSO only has three control parameters: inertia weight, cognitive acceleration coefficient, and social acceleration coefficient.

© The Author(s), under exclusive license to Springer Nature Singapore Pte Ltd. 2024
B. Luo et al. (Eds.): ICONIP 2023, LNCS 14449, pp. 277–287, 2024.
https://doi.org/10.1007/978-981-99-8067-3_21

A small change in any of these three parameters will bring about a difference in algorithm performance as shown in [3,4].

There are four main strategies to improve the standard PSO: modifying the control parameters of the PSO, mixing the PSO with GA, differential evolution algorithm and other meta-heuristic algorithms, collaboration and multi-swarm technology [5]. For the inertia weights in the PSO, literature [6] proposed a method to represent inertia weights with random values. Because in some applications, it is not easy to predict the required The size of the inertia weight, so this method is suitable for solving the dynamic environment. [7] proposed a linearly changing inertial weight, which can endow particles with the ability of individual optimization and collective optimization in different iteration periods. For the acceleration coefficients in the PSO, [8] proposed a hierarchical particle swarm algorithm with time-varying acceleration coefficients, using the number of iterations Dynamically adjust the particle swarm acceleration coefficient to improve the development and search capabilities of particles in the population. Literature [9] uses a mixture of PSO and GA to solve multimodal problems. Literature [10] divides the population into a master population and multiple subordinate populations, constituting a multi-group cooperative PSO.

Based on the PSO and SA, IPSO-SA is proposed in this paper to be used in cloud computing scenarios, with a multi-objective optimization considering both cloud computing execution time and resource cost. It is implemented from the following aspects:

1. The Tent chaotic map is used to initialize the particle swarm population to improve the traversal of the population.
2. A non-convex function decreasing function was constructed to modify the inertia weights as iterations were performed, adjusting the optimization-seeking ability of the particles at different times.
3. Hybrid Simulated Annealing algorithm is adopted, using the Metropolis criterion to disturb the population, thus improving the particle swarm velocity update equation to avoid particles falling into local optimal solutions.

It is worth pointing out that in our previous work, we also used Tent chaos mapping to improve the algorithm. In addition, the Cauchy mutation was introduced to realize the perturbation.

2   Problem Description

The cloud computing task scheduling model can be abstracted as transferring a large number of tasks to the data center and allocating them to different VMs in the resource pool through task execution resource allocation strategies. An appropriate allocation strategy can not only reduce the cost of cloud service providers, but also improve user satisfaction by reducing waiting time. A mathematical model for multi-objective optimization is established for task scheduling.

2.1 System Model

A virtual machine (VM) must be deployed on a PM in the Data Center. In this paper, we suppose that the number of VMs is n, $VM = \{vm_1, vm_2, vm_3, \ldots, vm_n\}$. The deployment and operation of each VM node is independent of each other. The VM nodes also have resources such as CPU, memory, storage, GPU and bandwidth. In this paper, we are mainly concerned with the computing and bandwidth resources of VMs. For example, $vm_j =< VMe_j, VMt_j >$ and $(j = 1, 2, 3, \ldots, n)$, where VMe and VMt are the computing power and bandwidth resources.

Each task is assigned to a VM for processing. Task scheduling sends each task to a selected VM for best performance. Assuming the total number of pending tasks is m, so the set of tasks is $T = \{t_1, t_2, t_3, \ldots, t_m\}$. In addition, $t_i =< Tl_i, Td_i >$ and $(i = 1, 2, 3, \ldots, m)$. Where Tl_i is the task length of t_i, which is proportional to the computation time of tasks. Td_i is the amount of data to be transferred for t_i, which is proportional to the transfer time of the task. Let x_{ij} be a binary variable indicating whether t_i is assigned on vm_j, the definition rules are as follows:

$$x_{ij} = \begin{cases} 1, t_i \text{ assign to } vm_j, \\ 0, \text{other.} \end{cases}$$

In the above equation, the constraint on x_{ij} is:

$$\sum_{j=1}^{n} x_{ij} = 1.$$

2.2 Fitness Function

The optimization objective of this paper is to minimize task execution time and execution cost. Therefore the total computation time, the total transmission time, the maximum end time and the execution cost of the task need to be defined. The details are as follows.

1. *Total Computation Time of Tasks.* According to the allocation strategy, t_i is assigned to the vm_j to perform computing tasks. Therefore, the computation time of the task is defined as follows:

$$ET_i = \sum_{j=1}^{n} x_{ij} \times \frac{Tl_i}{VMe_j}. \tag{1}$$

The total calculation time of a task is the sum of the time required to complete the task calculation by assigning a single task to a VM, so it is defined as follows:

$$ET = \sum_{i=1}^{m} ET_i. \tag{2}$$

Where, n is the total number of VMs and m is the total amount of tasks. The following is similar and will not be repeatedly defined.

2. *Total Transfer Time of Tasks.* Similar to the execution time, the transmission time of task data is related to the bandwidth resource of the VM. Therefore, the transmission time of the t_i is defined as follows:

$$TT_i = \sum_{j=1}^{n} x_{ij} \times \frac{Td_i}{VMt_j}. \tag{3}$$

The total task transfer time is the sum of all the individual task transfer times in the task sequence and is therefore defined as follows:

$$TT = \sum_{i=1}^{m} TT_i. \tag{4}$$

3. *The Maximum Ending Time of Tasks.* The time required for a complete cloud service is equal to the release time of the last VM in the cloud service, and the release time of a VM is:

$$T_j = \sum_{i=1}^{m} x_{ij} \times \left(\frac{TL_i}{VMe_j} + \frac{Td_i}{VMt_j} \right), (j = 1, 2, 3 \ldots, n). \tag{5}$$

Therefore, the maximum end time of the cloud service is:

$$T = \max (T_1, T_2, T_3, \ldots, T_n). \tag{6}$$

4. *Resource Occupation Cost.* Since the task set is accepted by the virtual node, the resource occupation cost is generated, so the following equation is defined:

$$C = (ET + TT) \times q. \tag{7}$$

The equation, q is the cost coefficient of cloud service per unit time.

Based on the above optimization objectives and problem description, the adaptation function constructed in this paper is as follows.

$$F = \alpha_1 f_1(x) + \alpha_2 f_2(x), \tag{8}$$

where $f_1(x)$ is the minimized task execution time, and $f_2(x)$ is the minimized resource cost, described as follows:

$$f_1(x) = \min(T), f_2(x) = \min(C). \tag{9}$$

Furthermore, in Eq. (8), α_1 and α_2 are the weight coefficients of the objective function. The difference between the weight coefficients determines the optimization focus of the resource allocation strategies.

3   Proposed IPSO-SA Algorithm

Based on PSO, this paper proposes IPSO-SA. This section is mainly concerned with the encoding and decoding of particles, and three major improvement strategies. These are Tent mapping initial population, non-linear dynamic adaptive inertia weights, and hybrid simulated annealing algorithm to improve the particle swarm update formulation, respectively.

3.1   Encoder and Decoder

Coding is used to map the position of particles to the solution space. In this paper, discretizing continuous position information is achieved through the minimal position principle (SPV) mentioned in [11], which assumes that the position information of the generation i particle is:

$$X_i = (x_1, x_2, \ldots, x_n).$$

The particles encoded using SPV is:

$$\lfloor |X_i| \rfloor = (\lfloor |x_1| \rfloor, \lfloor |x_2| \rfloor, \ldots, \lfloor |x_n| \rfloor).$$

For example, suppose there are 10 tasks are assigned to 5 VMs for processing. If the position information of the particle d in the iteration i is $\{2.34, 0.56, -4.3, 1.2, -1.56, 3.16, 4.56, -0.13, 2.09, -1.3\}$, the encoded position information is $\{2, 0, 4, 1, 1, 3, 4, 0, 2, 1\}$, and the corresponding allocation strategy is shown in the Table 1.

Table 1. Task scheduling strategy.

TaskID	1 2 3 4 5 6 7 8 9 10
VMID	2 0 4 1 1 3 4 0 2 1

3.2   Algorithm Improvements

The standard PSO algorithm defines the equation for updating the velocity and position of population particles [1]. Particle i updates its velocity and position according to Eq. (10) and (11) in the $t + 1$ iteration.

$$v_{in}^{t+1} = \omega v_{in}^t + c_1 \times r_1 \times \left(p_{in} - x_{in}^t\right) + c_2 \times r_2 \times \left(g_{in} - x_{in}^t\right), \tag{10}$$

$$x_{in}^{t+1} = x_{in}^t + v_{in}^{t+1}. \tag{11}$$

Where c_1 and c_2 are the cognitive acceleration coefficient and social acceleration coefficient, and r_1 and r_2 are two uniform random values generated within $[0, 1]$ interval. p_{in} is the individual optimal value of the current particle, and g_{in} is the overall optimal value of the population where the particle is located.

In the cloud scenario, PSO algorithm has the defects of unreasonable population initialization, easy convergence and premature, low convergence accuracy, and easy to fall into individual optimal solution [12]. This paper improves from the following aspects.

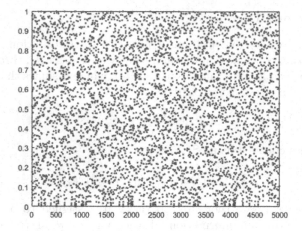

Fig. 1. Tent Chaos Map Chaos Value Distribution.

1. *Tent Map Initialization Population.* In the initial stage of PSO, the usual way is to use random functions to generate the initial information of particles. However, in the cloud scenario, it is necessary to use each VM as evenly as possible. At this time, the above method becomes less applicable. Therefore, in this paper, tent map is used to initialize the position and velocity of particles, as described in the following equation:

$$x_{n+1} = \begin{cases} \frac{x_n}{\alpha} & , (0 \leq x_n < \alpha), \\ \frac{1-x_n}{1-\alpha} & , (\alpha < x_n \leq 1). \end{cases}$$

Figure 1 shows the chaos values generated by the Tent Map. As can be seen from the figure, the sequence generated by the Tent Map is well distributed and random.

2. *Inertial Weights.* The ω in the PSO algorithm is called the inertia weight and is usually set to a fixed value. But, the results of the literature [12] show that when the number of iterations increases, taking a fixed-value solution leads to an amplification of many of the details of the problem that need to be solved. In addition, [12] indicates that, when $0 < \omega < 1$, particles gradually converge with the increase of iteration times. Larger ω are beneficial for finding the overall optimal solution. Smaller ω are beneficial for finding individual optimal solutions. According to the fitness function in this paper, we can know that the optimization problem in this paper belongs to the minimum value problem, so ω should gradually decrease with the increase of the number of iterations. Based on the above theory, this article adopts a nonlinear decreasing function to optimize ω. The value of ω can be changed adaptively during the iteration process. The improved nonlinear adaptive inertia weight iteration equation is as follows.

$$\omega = \omega_{\min} + (\omega_{\max} - \omega_{\min}) \times e^{\left(-\frac{2 \times t}{T_{\max}}\right)}. \tag{12}$$

In the Eq. (12), $\omega_{\max}$ is the maximum inertia weight. $\omega_{\min}$ is the minimum inertia weight. t is the current iteration number, and $T_{\max}$ is the maximum iteration number.

3. *Metropolis Criterion.* The Metropolis criterion is one of the core factor of SA algorithms [13], which does not use completely deterministic rules but chooses to accept new states with probability. In the t-th iteration, the algorithm will randomly generate a new solution g'_{in} in the neighborhood of all optimal g_{in} obtained by the particle swarm algorithm, and then determine whether to accept the solution through the Metropolis criterion. The Metropolis criteria are as follows:

$$\Delta F = F\left(g'_{in}\right) - F\left(g_{in}\right),\tag{13}$$

$$p\left(g_{in} \to g'_{in}\right) = \begin{cases} 1 & , \Delta F < 0, \\ e^{-\frac{\Delta F}{T}} & , \Delta F > 0. \end{cases}\tag{14}$$

Where ΔF is the difference value of fitness function, T is the current temperature, which is changed according to the following equation.

$$T_0 = -\frac{F\left(g_{0n}\right)}{\log(0.2)},\tag{15}$$
$$T = K \times T_0.$$

Where T_0 is the initial temperature, K is the cooling coefficient. Using the Metropolis criterion, we improve Eq. (11) as follows.

$$\begin{aligned} v_{in}^{t+1} =& \omega \times v_{in}^t + c_1 \times r_1 \times \left(p_{in} - x_{in}^t\right) \\ &+ c_2 \times r_2 \times \left(g_{in} - x_{in}^t\right) \\ &+ c_3 \times r_3 \times \left(g'_{in} - x_{in}^t\right). \end{aligned}\tag{16}$$

Where, c_1, c_2 and c_3 are the particle acceleration coefficients. The constraint conditions for using new solutions generated within the neighborhood as perturbed particles are:

$$p\left(g_{in} \to g'_{in}\right) > \mathrm{rand}(0, 1).$$

In addition, a contraction factor is introduced to ensure the contractility of the population particles, and the particle position is updated according to Eq. (17).

$$\chi = \frac{2}{\left|2 - C - \sqrt{C^2 - 4 \times C}\right|},\tag{17}$$
$$x_{in}^{t+1} = x_{in}^t + \chi v_{in}^{t+1}.$$

Where, $C = c_1 + c_2 + c_3$.

4 Simulation Results

In this paper the CloudSim software is adopted for simulation experiments, which is developed by [14]. Table 2 shows the experimental parameter configuration.

Table 2. VMs configuration table

Parameter	Value
Processing speed of VMs/MIPS	[200,600]
Bandwidth of VMs/Mbps	[1000,2500]
Memory of VMs/GB	1.70
Number of VMs	20
Number of tasks	200
Task length	[25000,250000]
Task data volume	[100,600]

In addition, the data for the VMs and tasks to be used in this paper were randomly generated based on the Table 2.

This paper uses the standard PSO, IPSO_1 in [8] and IPSO_2 [15], for comparison experiments with the proposed IPSO-SA. The strategy in [8] uses an adaptive acceleration factor, and [15] uses an inertial weight curve descent strategy. The population size is set to 25 and the number of iterations is 1000 for all the above four algorithms. The values of other parameters are referred to Table 3. In addition, in this paper, the parameter $q = 1.0$ in Eq. (7), $\alpha_1 = 0.7$ and $\alpha_2 = 0.3$ in Eq. (8).

Table 3. Algorithm parameter

PSO	inertia weight	0.9
	learning factor	$c_1 = c_2 = 2.05$
IPSO_1	inertia weight	0.9
	learning factor	$c_{max} = 2.5, c_{min} = 0.5$
IPSO_2	inertia weight	$\omega_{max} = 0.9, \omega_{min} = 0.2$
	learning factor	$c_1 = c_2 = 2.05$
	Inertia Curve Parameters	-0.95
IPSO-SA	inertia weight	$\omega_{max} = 0.9, \omega_{min} = 0.2$
	learning factor	$c_1 = c_2 = 2.05, c_3 = 0.5$
	The initial temperature	$T = 1000000$
	cooling coefficient	$K = 0.998$

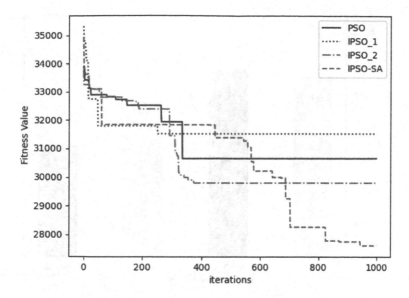

Fig. 2. Comparison of algorithm performance.

The comparison experiments are performed several times under the above parameter conditions. The algorithm performance comparison is shown in Fig. 2.

Fifty comparison experiments were conducted and the convergence accuracy of each algorithm is shown in the Table 4.

Table 4. Comparison of algorithm convergence accuracy

Algorithm	Maximum Fitness	Minimum Fitness	Average Fitness
PSO	31969.61723	29618.40511	31128.05389
IPSO_1	32276.67126	31023.48934	31635.17231
IPSO_2	31776.67126	27705.92521	29512.43004
IPSO-SA	29749.45768	23275.83437	26819.8504

According to Eq. 8, the optimisation objective of this paper lies in minimising the execution time and execution resources of the task. Therefore, the smaller the fitness value, the better the scheduling strategy.

Based on the results of the above-mentioned multiple comparison experiments, the IPSO-SA algorithm, after introducing simulated annealing perturbed particles, better avoids the defects of premature convergence of the PSO. In terms of algorithm convergence accuracy, IPSO-SA improves 13.8% over PSO. It is 15.2% better than IPSO_1. The improvement over IPSO_2 is 9.1%. Figure 3 shows a comparison of the average fitness values of the four algorithms.

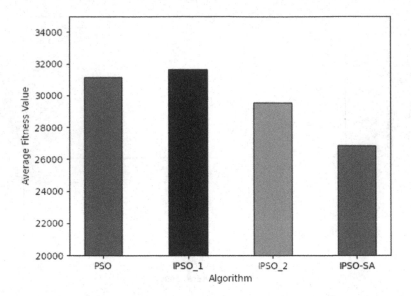

Fig. 3. Comparison of average fitness values.

Combining the analysis results in Fig. 2 and Table 4, it can be concluded that the IPSO-SA algorithm have better convergence accuracy and at the same time, and reducing the probability of premature convergence.

5   Conclusion

In this paper, we propose the improved PSO, which is used to solve the task scheduling problem in cloud data centers. Firstly, the initial population is made ergodic by Tent chaotic mapping. Then, we constructed a non-convex function decreasing function, so that the inertia weights can be adaptively modified with iterations, to achieve the purpose of adjusting the optimization ability of particles in different periods. Finally, the probability of premature convergence of the algorithm is reduced by using improved velocity and position update formulas, introducing the Metropolis criterion for generating perturbed particles, and modifying the perturbation triggering conditions. Comparative experimental results show that the IPSO-SA proposed in this paper has high convergence accuracy, and the particles do not remain in a locally optimal solution for long.

In future work, we will consider improving the convergence speed of the algorithm as a new optimization goal. Meanwhile, we consider applying the swarm intelligence algorithm to more application scenarios.

References

1. Kennedy, J., Eberhart, R.: Particle swarm optimization. In: Proceedings of ICNN 1995-International Conference on Neural Networks, vol. 4. IEEE (1995)

2. Garnier, S., Gautrais, J., Theraulaz, G.: The biological principles of swarm intelligence. Swarm Intell. **1**, 3–31 (2007)
3. Eltamaly, A.M.: A novel strategy for optimal PSO control parameters determination for PV energy systems. Sustainability **13**(2), 1008 (2021)
4. Harrison, K.R., Engelbrecht, A.P., Ombuki-Berman, B.M.: Optimal parameter regions and the time-dependence of control parameter values for the particle swarm optimization algorithm. Swarm Evol. Comput. **41**, 20–35 (2018)
5. Shami, T.M., et al.: Particle swarm optimization: a comprehensive survey. IEEE Access **10**, 10031–10061 (2022)
6. Li, M., et al.: A multi-information fusion "triple variables with iteration" inertia weight PSO algorithm and its application. Appl. Soft Comput. **84**, 105677 (2019)
7. Shi, Y., Eberhart, R.C.: Empirical study of particle swarm optimization. In: Proceedings of the 1999 Congress on Evolutionary Computation-CEC99 (Cat. No. 99TH8406), vol. 3. IEEE (1999)
8. Ratnaweera, A., Halgamuge, S.K., Watson, H.C.: Self-organizing hierarchical particle swarm optimizer with time-varying acceleration coefficients. IEEE Trans. Evol. Comput. **8**(3), 240–255 (2004)
9. Kao, Y.-T., Zahara, E.: A hybrid genetic algorithm and particle swarm optimization for multimodal functions. Appl. Soft Comput. **8**(2), 849–857 (2008)
10. Niu, B., et al.: MCPSO: a multi-swarm cooperative particle swarm optimizer. Appl. Math. Comput. **185**(2), 1050–1062 (2007)
11. Alguliyev, R.M., Imamverdiyev, Y.N., Abdullayeva, F.J.: PSO-based load balancing method in cloud computing. Autom. Control. Comput. Sci. **53**, 45–55 (2019)
12. Parsopoulos, K.E., et al.: Improving particle swarm optimizer by function "stretching". In: Hadjisavvas, N., Pardalos, P.M. (eds.) Advances in Convex Analysis and Global Optimization Nonconvex Optimization and Its Applications, vol. 54, pp. 445–457. Springer, Boston (2001). https://doi.org/10.1007/978-1-4613-0279-7_28. Ch 3
13. Van Laarhoven, P.J.M., et al.: Simulated Annealing. Springer, Dordrecht (1987). https://doi.org/10.1007/978-94-015-7744-1
14. Calheiros, R.N., et al.: CloudSim: a toolkit for modeling and simulation of cloud computing environments and evaluation of resource provisioning algorithms. Softw. Pract. Exp. **41**(1), 23–50 (2011)
15. Lei, K., Qiu, Y., He, Y.: A new adaptive well-chosen inertia weight strategy to automatically harmonize global and local search ability in particle swarm optimization. In: 2006 1st International Symposium on Systems and Control in Aerospace and Astronautics. IEEE (2006)

Traffic Signal Optimization at T-Shaped Intersections Based on Deep Q Networks

Wenlong Ni[1,2](✉), Chuanzhuang Li[2], Peng Wang[2], and Zehong Li[2]

[1] School of Computer Information Engineering, JiangXi Normal University, NanChang, China
[2] School of Digital Industry, JiangXi Normal University, ShangRao, China
{wni,lichuanzhuang,peng.wang,Zehong.Li}@jxnu.edu.cn

Abstract. In this paper traffic signal control strategies for T-shaped intersections in urban road networks using deep Q network (DQN) algorithms are proposed. Different DQN networks and dynamic time aggregation were used for decision-makings. The effectiveness of various strategies under different traffic conditions are checked using the Simulation of Urban Mobility (SUMO) software. The simulation results showed that the strategy combining the Dueling DQN method and dynamic time aggregation significantly improved vehicle throughput. Compared with DQN and fixed-time methods, this strategy can reduce the average travel time by up to 43% in low-traffic periods and up to 15% in high-traffic periods. This paper demonstrated the significant advantages of applying Dueling DQN in traffic signal control strategies for urban road networks.

Keywords: reinforcement learning · deep Q network · optimal control · traffic signal

1 Introduction

Traffic control systems play a key role in optimizing urban traffic efficiency and dealing with traffic congestion and special needs. However, due to the uncertainty and randomness of traffic intersections, people have been designing and optimizing intelligent traffic decision systems to solve these problems in the past few decades. These systems typically collect real-time information such as the number of vehicles, traffic rate, queue length, etc. for analysis by deploying road detection equipment [1–4]. With the rapid development of traffic intelligence, the establishment of on-board equipment and road detection facilities has greatly reduced costs and provided favorable conditions for improving traffic control systems. Traffic signal control types can be generally divided into timing control [5], fixed period responsive control method [6] and aperiodic real-time control method [7].

This work was supported in part by JiangXi Education Department under Grant No. GJJ191688.

This paper presents a novel adaptive traffic signal control approach using Deep Reinforcement Learning (DRL). This method integrates techniques like DQN and Dueling DQN while embedding a temporal dimension into its actions. With real-time road data, the system optimally controls traffic signals, significantly enhancing urban traffic efficiency. Four experiments underscore its efficacy in alleviating traffic congestion and promoting smoother traffic flow.

The second section introduces the relevant work and theoretical background. The third section describes the adaptive traffic signal control strategy, including the choice of reward function, state and action space, and the method and training process of DQN model. The fourth section verifies the effectiveness of the strategy through simulation experiments, and verifies the effectiveness of the dynamic timing combined with Dueling DQN method. The fifth part summarizes the content of the article and puts forward the future research direction.

2   Related Research

With the development of computer technology, data communication and onboard equipment, urban traffic flow statistics have changed from traditional manual methods to network statistics, which enables us to analyze road conditions and carry out data-driven traffic control with large amounts of data [8]. A regional detection platform can be built to provide reasonable basis for decision-making [9,10]. In this context, adaptive decision-making and reinforcement learning(RL) have made remarkable achievements in regional traffic research [11]. RL solves the shortcomings of Q-Learning in dealing with large-scale intersections and gives play to its advantages in continuous decision making [12]. In addition, by abstracting signal control lights as agents and making decisions on the current state in the Markov decision process, the control strategy based on RL realizes true closed-loop control [13]. This strategy can detect and respond to traffic flow changes in real time, and expand all impact factors within the control range.

Watkins et al. [14] proved that the Q-Learning algorithm can converge to the optimal Q function. Thorpe [15] applied Q-Learning algorithm to traffic control. Mikami et al. [16] introduced the dimension problem of RL [17] in traffic signal control. Mnih et al. [18] proposed DQN algorithm by combining deep neural network and Q-Learning algorithm. Li et al. [19] used deep learning technology to study and improve the single intersection problem. However, the traditional DQN architecture maximizes the target network when it is updated, resulting in the overestimation of action value. In order to avoid overestimation, Hasselt et al. [20] used double Q network to separate actions from strategies. Wang et al. [21] proposed Dueling DQN architecture for estimating state value function and action value, reducing the range of Q values and ensuring algorithm stability.

3   System Description

In this paper, the traffic signal control problem is abstracted as a Markov decision process, and the vehicle information provided by the pavement lane detector is used as the state input of DQN to realize the mapping of state and action, so as to optimize the control strategy.

3.1   Traffic Environment

In this paper, the common T-shaped intersection of the urban network is taken as an example, which can be seen from Fig. 1.

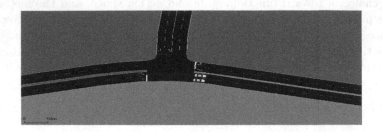

Fig. 1. An example of T-shaped intersection.

The whole road information consists of three inbound directions (I1, I2, I3) and three outbound directions (O1, O2, O3). Each inbound lane is installed with a regional detector to collect vehicle information. The node connecting the road network is provided with a traffic signal light that controls the phase of all lanes, through which vehicle collaborative control can be achieved. T-shaped intersection mapping Fig. 2 is as follows:

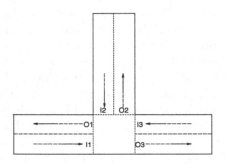

Fig. 2. Road inbound and outbound diagram.

By default, intelligent detection devices are installed in all lanes. Collaborative control of traffic flow is mainly controlled by traffic lights. All phases are shown in Fig. 3:

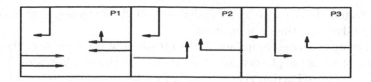

Fig. 3. Phase diagram of signal control.

3.2 DQN Model

The elements of DQN model include: states, actions, rewards, strategies, and deep neural networks, in the concrete definition, which can be represented a tuple (O, A, R, O_-), where O represents the state vector, A represents the action, R represents the reward and O_- is the next state vector after the end of the action. Strategy (π) is a function that maps a given state (O) to an action (A), while deep neural networks are used to estimate the Q function $(Q(O, A))$, which represents the expected reward for taking action A in state O.

State Space. According to the T-shaped intersection situation, the information that can be observed by the agent is the area detector placed on all inbound lanes. Therefore, the state space is a state vector composed of 6 different groups of inbound lane information, which is represented by $O = [O_1, O_2, ..., O_n]$, where O_n represents the traffic flow state of the n-th inbound lane. The state of a certain traffic flow is represented by $O_n = \{s_1, s_2, ..., s_i\}$, where i represents the information of the i-th vehicle in the current lane. In order to represent the detailed information of the lane, a binary group (T, S) is formed by obtaining the vehicle type T and the current speed S, for example, if the passenger car type number is 10 and its current speed is 5.0 km/h, the binary group is (10, 5).

Taking the inbound lane I3 as an example, Fig. 4 is part of the situation of vehicles in operation, Fig. 4(a) represents the real state of traffic flow on the road, and Fig. 4(b) represents the position distribution and speed matrix of vehicles in the current traffic flow.

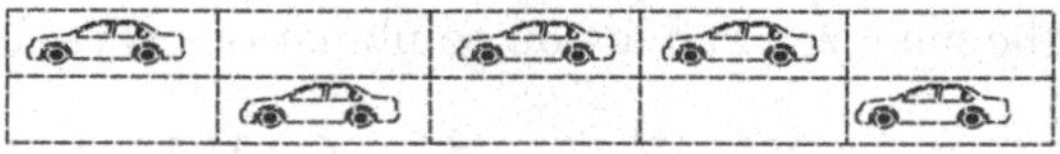

(a) Vehicle distribution.

(1,0.5)	(1,5)	(1,5)	0	0
(1,5)	(1,10)	0	0	0

(b) Vehicle position velocity matrix.

Fig. 4. Road element acquisition diagram.

Action Space. In the selection of action, two kinds of strategies are used: fixed timing and dynamic time combination.

First, using the fixed timing strategy to realize the basic phase of the system. The action space is the pre-defined phase scheme in the road network. The value range of the selected action space is {0, 1, 2} when fixed timing, and the phase representation is shown in Fig. 3. For example, when phase i is selected, the phase numbered $2i$ is switched for execution, while the phase numbered $2i + 1$ is executed passively. All phase schemes are shown in Table 1. The setting of control information is mainly composed of four parameters, indicating the access rights of each lane at this phase, where G represents that vehicles can pass through, g represents only straight or right turns are allowed, r represents the red light, and y represents the yellow light.

Table 1. Phase information.

Phase numbering	Phase combination
0	grggGGgGGrg
1	grggyygyyrg
2	grggrrgrrGg
3	grggrrgrryg
4	gGggrrgrrrg
5	gyggrrgrrrg

The dynamic time combination and adaptive decision-making make full use of the approximation ability of DRL to realize real-time learning and optimization of traffic signal timing. It dynamically obtains the traffic status at the intersection and adjusts the timing of the signal light according to the current traffic status, so as to optimizing the traffic flow.

Assuming that the maximum duration of a green light is MAX_G and the minimum duration of a green light is MIN_G, and the maximum duration of a yellow light is MAX_Y and the minimum duration of a yellow light is MIN_Y, the equation for the number of all action combinations is as follows:

$$totalPhaseCount = 3(MAX\_G - MIN\_G)(MAX\_Y - MIN\_Y) \quad (1)$$

Reward Function. Traffic metrics can effectively reflect the operation of vehicles and the status of traffic conditions. Metrics include the waiting number of vehicles, the waiting length of lanes, and the average waiting time.

In the context of traffic control systems, N_{t-1} and W_{t-1}^i are commonly used to represent specific traffic parameters. N_{t-1} typically represents the total number N of vehicles waiting in queue at time $t - 1$. It is a count of all vehicles that are not able to proceed through the intersection due to traffic controls (e.g., red lights) or other impediments (e.g., traffic congestion). W_{t-1}^i typically represents the accumulated waiting time W for vehicle i at time $t-1$. This is the total time that vehicle i has been waiting in queue. The index i indicates that this metric is calculated for each vehicle individually.

These parameters can provide a measure of traffic congestion at the intersection. By monitoring changes in these parameters over time, traffic control systems can adjust signal timings and other control strategies to improve traffic flow and reduce congestion. The equation for calculating the average waiting time of vehicles at the intersection is as follows:

$$AVG\_W^t = \frac{1}{N_t} \sum_{i}^{N_t} W_i^t \qquad (2)$$

The equation shows that a good phase will reduce the average waiting time of the intersection. The equation of reward value function is as follows:

$$R_t = AVG\_W^{t-1} - AVG\_W^t \qquad (3)$$

As can be seen from Eq. (3), it reflects the effect of traffic signal control strategies by using the average waiting time difference of two moments as the reward, and takes reducing the average waiting time as the optimization objective to reward effective strategies positively and inhibit ineffective ones.

Dueling DQN. As an improved DQN architecture, Dueling DQN is able to more efficiently learn and estimate expected future rewards for taking a particular action in a given state. The architecture diagram is shown in Fig. 5:

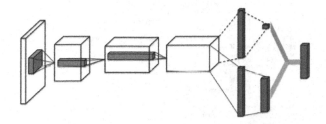

Fig. 5. The Dueling DQN architecture diagram.

The key idea of Dueling DQN is to decompose Q value function into state value function and action advantage function, so as to realize action selection based on value estimation. The network structure consists of two parts: the state value network and the action advantage network, which respectively estimate the expected value of the current state and the advantage of each action over the state. In the implementation process, average is used instead of maximization operation, and the equation is as follows:

$$Q_{\eta,\alpha,\beta}(s,a) = V_{\eta,\alpha}(s) + A_{\eta,\beta}(s,a) - \frac{1}{|A|}\sum_{a'} A_{\eta,\beta}(s,a') \tag{4}$$

In this equation, $V_{\eta,\alpha}(s)$ is the state value function, and $A_{\eta,\beta}(s,a)$ is the advantage function of taking different actions in this state, indicating the difference of taking different actions. η is the network parameter shared by the state value function and the advantage function. α and β are parameters of state value function and advantage function, respectively. The average value of Q is subtracted from the advantage function $A(s,a)$ when calculating the value of Q, making the network pay more attention to the relative advantage of the action.

Pseudo Code. The agent needs to obtain the T-shaped intersection state in the environment and get the maximum accumulable reward for each decision it makes. The pseudo-code of the training process is shown in Fig. 6.

Algorithm 1 DQN and Dueling DQN for Traffic Signal Control

```
 1:  Randomly initialize the weights of the DQN or Dueling DQN model
 2:  Initialize the capacity of the experience replay memory M
 3:  for each episode = 1 to E do
 4:      Initialize the state s
 5:      Preheat the system for some time
 6:      for each step t = 1 to T do
 7:          With probability ε, select a random action a
 8:          Otherwise select a = arg max_{a'} Q(s,a')
 9:          Execute action a, record the current average waiting time of the
            vehicles as pre_avg_time
10:          if using a specific time configuration then
11:              Set the green light duration G_t based on action a
12:          else
13:              Set the default green light duration G_t = 30s
14:          end if
15:          Record the current average waiting time of the vehicles as cur_avg_time
16:          Compute the reward r = pre_avg_time − cur_avg_time
17:          Observe the reward r and the new state s'
18:          Store the transition tuple (s, a, r, s') in M
19:          if replay memory is full then
20:              Randomly sample a batch of transitions (s, a, r, s') from M
21:              if method = DQN then
22:                  Compute the Q-learning target y = r + γ max_{a'} Q(s',a')
23:              else if method = Dueling DQN then
24:                  Split Q(s,a) into value V(s) and advantage A(s,a)
25:                  Compute the Q-learning target
26:                      y = r + γ(V(s') + A(s',a') − max_{a'} A(s',a'))
27:              end if
28:              Gradient descent minimize the loss (y − Q(s,a))²
29:          end if
30:          s = s'
31:          pre_avg_time = cur_avg_time
32:      end for
33: end for
```

Fig. 6. Pseudo-code of model training process.

By clearly separating state value from action advantage, Dueling DQN is able to more effectively estimate the expected future reward for taking a particular action in a given state. In the traffic signal control problem, when the change of signal light status does not significantly affect the traffic flow, such as sparse or excessive traffic flow, the structure of Dueling DQN makes the intelligent body more focused on learning valuable environmental states, rather than choosing which action is the best under the state.

4 Simulation Result

To verify the traffic efficiency of the proposed method, this paper uses SUMO software to control traffic signals and OpenStreetMap (OSM) to obtain actual road network parameters. OSM can download real road network data to realize the operation effect of the vehicle in the real environment.

4.1 Parameters

In the simulation process, the vehicle arrival rate is based on the traffic volume every 60 s, and its arrival time conforms to the exponential distribution. The end condition is that 6,000 vehicles have passed through the T-shaped intersection. Vehicle arrival rates can help manage traffic flow patterns, predict future traffic conditions, and evaluate different traffic management strategies.

The model training parameters are shown in Table 2.

Table 2. Training parameters.

Parameter	Value
Total number of iterations	500
Total number of vehicles	6000
Maximum time limit	20000
Default duration of green light	30
Default duration of yellow light	3
Maximum duration of the same phase	64
Preheat second	200
Learning rate	0.0001
Exploration rate	0.05
Experience pool size	2000
Sample batch size	128
Target network update period	200
Attenuation factor	0.8

4.2 Numerical Analysis

In this paper, the adaptive control scheme introduced a regional detector to obtain pavement conditions in real time, and made dynamic decisions according to the current optimal phase. All experiments set 6000 vehicles as the end condition, while maintaining the same neural network structure, to evaluate the differences of various performance metrics under different combinations of timing strategies and decision methods. By focusing on the comparison between

Dueling DQN and DQN under different action selection conditions, simulation tests are conducted from multiple dimensions to comprehensively compare the effects of the strategies. All the policies are shown in Table 3.

Table 3. Experimental timing methods and strategies.

Number	Timing strategy	Decision-making mode
Scheme 1	Fixed timing	Predefined decision
Scheme 2	Fixed timing	DQN
Scheme 3	Fixed timing	Dueling DQN
Scheme 4	Dynamic time combination	DQN
Scheme 5	Dynamic time combination	Dueling DQN

As can be seen from Fig. 7(a), under different vehicle arrival rates, the average waiting time of Scheme 1 increases rapidly with the arrival speed of vehicles. Among the two schemes of dynamic timing, Scheme 5 has the lowest average waiting time in the whole simulation process, indicating that vehicles can pass quickly at each selected phase and time, so as to saving more time.

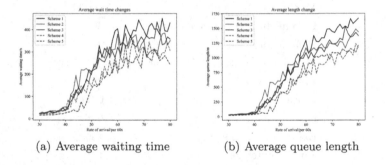

(a) Average waiting time   (b) Average queue length

Fig. 7. Average wait time and queue length change graph.

The average queue length represents the overall queuing situation of the road at the T-shaped intersection. In Fig. 7(b), under different arrival rates, the average queue length of Scheme 1 is the highest, while Scheme 2 and Scheme 3 with adaptive decision-making can effectively reduce the queue length, which also indicates that the decision made by the agent can choose to let the vehicle pass as soon as possible. The use of Schemes 4 and 5 has more time combinations, and also reduces the queue length to a greater extent, indicating the advantages of using this schemes.

The travel time and time loss of vehicles reflect the overall passing time of vehicles in the road network and the time loss when the speed is lower than the ideal, respectively. These two metrics can effectively reflect the traffic condition of the intersection. As shown in Fig. 8, Scheme 5 outperforms the other approaches on two metrics.

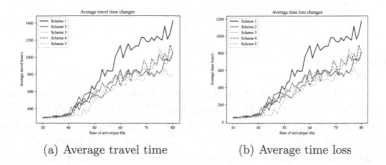

(a) Average travel time   (b) Average time loss

Fig. 8. Graph of average travel time and time loss as arrival rate increases.

From the comparison of the above metrics, it can be seen that compared with dynamic timing combination, adaptive decision-making has more obvious advantages. In terms of decision-making selection, Dueling DQN combined with more time combinations is superior to DQN.

Based on the results in Table 4, the performance of different schemes can be observed under varying vehicle arrival rates and quantities. When the number of vehicles is 4000, 8000, and 14000, Scheme 4 reduces the travel time by 10%, 43%, and 33% respectively compared to Scheme 2. However, at an arrival rate of 75, the average travel time of Scheme 4 is actually longer than that of Scheme 2 in some cases, increasing by 5%, 24%, and 1% respectively. This phenomenon may be attributed to the fact that the dynamic timing strategy used in Scheme 4 is not always superior to the fixed timing strategy under high traffic conditions. In high traffic, the changes in traffic flow may be more complex and unpredictable, and the dynamic timing strategy may not be able to adapt to these rapid changes in time. Conversely, the fixed timing strategy, although more conservative, may be more stable and reliable in some high traffic situations. Therefore, the performance of Scheme 4 may not be as good as Scheme 2 in some high traffic scenarios.

Meanwhile, Table 4 also demonstrates the advantage of Scheme 5 over Scheme 2. By combining more time combinations, Dueling DQN can reduce travel time by 17%, 43%, and 42% respectively when the number of vehicles is 4000, 8000, and 14000. During peak hours, when the T-shaped intersection is fully occupied, Scheme 5, with more choices in time combinations, plays a good role in adaptive decision-making, reducing the average travel time of vehicles by 2%, 8%, and 15%.

These results highlight the potential of dynamic timing and Dueling DQN in traffic signal control but also reveal the importance of choosing the appropriate strategy under different traffic conditions.

Table 4. Test results of signal control under different schemes.

Arrival rate	Vehicle number	Scheme 2 Average travel time/s	Scheme 4 Average travel time/s	Scheme 5 Average travel time/s	Scheme 4 Decline ratio/%	Scheme 5 Decline ratio/%
50	4000	547.9	493	450.0	10%	17%
	8000	926.2	531	526.8	43%	43%
	14000	1186.0	792	682.5	33%	42%
75	4000	683.0	715	666.3	-5%	2%
	8000	957.3	1190	879.2	-24%	8%
	14000	1531.1	1547	1305.7	-1%	15%

5   Conclusion

In this study, a novel adaptive traffic signal control strategy has been introduced, utilizing the Dueling DQN algorithm. The core strength of the Dueling DQN lies in its unique ability to separate state values and action advantages, allowing for a more precise estimation of Q values. This innovative approach has been meticulously applied to minimize the average waiting time for vehicles, serving as the reward function in the model. Through extensive experimentation, it has been confirmed that the Dueling DQN-based approach significantly enhances traffic flow and alleviates congestion. The results demonstrate not only the effectiveness of the method but also its potential applicability in real-world traffic management. Looking to the future, the aim is to extend these techniques to more intricate traffic scenarios, such as multi-T-shaped intersections, and to integrate vehicle priority considerations. This will pave the way for a more coordinated and efficient traffic management system, further showcasing the robust performance of the Dueling DQN.

References

1. Liang, X., Du, X., Wang, G., Han, Z.: A deep reinforcement learning network for traffic light cycle control. IEEE Trans. Veh. Technol. **68**(2), 1243–1253 (2019)
2. Chin, Y.K., Bolong, N., Yang, S.S., Teo, K.: Exploring q-learning optimization in traffic signal timing plan management. In: Third International Conference on Computational Intelligence (2011)
3. Abdoos, M., Mozayani, N., Bazzan, A.: Holonic multi-agent system for traffic signals control. Eng. Appl. Artif. Intell. **26**(5–6), 1575–1587 (2013)
4. Balaji, P.G., German, X., Srinivasan, D.: Urban traffic signal control using reinforcement learning agents. IET Intel. Transport Syst. **4**(3), 177–188 (2010)
5. Varaiya, P.: The max-pressure controller for arbitrary networks of signalized intersections. In: Ukkusuri, S., Ozbay, K. (eds.) Advances in Dynamic Network Modeling in Complex Transportation Systems. Complex Networks and Dynamic Systems, vol. 2. Springer, New York (2013). https://doi.org/10.1007/978-1-4614-6243-9_2
6. Bazzan, A.: Opportunities for multiagent systems and multiagent reinforcement learning in traffic control. Auton. Agent. Multi-Agent Syst. **18**(3), 342 (2009)
7. Abdulhai, B., Pringle, P.: Autonomous multiagent reinforcement learning5 GC urban traffic control. In: Annual Transportation Research Board Meeting (2003)
8. Hamilton, A., Waterson, B., Cherrett, T., Robinson, A., Snell, I.: The evolution of urban traffic control: changing policy and technology. Transp. Plan. Technol. **36**, 24–43 (2013)

9. Zhang, J., Wang, F.Y., Wang, K., Lin, W.H., Xu, X., Chen, C.: Data-driven intelligent transportation systems: a survey. IEEE Trans. Intell. Transp. Syst. **12**(4), 1624–1639 (2011)
10. Wu, X., Liu, H.X.: Using high-resolution event-based data for traffic modeling and control: an overview. Transp. Res. Part C Emerg. Technol. **42**(2), 28–43 (2014)
11. El-Tantawy, S., Abdulhai, B.: Towards multi-agent reinforcement learning for integrated network of optimal traffic controllers (Marlin-OTC). Transp. Lett. **2**(2), 89–110 (2010)
12. Genders, W., Razavi, S.: Using a deep reinforcement learning agent for traffic signal control (2016)
13. Abdulhai, B., Pringle, R., Karakoulas, G.J.: Reinforcement learning for true adaptive traffic signal control. J. Transp. Eng. **129**(3), 278–285 (2003)
14. Watkins, C., Dayan, P.: Q-learning. In: Machine Learning (1992)
15. Thorpe, T.L.: Vehicle traffic light control using SARSA (1997)
16. Mikami, S., Kakazu, Y.: Genetic reinforcement learning for cooperative traffic signal control. In: IEEE Conference on Evolutionary Computation, IEEE World Congress on Computational Intelligence (1994)
17. Volodymyr, M., et al.: Human-level control through deep reinforcement learning. Nature **518**(7540), 529–533 (2015)
18. Mnih, V., et al.: Playing Atari with deep reinforcement learning. Computer Science (2013)
19. Li, L., Lv, Y., Wang, F.Y.: Traffic signal timing via deep reinforcement learning. IEEE/CAA J. Automatica Sinica **3**, 247–254 (2016)
20. Hasselt, H.V., Guez, A., Silver, D.: Deep reinforcement learning with double q-learning. In: National Conference on Artificial Intelligence (2016)
21. Freitas, N.D., Lanctot, M., Hasselt, H.V., Hessel, M., Schaul, T., Wang, Z.: Dueling network architectures for deep reinforcement learning. In: International Conference on Machine Learning (2016)

A Multi-task Framework for Solving Multimodal Multiobjective Optimization Problems

Xinyi Wu[✉], Fei Ming, and Wenyin Gong

School of Computer Science, China University of Geosciences, Wuhan, WH 430074, China
m18392150842@163.com, {feiming,wygong}@cug.edu.cn

Abstract. In multimodal multiobjective optimization problems, there may have more than one Pareto optimal solution corresponding to the same objective vector. The key is to find solutions converged and well-distributed. Even though the existing evolutionary multimodal multiobjective algorithms have taken both the distance in the decision space and objective space into consideration, most of them still focus on convergence property. This may omit some regions difficult to search in the decision space during the process of converging to the Pareto front. In order to resolve this problem and maintain the diversity in the whole process, we propose a differential evolutionary algorithm in a muti-task framework (MT-MMEA). This framework uses an ε-based auxiliary task only concerning the diversity in decision space and provides well-distributed individuals to the main task by knowledge transfer method. The main task evolves using a non-dominated sorting strategy and outputs the final population as the result. MT-MMEA is comprehensively tested on two MMOP benchmarks and compared with six state-of-the-art algorithms. The results show that our algorithm has a superior performance in solving these problems.

Keywords: Multimodal multiobjective optimization · Multi-task framework · ε dominance

1 Introduction

In the real world, a problem with two or more conflicting objectives that need to be optimized at the same time can be formulated as a multiobjective optimization problem (MOP). For this kind of problem, it is unrealistic to find one single solution optimizing all objectives, thus finding a set of trade-off solutions (known as the Pareto optimal solution set) among all the conflicting objectives is in need so that more choices can be provided. Due to the population-based search property, multiobjective evolutionary algorithms (MOEAs) have become a popular and effective approach for solving MOPs. They have the inherent ability to obtain a population of solutions in a single run.

Recently, a special kind of MOP with two or more distinct Pareto optimal solutions corresponding to very similar Pareto front positions, named multimodal multiobjective optimization problem (MMOP), has drawn great attention

© The Author(s), under exclusive license to Springer Nature Singapore Pte Ltd. 2024
B. Luo et al. (Eds.): ICONIP 2023, LNCS 14449, pp. 300–313, 2024.
https://doi.org/10.1007/978-981-99-8067-3_23

to engineering researchers. These problems are frequently encountered in many real-world applications such as flow-shop scheduling [1], feature selection [2], and 0–1 knapsack problems [3]. Figure 1a shows a simple example of a two-variable two-objective MMOP. In Fig. 1a, there are three distinct Pareto optimal sets (PSs) in the decision space and they all correspond to the same Pareto front (PF) in the objective space. When solving this kind of problem, it is essential to obtain as many equivalent Pareto solutions as possible. Reserving a full set of equivalent solutions can provide more information and help researchers to understand the underlying properties of a problem. Furthermore, in a real-world application, if a chosen solution is hard to implement or infeasible in certain scenarios, decision-makers can have other backup plans so that similar results can be achieved.   Since more equivalent PSs in the decision space need to be

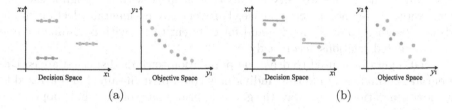

(a)   (b)

Fig. 1. (a) illustrates a general MMOP. (b) shows a problem that may be encountered during the process of evolving.

reserved when solving MMOPs, the existing MOEAs are not applicable. Up to now, many effective and efficient multimodal multiobjective optimization algorithms (MMOEAs) have been proposed for solving MMOPs. For example, Omnioptimizer [4] checks the crowding distance in the decision space; DN_NSGAII [5] adopts a niching method during mating selection and chooses the best solutions in a specific range; TriMOEA&TAR [6] uses two archives (convergence and diversity archive) and recombination strategies; MMOEA-WI [7] involves a modified crowding distance based on the additive ε indicator to evaluate the potential convergence of the population. These methods are based on the existing MOEAs and use niching-based approaches including fitness sharing, special crowding distance, and clustering. They have been proven effective in several MMOP test benchmarks and can be adjusted to different application problems.

Liu in [8] suggests there exists a situation when the difficulty in finding each equivalent PSs is not the same. In this way, how to tackle the imbalance between convergence and diversity in the decision space poses a challenge. One optimal Pareto set may be much more difficult to find than another. This appears to be a challenge for the algorithms above, mainly based on convergence-first selection criteria. To be specific, if some optimal solutions are easy to find, the population will discover them quickly, and then other individuals will converge towards the same Pareto set in hot pursuit. This results in the loss of some regions and deteriorates the diversity in decision space severely. Figure 1b gives a specific example. When evolving, it is hard for the grey points with slow convergence speed to dominate the red and orange points around other Pareto sets. For the next generation, if only six points are wanted, the grey points will be all sifted out due to poor performance over other non-dominated points. In the end, only the rest of the two PSs remain and the solutions obtained are incomplete. To resolve this challenge, we adopt the multitask framework and propose a novel algorithm (MT-MMEA) to attain complete and evenly-distributed PSs. The main contributions of this study are as follows:

1. A multi-task framework is designed for solving MMOPs to better balance the diversity in decision space and objective space. In this paper, two tasks are specifically designed. A main task is based on regular non-dominated sorting, while an auxiliary task is designed with a special ε-relaxation dominate relation that takes the solutions in the specific range as equivalent solutions. The relaxation can maximize the diversity in the decision space with a negligible sacrifice of convergence in the objective space. Different environmental selection methods are designed to meet the demand for different tasks, with ε-dominance and non-dominated relations accordingly.

2. DE operator is used to drive the population towards dispersed areas. Differential information of multiple individuals could be maintained and utilized to the next generation. In this way, the general performance of the whole population is increased.

3. Systematic experiments are carried out on two MMOP benchmark sets with twenty test problems to verify the effectiveness of our approach. Compared with six other state-of-the-art MMOEAs, the experimental results show that the solutions obtained by MT-MMEA are better than the other algorithms.

The remainder of this article is organized as follows. The related works are explained in Sect. 2. The details of MT-MMEA are described in Sect. 3. Section 4 presents the experimental study. Section 5 concludes this paper.

2   Literature Review

For MMOPs, maintaining convergence and diversity in both decision space and objective space is crucial for a population. Much previous work has proposed effective techniques to preserve the diversity of solutions in decision space. For the most commonly used method, niching has been widely applied to prevent diversity loss for the population by dividing it into several niches and comparing the closest individuals rather than the whole set. For example, Liang et al. [5] proposed DN_NSGAII, which uses a niching method during mating selection. Liu et al. [9] suggested DNEA, a double-niched evolutionary algorithm for specific. It applies a niche-sharing method to both the objective and decision space. Lin et al. [10] proposed MMOEADE using a dual clustering-based method. Liu et al. [6] designed a TriMOEA&TAR combining two archives, the convergence archive (CA) and the diversity archive (DA), to focus on convergence and diversity respectively. And they cooperate to achieve a balance. However, one shortage of the niching method is the demand for an additional parameter, named niching size for specific. The setting of this parameter will affect the performance of the above algorithms severely, especially when tackling practical problems without any prior knowledge. Improper division of the niches will affect the searchability of an algorithm in the early stage and even let the population fall into local optimization.

Differential evolution (DE) [11,12] is a simple but powerful stochastic search technique for solving global optimization problems. Exploration in using DE to solve MMOPs becomes more and more popular in recent years. MMODE [13] used crowding distance based on decision space to select an elite mutation pool to maintain population diversity. Liang et al. proposed a differential evolution based on a clustering-based special crowding distance method (MMODE_CSCD) [15] to measure the crowding distance more accurately. Yue et al. [14] developed a differential evolution using improved crowding distance, named MMODE_ICD in which all the selected solutions are taken into account instead of only considering the ones in the same Pareto front.

In addition to sub-population-based methods, a few multi-population-based frameworks have also been brought up. A novel coevolutionary framework [16] proposed by Li et al. introduces two archives for convergence and diversity. The diversity archive utilizes the local convergence indicator and ε-dominance-based method to obtain global and local PSs. Li et al. [17] used a grid search-based multi-population particle swarm optimization algorithm named GSMPSO-MM. Multi-populations based on the k-means clustering method are adopted to locate more equivalent PS in decision space, and a grid is applied to explore high-quality solutions in decision space in GSMPSO-MM. Ming et al. [18] adopted a coevolutionary algorithm CMMO with a convergence-first population that aims at pursuing a solution set well distributed on both the Pareto front and Pareto set assisted by a convergence-relaxed population.

Much research has been done on MMOEAs, and researchers are mostly contributing to the balance between crowding distance, or special crowding distance, and convergence in the objective space. However, there exists a trade-off in the convergence of Pareto solutions towards the true PSs and the well distribution in decision space. And the general MMOEAs fail to attain even distribution when there is a need to reach a compromise between the above two. To address these issues, a multi-task framework is developed for MMOPs to improve the exploration and exploitation ability of MMOEAs.

3   Proposed Method

In this section, the detail of our proposed algorithm is introduced and amplified. First, the overall framework is presented. Then, the details of the sorting mechanism for ε-based dominance and the corresponding diversity-enhanced environmental selection strategy are described.

3.1 Multitask Framework

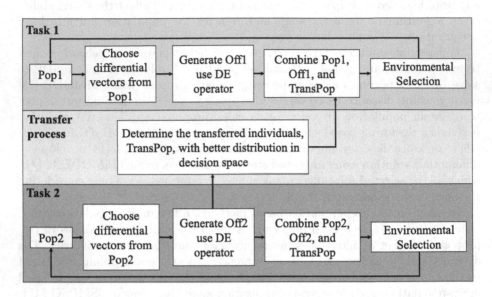

Fig. 2. General flow chart of MT-MMEA.

Figure 2 shows the general flow chart of MT-MMEA. There are two populations aiming at two different tasks, the main task for solving MMOPs and the auxiliary task for solving the corresponding problems with ε-dominance relaxation. With the range of ε, the auxiliary task can give priority to some inferior solutions with better distributions. In this way, the convergence speed of Pop_2 will slow down, and the decision space can be fully explored by the auxiliary task. At the same time, with the exploration of the auxiliary task, the solutions obtained in each iteration with good diversity in decision space are deemed as useful information and transferred to the main task to assist the normal population to get the knowledge of the whole decision space. In this way, Pop_1 has the possibility of reserving the individuals from a new Pareto set provided by Pop_2. If so, the diversity of the Pop_2 will be increased. And it will be beneficial, especially for the MMOPs with imbalanced search difficulties.

The pseudocode of the general framework of MT-MMOEA is given in Algorithm 1. First, two initial populations, Pop_1, and Pop_2, are generated by randomly sampling from the whole decision space. Then, the two populations are evaluated by two different sorting mechanisms: commonly used non-dominated sorting, and ε-relaxation dominance sorting. In each iteration, the mating differential vectors are selected by binary tournament selection based on crowding distances and fitness values. DE operator is performed for the reproduction of two offsprings, Off_1, and Off_2. During this process, for each individual, the

DE/rand/1/bin mutation vector v is used as follows:

$$\mathbf{v} = \mathbf{x_{r1}} + F \times (x_{r2} - x_{r3}) \tag{1}$$

where x_{r1} is a random individual chosen from the population, x_{r2} and x_{r3} are the random individuals from the corresponding mating pool that are different with each other. They are not equal to the original individual x. After the mutation operation, a simple binary crossover is used and implemented as follows:

$$u = \begin{cases} \mathbf{v}, & \text{if } r{\leq}Cr \\ \mathbf{r}, & \text{otherwise} \end{cases} \tag{2}$$

where Cr is the crossover rate, u is the trail vector generated. Then the final decision will be made on x and u based on their values in objective space.

The crowding distance is taken as the first criterion during mating selection. In this way, the individual with a less crowded space distance will have more probability to be chosen, and their genes are more likely to be retained in the next generation. The whole evolving process will be moving towards a population with a more diverse distribution in the space. In line 9, after two populations breed, the points in Pop_2 and Off_2 will be evaluated and sorted by their crowding distances. The top half of them will be selected as useful transfer knowledge and merged to Pop_1. To select the next generation population, an environmental selection strategy is designed, including the diversity-first criterion and ε-relaxation dominance strategy. The above procedures in lines 5–11 will be executed repeatedly until the stopping criterion is met. In the end, Pop_1 is returned as the final approximate set.

Algorithm 1: Framework of MT-MMEA

Input : Population size N, maximum evaluation times FE_{max}, ε-relaxation value ε

Output: Population Pop_1

1 Initial Population Pop_1, Pop_2 and $FE = 0$
2 Evaluate Pop_1 based on non-dominated sorting
3 Evaluate Pop_2 based on ε-dominated sorting
4 **while** $FE \leq FE_{max}$ **do**
5      $MatingPool_1 = TournamentSelection(Pop_1)$
6      $MatingPool_2 = TournamentSelection(Pop_2)$
7      $Off_1 = DEReproduction(Pop_1, MatingPool_1)$
8      $Off_2 = DEReproduction(Pop_2, MatingPool_2)$
9      $P_{transfer} \leftarrow$ transfer solutions with better crowding distance from Pop_2, Off_2
10     $Pop_1 = EnvironmentalSelection(Pop_1, Off_1, P_{transfer})$
11     $Pop_2 = EnvironmentalSelection(Pop_2, Off_2)$
12 **end**

3.2 ϵ-Dominance Sorting

Most dominated-based evolutionary algorithms are stuck to the convergence-first principle. In some cases, it will inevitably discard some regions with slow convergence speed. To address the shortage above, a relaxation of the strict dominance concept based on an extension of the ε-dominance criterion to reserve some potential possible solutions is suggested. The main idea of this relaxation is to maintain good distribution in the decision space with the cost of slight deterioration of performance in the objective space. This technique improves the diversity of the solutions and the convergence of the optimization algorithm.

In each evaluation process, we first sort the solutions by Pareto-dominated relationship. And if a dominated solution is within a tolerably small range of any other solutions dominating it, the dominance relation will be eliminated and they will be reassigned as a non-dominated solution to those solutions as well. In other words, we apply a "relaxation" condition of dominance, thus reserving a new set of solutions that may be discarded because they are dominated by other solutions.

3.3 Environmental Selection

In multimodal multiobjective problems, we need to choose solutions with great diversity in the decision and objective spaces as well as convergence from the offspring, population and transferred population for the main task. Algorithm 2 shows the pseudocode of the modified environmental selection procedure. In our algorithm, the population input refers to the population, offspring, and transfer solutions for the main task. And ε with the value of zero refers to the main task. Firstly, a combined population is sorted based on the non-dominated relation. If the number of non-dominated solutions is larger than population size N, they will be truncated according to the crowding distances. Oppositely, if the number is less than N, the solutions close to the top half layers will be selected, and the best-distributed solutions will be chosen for the next generation. This procedure guarantees that the diversity is preserved to the maximum extent with little damage to convergence, for the reason that some crowded solutions from the first layer may be weeded out. And this will not damage the overall performance of the algorithm, because of the fast convergence speed of the DE operator. For the auxiliary task, the population to be selected contains the population itself and the offspring. They will be sorted by ε-dominance. The fitness value is the sum of the number of domination solutions and normalized distances in decision space. Next, the solutions with the best fitness values will be selected for the next generation.

4 Experiments

4.1 Experimental Methodology

To clearly interpret the mechanism and examine the effectiveness of the proposed MT-MMEA, we evaluate and compare it with other six state-of-the-art MMEAs:

Algorithm 2: Environmental selection

Input : Population to be selected P, number of selected population N,
ε-dominance relaxation value ε
Output: Population P_{next}

1 **if** $\varepsilon == 0$ **then**
2   C
3 **else**
4   a
5 **end**
6 lculate the Front number FNo and max front number $MaxFNo$ of N based on non-dominated sorting
7 Select individuals with $FNo \leq MaxFNo$, and calculate the crowding distance.
8 Select the N solution with best crowding distance as P_{next} Calculate the fitness values of P based on ε-dominance sorting
9 Select N individuals with best fitness value as P_{next}

1) DN_NSGAII [5]; 2) MMEAWI [7]; 3) MMOEADC [10]; 4) MO_Ring_PSO_SCD [19]; 5) TriMOEA&TAR [6]; and 6) CMMO [18].

In our experiments, we assess the proposed algorithm's effectiveness by using twenty typical multimodal multiobjective test problems, including eight MMF problems (MMF1-8) [5], and twelve IDMP problems [8]. For MMF test problems, the difficulty in finding each equivalent Pareto optimal solution is the same. The complexity of finding equivalent PSs is different on IDMP test problems.

For MMFs, the population size is specified as 200 with the termination condition of 40 000 solution evaluations. For, IDMPs, the parameters are set as the original literature. The ε value of our algorithm is set to 0.18. If the ε value is too small, the difference between two tasks will be vague. Whereas, if it is set too large, a number of useless results will be remained. The crowding factor in DN_NSGAII is set to 5. and the indicator parameter for MMEAWI is set to 0.05.

4.2 Performance Metrics

To fairly compare the results of all the algorithms, we use typical indicators to evaluate the solution quality set: IGD and IGDX [5] for the objective space and decision space, respectively. For both of them, a smaller value indicates more desirable results obtained.

4.3 Ablation Study

To verify the effectiveness of the individual components of our proposed MT-MMEA, we conduct an ablation analysis on the multi-task framework, DE operator, and environmental selection mechanism.

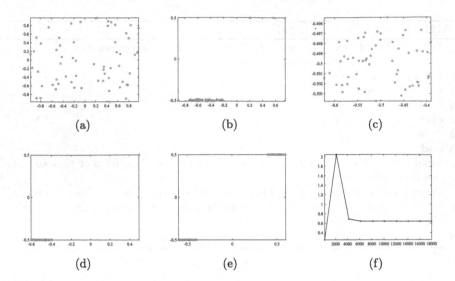

Fig. 3. (a)-(e) shows the evolutionary process in decision space using MT-MMEA on IDMPM2T4 test problem. The individuals transferred help the main population regain the PS in the upper right. (f) shows the change of indicator IGDX during the process.

Multi-task Framework. To show the superiority of the multi-task framework and the effectiveness of transfer knowledge, we modify the task and compare their performance. Task 1 is the same framework while the individuals with better fitness values are deemed as useful information and transferred to help the main task. In task 2, the auxiliary population uses the same task as the main population searching for non-dominated solutions, and the transfer segment remains the same. Task 3 only utilizes the main task for the whole process. These tasks are compared and the results are shown in Table 1. As illustrated in the table, the original algorithm outperforms all the other formats. With the auxiliary task of selecting well-distributed solutions and providing useful information, the main population can find a full set of Pareto optimal solutions in the entire decision space. And the distribution in the objective space is improved at the same time. Figure 3 shows a specific case of the evolutionary process. In the beginning, the population search in two directions. Soon the upper solutions are eliminated due to the low convergence rate. However, the other task remains and transfers the missing part, helping the main population acquire well-distributed and converged PSs in Fig. 4f. Figure 4g demonstrates the change of IGDX during this process.

Environmental Selection. Elite strategy is used instead of choosing the most diverse individuals among the half-upper layer. When the number of non-dominated solutions for the next generation is less than the population size. The experiments are conducted on 20 test benchmark problems mentioned above.

Table 1. Statistical comparison results between different tasks.

	IGD			IGDX		
	+/ − / ≈	Average ranking	Final Ranking	+/ − / ≈	Average ranking	Final Ranking
Original	baseline	1.7	1	baseline	1.6	1
Task1	0/5/15	3.1	3	3/9/8	2.4	2
Task2	0/14/6	3.7	4	4/13/3	2.8	3
Task3	2/3/15	2.1	2	0/8/12	3.4	4

It can be seen from Table 2 that the performance deteriorates severely when environmental strategy changes.

Table 2. Statistical comparison results between different environmental selections.

	IGD			IGDX		
	+/ − / ≈	Average ranking	Final Ranking	+/ − / ≈	Average ranking	Final Ranking
Original	baseline	1.1	1	baseline	1.3	1
ES1	0/14/6	1.9	2	4/13/3	1.7	2

DE Operator. We replace DE operators with GA to demonstrate their effect on populations. As shown in the bottom part of Table 3, Both the IGD and IGDX get influenced by the operator. We guess that the DE operator could make full use of the information of better-distributed solutions by selecting differential vectors and directing the population to a more widespread distribution.

Table 3. Statistical comparison results between different operations.

	IGD			IGDX		
	+/ − / ≈	Average ranking	Final Ranking	+/ − / ≈	Average ranking	Final Ranking
Original	baseline	1.15	1	baseline	1.1	1
GA	3/17/1	1.85	2	2/17/1	1.9	2

4.4   Comparison with Other Algorithms

The comparison results of IGD and IGDX values between MT-MMEA and the six state-of-the-art MMOEAs are shown in Table 4 and Table 5, respectively. The best results are in gray shade. In Table 4, we can observe that MT-MMEA obtained better values for the IGD indicator than the other six algorithms in 17 of the 20 test functions. Compared with MO_Ring_PSO_SCD, the IGD value of our algorithm on MMF2, MMF3, and MMF8 is slightly lower. But it still ranks second. On MMF7, MT-MMEA is ranked the same as the best performance algorithm CMMO. It can be seen from Table 5 that MT-MMEA achieves the highest values on 11 test functions. Although the performance on the MMF test is not ideal, it is only second to CMMO on the six test functions, and the performance on IDMP is better than those of the other algorithms. Combined with Tables 4 and 5, the results show that the MT-MMEA proposed in this paper achieves great performance on IDMP, and has relatively competitive performance on MMF benchmarks. It proves that the algorithm can maintain some hard-to-search Pareto set to some extent, and thus increase the distribution in both the decision space and objective space.

To graphically interpret these experiment results, typical simulation results of the distribution of the PSs obtained by seven algorithms when solving the IDMPM3T4 problem are shown in Fig. 4. It is evident that our method has much more distributed solutions than the others when solving imbalance conver-

Table 4. Mean IGD values obtained from different MMEAs over 30 runs.

Problem	DN_NSGAII	MMEAWI	MMOEADC	MO_Ring_PSO_SCD	TriMOEA&TAR	CMMO	MT-MMEA
IDMPM2T1	1.2721e+0 (4.20e-4) −	1.2713e+0 (1.26e-3) −	1.2714e+0 (1.93e-3) −	1.2696e+0 (6.97e-4) −	1.2728e+0 (2.86e-5) −	1.2725e+0 (3.98e-4) −	1.2611e+0 (3.27e-3)
IDMPM2T2	1.2720e+0 (6.59e-4) −	1.2712e+0 (1.20e-3) −	1.2718e+0 (7.43e-4) −	1.2712e+0 (5.56e-4) −	1.2728e+0 (8.52e-5) −	1.2726e+0 (3.50e-4) −	1.2643e+0 (2.14e-3)
IDMPM2T3	1.2718e+0 (6.06e-4) −	1.2708e+0 (1.06e-3) −	1.2711e+0 (7.41e-4) −	1.2707e+0 (6.15e-4) −	1.2727e+0 (4.18e-4) −	1.2719e+0 (6.28e-4) −	1.2641e+0 (2.62e-3)
IDMPM2T4	1.2724e+0 (3.57e-4) −	1.2723e+0 (4.77e-4) −	1.2718e+0 (7.23e-4) −	1.2711e+0 (5.45e-4) −	1.2728e+0 (9.32e-6) −	1.2727e+0 (5.39e-5) −	1.2668e+0 (2.14e-3)
IDMPM3T1	1.5271e+0 (3.47e-3) −	1.5255e+0 (3.65e-3) −	1.5230e+0 (3.52e-3) −	1.5212e+0 (3.40e-3) −	1.5276e+0 (5.40e-3) −	1.5275e+0 (2.81e-3) −	1.4756e+0 (1.11e-2)
IDMPM3T2	1.5270e+0 (3.41e-3) −	1.5262e+0 (3.27e-3) −	1.5231e+0 (2.33e-3) −	1.5269e+0 (1.84e-3) −	1.5293e+0 (4.60e-3) −	1.5276e+0 (2.31e-3) −	1.4951e+0 (1.04e-2)
IDMPM3T3	1.5266e+0 (2.74e-3) −	1.5253e+0 (2.44e-3) −	1.5216e+0 (3.15e-3) −	1.5254e+0 (2.65e-3) −	1.5278e+0 (5.11e-3) −	1.5273e+0 (3.06e-3) −	1.4852e+0 (1.27e-2)
IDMPM3T4	1.5272e+0 (3.57e-3) −	1.5262e+0 (3.88e-3) −	1.5259e+0 (2.27e-3) −	1.5259e+0 (2.31e-3) −	1.5264e+0 (5.13e-3) −	1.5278e+0 (2.97e-3) −	1.5049e+0 (6.27e-3)
IDMPM4T1	1.7451e+0 (4.87e-3) −	1.7507e+0 (2.83e-3) −	1.7471e+0 (4.76e-3) −	1.7261e+0 (4.81e-3) −	1.7584e+0 (6.83e-3) −	1.7522e+0 (3.54e-3) −	1.6797e+0 (1.12e-2)
IDMPM4T2	1.7496e+0 (2.88e-3) −	1.7505e+0 (3.22e-3) −	1.7488e+0 (3.91e-3) −	1.7399e+0 (2.71e-3) −	1.7609e+0 (4.87e-3) −	1.7514e+0 (3.43e-3) −	1.7021e+0 (8.45e-3)
IDMPM4T3	1.7459e+0 (7.06e-3) −	1.7487e+0 (2.64e-3) −	1.7453e+0 (3.93e-3) −	1.7404e+0 (3.38e-3) −	1.7565e+0 (1.25e-2) −	1.7514e+0 (4.07e-3) −	1.6885e+0 (1.17e-2)
IDMPM4T4	1.7476e+0 (6.44e-3) −	1.7496e+0 (3.24e-3) −	1.7483e+0 (5.75e-3) −	1.7405e+0 (4.37e-3) −	1.7592e+0 (6.72e-3) −	1.7514e+0 (3.60e-3) −	1.7015e+0 (1.06e-2)
+/−/≈	0/12/0	0/12/0	0/12/0	0/12/0	0/12/0	0/12/0	
MMF1	1.2583e+0 (8.54e-5) −	1.2584e+0 (2.94e-4) −	1.2587e+0 (1.03e-3) −	1.2590e+0 (7.84e-4) −	1.2594e+0 (3.49e-3) −	1.2584e+0 (1.61e-4) −	1.2579e+0 (2.74e-4)
MMF2	6.1513e-1 (1.62e-3) ≈	6.1541e-1 (8.39e-4) −	6.1372e-1 (4.66e-3) −	4.9431e-1 (4.35e-2) +	5.9727e-1 (3.86e-2) ≈	6.1548e-1 (7.80e-4) −	6.1491e-1 (8.91e-4)
MMF3	4.6753e-1 (9.19e-4) −	4.6751e-1 (7.62e-4) ≈	4.6656e-1 (2.13e-3) −	4.4758e-1 (5.15e-3) +	4.6567e-1 (5.17e-3) −	4.6730e-1 (1.02e-3) −	4.6723e-1 (7.61e-4)
MMF4	6.140e-1 (3.97e-2) −	6.0741e-1 (4.16e-2) −	6.1114e-1 (4.08e-2) −	6.0283e-1 (3.69e-2) −	6.3112e-1 (2.44e-2) −	6.0484e-1 (4.08e-2) −	5.5029e-1 (5.81e-2)
MMF5	1.6839e+0 (4.30e-2) −	1.6761e+0 (5.07e-2) −	1.6641e+0 (5.54e-2) −	1.6436e+0 (5.46e-2) −	1.6635e+0 (5.78e-2) −	1.6667e+0 (5.98e-2) −	1.6127e+0 (5.77e-2)
MMF6	1.3831e+0 (5.99e-3) −	1.3804e+0 (7.19e-3) −	1.3796e+0 (7.26e-3) −	1.3725e+0 (5.76e-3) −	1.3817e+0 (7.22e-3) −	1.3807e+0 (5.71e-3) −	1.3718e+0 (6.63e-3)
MMF7	1.0510e+0 (3.69e-5) −	1.0511e+0 (1.48e-4) −	1.0515e+0 (1.27e-3) −	1.0516e+0 (6.14e-4) −	1.0509e+0 (3.98e-4) ≈	1.0511e+0 (1.08e-4) −	1.0509e+0 (8.37e-5)
MMF8	3.7217e+0 (1.93e-4) −	3.7212e+0 (3.70e-4) −	3.7202e+0 (7.29e-3) −	2.9263e+0 (5.93e-1) +	3.7223e+0 (1.60e-4) −	3.7216e+0 (2.84e-4) −	3.6939e+0 (6.39e-2)
+/−/≈	0/18/2	0/19/1	0/18/2	3/17/1	1/17/2	0/19/1	

Table 5. Mean IGDX values obtained from different MMEAs over 30 runs.

Problem	DN_NSGAII	MMEAWI	MMOEADC	MO_Ring_PSO_SCD	TriMOEA&TAR	CMMO	MT-MMEA
IDMPM2T1	1.8166e+0 (5.35e-1) −	1.2522e+0 (7.11e-1) −	7.8009e-1 (4.23e-1) −	1.5007e+0 (6.87e-1) −	1.9110e+0 (4.30e-1) −	1.1113e+0 (6.76e-1) −	6.4073e-1 (1.45e-3)
IDMPM2T2	1.8165e+0 (5.35e-1) −	1.0169e+0 (6.35e-1) ≈	7.8195e-1 (4.31e-1) −	7.0354e-1 (2.55e-1) −	1.7225e+0 (6.07e-1) −	9.6984e-1 (6.07e-1) ≈	6.4077e-1 (1.09e-3)
IDMPM2T3	1.3173e+0 (6.99e-1) −	6.7260e-1 (1.29e-3) −	8.5586e-1 (4.77e-1) −	6.7562e-1 (6.66e-3) −	1.6921e+0 (6.07e-1) −	8.1013e-1 (4.21e-1) −	6.6997e-1 (1.88e-3)
IDMPM2T4	1.7225e+0 (6.07e-1) −	2.0050e+0 (2.58e-1) −	1.6289e+0 (6.58e-1) −	9.3610e-1 (5.68e-1) −	1.9577e+0 (3.58e-1) −	1.5814e+0 (6.77e-1) −	1.1577e+0 (6.92e-1)
IDMPM3T1	2.0469e+0 (4.58e-1) −	1.0396e+0 (9.84e-2) −	9.9305e-1 (5.80e-2) −	1.2651e+0 (3.68e-1) −	2.0256e+0 (5.52e-1) −	1.0409e+0 (9.76e-2) −	9.8627e-1 (9.12e-3)
IDMPM3T2	1.5587e+0 (5.88e-1) −	1.0400e+0 (1.00e-1) −	1.0110e+0 (1.22e-1) −	1.0354e+0 (1.83e-2) −	1.7406e+0 (4.97e-1) −	1.0689e+0 (1.87e-1) −	9.9157e-1 (8.97e-3)
IDMPM3T3	1.7975e+0 (5.70e-1) −	1.0728e+0 (1.36e-1) −	1.0676e+0 (1.79e-1) −	1.0404e+0 (5.09e-3) −	1.7622e+0 (5.36e-1) −	1.1211e+0 (2.22e-1) −	1.0132e+0 (7.42e-3)
IDMPM3T4	2.0652e+0 (6.21e-1) −	1.6165e+0 (3.86e-1) −	1.3605e+0 (3.55e-1) −	1.0531e+0 (1.08e-1) −	2.0429e+0 (6.14e-1) −	1.2482e+0 (3.41e-1) −	1.0148e+0 (9.69e-2)
IDMPM4T1	3.0084e+0 (1.42e-1) −	1.3387e+0 (1.91e-1) ≈	1.4901e+0 (4.13e-1) ≈	2.4712e+0 (6.25e-1) −	2.9825e+0 (1.96e-1) −	2.3285e+0 (5.87e-1) −	1.6462e+0 (4.64e-1)
IDMPM4T2	2.7114e+0 (5.17e-1) −	1.7688e+0 (6.77e-1) −	1.5139e+0 (4.47e-1) −	1.3460e+0 (1.85e-1) +	2.6620e+0 (5.20e-1) −	1.7428e+0 (5.33e-1) −	1.4608e+0 (3.92e-1)
IDMPM4T3	2.4840e+0 (5.83e-1) −	1.4032e+0 (1.47e-1) −	1.3580e+0 (2.24e-1) −	1.3032e+0 (5.83e-3) −	2.2771e+0 (5.40e-1) −	1.5746e+0 (4.75e-1) −	1.2831e+0 (4.78e-3)
IDMPM4T4	2.6878e+0 (5.52e-1) −	1.5110e+0 (1.39e-1) −	1.5206e+0 (4.63e-1) −	1.2935e+0 (1.01e-2) −	2.5146e+0 (6.80e-1) −	1.5041e+0 (4.42e-1) −	1.2739e+0 (5.22e-2)
+/−/≈	0/12/0	0/10/2	0/10/2	2/10/0	0/12/0	0/11/1	
MMF1	4.0661e-2 (1.01e-3) −	3.9619e-2 (2.49e-3) ≈	4.4639e-2 (1.52e-3) −	6.7804e-2 (1.15e-2) −	5.0319e-2 (3.95e-3) −	3.0834e-2 (5.60e-4) +	3.9444e-2 (7.38e-4)
MMF2	4.6917e-2 (4.06e-2) −	1.1396e-2 (2.35e-3) −	6.3879e-3 (1.64e-3) −	6.3879e-2 (3.58e-2) −	5.9082e-2 (4.45e-2) −	1.1591e-2 (3.29e-3) −	6.3019e-3 (9.16e-4)
MMF3	3.3942e-2 (2.81e-2) −	9.6408e-3 (2.29e-3) −	8.2615e-3 (1.42e-3) −	4.5576e-2 (2.07e-2) −	4.5934e-2 (2.66e-2) −	9.6971e-3 (1.58e-3) −	5.8586e-3 (4.27e-4)
MMF4	2.6560e-2 (2.16e-3) +	2.3657e-2 (8.01e-4) +	2.1411e-2 (1.76e-3) +	6.4747e-2 (2.01e-2) −	1.0630e-1 (1.59e-1) ≈	3.2101e-2 (5.36e-4) +	3.0313e-2 (8.76e-4)
MMF5	4.9728e-2 (2.75e-3) −	4.3842e-2 (1.77e-3) −	4.7049e-2 (1.02e-3) −	8.4888e-2 (9.81e-3) −	5.9267e-2 (1.66e-2) −	3.2101e-2 (5.36e-4) +	3.7436e-2 (5.53e-4)
MMF6	7.1016e-2 (3.22e-3) −	6.1801e-2 (1.93e-3) −	6.8064e-2 (2.46e-3) −	1.0243e-1 (1.15e-2) −	7.4945e-2 (4.36e-3) −	5.2591e-2 (1.48e-3) +	6.0250e-2 (9.45e-4)
MMF7	2.2096e-2 (1.51e-3) +	2.3340e-2 (1.05e-3) −	2.4079e-2 (3.86e-3) +	7.1156e-2 (5.86e-2) −	3.7043e-2 (2.66e-2) −	1.9106e-2 (6.40e-4) +	2.3885e-2 (2.40e-3)
MMF8	7.0959e-2 (1.75e-2) +	5.0441e-2 (4.61e-3) +	4.6484e-2 (5.19e-3) +	7.1156e-1 (5.86e-2) −	4.4650e-1 (9.32e-2) −	3.8993e-2 (3.11e-3) +	7.5004e-2 (2.41e-3)
+/−/≈	3/17/0	3/14/3	3/15/2	2/18/0	0/18/2	6/13/0	

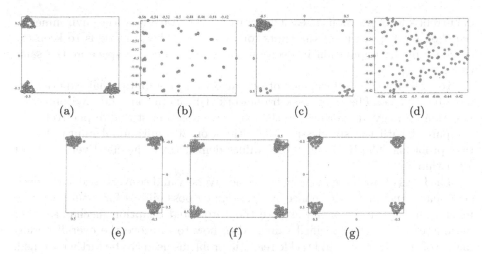

Fig. 4. Visualization of the solution sets obtained on IDMPM4T3 test problem. Results are shown in the decision space. (a) MMEAWI. (b) TriMOEA&TAR. (c)) MO_Ring_PSO_SCD. (d) DN_NSGAII. (e) MMOEADC. (f) CMMO. (g) MT-MMEA.

gence problems. For specific, TriMOEA&TAR and DN_NSGAII have the worst distribution in that they can only find one Pareto set in total. Even though in DN_NSGAII, the crowding distance is calculated purely based on the decision variables, it has no diversity reservation method, so that the points in the slow convergence-rate Pareto set will be omitted. As for MMEAWI and MMOEADC, they are better than the former two. They can find multiple PSs and preserve even distribution in each part. Unfortunately, there is still some fragments missing. MO_Ring_PSO_SCD finds every equivalent Pareto optimal set, but the scope of PS attenuates where solutions are much more difficult to find. CMMO and MT-MMEA are the best solvers for IDMP problems, and what is interesting is that they are both based on a multi-population framework, and use an ε relaxation method. However, the distribution of CMMO on this problem is worse than MT-MMEA in both decision space and objective space, which can also be certified in Tables 4 and 5. This may because the design of the two tasks is totally different, as well as the transfer knowledge and transfer strategy.

From the above discussion, we can conclude that our algorithm is capable to solve different MMOPs and especially effective in the imbalanced test problems in IDMP.

5 Conclusion

In this paper, we propose a novel algorithm MT-MMEA integrated with a multi-task framework for multimodal multiobjective optimization. We introduce an ε-dominance relaxation in the auxiliary task to help the main task get various options. DE operator is used for the fast-convergence speed and full utilization of current known positions in each generation. Then we adjust the environmental selection strategy to meet the traits for each task. The purpose is to keep the diversity of the population in decision space and objective space at the same time.

To evaluate the performance of our algorithm, computational experiments have been made. The multi-task framework, DE operation, and environmental selection strategy are verified by the extensive ablation study, respectively. We compare it with the six classical algorithms on 20 multimodal multiobjective task problems. The IGD and IGDX values demonstrate the effectiveness of our algorithm.

The focus of this paper is to balance the diversity and convergence in the decision space, and the performance suggests many possibilities for future achievements. Subsequent work can explore the automated adaption mechanism over parameter ε over the specified value. Also, how to improve the overall performance of this algorithm and tackle realistic problems needs to be further studied.

References

1. Han, Y., Gong, D., Jin, Y., Pan, Q.: Evolutionary multiobjective blocking lot-streaming flow shop scheduling with machine breakdowns. IEEE Trans. Cybern. **49**(1), 184–197 (2017)
2. Yue, C.T., Liang, J.J., Qu, B.Y., Yu, K.J., Song, H.: Multimodal multiobjective optimization in feature selection. In: 2019 IEEE Congress on Evolutionary Computation (CEC), pp. 302–309. IEEE (2019)
3. Jaszkiewicz, A.: On the performance of multiple-objective genetic local search on the 0/1 knapsack problem-a comparative experiment. IEEE Trans. Evol. Comput. **6**(4), 402–412 (2002)
4. Deb, K., Tiwari, S.: Omni-optimizer: a generic evolutionary algorithm for single and multi-objective optimization. Eur. J. Oper. Res. **185**(3), 1062–1087 (2008)
5. Liang, J.J., Yue, C.T., Qu, B.Y.: Multimodal multi-objective optimization: a preliminary study. In: 2016 IEEE Congress on Evolutionary Computation (CEC), pp. 2454–2461. IEEE. (2016)
6. Liu, Y., Yen, G.G., Gong, D.: A multimodal multiobjective evolutionary algorithm using two-archive and recombination strategies. IEEE Trans. Evol. Comput. **23**(4), 660–674 (2018)
7. Li, W., Zhang, T., Wang, R., Ishibuchi, H.: Weighted indicator-based evolutionary algorithm for multimodal multiobjective optimization. IEEE Trans. Evol. Comput. **25**(6), 1064–1078 (2021)
8. Liu, Y., Ishibuchi, H., Yen, G.G., Nojima, Y., Masuyama, N.: Handling imbalance between convergence and diversity in the decision space in evolutionary multimodal multiobjective optimization. IEEE Trans. Evol. Comput. **24**(3), 551–565 (2019)
9. Liu, Y., Ishibuchi, H., Nojima, Y., Masuyama, N., Shang, K.: A double-niched evolutionary algorithm and its behavior on polygon-based problems. In: Auger, A., Fonseca, C.M., Lourenço, N., Machado, P., Paquete, L., Whitley, D. (eds.) PPSN 2018. LNCS, vol. 11101, pp. 262–273. Springer, Cham (2018). https://doi.org/10.1007/978-3-319-99253-2_21
10. Lin, Q., Lin, W., Zhu, Z., Gong, M., Li, J., Coello, C.A.C.: Multimodal multiobjective evolutionary optimization with dual clustering in the decision and objective spaces. IEEE Trans. Evol. Comput. **25**(1), 130–144 (2020)

11. Vesterstrom, J., Thomsen, R.: A comparative study of differential evolution, particle swarm optimization, and evolutionary algorithms on numerical benchmark problems. In: Proceedings of the 2004 Congress on Evolutionary Computation (IEEE Cat. No.04TH8753), vol. 2, pp. 1980–1987, Portland, OR, USA (2004)

12. Das, S., Suganthan, P.N.: Differential evolution: a survey of the state-of-the-art. IEEE Trans. Evol. Comput. **15**(1), 4–31 (2010)

13. Liang, J., et al.: Multimodal multiobjective optimization with differential evolution. Swarm Evol. Comput. **44**, 1028–1059 (2019)

14. Yue, C., et al.: Differential evolution using improved crowding distance for multimodal multiobjective optimization. Swarm Evol. Comput. **62**, 100849 (2021)

15. Liang, J., et al.: A clustering-based differential evolution algorithm for solving multimodal multi-objective optimization problems. Swarm Evol. Comput. **60**, 100788 (2021)

16. Li, W., Yao, X., Li, K., Wang, R., Zhang, T., Wang, L.: Coevolutionary Framework for Generalized Multimodal Multi-objective Optimization. arXiv preprint, arXiv:2212.01219 (2022)

17. Li, G., Wang, W., Zhang, W., Wang, Z., Tu, H., You, W.: Grid search based multi-population particle swarm optimization algorithm for multimodal multi-objective optimization. Swarm Evol. Comput. **62**, 100843 (2021)

18. Ming, F., Gong, W., Wang, L., Gao, L.: Balancing convergence and diversity in objective and decision spaces for multimodal multi-objective optimization. IEEE Trans. Emerg. Top. Comput. Intell. **7**, 474–486 (2022)

19. Liang, J., Guo, Q., Yue, C., Qu, B., Yu, K.: A self-organizing multi-objective particle swarm optimization algorithm for multimodal multi-objective problems. In: Tan, Y., Shi, Y., Tang, Q. (eds.) ICSI 2018. LNCS, vol. 10941, pp. 550–560. Springer, Cham (2018). https://doi.org/10.1007/978-3-319-93815-8_52

Domain Generalized Object Detection with Triple Graph Reasoning Network

Zhijie Rao[1,2], Luyao Tang[1,2], Yue Huang[1,2(✉)], and Xinghao Ding[1,2]

[1] School of Informatics, Xiamen University, Xiamen, China
huangyue05@gmail.com
[2] Institute of Artificial Intelligent, Xiamen University, Xiamen, China

Abstract. Recent advances in Domain Adaptive Object Detection (DAOD) have vastly restrained the performance degradation caused by distribution shift. However, DAOD relies on the strong assumption of accessible target domain during the learning procedure, which is tough to be satisfied in real-world applications. Domain Generalized Object Detection (DGOD) aims to generalize the detector trained on the source domains directing to an unknown target domain without accessing the target data. Thus it is a much more challenged problem and very few contributions have been reported. Extracting domain-invariant information is the key problem of domain generalization. Considering that the topological structure of objects does not change with the domain, we present a general DGOD framework, Triple Graph Reasoning Network (TGRN) to uncover and model the structure of objects. The proposed TGRN models the topological relations of foregrounds via building refined sparse graphs on both pixel-level and semantic-level. Meanwhile, a bipartite graph is created to capture structural consistency of instances across domain, implicitly enabling distribution alignment. Experiments on our newly constructed datasets verify the effectiveness of the proposed TGRN. Codes and datasets are available at https://github.com/zjrao/tgrn.

Keywords: Domain generalization · Object detection · Graph convolutional Network · Transfer learning

1 Introduction

Object detection is a crucial and challenging topic in computer vision, where the goal is to precisely recognize and localize objects of interest in images. Current detectors based on deep convolutional neural network have reaped extraordinary achievements in coping with identical distributed data. Nevertheless, the training and test sets may be sampled from disparate styles, lighting, or environments, *e.g.*, *photo* vs. *sketch*, *sunny* vs. *foggy*, a phenomenon known as domain shift [18]. Domain shift typically lead to severe degradation of the detector performance.

The work was supported in part by the National Natural Science Foundation of China under Grant 82172033, U19B2031, 61971369, 52105126, 82272071, 62271430, and the Fundamental Research Funds for the Central Universities 20720230104.

Domain Adaptive Object Detection (DAOD) acts as a credible solution to tackle this issue by knowledge transfer and distribution alignment [3,14]. Unfortunately, DAOD relies on the prerequisite that the source and target domains can be jointly trained, which is arduous to be satisfied in real-world scenarios. For example, collecting data under all weather conditions in the autonomous driving task or asking for private data with digital copyright. On the other hand, research on Domain Generalization (DG) has evolved considerably for classification and segmentation tasks [15,17], where the goal is to learn a well-generalized model without accessing target domain during the training phase. Regrettably, most of their methods are custom-made, resulting in a difficult way to transfer or integrate into the detection pipelines. To this end, we delve into a new problem, Domain Generalized Object Detection (DGOD), which has not yet been thoroughly explored. The objective of DGOD is to learn a robust detector with multiple source domains and then apply it directly to an unseen target domain.

Extracting domain-invariant information is one of the effective means to deal with cross-domain issues. To this end, many researchers resorted to adversarial learning or consistency constraints to motivate models to learn domain-invariant representations [3–5,9]. However, such approaches can easily overfit the model to the source domains and thus make it difficult to generalize to unknown domains. Considering that the topology of an object is a kind of information that cannot be easily changed, our goal is to encourage the model to focus on the holistic morphology of the object through explicit structural modeling.

To achieve this goal, we present a novel framework named Triple Graph Reasoning Network (TGRN), to model the topological relation of the foreground at three levels. The proposed TGRN mainly consists of three modules, namely, Pixel-level Graph Reasoning Module (PGRM), Instance-level Graph Reasoning Module (IGRM) and Cross-domain Graph Reasoning Module (CGRM). PGRM works on low-level feature map extracted by shallow convolutional layer. Through similarity filtering, a sparse graph is created to aggregate the pixel points belonging to the foreground and background. To further contour the morphology of foregrounds, we introduce distance restriction to remove redundant links. IGRM constructs a semantic-level graph to associate instances belonging to the same category and fuse information from different regions, prompting the network to exploit those hard regions. To mitigate the affect of domain shift, CGRM implicitly performs distribution alignment by virtue of capturing cross-domain structural consistency. Specifically, a bipartite graph is built to model the pairwise relationships of inter-class instances across domains. Meanwhile, we develop a prototype-based re-weighting technique to match reliable pairs.

Due to the current lack of datasets for DGOD, we construct two datasets. **PCWS** contains Pascal VOC2007 [19] and three domains collected and annotated by us, which differ in image style. **BDD-DG** is created based on BDD100k [20] and simulates four different weather conditions in autonomous driving. The main contributions can be summarized as follows:

- The proposed challenged DGOD problem has the ability of generalizing the detector trained on source domains directly to an unknown target domain, which has not been fully explored yet.
- To boost the out-of-domain generalizability, we present a novel and general framework to model the topological relations of objects at three levels.
- Two datasets are constructed to facilitate the study of DGOD, named PCWS and BDD-DG, reflecting domain shift in image style and weather condition respectively. The datasets will be available soon to benefit the DG community.
- Experimental results demonstrate that our approach outperforms existing methods by a large margin.

2 Related Work

2.1 Domain Adaptive Object Detection

Domain Adaptive Object Detection is an effective solution for domain shift and numerous algorithms have been developed [29]. To give a few examples, DA-Faster [3] is the first study for DAOD, which proposes a framework based on adversarial learning to align source and target domains at the global and instance levels. SWDA [4] proposes a strong and weak alignment strategy for the problem of inconsistent degree of domain shift, as a way to give stronger constraint on foreground features. Chen et al. [5] propose a novel framework HTCN to alleviate the contradiction between transferability and discriminability, thus enhancing the recognition of foreground regions. Zhou et al. [6] devise a multi-granularity alignment network to cope with the inconsistency of foreground features at different scales. In addition, some studies achieves promising results with self-training by using pseudo-label [30–32]. Although DAOD has shown promising results, its requirement of acquiring the target domain during the training phase is tough to meet in realistic scenarios.

2.2 Domain Generalization

Domain Generalization aims to use the visible source domains to train a generic and robust model for direct deployment to an unseen target domain. The mainstream approaches for DG fall into three directions, domain augmentation, domain alignment and learning strategies [7]. To name a few, L2A-OT [8] uses adversarial learning to generate new image-wise samples. MMLD [9] discards domain labels and assigns pseudo-domain labels to samples and performs domain alignment by a self-discovery strategy. JiGen [10] designs a puzzle classification task for self-supervised learning and MetaReg [34] proposes a meta-learning based regularization technique. Huang et al. [15] propose an augmentation method based on frequency domain transfer for the segmentation task. Other DG methods [33,35] have also yielded promising results. However, most of these approaches are tailored for classification or segmentation tasks and are not easy to transfer to detection task.

2.3 Domain Generalized Object Detection

Domain Generalized Object Detection attempts to utilize the paradigm of DG to learn a cross-domain robust detector, which is currently less studied. Liu et al. [12] design a novel framework for underwater object detection that uses style transfer to convert images to other water qualities and aligns their features by adversarial training. Zhang et al. [13] develop Region-Aware Proposal reweighTing (RAPT) to eliminate statistical dependences between relevant features and irrelevant features for pedestrian detection. However, their methods are designed for specialized scenarios, hindering the versatility and transferability. Seemakurthy et al. [2] propose a general framework that leverages auxiliary domain discriminators and entropy regularization to learn domain-invariant features in an adversarial learning manner, which is difficult to train and tends to make the model source-biased. Moreover, they do not consider the limitation of the local field of view of the CNNs-based detectors.

3 Methodology

Here, we give the problem description of DGOD. Suppose we have D domains $\mathcal{D} = \{X_d, (Y_d, B_d)\}_{d=1}^{D}$ where X is input image, and Y, B are the corresponding category and bounding box. Note that all domains share the same label space. The objective of DGOD is to minimize the empirical risk of the model over all domains using visible source domains. Specifically, we have source domains $\mathcal{D}_s = \{X_d, (Y_d, B_d)\}_{d=1}^{D-1}$ available for training and a target domain $\mathcal{D}_t = \{X_d\}_{d=D}$ for testing.

The overall architecture of TGRN is shown in Fig. 1, which is mainly composed of PGRM, IGRM and CGRM. Our method is based on Faster-RCNN framework [16]. At each iteration, two randomly sampled images from different source domains are fed into the network. Then PGRM and IGRM model the topological relationships of foregrounds at pixel-level and semantic-level, respectively, to extend the horizon and improve the reasoning ability. CGRM builds pairwise links between instances across domains based on structural consistency, allowing the network to automatically mine domain-invariant information in the process of learning such correspondences.

3.1 Pixel-Level Graph Reasoning Module

Many domain augmentation-driven DG methods manipulate the statistics of low-level features to reduce the sensitivity of the network to domain change [11,21]. However, their methods fail to distinguish the foreground from the background. The proposed PGRM naturally clusters pixels in the foreground or the background by similarity filtering and further does detailed topological modeling for foreground pixels to expand the field of view.

Let $F \in \mathbb{R}^{C \times H \times W}$ denotes the feature map of one image extracted from shallow layer of the backbone network, where C, H, W are the number of channels,

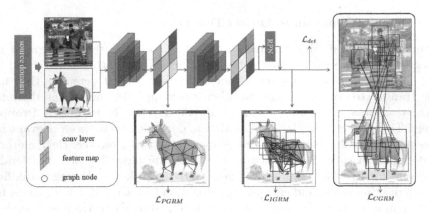

Fig. 1. The overall architecture of TGRN, which mainly consists of three modules, *i.e.*, PGRM, IGRM and CGRM. $\mathcal{L}_{det}$ denotes the detection loss that contains classification and bounding box regression losses.

the height and width of feature map, respectively. We are going to build a graph $\mathcal{G}^{\mathcal{P}} = \{\mathcal{V}^{\mathcal{P}}, \mathcal{A}^{\mathcal{P}}\}$. $\mathcal{V}^{\mathcal{P}}$ denotes the set of nodes (each pixel point in F corresponds to a node). $\mathcal{A}^{\mathcal{P}}$ is a weighted adjacency matrix representing the edges between nodes.

Connecting all nodes is obviously not make sense. To obtain a refined sparse graph, we perform a series of measures to remove redundant edges. Depending on the annotated bounding box, the pixels of background are separated out and their edges are simply discarded (set the weight to 0). For the pixels of foreground, only connections between the same category are allowed to be preserved. Meanwhile, the pixels in the bounding box not all belong to foreground object. To isolate the object from environment, we utilize cosine function to measure the similarity and truncate the connections with low similarity. Specifically, the weight of an edge connecting two nodes i and j is formulated as,

$$\mathcal{A}^{\mathcal{P}}_{ij} = \mathbb{1} \cdot \sigma(\cos(F_i, F_j) > 0.5) \cdot \cos(F_i, F_j), \tag{1}$$

where $\sigma(\cdot)$ is an indicator function which is 1 if $(\cdot)$ is true and 0 otherwise. $\mathbb{1}$ is a binary function that indicates whether the two nodes belong to the same category. The category label of a pixel is assigned by bounding box and object label.

The connections between foregrounds and backgrounds are severed by careful pruning, but the structural representations of objects remain blurred. Global linking introduce numerous unnecessary and even harmful links, *e.g.*, tail and hair, resulting in confusion of semantic information and unclear topological relations. Therefore, we impose distance restriction to eliminate those connections that are too far away. Specifically, the Euclidean distance of two nodes is $E_{ij} = (a_i - a_j)^2 + (b_i - b_j)^2$, here a, b represent the 2D coordinates of a node on the feature map. Only connections with distances less than threshold τ can be retained,

$$\mathcal{A}_{ij}^{\mathcal{P}} = \sigma(E_{ij} < \tau) \cdot \mathcal{A}_{ij}^{\mathcal{P}}. \tag{2}$$

We set τ to 50 in experiments. Note that the distance restriction is not contradictory to the original intention of establishing long-range associations, since the value of τ is much larger than the size of conventional convolution kernel.

Once the graph $\mathcal{G}^{\mathcal{P}} = \{\mathcal{V}^{\mathcal{P}}, \mathcal{A}^{\mathcal{P}}\}$ is established, it is fed into a Graph Convolutional Network(GCN) [22]. The features are processed recursively by the graph convolutional layers, which is defined as $f^{(l+1)} = ReLU(A^* f^{(l)} W^{(l)})$, $1 \leq l < L$, where A^* is adjacency matrix, $f^{(l)}$ and $W^{(l)}$ are the hidden features and parameters of the l-th layer. The last (L-th) linear layer outputs the softmax classification results. Finally the training loss is defined as,

$$\mathcal{L}_{PGRM} = -\frac{1}{N|\mathcal{V}^{\mathcal{P}}|} \sum^{N} \sum^{\mathcal{V}^{\mathcal{P}}} \hat{y}_1 \cdot log(GCN_1(\mathcal{G}^{\mathcal{P}})), \tag{3}$$

where $\hat{y}_1$ is the category label of the node and N is the batchsize, as mentioned above, $N = 2$.

3.2   Instance-Level Graph Reasoning Module

Instance-level features refer to high-level features based on ROI-Pooling, which contain rich semantic information for the final decision-layer inference. It is challenging to search for foreground regions from complex environments and overlapping objects. To address this problem, some studies resort to relational binding of foregrounds and backgrounds with a view to using environmental elements to reason targets of interest [23,24]. Despite the success achieved, it is not applicable to DGOD, where the target domain is agnostic, leading to the uncontrollable environment variable. On the other hand, various parts of an object are often integrated with different environments, so the degree of domain shift and the difficulty of identification vary. The topological relations between the parts of an object do not change easily. Taking advantage of this, IGRM incorporates the semantic information of various regions by modeling the relationship of foreground instances, endowing the network with the ability to reason from easy to difficult and from local to full view.

Suppose $I \in \mathbb{R}^{M \times L}$ denotes the instance-level features of an image extracted by RPN, where M, L represent the number and feature dimension of the instance. We firstly create a graph $\mathcal{G}^{\mathcal{I}} = \{\mathcal{V}^{\mathcal{I}}, \mathcal{A}^{\mathcal{I}}\}$, where each instance corresponds to a node. We establish strong correlations for instance features belonging to the same category, forcing the model to discriminate features that incorporate information from multiple regions. Specifically, the weight of an edge between any two nodes is defined as,

$$\mathcal{A}_{ij}^{\mathcal{I}} = \mathbb{1} \cdot [1.0 - cos(I_i, I_j)], \tag{4}$$

where $\mathbb{1}$ denotes whether the two nodes belong to the same category. The reason we make the weights inversely proportional to the similarity is to allow regions with greater differentiation to produce stronger connections so that the model can reason from the easy parts to the hard ones. Note that here, unlike PGRM, we encourage the establishment of associations between instances at ultra-long distances, due to the fact that instance-level features have more redundant information compared to pixel-level features and do not lead to crashing semantic confusion. Finally, we similarly feed the graph $\mathcal{G}^{\mathcal{I}}$ into a GCN and the training loss is formulated as,

$$\mathcal{L}_{IGRM} = -\frac{1}{N|\mathcal{V}^{\mathcal{I}}|} \sum^{N} \sum^{\mathcal{V}^{\mathcal{I}}} \hat{y}_2 \cdot log(GCN_2(\mathcal{G}^{\mathcal{I}})), \tag{5}$$

where $\hat{y}_2$ denotes the category label of the instance assigned by RPN.

3.3   Cross-Domain Graph Reasoning Module

For cross-domain instances, many studies resort to learning domain-invariant features by domain alignment with adversarial training [3,4]. However, adversarial training is difficult to converge and excessive domain alignment will lead to source bias. The proposed CGRM learns topology-invariant features by mining structural interactions across domains, which achieves domain alignment implicitly.

At each iteration, two images from different source domains are fed into the network. Suppose $I_1, I_2 \in \mathbb{R}^{M \times L}$ represent their instance features. Following DBGL [1], we construct a bipartite graph $\mathcal{G}^{\mathcal{C}} = \{\mathcal{V}^{\mathcal{C}}, \mathcal{A}^{\mathcal{C}}\}$. $\mathcal{V}^{\mathcal{C}} = \{\mathcal{V}_1, \mathcal{V}_2\}$, where $\mathcal{V}_1, \mathcal{V}_2$ correspond to the instances of the two images respectively. And the adjacency matrix is

$$\mathcal{A}^{\mathcal{C}} = \begin{bmatrix} \mathbf{0}_{M \times M} & \widetilde{A}_{M \times M} \\ \widetilde{A}^T_{M \times M} & \mathbf{0}_{M \times M} \end{bmatrix}, \tag{6}$$

where $\mathbf{0}$ indicates zero matrix. $\widetilde{A}$ denotes the weight matrix of the edges between the instances of the two images and $\widetilde{A}^T$ is the transpose matrix. For instances i and j from two images, the weight of their edge is $\widetilde{A}_{ij} = cos(I_1^i, I_2^j)$. However, such a measure is not accurate due to the asymmetry of domain. To ensure structural consistency across domains, we need to assign the greatest weight to the instances with the closest semantics. Therefore, we introduce Prototype Distance Re-weighting (PDR) to capture reliable pairs.

Prototype Distance Re-weighting. We first define a prototype for each category in each image,

$$p_n^k = \frac{1}{|I_n^q \in k|} \sum_{I_n^q \in k} I_n^q, \tag{7}$$

where $n \in \{1, 2\}$ and $k \in \{1, 2, ..., K\}$ denote the image index and class number, and p_n^k is the prototype of the class k in the n-th image and I_n^q denotes the q-th instance in the n-th image. Then we can calculate the similarity between each instance and its corresponding prototype,

$$\hat{S}_n^q = cos(I_n^q, p_n^k), I_n^q \in k, \tag{8}$$

Due to the asymmetry of the domain, the similarity of instances with the same semantic information and their prototypes are different, which does not facilitate comparison. Therefore, we further normalize the similarity,

$$S_n^q = \frac{\hat{S}_n^q}{max(\hat{S}_n^0, \hat{S}_n^1, ..., \hat{S}_n^q, ..., \hat{S}_n^Q)}, \{I_n^0, I_n^1, ..., I_n^Q\} \in k, \tag{9}$$

where $max(\cdot)$ denotes the maximum value function. S, $i.e.$, Prototype Distance, describes the distance between an instance and its corresponding prototype. If two instances from different domains have similar semantic information, then their prototype distances are also similar. Based on this, we reweight the cosine similarity,

$$\widetilde{A}_{ij} = (1.0 - |S_1^i - S_2^j|) \cdot cos(I_1^i, I_2^j) \tag{10}$$

Then we get the adjacency matrix $\mathcal{A}^\mathcal{C}$. Finally the graph $\mathcal{G}^\mathcal{C}$ is fed into a GCN and trained with the loss function,

$$\mathcal{L}_{CGRM} = -\frac{1}{|\mathcal{V}^\mathcal{C}|} \sum^{\mathcal{V}^\mathcal{C}} \hat{y}_2 \cdot log(GCN_3(\mathcal{G}^\mathcal{C})), \tag{11}$$

where $\hat{y}_2$ denotes the category label of the instance.

3.4   Overall Objective

The basic optimize objective of Faster-RCNN is defined as $\mathcal{L}_{det}$, including the classification and bounding box regression losses. Then the three proposed modules are trained jointly with the backbone. The overall objection function of TGRN is formulated as,

$$\mathcal{L}_{TGRN} = \mathcal{L}_{det} + \lambda(\mathcal{L}_{PGRM} + \mathcal{L}_{IGRM} + \mathcal{L}_{CGRM}) \tag{12}$$

where λ is the hyper-parameter.

4   Datasets

As no suitable dataset for DGOD, we construct two datasets for experiments. Figure 2 shows several examples. **PCWS** is composed of four domains, **P**ascal VOC2007 [19], **C**lipart22, **W**atercolor22 and **S**ketch22, where the data of the last three domains are collected and annotated by us. Pascal VOC2007 is a real-world image benchmark for object detection that includes 5011 images for training and 4952 images for testing. Clipart22, Watercolor22 and Sketch22, as their names suggest, are respectively clipart, watercolor and sketch styles. They each contain 2000, 1050, 1000 images and a total of 9717 instance-level annotations. All three domains share the same 20 category labels as Pascal VOC2007. **BDD-DG** is created based on BDD100k [20] and contains four domains: clear, rainy, dark, and foggy. BDD100k is a standard benchmark for autonomous driving scenarios, covering different weather conditions and time periods. We sampled images from clear, rainy and dark to form the three domains. Then another domain, foggy, is obtained by using the fogging algorithm [36]. Each domain contains 2000 images and shares 7 categories.

(a) Clipart22   (b) Watercolor22   (c) Sketch22   (d) BDD-DG

Fig. 2. (a)-(c) Examples of three domains collected by us. (d) Examples of BDD-DG. **Top Left**: clear. **Top Right**: rainy. **Bottom Left**: dark. **Bottom Right**: foggy.

5   Experiments

5.1   Implementation Details

Training Details. Our method is based on Faster-RCNN framework [16] with VGG-16 [25] backbone, which is pre-trained on ImageNet [26]. At each iteration, we randomly select two source domains and take one image each to feed into the network, *i.e.*, batchsize is 2. The shorter side of each image is resized to 600. We set the training epoch to 7. The network is optimized by stochastic gradient descent(SGD) with a momentum of 0.9 and the initial learning rate is set to 0.001, which is decreased to 0.0001 after 5 epochs. The hyper-parameter λ is fixed to 0.2. We report mean average precision (mAP) with a IoU threshold of 0.5. All experiments are implemented by the Pytorch framework and trained with a TITAN RTX GPU.

Evaluation Protocol. Following DG studies [17,27], we conduct the leave-one-domain out evaluation protocol. One domain is selected as the target domain while the remaining domains are source domains for training.

Table 1. Results on PCWS (%)

Method	Pascal VOC	Clipart22	Watercolor22	Sketch22	Avg
Faster-RCNN	53.9	50.9	63.7	66.4	58.7
DA-Faster	53.5	50.2	64.3	67.4	58.9
Ent-Reg	52.8	50.7	64.1	68.3	59.0
MixStyle	52.9	52.9	64.5	67.2	59.4
Seemakurthy	52.2	51.5	64.7	67.6	59.0
TGRN(ours)	**54.1**	**54.7**	**65.6**	**69.7**	**61.0**

Table 2. Results on BDD-DG (%)

Method	clear	rainy	foggy	dark	Avg
Faster-RCNN	32.2	31.8	21.5	25.2	27.7
DA-Faster	33.1	**34.7**	22.5	24.7	28.8
Ent-Reg	33.0	33.7	23.5	24.6	28.7
MixStyle	30.9	29.7	22.1	23.7	26.6
Seemakurthy	32.2	33.6	23.2	**26.5**	28.9
TGRN(ours)	**33.2**	34.0	**23.9**	25.8	**29.2**

5.2 Baselines

Following DG studies [7], we aggregate source domains and train with **Faster-RCNN** [16] as a strong baseline. For a comprehensive assessment, we implement several methods for comparison. **DA-Faster** [3] is an effective method for DAOD, which eliminates domain shift by aligning global and instance features of source and target domains. In this study, we use this method to align only the source domains. **Ent-Reg** [28] is a regularization method for DG used to solve the problem of unstable class condition distribution in adversarial learning. **MixStyle** [11] is a plug-and-play feature augmentation-based DG method that obtains new domain information by mixing the statistics of features. **Seemakurthy** et al. propose a general detection framework which integrates domain invariant learning and entropy regularization.

Table 3. Ablation Results on PCWS

PGRM	IGRM	CGRM	PCWS
-	-	-	58.7
✓	-	-	59.6
-	✓	-	59.3
-	-	✓	59.1
✓	✓	-	60.7
✓	-	✓	60.3
-	✓	✓	60.2
✓	✓	✓	61.0

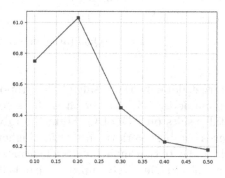

Fig. 3. Effect of hyper-parameter λ

5.3 Main Results

The experimental results on PCWS are shown in Table 1. We can observe that our method outperforms all the baseline methods by a large margin and improves over Faster-RCNN by 2.3% on average. In particular, when Clipart22 is used as

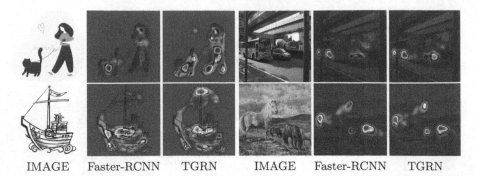

IMAGE   Faster-RCNN   TGRN   IMAGE   Faster-RCNN   TGRN

Fig. 4. Visualization of attention map on PCWS

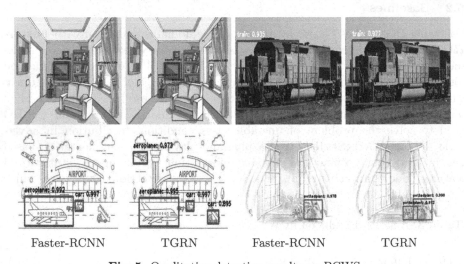

Faster-RCNN   TGRN   Faster-RCNN   TGRN

Fig. 5. Qualitative detection results on PCWS

the target domain, the result of the proposed TGRN is much higher than the
other methods, 1.8% higher than the second place. Table 2 shows the experimental results on BDD-DG. From the table, we can see that our method still
achieves the best performance, 1.5% higher than faster-rcnn on average. These
results indicate that modeling topological relations can enhance the robustness of
the model and effectively improve the out-of-domain generalization performance.

5.4 Empirical Analysis

Ablation Study. We conduct a series of ablation experiments on PCWS to
investigate the role of the three modules. As shown in Table 3, we can see that
adding any of the modules can improve on the baseline. At the same time, removing any of the modules results in less than optimal performance. It demonstrates
that each module plays a positive role.

Effect of λ. We investigate the effect of hyper-parameter λ. The results are shown in Fig. 3. We set λ to 0.1, 0.2, 0.3, 0.4 and 0.5, respectively, and report the average results on PCWS. It can be observed that the best result is obtained when λ is 0.2. The results decay as λ gradually increases.

Visualization of Attention Map. To study the deep network activations learned by Faster-RCNN and TGRN. We visualize their attention heat maps. As shown in Fig. 4, we can see that Faster-RCNN tends to focus on partial easy regions. In contrast, the proposed TGRN is able to capture relatively hard regions and overall topological information of objects. For example, TGRN 'sees' the mast of the ship, which is a hard region to detect, illustrating its topological reasoning ability.

Qualitative Detection Results. Figure 5 illustrates the example of detection results on PCWS. We can see that our method consistently outperforms Faster-RCNN. TGRN detects the whole train as well as those hard regions such as car, potted plant, etc. TGRN is able to detect the whole profile of foregrounds, relying on the global topology modeling capability of PGRM and IGRM. Since CGRM explores cross-domain structural similarity, TGRN is able to recognize those regions with large domain shift.

6   Conclusion

We delve into a more practical and challenging topic, DGOD, which aims to generalize the detector trained with source domains to an unseen target domain. Meanwhile, we propose a new framework, TGRN, which models the topology of foregrounds to obtain long-range perception and topological reasoning capabilities to improve out-of-domain detection performance. Moreover, two datasets are constructed to facilitate the study of DGOD. Experimental results demonstrate the effectiveness of the proposed TGRN.

References

1. Chen, C., Li, J., Zheng, Z., Huang, Y., Ding, X., Yu, Y.: Dual bipartite graph learning: a general approach for domain adaptive object detection. In: ICCV (2021)
2. Seemakurthy, K., Fox, C., Aptoula, E., Bosilj, P.: Domain generalisation for object detection. arXiv preprint arXiv:2203.05294 (2022)
3. Chen, Y., Li, W., Sakaridis, C., Dai, D., Van Gool, L.: Domain adaptive faster R-CNN for object detection in the wild. In: CVPR, pp. 3339–3348 (2018)
4. Saito, K., Ushiku, Y., Harada, T., Saenko, K.: Strong-weak distribution alignment for adaptive object detection. In: CVPR, pp. 6956–6965 (2019)
5. Chen, C., Zheng, Z., Ding, X., Huang, Y., Dou, Q.: Harmonizing transferability and discriminability for adapting object detectors. In: CVPR, pp. 8869–8878 (2020)
6. Zhou, W., Du, D., Zhang, L., Luo, T., Wu, Y.: Multi-granularity alignment domain adaptation for object detection. In: CVPR, pp. 9581–9590 (2022)
7. Wang, J., et al.: Generalizing to unseen domains: a survey on domain generalization. IEEE Trans. Knowl. Data Eng. (2022)

8. Zhou, K., Yang, Y., Hospedales, T., Xiang, T.: Learning to generate novel domains for domain generalization. In: Vedaldi, A., Bischof, H., Brox, T., Frahm, J.-M. (eds.) ECCV 2020. LNCS, vol. 12361, pp. 561–578. Springer, Cham (2020). https://doi.org/10.1007/978-3-030-58517-4_33

9. Matsuura, T., Harada, T.: Domain generalization using a mixture of multiple latent domains. In: AAAI, vol. 34, no. 07, pp. 11749–11756 (2020)

10. Carlucci, F.M., D'Innocente, A., Bucci, S., Caputo, B., Tommasi, T.: Domain generalization by solving jigsaw puzzles. In: CVPR, pp. 2229–2238 (2019)

11. Zhou, K., Yang, Y., Qiao, Y., Xiang, T.: Domain generalization with mixstyle. In: ICLR (2021)

12. Liu, H., Song, P., Ding, R.: Towards domain generalization in underwater object detection. In: ICIP, pp. 1971–1975. IEEE (2020)

13. Zhang, X., et al.: Towards domain generalization in object detection. arXiv preprint arXiv:2203.14387 (2022)

14. Ben-David, S., Blitzer, J., Crammer, K., Kulesza, A., Pereira, F., Vaughan, J.W.: A theory of learning from different domains. Mach. Learn. **79**(1), 151–175 (2010)

15. Huang, J., Guan, D., Xiao, A., Lu, S.: FSDR: frequency space domain randomization for domain generalization. In: CVPR, pp. 6891–6902 (2021)

16. Ren, S., He, K., Girshick, R., Sun, J.: Faster R-CNN: towards real-time object detection with region proposal networks. In: NeurIPS, vol. 28 (2015)

17. Li, D., Yang, Y., Song, Y.-Z., Hospedales, T.M.: Deeper, broader and artier domain generalization. In: ICCV, pp. 5542–5550 (2017)

18. Quinonero-Candela, J., Sugiyama, M., Schwaighofer, A., Lawrence, N.D.: Dataset Shift in Machine Learning. MIT Press, Cambridge (2008)

19. Everingham, M., Van Gool, L., Williams, C.K., Winn, J., Zisserman, A.: The pascal visual object classes (VOC) challenge. IJCV **88**(2), 303–338 (2010)

20. Yu, F., et al.: BDD100K: a diverse driving dataset for heterogeneous multitask learning. In: CVPR, pp. 2636–2645 (2020)

21. Li, X., Dai, Y., Ge, Y., Liu, J., Shan, Y., Duan, L.: Uncertainty modeling for out-of-distribution generalization. In: ICLR (2022)

22. Welling, M., Kipf, T.N.: Semi-supervised classification with graph convolutional networks. In: ICLR (2016)

23. Xu, H., Jiang, C., Liang, X., Li, Z.: Spatial-aware graph relation network for large-scale object detection. In: CVPR, pp. 9298–9307 (2019)

24. Li, Z., Du, X., Cao, Y.: GAR: graph assisted reasoning for object detection. In: WACV, pp. 1295–1304 (2020)

25. Simonyan, K., Zisserman, A.: Very deep convolutional networks for large-scale image recognition. arXiv preprint arXiv:1409.1556 (2014)

26. Krizhevsky, A., Sutskever, I., Hinton, G.E.: Imagenet classification with deep convolutional neural networks. In: NeurIPS, vol. 25 (2012)

27. Zhou, K., Yang, Y., Hospedales, T., Xiang, T.: Deep domain-adversarial image generation for domain generalisation. In: AAAI, vol. 34, no. 07, pp. 13025–13032 (2020)

28. Zhao, S., Gong, M., Liu, T., Fu, H., Tao, D.: Domain generalization via entropy regularization. In: NeurIPS, vol. 33, pp. 16096–16107 (2020)

29. Weiss, K., Khoshgoftaar, T.M., Wang, D.: A survey of transfer learning. J. Big data **3**(1), 1–40 (2016)

30. Inoue, N., Furuta, R., Yamasaki, T., Aizawa, K.: Cross-domain weakly-supervised object detection through progressive domain adaptation. In: CVPR, pp. 5001–5009 (2018)

31. Kim, S., Choi, J., Kim, T., Kim, C.: Self-training and adversarial background regularization for unsupervised domain adaptive one-stage object detection. In: ICCV, pp. 6092–6101 (2019)
32. Zhao, G., Li, G., Xu, R., Lin, L.: Collaborative training between region proposal localization and classification for domain adaptive object detection. In: Vedaldi, A., Bischof, H., Brox, T., Frahm, J.-M. (eds.) ECCV 2020. LNCS, vol. 12363, pp. 86–102. Springer, Cham (2020). https://doi.org/10.1007/978-3-030-58523-5_6
33. Motiian, S., Piccirilli, M., Adjeroh, D.A., Doretto, G.: Unified deep supervised domain adaptation and generalization. In: ICCV, pp. 5715–5725 (2017)
34. Balaji, Y., Sankaranarayanan, S., Chellappa, R.: Metareg: towards domain generalization using meta-regularization. In: NeurIPS, vol. 31 (2018)
35. Nam, H., Lee, H., Park, J., Yoon, W., Yoo, D.: Reducing domain gap by reducing style bias. In: CVPR, pp. 8690–8699 (2021)
36. Koschmieder, H.: Theorie der horizontalen sichtweite, Beitrage zur Physik der freien Atmosphare, pp. 33–53 (1924)

RPUC: Semi-supervised 3D Biomedical Image Segmentation Through Rectified Pyramid Unsupervised Consistency

Xiaogen Zhou, Zhiqiang Li, and Tong Tong[✉]

College of Physics and Information Engineering, Fuzhou University,
Fuzhou, People's Republic of China
ttraveltong@gmail.com

Abstract. Deep learning models have demonstrated remarkable performance in various biomedical image segmentation tasks. However, their reliance on a large amount of labeled data for training poses challenges as acquiring well-annotated data is expensive and time-consuming. To address this issue, semi-supervised learning (SSL) has emerged as a potential solution to leverage abundant unlabeled data. In this paper, we propose a simple yet effective consistency regularization scheme called Rectified Pyramid Unsupervised Consistency (RPUC) for semi-supervised 3D biomedical image segmentation. Our RPUC adopts a pyramid-like structure by incorporating three segmentation networks. To fully exploit the available unlabeled data, we introduce a novel pyramid unsupervised consistency (PUC) loss, which enforces consistency among the outputs of the three segmentation models and facilitates the transfer of cyclic knowledge. Additionally, we perturb the inputs of the three networks with varying ratios of Gaussian noise to enhance the consistency of unlabeled data outputs. Furthermore, three pseudo labels are generated from the outputs of the three segmentation networks, providing additional supervision during training. Experimental results demonstrate that our proposed RPUC achieves state-of-the-art performance in semi-supervised segmentation on two publicly available 3D biomedical image datasets.

Keywords: Semi-supervised learning · Rectified pyramid consistency

1 Introduction

Magnetic Resonance Imaging (MRI), a non-radiative imaging method, is crucial in clinical practice. Quantitative MRI characteristics are associated with brain tumor histology and genetics. Radiogenomics, or radiomics, is gaining prominence, involving image acquisition, data preprocessing, tumor delineation, radiomic feature extraction, selection, and classification modeling. Multimodal MRI is the gold standard for brain tumor and atrial fibrillation diagnosis, surpassing CT scans. Brain tumors include primary, brain-derived, and metastatic tumors. Gliomas, the most common primary tumors, account for over 50% of central nervous tumors. High-grade gliomas have a two-year median survival, while low-grade gliomas progress and become fatal. Various MRI

This work was partially supported by the National Natural Science Foundation of China under Grants 62171133.

modalities (T2-weighted FLAIR, T1-weighted, contrast-enhanced, and T2-weighted scans) highlight gliomas' different properties [2]. Atrial fibrillation, a prevalent cardiac arrhythmia, is common in the elderly. Late gadolinium-enhanced MRI visualizes cardiac scars. Left atrium and scar tissue segmentation from LGE-MRI aids diagnosis and treatment planning. Manual annotation is time-consuming and subjective, necessitating automated methods, but faces challenges due to image quality, variable shapes, thin boundaries, and tissue noise [10].

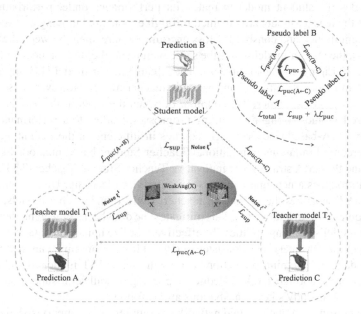

Fig. 1. Illustration of the architecture of the proposed RPUC. It comprises three segmentation networks with distinct initializations. Specifically, the whole network structure looks like a pyramid. A novel PUC loss is presented to enforce consistency among three segmentation networks for unlabeled data input, and their inputs are perturbed with varying ratios of Gaussian noise.

Automatic segmentation of brain tumors and the left atrium (LA) from various MRI modalities is crucial for assessing treatment effectiveness. While deep convolutional neural networks (DCNNs) have shown promising results in biomedical image segmentation [5,9,16,18–23], their performance heavily relies on a large volume of annotated data. However, the acquisition and annotation of such extensive datasets are time-consuming and expensive, particularly in the biomedical image domain, where annotations demand expert diagnostic skills. To address this challenge, substantial research has been dedicated to leveraging additional information from unlabeled data.

Recently, various deep convolutional neural network-based approaches have been proposed to tackle the challenge of limited labeled data in biomedical image segmentation, aiming to reduce labeling costs while achieving impressive performance. For instance, self-supervised learning methods [24] utilize unlabeled data to train a model in a self-supervised manner, enabling it to learn essential information for knowledge transfer. Semi-supervised learning (SSL) methods [11] leverage both limited labeled data and a large amount of unlabeled data to generate high-confidence segmentation predictions. Weakly supervised learning frameworks [3] learn from bounding boxes, scribbles, or image-level tags instead of pixel-wise annotations, thereby alleviating the burden of manual annotation. In this study, we adopt a semi-supervised learning app-roach to achieve 3D biomedical image segmentation using a reduced amount of labeled data and a vast pool of unlabeled data.

The consistency-based model [11,14] is a common structure in semi-supervised learning (SSL), often employing a Teacher-Student framework. Typically, the teacher model guides the student model to match its performance under perturbations. The weights of the teacher model are obtained through exponential moving average (EMA) of the student model, and consistency loss encourages agreement between their predic-tions. In [4], the teacher model shares the same weights as the student model, effectively setting the averaging coefficient to zero. The Mean Teacher model [11] applies EMA to the student model, resulting in an ensemble teacher model. However, these methods suffer from a limitation where the coupling between the roles in the Teacher-Student architecture becomes stronger as training progresses, leading to a performance bottle-neck as the EMA-based teacher model becomes insufficient for the student model.

To overcome limitations in traditional Teacher-Student-based methods like weight coupling and the EMA strategy, our innovative Teacher-Student-Teacher (TST) network structure introduces a new approach to semi-supervised learning. Unlike conventional methods, TST separates the weights of teacher, student, and teacher models, allowing each to acquire its own knowledge. This mitigates the drawbacks of weight coupling and boosts the model's learning abilities. To effectively utilize unlabeled data, we introduce a novel Pyramid Unsupervised Consistency (PUC) loss, promoting consistency among unlabeled data and aligning predictions across the three TST models. PUC also aids the transfer of pseudo-labeled knowledge, improving overall semi-supervised learning performance. Our study's key contributions are as follows:

(1) We present a unique pyramid network structure for semi-supervised 3D biomed-ical image segmentation, addressing limitations in existing Teacher-Student models by adding a teacher model.

(2) To ensure consistency among unlabeled data and facilitate knowledge transfer among three segmentation models, we introduce a novel pyramid unsupervised consis-tency loss.

(3) We conducted extensive experiments on two public 3D MRI datasets, using established metrics like the Dice coefficient. Our results demonstrate significant improvements in semi-supervised learning, confirming its feasibility and superiority.

2   Method

2.1   Overview Architecture

Figure 1 presents an overview of our RPUC architecture. In contrast to most exist-ing consistency-based semi-supervised learning (SSL) methods, our RPUC deviates from the conventional Teacher-Student structure by incorporating the joint learning of three segmentation models. The RPUC architecture, depicted as a pyramid-like network structure, consists of three segmentation models. To ensure consistency among these models when processing unlabeled data inputs, we introduce a novel pyramid unsupervised consistency (PUC) loss. Furthermore, we apply Gaussian noise to perturb the inputs. The subsequent sections will provide detailed explanations of each component.

2.2 Rectified Unsupervised Pyramid Consistency

Data Definition. In our semi-supervised biomedical image segmentation task, the training set X consists of a total of N volumes, including M labeled data and $N - M$ unlabeled input volumes, wherein, $M \gg N - M$. We denote the labeled set as $X^l = \{(x_i, y_i)\}_{i=1}^{M}$ and the unlabeled set as $X^u = \{x_i\}_{i=M+1}^{N}$, specifically, $X = X^l \cup X^u$. For the 3D MRI volumes, $x_i \in \mathcal{R}^{H \times W \times D}$ represents the input image, and $y_i \in \{0, 1\}^{H \times W \times D}$ is the ground truth segmentation volume. The general semi-supervised segmentation learning task can be formulated to learn the network parameters θ by optimizing the following equation:

$$\min_{\theta} = \sum_{i=1}^{M} \mathcal{L}_{sup}(f(x_i; \theta), y_i) + \lambda \mathcal{L}_{puc}(\theta, X^l, X^u) \tag{1}$$

Here, $\mathcal{L}_{sup}$ represents the supervised loss function, $\mathcal{L}_{puc}$ represents the regularization unsupervised loss function, and $f(.)$ represents the segmentation network with θ denoting the model weights. λ is a weighting factor that controls the strength of the regularization. The first term in the loss function integrates the cross-entropy loss and the Dice coefficient loss function, evaluating the correctness of the network's output on labeled input volumes only. The second term is optimized using the regularization unsupervised loss, which leverages both labeled and unlabeled data.

Pyramid Network Structure. The RPUC consists of one student model and two teacher models, forming a pyramid-shaped architecture. As we know, the pyramid structure provides geometric stability. Additionally, we propose a novel pyramid unsupervised consistency (PUC) loss to ensure consistency and transfer cyclical knowledge among the three segmentation models for unlabeled data. Similar to various semi-supervised learning (SSL) approaches [6,14], our method incorporates perturbations such as weak data augmentation and Gaussian noise to enhance the consistency of the unlabeled data output and fully leverage the available unlabeled data. The models can be defined as follows:

$$T_1 = f(X^l; X^u; \theta_1) \tag{2}$$

$$S = f(X^l; X^u; \theta_2) \tag{3}$$

$$T_2 = f(X^l; X^u; \theta_3) \tag{4}$$

$$X_{output}^{rpuc} = \max(T_1, S, T_2) \tag{5}$$

Here, T_1 denotes the first teacher model, S denotes the student model, and T_2 denotes the second teacher model. These models have the same network structure, but their weights, θ_1, θ_2, and θ_3, are initialized differently. X_{output}^{rpuc} denotes the final semi-supervised learning segmentation prediction of our proposed RPUC, obtained by selecting the maximum output from the three segmentation models (T_1, S, and T_2). The logical illustration of the proposed method is provided below:

$$(X^l, X^u) \rightarrow \text{WeakAug}(X^l, X^u) \rightarrow \text{Gaussian}(\text{WeakAug}(X^l, X^u), \xi^i)$$

$$\rightarrow \begin{cases} f(\theta_1; X^l; X^u) \rightarrow Y_1 \rightarrow P_1, \\ f(\theta_2; X^l; X^u) \rightarrow Y_2 \rightarrow P_2, \\ f(\theta_3; X^l; X^u) \rightarrow Y_3 \rightarrow P_3. \end{cases} \tag{6}$$

$$P_i = \text{argmax}(f(\text{WeakAug}(X^u); \theta_i), \xi^i) \tag{7}$$

where ξ^i denotes the perturbation of Gaussian noise rate for the i_{th} segmentation network's input. Y_i represents the i_{th} segmentation prediction, and P_i denotes the i_{th} predicted one-hot label map, referred to as a pseudo label, specifically, $i \in \{1, 2, 3\}$. P_1 represents the pseudo label generated by the first teacher model (see Fig. 1 pseudo label A), P_2 represents the pseudo label generated by the student model (see Fig. 1 pseudo label B), and P_3 represents the pseudo label generated by the second teacher model (see Fig. 1 pseudo label C). The pseudo label is utilized as an additional supervision signal during the training phase to guide the other two segmentation models.

2.3 Loss Functions

Supervised Loss: Our proposed RPUC is trained by minimizing a supervised loss $\mathcal{L}_{sup}$ among the three segmentation models using limited labeled data. The formulation of the supervised loss $\mathcal{L}_{sup}$ is as follows:

$$\mathcal{L}_{sup} = \frac{1}{|D^L|} \sum_{X \in D^L} \frac{1}{w \times h} \sum_{i=0}^{w \times h} (\mathcal{L}_{ce}(Y i^k, p_i^k) + \mathcal{L}_{dice}(Y i^k, p_i^k)), \tag{8}$$

Here, $\mathcal{L}_{ce}$ represents the cross-entropy loss function, and $\mathcal{L}_{dice}$ represents the Dice coefficient loss function. p_i^k denotes the ground truth for the k-th model (where k is in the range of $[1, 2, 3]$). w and h represent the width and height of the input image.

Pyramid Unsupervised Consistency Loss: The proposed pyramid unsupervised consistency (PUC) loss is tri-directional and has a pyramid structure (see Fig. 1). The PUC loss $\mathcal{L}_{puc}$ on the unlabeled data is defined as follows:

$$\mathcal{L}_{puc} = \frac{1}{|D^U|} \sum_{X \in D^U} \frac{1}{w \times h} \sum_{i=0}^{w \times h} (\mathcal{L}_{ce}(Y i^1, p_i^2) + \mathcal{L}_{ce}(Y i^2, p_i^3) + \mathcal{L}_{ce}(Y i^3, p_i^1)), \tag{9}$$

The total loss is defined as:

$$\mathcal{L}_{total} = \mathcal{L}_{sup} + \lambda \mathcal{L}_{puc} \tag{10}$$

where $\mathcal{L}_{sup}$ is a hybrid loss that combines cross-entropy and Dice losses. Following [11], we utilize a time-dependent Gaussian warming-up function $\lambda(t) = e^{-5(1-\frac{t}{tmax})^2}$ to regulate the balance between the supervised loss and the unsupervised consistency loss. Here, t represents the current training step, and t_{max} is the maximum training step.

3 Experiments and Results

3.1 Dataset

To evaluate our RPUC, we conducted experiments on two publicly available biomedical image datasets as follows.

Table 1. Quantitative results of brain tumor semi-supervised segmentation between our proposed RPUC and the other methods on the BraTS2019 dataset.

Methods	# Scan Used		Metrics			
	Labeled	Unlabeled	Dice ↑	JI ↑	95HD	ASD ↓
			(%)	(%)	(voxel)	(voxel)
V-Net [8]	25(10%)	0	80.09	68.12	22.43	7.53
V-Net [8]	50(20%)	0	83.58	76.06	22.09	7.33
V-Net [8]	250(All)	0	88.51	83.82	7.52	1.81
MT [11]	25(10%)	225	81.7	75.12	13.28	3.56
EM [12]	25(10%)	225	82.35	75.75	14.7	3.68
UAMT [14]	25(10%)	225	83.93	5.43	15.81	3.27
DAN [15]	25(10%)	225	82.5	73.48	15.11	3.79
DTC [6]	25(10%)	225	85.06	76.55	14.47	3.74
CPS [1]	25(10%)	225	85.03	77.53	10.27	3.26
DTSC-Net [17]	25(10%)	225	85.12	**77.83**	10.15	3.16
RPUC	25(10%)	225	**85.62**	77.53	**9.26**	**2.7**
MT [11]	50(20%)	200	85.03	78.72	11.8	1.89
EM [12]	50(20%)	200	84.82	80.21	12.37	3.21
UAMT [14]	50(20%)	200	85.05	81.14	12.31	3.03
DAN [15]	50(20%)	200	84.63	81.21	8.96	2.34
DTC [6]	50(20%)	200	86.37	81.65	8.97	2.1
CPS [1]	50(20%)	200	87.12	81.48	9.48	2.16
DTSC-Net [17]	50(20%)	200	87.22	81.58	8.41	2.13
RPUC	50(20%)	200	**87.68**	**82.51**	**7.95**	**1.83**

BraTS2019 Dataset: The brain tumor image dataset used in our experiments is the BraTS2019 dataset [7]. It comprises 335 scans, each consisting of four modalities (FLAIR, T1, T1ce, and T2), with an isotropic resolution of 1 mm^3. The scans were randomly divided into 250, 25, and 60 scans for training, validation, and testing, respectively.

Left Atrium Dataset: The left atrium dataset, consisting of gadolinium-enhanced MRI scans, is part of the 2018 Atrial Segmentation Challenge [13]. It includes 100 gadolinium-enhanced MR imaging scans along with corresponding left atrium (LA) segmentation masks for training and validation. The images have an isotropic resolution of $0.625 \times 0.625 \times 0.625 \text{ mm}^3$. We partitioned the 100 scans into 80 samples for training, 10 samples for validation, and 10 samples for testing.

Table 2. Quantitative results of left atrium semi-supervised segmentation between our proposed RPUC and the other methods on the LA2018 dataset.

Methods	# Scan Used		Metrics			
	Labeled	Unlabeled	Dice ↑	JI ↑	95HD	ASD
			(%)	(%)	(voxel)	(voxel)
V-Net [8]	8(10%)	0	79.99	68.12	21.11	5.48
V-Net [8]	16(20%)	0	86.03	76.06	14.26	3.51
V-Net [8]	80(All)	0	91.14	83.82	5.75	1.52
MT [11]	8(10%)	72	85.54	75.12	13.29	3.77
EM [12]	8(10%)	72	85.91	75.75	12.67	3.31
UAMT [14]	8(10%)	72	81.89	71.23	15.81	3.8
DAN [15]	8(10%)	72	84.25	73.48	13.84	3.36
DTC [6]	8(10%)	72	86.57	76.55	14.47	3.74
CPS [1]	8(10%)	72	87.32	77.23	10.70	2.86
DTSC-Net [17]	8(10%)	72	87.68	77.53	10.26	2.70
RPUC	8(10%)	72	**88.28**	**78.43**	**10.10**	**2.60**
MT [11]	16(20%)	64	88.23	78.72	10.64	2.73
EM [12]	16(20%)	64	88.45	80.21	14.14	3.72
UAMT [14]	16(20%)	64	88.88	81.24	7.51	2.26
DAN [15]	16(20%)	64	87.52	81.31	9.01	2.42
DTC [6]	16(20%)	64	89.42	81.35	7.32	2.1
CPS [1]	16(20%)	64	89.61	80.98	7.61	3.26
DTSC-Net [17]	16(20%)	64	90.15	82.11	**7.23**	1.83
RPUC	16(20%)	64	**91.06**	**82.51**	7.33	**1.72**

3.2 Implementation Details and Evaluation Metrics

In this study, all approaches were implemented using PyTorch on an Ubuntu 18.04 desktop with an NVIDIA GTX 2080TI GPU. The V-Net [8] was used as the backbone of the SSL segmentation network. To ensure a fair comparison, all comparison methods employed the same backbone network. The dropout rate was set to 0.3. We utilized the SGD optimizer with a weight decay of $1e^{-4}$ and momentum of 0.9, along with Eq. 10 as the final loss function. To adjust the learning rate, we employed the poly learning rate scheme, where the initial learning rate (LR) was multiplied by $(1 - \frac{t}{t_{max}})^{0.9}$, with LR set to 0.1 and t_{max} set to 60000 for the BraTS2019 dataset. The batch size was fixed at 4 for all the compared approaches, with half of the labeled data and the other half unlabeled data. This study took randomly cropped patches as input, with patch size set to $96 \times 96 \times 96$ for the BraTS2019 dataset and $112 \times 112 \times 80$ for the LA2018 dataset. Random cropping, flipping, and rotation techniques were employed to augment the training dataset and mitigate overfitting. During the testing phase, four metrics were used to quantitatively evaluate the segmentation performance: Dice, Jaccard Index (JI), 95% Hausdorff Distance (HD95), and Average Surface Distance (ASD).

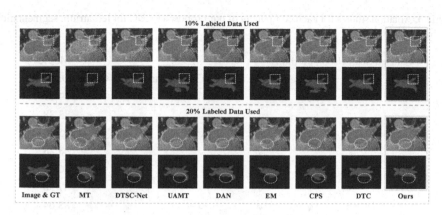

Fig. 2. Visualization of segmentation results of left atrium images obtained by our proposed RPUC and the other methods.

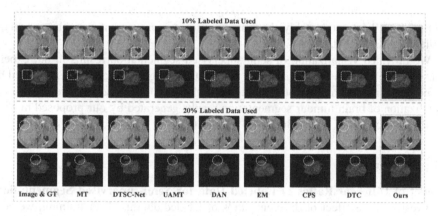

Fig. 3. Visualization of segmentation results of brain tumor MR images obtained by our proposed RPUC and the other methods.

3.3 Comparison with Other Semi-supervised Methods

We compared the semi-supervised segmentation results of our proposed method with other SSL methods, including MT [11], EM [12], UAMT [14], CPS [1], DAN [15], DTC [6], DTSC-Net [17], and a fully supervised learning method, V-Net [8]. The implementations of these approaches have been publicly released, allowing us to conduct fair comparisons on the same datasets. Table 1 and Table 2 present the quantitative comparisons between our proposed RPUC and the other models using 10% and 20% labeled training samples on two benchmark datasets.

Our proposed method outperformed all compared methods in terms of Dice coefficient and Average Surface Distance (ASD), particularly with 20% labeled data. It demonstrated an improvement of approximately 3.92% and 2.65% in Dice score compared to MT [11] with a student-Teacher structure, as well as a boost of around 0.5% and 0.46% in Dice score compared to DTSC-Net [17] with a dual-teacher structure on

Table 3. Ablation studies for each component of our proposed RPUC for left atrium segmentation on the LA2018 dataset.

# Labeled Scan Used	Methods				Metrics
	Baseline	WeakAug.	Gaussian noise	PUC loss	Dice (%)
10%	✓				79.99
	✓	✓			83.88
	✓	✓	✓		85.38
	✓	✓	✓	✓	**88.28**
20%	✓				86.03
	✓	✓			87.56
	✓	✓	✓		88.64
	✓	✓	✓	✓	**91.06**

the BraTS2019 dataset. Notably, the most significant improvement was achieved on the challenging LA2018 dataset.

We provided visualization of semi-supervised segmentation results in our experiments. Figure 2 shows the visualization of left atrium segmentation results obtained by our proposed method and the other methods on the LA2018 dataset. Figure 3 shows the visualization of brain tumour segmentation results generated by our proposed method and the other methods on the BraTS2019 dataset.

3.4   Ablation Study

Impact of Different Ratios of Labeled Data for Semi-supervised Segmentation: To investigate the effect of different ratios of labeled data used for training, we conducted ablation studies under different ratios of labeled data on the BraTS2019 and LA2018 datasets. The semi-supervised segmentation performance with different ratios (i.e., labeled ratios are from 10% to 100%) of labeled data on the two datasets. It is obvious that both CPS [1] and our proposed RPUC outperform the fully supervised V-Net [8] by a huge margin on the BraTS2019 dataset, especially when the labeled ratios were from 10% to 50%. It demonstrates that SSL methods can obtain remarkable performance using limited labeled data. Meanwhile, our RPUC consistently performed better than CPS [1] in different labeled ratio settings, indicating that our proposed method can leverage abundant unlabeled data and improve results. Furthermore, it can be observed that when increasing the labeled ratio to 50%, our proposed RPUC very close results to 100% labled data of the supervised learning V-Net [8]. These results indicate that our RPUC has the potential to achieve more accurate semi-supervised segmentation results with only limited labeled data than other methods, which is desirable for reducing the annotation cost in clinical practice.

Impact of the Number of Teacher Models in the Pyramid Structure: To validate the effectiveness of our proposed pyramid structure, we conducted an investigation to evaluate the impact of different numbers of teacher models, ranging from zero to nine,

Table 4. Ablation studies for different the number of teacher models used in our RPUC structure on the BraTS2019 dataset.

Method/with different number of teacher models	10% Labeled data Dice (%)	20% Labeled data Dice (%)
zero teacher model	80.09	83.58
one teacher model	81.81	85.03
two teacher models (Ours)	**85.62**	**87.68**
three teacher models	82.53	83.82
four teacher models	83.19	82.78
five teacher models	82.54	83.67
six teacher models	79.87	83.14
seven teacher models	82.72	83.26
eight teacher models	80.83	82.34
nine teacher models	82.64	81.97

on our proposed RPUC method. The performance of our proposed RPUC with varying numbers of teacher models in the pyramid structure was evaluated on the BraTS2019 dataset, and the results are summarized in Table 4. By analyzing the effectiveness of our Teacher-Student-Teacher scheme, we observed that the number of teacher models has an impact on the results. Notably, the best performance was achieved when two teacher models were employed, aligning with our proposed Teacher-Student-Teacher structure. The corresponding Dice scores ranged from 80.09% to 85.62% when using 10% labeled data (as shown in Table 4). However, an increase in the number of teacher models beyond two resulted in a decrease in Jaccard Index scores. These findings demonstrate the superiority of our proposed pyramid network structure strategy. The pyramid network structure offers advantages in two aspects. Firstly, it provides a geometrically stable architecture. Secondly, the proposed pyramid unsupervised consistency (PUC) loss ensures consistency among the outputs of the three segmentation models for unlabeled data, facilitating the transfer of cyclical knowledge between the three models.

Impact of Perturbation of Weak Data Augmentation and Gaussian Noise: Ablation studies were conducted to investigate the impact of perturbation of weak data augmentation and Gaussian noise on semi-supervised segmentation. The results are presented in Table 3. Compared to the approach without the weak data augmentation strategy and perturbation of Gaussian noise, our proposed RPUC achieved more accurate left atrium segmentation results compared to the Baseline. These findings demonstrate the beneficial effects of perturbing weak data augmentation and introducing Gaussian noise in 3D semi-supervised medical image segmentation.

Impact of the PUC Loss in Our RPUC: We have investigated the influence of incorporating the pyramid unsupervised consistency (PUC) loss in our method. As presented in Table 3, it is evident that our proposed method with the PUC loss outperforms the other cases in terms of the Dice score. Specifically, compared to the case without the PUC loss ('Baseline+WeakAug.+Gaussian noise'), our method with the PUC loss shows an improvement of 3.96% and 2.91% in Dice when the labeled ratio is 10% and 20%, respectively. These results highlight the effectiveness of the PUC loss in enhancing the SSL segmentation task within our proposed RPUC framework.

4   Conclusion

In this study, we introduce an innovative pyramid network structure for semi-supervised 3D biomedical image segmentation. Our approach has two key components. First, we propose a unique pyramid network that generates pyramid pseudo labels and facilitates knowledge transfer between three segmentation models. Second, we introduce a novel pyramid unsupervised consistency loss to ensure consistency among the outputs of unlabeled data from these models during training. Our experiments on two publicly available biomedical image datasets show that our method outperforms current state-of-the-art approaches both visually and quantitatively. Future work will concentrate on semi-supervised domain adaptation segmentation for multimodal images.

References

1. Chen, X., Yuan, Y., Zeng, G., Wang, J.: Semi-supervised semantic segmentation with cross pseudo supervision. In: Proceedings of the IEEE/CVF Conference on Computer Vision and Pattern Recognition, pp. 2613–2622 (2021)
2. Fountas, K., Kapsalaki, E.Z.: Epilepsy Surgery and Intrinsic Brain Tumor Surgery. Springer, Cham (2019). https://doi.org/10.1007/978-3-319-95918-4
3. Gao, F., et al.: Segmentation only uses sparse annotations: unified weakly and semi-supervised learning in medical images. Med. Image Anal. **80**, 102515 (2022)
4. Laine, S., Aila, T.: Temporal ensembling for semi-supervised learning. arXiv preprint arXiv:1610.02242 (2016)
5. Lin, X., et al.: A super-resolution guided network for improving automated thyroid nodule segmentation. Comput. Methods Programs Biomed. **227**, 107186 (2022)
6. Luo, X., Chen, J., Song, T., Wang, G.: Semi-supervised medical image segmentation through dual-task consistency. In: Proceedings of the AAAI Conference on Artificial Intelligence, vol. 35, pp. 8801–8809 (2021)
7. Menze, B.H., et al.: The multimodal brain tumor image segmentation benchmark (Brats). IEEE Trans. Med. Imaging **34**(10), 1993–2024 (2014)
8. Milletari, F., Navab, N., Ahmadi, S.A.: V-net: fully convolutional neural networks for volumetric medical image segmentation. In: 2016 Fourth International Conference on 3D Vision (3DV), pp. 565–571. IEEE (2016)
9. Nie, X., et al.: N-net: a novel dense fully convolutional neural network for thyroid nodule segmentation. Front. Neurosci. **16**, 872601 (2022)
10. Njoku, A., et al.: Left atrial volume predicts atrial fibrillation recurrence after radiofrequency ablation: a meta-analysis. EP Europace **20**(1), 33–42 (2018)
11. Tarvainen, A., Valpola, H.: Mean teachers are better role models: weight-averaged consistency targets improve semi-supervised deep learning results. In: Advances in Neural Information Processing Systems, vol. 30 (2017)
12. Vu, T.H., Jain, H., Bucher, M., Cord, M., Pérez, P.: Advent: adversarial entropy minimization for domain adaptation in semantic segmentation. In: Proceedings of the IEEE/CVF Conference on Computer Vision and Pattern Recognition, pp. 2517–2526 (2019)

13. Xiong, Z., et al.: A global benchmark of algorithms for segmenting the left atrium from late gadolinium-enhanced cardiac magnetic resonance imaging. Med. Image Anal. **67**, 101832 (2021)
14. Yu, L., Wang, S., Li, X., Fu, C.-W., Heng, P.-A.: Uncertainty-aware self-ensembling model for semi-supervised 3D left atrium segmentation. In: Shen, D., et al. (eds.) MICCAI 2019. LNCS, vol. 11765, pp. 605–613. Springer, Cham (2019). https://doi.org/10.1007/978-3-030-32245-8_67
15. Zhang, Y., Yang, L., Chen, J., Fredericksen, M., Hughes, D.P., Chen, D.Z.: Deep adversarial networks for biomedical image segmentation utilizing unannotated images. In: Descoteaux, M., Maier-Hein, L., Franz, A., Jannin, P., Collins, D.L., Duchesne, S. (eds.) MICCAI 2017. LNCS, vol. 10435, pp. 408–416. Springer, Cham (2017). https://doi.org/10.1007/978-3-319-66179-7_47
16. Zheng, H., Zhou, X., Li, J., Gao, Q., Tong, T.: White blood cell segmentation based on visual attention mechanism and model fitting. In: 2020 International Conference on Computer Engineering and Intelligent Control (ICCEIC), pp. 47–50. IEEE (2020)
17. Zhou, X., Li, Z., Tong, T.: DTSC-net: semi-supervised 3d biomedical image segmentation through dual-teacher simplified consistency. In: 2022 IEEE International Conference on Bioinformatics and Biomedicine (BIBM), pp. 1429–1434. IEEE (2022)
18. Zhou, X., Li, Z., Tong, T.: DM-Net: a dual-model network for automated biomedical image diagnosis. In: Tang, H. (eds.) Research in Computational Molecular Biology. RECOMB 2023. LNCS, vol. 13976, pp. 74–84. Springer, Cham (2023). https://doi.org/10.1007/978-3-031-29119-7_5
19. Zhou, X., et al.: CUSS-net: a cascaded unsupervised-based strategy and supervised network for biomedical image diagnosis and segmentation. IEEE J. Biomed. Health Inform. (2023)
20. Zhou, X., et al.: Leukocyte image segmentation based on adaptive histogram thresholding and contour detection. Curr. Bioinform. **15**(3), 187–195 (2020)
21. Zhou, X., et al.: H-net: a dual-decoder enhanced FCNN for automated biomedical image diagnosis. Inf. Sci. **613**, 575–590 (2022)
22. Zhou, X., Tong, T., Zhong, Z., Fan, H., Li, Z.: Saliency-CCE: exploiting colour contextual extractor and saliency-based biomedical image segmentation. Compute. Biol. Med. 106551 (2023)
23. Zhou, X., Wang, C., Li, Z., Zhang, F.: Adaptive histogram thresholding-based leukocyte image segmentation. In: Pan, J.-S., Li, J., Tsai, P.-W., Jain, L.C. (eds.) Advances in Intelligent Information Hiding and Multimedia Signal Processing. SIST, vol. 157, pp. 451–459. Springer, Singapore (2020). https://doi.org/10.1007/978-981-13-9710-3_47
24. Zhuang, X., Li, Y., Hu, Y., Ma, K., Yang, Y., Zheng, Y.: Self-supervised feature learning for 3D medical images by playing a Rubik's cube. In: Shen, D., et al. (eds.) MICCAI 2019. LNCS, vol. 11767, pp. 420–428. Springer, Cham (2019). https://doi.org/10.1007/978-3-030-32251-9_46

Cancellable Iris Recognition Scheme Based on Inversion Fusion and Local Ranking

Dongdong Zhao[1,2], Wentao Cheng[1], Jing Zhou[3], Hongmin Wang[4], and Huanhuan Li[5(✉)]

[1] School of Computer Science and Artificial Intelligence,
Wuhan University of Technology, Wuhan, Hubei, China
{zdd,cwt}@whut.edu.cn
[2] Chongqing Research Institute, Wuhan University of Technology, Chongqing, China
[3] School of Economics and Management, Dalian University of Technology,
Dalian, China
zj_0562@163.com
[4] College of Information Science and Technology, Bohai University, Jinzhou, China
wanghongmin@qymail.bhu.edu.cn
[5] China University of Geosciences, Wuhan, Hubei, China
julylhh@gmail.com

Abstract. Iris recognition has gained significant attention and application in real-life and financial scenarios in recent years due to its importance as a biometric data source. While many proposed solutions boast high recognition accuracy, one major concern remains the effective protection of users' iris data and prevention of privacy breaches. To address this issue, we propose an improved cancellable biometrics scheme based on the inversion fusion and local ranking strategy (IFCB), specifically targeting the vulnerability of the local ranking-based cancellable biometrics scheme (LRCB) to the ranking-inversion attack when recognition accuracy is high. The proposed method disrupts the original iris data by applying a random substitution string and rearranging blocks within each iris string. For every rearranged block, it is either inverted or kept unchanged. This combination of inversed and unchanged blocks, referred as inversion fusion, is then sorted to obtain rank values that are stored for subsequent matching. It is important to note that the inversion fusion step may lead to a loss of accuracy, which can be compensated by amplifying the iris data to improve accuracy. By utilizing a set of different random substitution strings, the rearranged iris strings are employed in both the inversion fusion and local ranking steps. A long iris template is generated and stored as the final protected iris template, forming the basis of the proposed IFCB method. Theoretical and experimental analyses demonstrate that the IFCB scheme effectively withstands rank-inversion attacks and achieves a favorable balance of accuracy, irreversibility, unlinkability, and revocability.

Keywords: Iris recognition · Privacy protection · Cancellable biometric · Template protection

B. Luo et al. (Eds.): ICONIP 2023, LNCS 14449, pp. 340–356, 2024.
https://doi.org/10.1007/978-981-99-8067-3_26

1 Introduction

In recent years, due to the development and maturity of biometric technologies, various modal biometric technologies have been applied to financial and livelihood scenarios, such as iris recognition, face recognition, and fingerprint recognition. Iris recognition is widely regarded as one of the most commonly utilized biometric features in the past decade. Additionally, most iris feature data remains constant throughout a person's lifetime. Once disclosed, the previous data cannot be revoked or replaced with new iris data, so it is important to protect the user's iris data.

According to the international standards ISO/IEC 24745 [1], iris template protection approaches should satisfy three basic characteristics:

Irreversibility: This means that it should be extremely difficult to recover the original iris template from the generated protected template.

Revocability: A compromised iris template should be revocable, meaning that if a previous template is leaked, a new iris template can be issued for the purpose of authentication.

Unlinkability: Iris templates from different applications of the same individual should not be matched crossly, which prevents a template leak in one application from exposing other applications to the threat.

Recently, a number of methods have been proposed to meet the above three requirements, which can be categorized into cancelable iris biometrics [2] and iris biometric cryptosystems [3]. This paper focuses on cancelable iris recognition, which involves transforming the original iris template into another domain using repeatable and non-invertible transformation while preserving recognition accuracy. However, several proposed methods [4–6] in the field of iris recognition and cancelable biometrics have faced challenges in satisfying the requirements of irreversibility, unlinkability, and revocability at the same time.

Zhao et al. [7] introduced a cancelable biometric recognition method called LRCB, which is based on local ranking. However, Ouda et al. [8] discovered a ranking-inversion attack against the LRCB scheme. This attack enabled the recovery of over 95% of the original iris template with the parameter setting achieving the highest recognition accuracy in the LRCB scheme.

To address the security issue of LRCB at high recognition accuracy, we introduce a new scheme called IFCB which includes LRCB along with inversion fusion, random permutation, and other operations. The core of IFCB involves dividing the iris template into certain blocks and comparing the decimal value with half of the maximum value the block can represent. If the decimal value is greater than half, the corresponding block is inversed, otherwise it remains unchanged. The inversed blocks are then combined with the unchanged blocks to enhance the irreversibility of IFCB. Even if an attacker obtains the value of each block after inversion fusion, there are two possible choices for each block, making it exponentially difficult to recover the original iris template. Theoretical and experimental analyses show that IFCB has good irreversibility, revocability, and unlinkability while being resistant to the ranking-inversion attack. In general, the contributions of our work are as follows:

– We improve the irreversibility of IFCB by combining the inversion fusion mechanism with local ranking and making it well resistant to the ranking-inversion attack.
– We demonstrate that IFCB provides the three main security features of irreversibility, revocability, and unlinkability through theoretical analysis and experiments. Furthermore, experimental results demonstrate that IFCB can achieve high recognition accuracy.

The rest of this paper is organized as follows: In Sect. 2, we review related work on cancelable biometric recognition methods. Section 3 presents a detailed introduction to IFCB, as well as a brief overview of LRCB. Section 4 discusses the security of IFCB and the improved attack method. In Sect. 5, we present the experimental results, including recognition accuracy and comparisons with other cancelable biometric recognition methods. Finally, we conclude this paper in Sect. 6 by summarizing the contributions of our work and discussing potential future directions for research in this area.

2   Related Work

In recent years, there have been numerous methods proposed to protect the privacy of iris biometrics. These methods can be broadly classified into two categories: iris biometric encryption and cancelable iris biometrics. The representative methods in the field of iris biometric encryption are fuzzy commitment [9], fuzzy vaults [10], fuzzy extractors [11]. In the field of cancelable iris recognition, many schemes have been proposed in recent years, among which the well-known ones are based on Bloom filters [12], locality sensitive hashing [13–15], and non-invertible approaches et al. [16–18].

The fuzzy commitment (FC) scheme was proposed by Juels and Wattenberg [9] in 1999 as a way to reflect the fuzziness of biometric features by using error correction codes for biometric representation. In 2006, Hao et al. [19] designed and implemented an iris encryption scheme, which achieved good recognition results with a low false rejection rate of 0.47% under the assumption that the false acceptance rate was 0. However, Kanade et al. [20] pointed out that this scheme was only effective in specific iris datasets [19] and performed poorly in other datasets. Furthermore, Kelkboom et al. [21] raised concerns about the security of fuzzy commitment, revealing that it is vulnerable to decodability attacks based on cross-matching.

The fuzzy vault (FV) scheme, proposed by Juels et al. [9] in 2006, is a popular and practical scheme in the field of biometric encryption. The scheme is built on the foundation of unordered set construction and is thus especially applicable to the recognition of unordered biometric features. However, it is shown in [22, 23] that fuzzy vault schemes are vulnerable to cross-matching attacks leading to compromise of user privacy and do not have good unlinkability.

Key generation schemes are a popular method in biometric encryption, but the main challenge is how to obtain stable, uniformly distributed, and regenerable keys from noisy biometric data. One solution proposed by Dodis et al. [11]

in 2008 is the concept of fuzzy extractors, which aim to extract approximately uniform random numbers R from the input regardless of any variations or errors in the input. This property makes fuzzy extractors well-suited for key generation. In 2008, Bringer et al. [24] attempted to extract secure sketches to protect the privacy of a given iris database. However, Blanton and Aliasgari [25] pointed out that existing fuzzy sketch constructions have security issues such as privacy leakage and vulnerability to cross-matching attacks if multiple sketches of an iris are revealed. Therefore, further research is required to address these concerns and develop secure and robust key generation schemes for biometric encryption.

Rathgeb et al. [12] proposed a cancelable iris recognition method based on Bloom filter. However, Hermans et al. [26] analyzed the security of this scheme and pointed out that the unlinkability of the Bloom filter scheme is problematic. The proposed linkability attack can differentiate two Bloom filters from the same iris template with more than 96% probability.

The idea behind locality sensitive hashing (LSH) is similar to a spatial domain transformation, it assumes that if two samples are similar in the original data space, they should show a high degree of similarity after being transformed individually by a hash function. Conversely, if two samples are not similar to each other, they should continue to exhibit dissimilarities after the transformation. Sadhya et al. [15] proposed a cancelable iris template generation method that utilizes random bit sampling. This approach retains the iris data as an iris template after performing a modulo operation by random bit sampling. Meanwhile, Yang et al. [27] proposed a cancelable biometric scheme based on feature-adaptive random projection. The technique is designed to effectively resist attack via record multiplicity by generating a projection matrix from bio-localized feature data. This matrix is discarded after each random projection to enhance security.

Ouda et al. [17] proposed an iris template protection scheme called Bioencoding that does not require tokens, which overcomes the issue of significant performance degradation due to token loss in previous schemes such as GREY-COMBO and BIN-COMBO [28]. However, the irreversibility of this approach was later questioned by Lacharme [29]. Zhao et al. [7] proposed a new iris template protection scheme based on local ranking, referred to as LRCB. Nevertheless, Ouda et al. [8] presented a ranking-inversion attack, which demonstrated that the irreversibility of the LRCB scheme still needs to be improved. Dwivedi et al. [18] proposed an iris template generation scheme based on look-up table for privacy protection. Their approach achieved irreversibility through the random selection of values in the look-up table, while ensuring revocability and unlinkability through substitution of the look-up table. However, their paper lacks an in-depth analysis of possible attacks, such as similarity attacks and attack via record multiplicity.

3   The Proposed Method

This section explains the IFCB method and gives a brief outline of the LRCB scheme process.

3.1   The LRCB Scheme

The process of the LRCB scheme [7] begins from XORing the original iris string x with an application-specific random binary string p to obtain the binary string t. This binary string t is then divided into m blocks with block size b. Each block is then converted from binary to decimal value. The converted blocks are then grouped based on group size d, resulting in g groups. Sorting is then performed within each group, and the rank values obtained are saved as the protected iris template r.

The security of the LRCB scheme relies mainly on the many-to-one mapping, which occurs during the sorting process within each group. When the rank value of a block within a group is known, there are multiple possible decimal values corresponding to this rank value. Ouda et al. [8] proposed a ranking-inversion attack that can recover most of the original iris data when recognition accuracy is high. To counter this attack, we improve the LRCB and propose the IFCB, which has been shown through analysis and experiments to offer better resistance to the ranking-inversion attack.

3.2   The IFCB Scheme

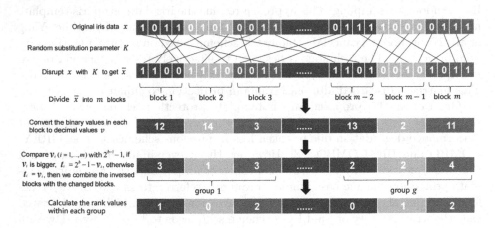

Fig. 1. An illustration of the proposed method ($b = 4$, $d = 3$)

The architecture of IFCB is illustrated in Fig. 1. When given an original iris string x of length n, the application generates a random substitution string K of equal length. The details of this process are as follows:

(1) The string K is an array of length n, expressed as $K = K_1...K_n$. For example, if K_1 equals 8, the data at the first position of x is exchanged with the data at the ninth position.
(2) K is used to disrupt x and obtain $\bar{x}$. Users in the same application use the same K.

(3) $\overline{x}$ is divided into m blocks: $\overline{x} = \overline{x}_1...\overline{x}_m$, each containing b bits. The binary values of each block in $\overline{x}$ are converted to decimal values v so that $v = v_1...v_m$.

(4) Corresponding decimal values $v_i (i = 1...m)$ are compared with $2^{b-1} - 1$. If v_i exceeds $2^{b-1} - 1$, then $t_i = 2^b - 1 - v_i$, otherwise $t_i = v_i$.

(5) t is divided into g groups, denoted as $T = T_1...T_g$, with each group containing d blocks and $m = g \times d$.

(6) Sorting is performed within each group and the rank values are recorded and stored as the final protected iris template r. If multiple blocks have the same decimal value in a group, they are sorted according to their positions in the group. For instance, the decimal values of T_1 in Fig. 1 are $\{3, 1, 3\}$, so the rank values should be $\{1, 0, 2\}$.

It should be noted that the length of the template needs to be increased to improve accuracy, as a shorter template may result in the loss of relevant information. We repeat the above process for one original iris template with 10 different random substitution strings $K^1...K^{10}$, yielding protected templates $r^1...r^{10}$. These templates are spliced into one long template r, expressed as $r = r^1||...||r^{10}$, with a length of $10 \times m$.

In the recognition phase, to determine whether two protected iris templates r_1 and r_2 are from the same user, we calculate the distance between the two templates according to:

$$Dis(r_1, r_2) = \sum_{i=1}^{10 \times m} [r_{1,i} \neq r_{2,i}] \tag{1}$$

Here, $r_{1,i}$ refers to the i-th block of r_1 and $r_{2,i}$ refers to the i-th block of r_2. The range of i is from 1 to $10 \times m$.

4   Security Analysis

4.1   Irreversibility

Our analysis of irreversibility assumes that the attacker already knows the random substitution string K and the final protected iris template r. In such a scenario, if the system satisfies the condition of irreversibility, the attacker will be unable to recover the original iris template x.

The Improved Attack Methods: When attempting to recover the original iris template x from the generated protected iris template r, the attacker must first recover the decimal value of the corresponding block from the result after local ranking. Therefore, the ranking inversion attack proposed by Ouda et al. [8] requires improvement.

Given that there is a inversion fusion operation in the generation of the protected iris template, the range of decimal values corresponding to the rank values in template r is altered. If the block size is b and the range of decimal values is $\Omega = \{0, ..., 2^{b-1} - 1\}$, then for example, $U_j = \{t_{(j-1) \times d+1}, ..., t_{j \times d}\}$ denotes a set of decimal values of group size d in t. The values of elements in U_j can be considered as a set of d random variables within the value domain Ω, denoted by $X_1, X_2, ..., X_d$, each of them has cumulative distribution function (CDF) $F_X(x)$ and probability density function (PDF) $f_X(x)$. We assume that these elements are arranged in ascending order, denoted by $X_{(1)} \leq X_{(2)} \leq ... \leq X_{(d)}$. The process of obtaining the approximate U_j' from the corresponding group of U_j in t is analogous to calculating the PDF of the k-th order statistic $X_{(k)}$ when sampling d items with replacement from the value range Ω. This calculation is performed as shown in formula (2) [8, 30].

$$
f_{X_{(k)}}(x) = \begin{cases}
\sum_{w=0}^{d-k} \binom{d}{w} [f_X(0)]^{d-w} [S_X(1)]^w, & x = 0 \\
\sum_{u=0}^{k-1} \sum_{w=0}^{d-k} \binom{d}{u, d-u-w, w} [F_X(x-1)]^u \times [f_X(x)]^{d-u-w} [S_X(x+1)]^w, & \\
& x = 1, 2, ..., 2^{b-1} - 2 \\
\sum_{u=0}^{k-1} \binom{d}{u} [F_X(2^{b-1} - 2)]^u [f_X(2^{b-1} - 1)]^{d-u}, & \\
& x = 2^{b-1} - 1
\end{cases}
\tag{2}
$$

Formula (2) uses $F_X(x)$ to denote the CDF, which calculates the probability that the obtained value is less than or equal to $x - 1$. $S_X(x)$ denotes the survival function (SF), which calculates the probability that the obtained value is greater than or equal to $x + 1$. Lastly, the PDF $f_X(x)$ calculates the probability that the obtained value is equal to x.

In the IFCB scheme, the decimal values in template t after inversion fusion are considered a set of random variables, where PDF $f_X(x) = \frac{1}{M}$, with $M = 2^{b-1}$ and $x = 0, 1, ..., M - 1$. Thus, the CDF and SF of X can be calculated as follows:

$$
F_X(x) = \begin{cases}
0 & x < 0 \\
\frac{\lfloor x+1 \rfloor}{M} & 0 \leq x < M - 1 \\
1 & x \geq M - 1
\end{cases}
\tag{3}
$$

$$
S_X(x) = \begin{cases}
1 & x \leq 0 \\
1 - \frac{\lfloor x+1 \rfloor}{M} & 0 < x \leq M - 1 \\
1 & x > M - 1
\end{cases}
\tag{4}
$$

After approximating the encoding t' of t, we invert t' to estimate the value v' of v. Each block in t' can either be kept unchanged or inverted by bit. The attacker has a 50% probability of choosing either method, as it is impossible to determine whether a block has been inverted or not during the encryption process. This randomness makes it difficult to estimate a value of v' that is similar to v. Once we have obtained v', we convert its decimal values to binary and use the random substitution string K to obtain an estimate x' of the original iris string x. Since the attacker is aware of K by default, this step does not enhance irreversibility.

To improve the effectiveness of the attack, we implement a majority voting mechanism. After obtaining the protected templates $r = r^1||...||r^{10}$, we perform the recovery operation described above on each $r^i(i = 1, ..., 10)$ to obtain 10 estimates $x'_1, ..., x'_{10}$ of the original iris string x. We count the number of 0 s and 1 s at each position of the 10 estimates $x'_1, ..., x'_{10}$, respectively denoted by a_0 and a_1. For each bit in x', we assign a value of 0 if a_0 is greater than a_1 at the corresponding position, and 1 otherwise. If $a_0 = a_1$, then randomly select 0 or 1. This generates an estimate x' of the original iris template x. Finally, we calculate the Hamming distance between x and x' using the following formula:

$$HD(x, x') = \frac{\sum_{i=1}^{n} x_i \oplus x'_i}{n} \times 100\% \tag{5}$$

where x_i denotes the i-th bit of the template x and x'_i is the i-th bit of the template x', n is the length of the original iris template. If $HD(x, x')$ is smaller, it means that x' is more similar to x and the attack is more effective.

The Number of Possibilities Corresponding to the Original Iris Template: In this section, the attacker needs to exhaustively search through G possible iris strings in order to obtain the correct registered iris template. If $v_i(i = 1...m)$ is greater than $2^{b-1} - 1$, then the value of v_i is changed to: $t_i = 2^b - 1 - v_i$. Otherwise, $t_i = v_i$. Thus, t is the template obtained by performing inversion fusion on v. As a result of this step, in practice, given r, the corresponding number of possibilities G_1 for w can be expressed as:

$$G_1 = 2^m = 2^{(n/b)} \tag{6}$$

After dividing the generated t into g groups denoted by $T = T_1...T_g$, each group $T_i = \{t_{(i-1) \times d+1}, ..., t_{i \times d}\}$ can be denoted by a group of decimal values $a_1, ..., a_d$, where each $a_i(i = 1...d)$ corresponds to a block in R_i. Sorting this group of decimal values results in a set of rank values $r_{(i-1) \times d+1}, ..., r_{i \times d}$. We assume that $a_1, ..., a_d$ are sorted in ascending order, the question then arises of finding the number of possible mappings from $r_{(i-1) \times d+1}, ..., r_{i \times d}$ to $a_1, ..., a_d$. To solve this problem, we use the idea of dynamic programming. First, we determine all possible cases for the largest decimal value a_d, and then derive them in turn. The final number of possible cases can be expressed as $f_d(2^{b-1})$, and the specific formula is as follows [7]:

$$f_d(2^{b-1}) = \sum_{j=1}^{2^{b-1}} f_{d-1}(j) \tag{7}$$

In the above formula, the range of possible values for a_d is reduced by half because of the inversion fusion operation previously performed. 2^{b-1} represents the number of possible values of a_d. The formula is obtained by first determining the value of a_d as $0, ..., 2^{b-1} - 1$, and then continuing recursively from there. The boundary conditions in the formula are defined as $f_1(i) = i(i = 1, ..., 2^{b-1})$ and $f_j(1) = 1(j = 1, ..., d)$. After calculation, we obtain:

$$f_d(2^{b-1}) = \binom{2^{b-1} + d - 1}{d} \tag{8}$$

Since the template r is divided into g groups, the number of possible mappings G_2 in this step is:

$$G_2 = \left(f_d(2^{b-1})\right)^g = \binom{2^{b-1} + d - 1}{d}^{n/(b \times d)} \tag{9}$$

Additionally, we can calculate the total number of all possible alternatives G corresponding to the original iris template x as follows:

$$G = G_1 \times G_2 = 2^{(n/b)} \times \binom{2^{b-1} + d - 1}{d}^{n/(b \times d)} \tag{10}$$

To represent the large value of G, $log_2 G$ is utilized in Table 1. By varying the parameters b and d, with b equal to 1, 2, 4, or 8, and d equal to 2, 4, 8, 16, 32, or 64, the calculated $log_2 G$ values are listed in Table 1. It should be noted that IFCB is not applicable when b equals 1, as the corresponding decimal value of the block can only be 0 or 1. After the inversion fusion operation of the block, a block with the original value of 1 will become 0, and the final rank value will then only be determined by the positions of the blocks within each group. As a result, all iris templates will be the same after the inversion and cannot be used for authentication. Table 1 demonstrates that the value of $log_2 G$ decreases as d increases. The maximum value of $log_2 G$ is achieved when $b = 8$ and $d = 2$. We can modify the values of b and d to achieve varying degrees of irreversibility. Regardless of the $log_2 G$ value in Table 1, the smallest G value is approximately equal to 2^{3772}, making it practically impossible for an attacker to recover the original iris template from over 2^{3772} different choices. When the attacker obtains and utilizes multiple protected iris templates, the value of G can potentially decrease. Due to the high computational complexity of the reverse process, determining the set of all possible original iris templates using multiple protected templates is a complex problem. Therefore, it can be considered that IFCB could achieve irreversibility.

Table 1. The values of $log_2 G$ under different b and d

b	d					
	2	4	8	16	32	64
1	N/A	N/A	N/A	N/A	N/A	N/A
2	9177.50	8092.07	7148.75	6427.99	5927.10	5601.79
4	9177.50	7914.45	6608.55	5424.54	4469.97	3772.35
8	9607.18	8794.25	7841.66	6804.07	5740.71	4713.85

4.2   Revocability

In the IFCB scheme, when a user's protected iris template is compromised, we can revoke the compromised iris template and regenerate a new one for subsequent authentication of the user. The specific steps are as follows:

(1) The iris template of the compromised user is removed from the iris database stored in the authentication server, thereby preventing any unauthorized access to the user's account.
(2) The iris data of the compromised user is re-captured and pre-processed, and a new random substitution string K' is generated. The extracted iris data is then transformed into a new protected iris template using the newly generated K'.

Only the iris feature data of users who have suffered a template leak is transformed with the new random substitution string K', and other users who have not suffered a leak or newly registered users continue to use the old random substitution string K for the template transformation.

Due to the difference between the new random substitution string K' and the old string K, the corresponding protected iris templates are quite different. Therefore, the newly generated iris template can be used for authentication for that user, and the system continues to ensure the authentication and privacy of the user's identity.

4.3   Unlinkability

Good unlinkability in the IFCB scheme refers to the guarantee that protected templates from different applications cannot be used to perform cross-matching, i.e., an attacker cannot determine whether iris templates are from the same user in different applications. To achieve this, the IFCB scheme uses two different random substitution strings K and K' to disrupt the same iris template x in two different applications, thereby generating two protected iris templates t and t'. The decimal values in t and t' can be considered as independent of each other, and hence the final rank values in the protected iris templates r and r' are also independent of each other. Therefore, the distance between r and r' would not be smaller than the distance between the protected templates generated from different irises.

Experimental results presented in Sect. 5.4 show that the distance distribution of intra-class and inter-class matching is very similar, indicating that it is difficult to distinguish between intra-class and inter-class when performing cross-matching between t and t', i.e., it is difficult to know if t and t' are from the same person. Furthermore, the framework proposed in paper [31] was used to evaluate the unlinkability of the IFCB scheme, which demonstrates that the IFCB scheme achieves promising unlinkability.

4.4   Security Attacks on IFCB

In this section, we analyze the reasons why the IFCB scheme can resist brute force attack and the attack via record multiplicity (ARM).

The brute force attack: The attacker tries to recover the original iris template x by exhaustively guessing all possible combinations. According to the derivations made in Sect. 4.1 while calculating G, the minimum number of all possible mappings from the protected template t to the original iris template x has reached 2^{3772}, which is an extremely large number. Therefore, it would be computationally infeasible and practically impossible for an attacker to determine the original iris template by using the brute-force attack since the sheer number of possibilities is astronomically high.

The ARM: Suppose that an attacker obtains multiple cancellable iris templates, denoted as $t_i(i = 1, ..., z)$, from the same original iris template x. If the attacker attempts to use them with the ARM, it is theoretically possible to recover a larger percentage of the original iris template compared to using only one cancellable iris template, given that multiple templates provide more information about the original iris template. In Sect. 5.3, the Hamming distance between the original iris template x and the obtained approximate iris templates $x'_i(i = 1, ..., z)$ was found to be approximately 50%, indicating that the probability of successful recovery for each bit on the original template x can be considered about 50%, denoted as p. Even with the majority voting mechanism, only when the number of correct recoveries on each bit is greater than the number of incorrect recoveries, then the recovery can be considered successful. However, since p is about 50%, the effect of using the majority voting mechanism is similar to random guessing of the original iris template. Therefore, the IFCB scheme could resist the ARM.

5   Experimental Results

This section presents a series of experiments carried out on the classical iris dataset CASIA-IrisV3-Interval [32] using the proposed scheme outlined in this paper. These experiments consist of parameter varying experiments, irreversibility and unlinkability evaluation experiments, and comparison experiments with several other classical schemes.

5.1   Experimental Setup

To conduct the experiments, we first pre-processed the dataset by segmenting the iris images using the USIT [33] system and then extracting iris features with the method proposed by Ma et al. [34]. This process yielded a binary string with a length of $20 \times 512 = 10240$ from each iris image. For the experiments, we used a total of 2639 iris images from 249 individuals, only using the images from the left eyes. To evaluate the performance of the proposed scheme, we used metrics such as the False Acceptance Rate (FAR), the False Rejection Rate (FRR), the Genuine Acceptance Rate (GAR), and the Equal Error Rate (EER). The irreversibility of the proposed scheme was evaluated using the Hamming distance (HD) between the original iris string x and the iris string x' obtained through attack, as mentioned in Sect. 5.3. In addition, we used the shift strategy with $sn = 10$ by default in the accuracy evaluation experiment.

5.2 Parameters Variation Experiment

This experiment analyzes the effect of varying the block size b and the group size d on the recognition accuracy. Additionally, the shifting strategy is implemented to enhance recognition results. The performance of IFCB with changes in block size b and group size d is shown in Table 2.

Table 2. The GAR(%) (FAR = 0.01%) of IFCB with different block size b and group size d

b	d					
	2	4	8	16	32	64
1	N/A	N/A	N/A	N/A	N/A	N/A
2	91.38	93.96	96.07	95.27	93.99	92.04
4	93.52	94.65	94.77	94.90	95.41	94.92
8	90.62	92.50	92.34	92.81	93.51	92.35

Table 2 shows that the IFCB scheme is not applicable when b equals 1, as explained in the Sect. 4.1. It is easy to find that the recognition accuracy of IFCB peaks at 96.07% when the block size b is 2 and the group size d is 8. Table 1 demonstrates that the corresponding value of G is about 2^{7148}, which indicates the irreversibility of IFCB.

5.3 Irreversibility Evaluation

For this part of the experiment, 50 participants were randomly chosen from the 249 test individuals, and one iris code per individual was selected at random for the attack experiment. The selected iris codes were transformed using IFCB to obtain the protected templates. The improved attack method proposed in Sect. 4.1 was then applied to attack the protected templates and obtain the approximate value x' of the original iris string x. The average Hamming distance (HD) of the 50 participants between x and x' was then calculated using formula (5). Table 3 compares the effectiveness of the IFCB and LRCB schemes on defending against the ranking-inversion attack.

Table 3. The HD(%) between the original iris template x and the recovered x' of LRCB and IFCB with different block size b and group size d

IFCB/LRCB		d					
		2	4	8	16	32	64
b	2	50.05/31.26	49.87/27.73	49.62/23.66	49.85/17.99	49.77/13.07	50.03/9.34
	4	49.94/38.28	49.82/36.50	49.70/34.27	50.81/31.10	50.41/27.29	49.76/22.96
	8	49.84/43.77	49.79/42.28	52.13/41.77	57.55/40.17	51.97/38.25	50.28/35.99

Throughout the variations of the b and d, the Hamming distance between the approximate string x' obtained by the attack and the original iris template x remained around 50%, indicating that the ranking-inversion attack is not effective against the proposed scheme. Essentially, the final result obtained is close to random guessing of the values of each position in x'. This outcome primarily stems from the inversion fusion operation performed during the generation of the protected iris templates, which greatly improves the irreversibility of IFCB.

5.4   Unlinkability Evaluation

In this section, we evaluate the unlinkability of IFCB by setting the experimental parameters to the case that the recognition accuracy is at its peak. Specifically, the block size b is set to 2 and group size d is set to 8.

Evaluating Unlinkability by Distance Distribution of Cross-Matching: We evaluate the unlinkability of IFCB through matching across two different applications and calculating intra-class and inter-class distances, and comparing the distributions of both.

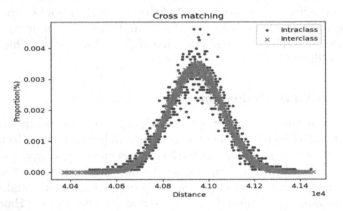

Fig. 2. The distance distribution of intra-class and inter-class cross-matching of the IFCB

We assume the existence of two applications c and e. Firstly, we use two different random substitution strings K_c and K_e to transform the original iris data from the two applications into respective protected templates. For the iris templates in application c, we randomly select one template per individual and perform cross-matching with the protected templates in application e. The templates are then categorized based on whether they come from the same individual. If the two templates are from the same individual, the comparison is referred to as intraclass cross-matching. Otherwise, the comparison is referred to as interclass cross-matching, and the matching distance is recorded.

The distribution of cross-matching distances using the proposed IFCB method is depicted in Fig. 2. The distance distributions of intra-class and inter-class cross-matching almost overlap. As a result, an attacker can hardly determine whether the templates in the two applications are from the same person or not. This indicates that IFCB could achieve the unlinkability.

Evaluating Unlinkability Through the Metric in [31]**:** In this section, we evaluate the unlinkability of IFCB using the framework proposed in [31]. In this framework, we use the local unlinkability metric $D_\leftrightarrow(s)$ to measure the unlinkability under a particular cross-matching score s. The range of $D_\leftrightarrow(s)$ is from 0 to 1, where a value closer to 0 indicates better local unlinkability under the corresponding matching score s, whereas a value closer to 1 denotes worse unlinkability. We use the global unlinkability metric $D_\leftrightarrow^{sys}$ to evaluate the unlinkability of the entire system. A value closer to 0 indicates better global unlinkability.

The evaluation results for the unlinkability of IFCB are presented in Fig. 3. The Mated curve represents the distance calculation between two templates derives from intraclass cross-matching, while the Non-Mated curve represents the distance calculation derives from inter class cross-matching. The global unlinkability $D_\leftrightarrow^{sys}$ of IFCB is 0.0107, approaching 0. Therefore, it can be concluded that IFCB could achieve promising unlinkability.

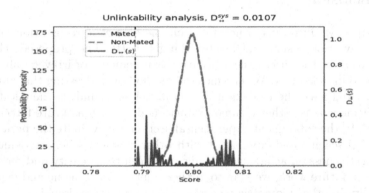

Fig. 3. Local unlinkability and global unlinkability evaluation of the IFCB

5.5   Comparative Experiments

In this experiment, we conduct a comparison between IFCB and three classical cancellable iris template protection methods. These methods include the double bloom filter-based scheme proposed by Ajish et al. [35], the scheme utilizing Randomized Bit Sampling proposed by Sadhya et al. [15], and the local ranking-based scheme proposed by Zhao et al. [7]. We use DBF, RBS, and LRCB to refer to these three schemes respectively.

Table 4. The best GAR(%) (FAR = 0.01%) and security comparison of several methods

Method	Parameter	GAR	Irreversibility	Revocability	Unlinkability
DBF [35]	$l = 32, w = 20$	94.28	$2^{129.97}$	True	0.56860
RBS [15]	$l = 400, k = 10, T = 0.5$	94.45	2^{1024}	True	0.17924
LRCB [7]	$b = 1, d = 64$	97.11	$2^{965.58}$	True	0.05433
IFCB	$b = 2, d = 8$	96.07	$2^{7148.75}$	True	0.01070

From Table 4, it can be observed that the proposed IFCB scheme achieves a higher accuracy of approximately 1.62%–1.79% compared to DBF and RBS. Additionally, IFCB demonstrates superior irreversibility and unlinkability when compared to both DBF and RBS schemes. Although the accuracy of IFCB is lower than that of LRCB, IFCB introduces certain modifications and improvements to the LRCB scheme. This enhances the irreversibility of IFCB and addresses the vulnerability of LRCB to ranking-inversion attack in high-accuracy recognition scenarios. Furthermore, IFCB exhibits better unlinkability than LRCB.

6   Conclusion

In this paper, we introduce a novel cancellable biometric recognition approach based on inversion fusion and local ranking (IFCB). The proposed scheme is designed to meet the three crucial security requirements of irreversibility, revocability, and unlinkability. We enhance the method by addressing the vulnerabilities associated with the ranking-inversion attack and analyze the effectiveness of the scheme against other common attack models, such as brute force attacks and ARM. In the subsequent experimental section, we evaluate the performance of the IFCB scheme and compare it with other classical schemes, demonstrating its effectiveness on achieving a balance between the security and recognition accuracy. In future work, we plan to further improve the scheme and explore its applicability in other biometric recognition and protection domains.

Acknowledgements. This work was partially supported by the National Natural Science Foundation of China (Grant No. 61806151), the Natural Science Foundation of Chongqing City (Grant No. CSTC2021JCYJ-MSXMX0002), and the National Key Research and Development Program (Grant No. 2022YFC3321102).

References

1. Secretary, I.: Information technology-security techniques-biometric information protection. International Organization for Standardization, Standard ISO/IEC 24745, 2011 (2011)

2. Patel, V.M., Ratha, N.K., Chellappa, R.: Cancelable biometrics: a review. IEEE Signal Process. Mag. **32**(5), 54–65 (2015)
3. Rathgeb, C., Uhl, A.: A survey on biometric cryptosystems and cancelable biometrics. EURASIP J. Inf. Secur. **2011**(1), 1–25 (2011)
4. Nandakumar, K., Jain, A.K.: Biometric template protection: bridging the performance gap between theory and practice. IEEE Signal Process. Mag. **32**(5), 88–100 (2015)
5. Natgunanathan, I., Mehmood, A., Xiang, Y., Beliakov, G., Yearwood, J.: Protection of privacy in biometric data. IEEE Access **4**, 880–892 (2016)
6. Lee, M.J., Jin, Z., Liang, S.N., Tistarelli, M.: Alignment-robust cancelable biometric scheme for iris verification. IEEE Trans. Inf. Forensics Secur. **17**, 3449–3464 (2022)
7. Zhao, D., Fang, S., Xiang, J., Tian, J., Xiong, S.: Iris template protection based on local ranking. Secur. Commun. Netw. **2018**, 1–9 (2018)
8. Ouda, O.: On the practicality of local ranking-based cancelable iris recognition. IEEE Access **9**, 86392–86403 (2021)
9. Juels, A., Wattenberg, M.: A fuzzy commitment scheme. In: Proceedings of the 6th ACM Conference on Computer and Communications Security, pp. 28–36 (1999)
10. Juels, A., Sudan, M.: A fuzzy vault scheme. Des. Codes Crypt. **38**, 237–257 (2006)
11. Dodis, Y., Reyzin, L., Smith, A.: Fuzzy extractors: how to generate strong keys from biometrics and other noisy data. In: Cachin, C., Camenisch, J.L. (eds.) EUROCRYPT 2004. LNCS, vol. 3027, pp. 523–540. Springer, Heidelberg (2004). https://doi.org/10.1007/978-3-540-24676-3_31
12. Rathgeb, C., Breitinger, F., Busch, C.: Alignment-free cancelable iris biometric templates based on adaptive bloom filters. In: 2013 international Conference on Biometrics (ICB), pp. 1–8. IEEE (2013)
13. Jin, Z., Hwang, J.Y., Lai, Y.L., Kim, S., Teoh, A.B.J.: Ranking-based locality sensitive hashing-enabled cancelable biometrics: index-of-max hashing. IEEE Trans. Inf. Forensics Secur. **13**(2), 393–407 (2017)
14. Lai, Y.L., et al.: Cancellable iris template generation based on indexing-first-one hashing. Pattern Recogn. **64**, 105–117 (2017)
15. Sadhya, D., Raman, B.: Generation of cancelable iris templates via randomized bit sampling. IEEE Trans. Inf. Forensics Secur. **14**(11), 2972–2986 (2019)
16. Ouda, O., Tsumura, N., Nakaguchi, T.: Tokenless cancelable biometrics scheme for protecting iris codes. In: 2010 20th International Conference on Pattern Recognition, pp. 882–885. IEEE (2010)
17. Ouda, O., Tsumura, N., Nakaguchi, T.: Bioencoding: a reliable tokenless cancelable biometrics scheme for protecting iriscodes. IEICE Trans. Inf. Syst. **93**(7), 1878–1888 (2010)
18. Dwivedi, R., Dey, S., Singh, R., Prasad, A.: A privacy-preserving cancelable iris template generation scheme using decimal encoding and look-up table mapping. Comput. Secur. **65**, 373–386 (2017)
19. Hao, F., Anderson, R., Daugman, J.: Combining crypto with biometrics effectively. IEEE Trans. Comput. **55**(9), 1081–1088 (2006)
20. Kanade, S., Petrovska-Delacr'etaz, D., Dorizzi, B.: Cancelable iris biometrics and using error correcting codes to reduce variability in biometric data. In: 2009 IEEE Conference on Computer Vision and Pattern Recognition, pp. 120–127. IEEE (2009)
21. Kelkboom, E.J., Breebaart, J., Kevenaar, T.A., Buhan, I., Veldhuis, R.N.: Preventing the decodability attack based cross-matching in a fuzzy commitment scheme. IEEE Trans. Inf. Forensics Secur. **6**(1), 107–121 (2010)

22. Scheirer, W.J., Boult, T.E.: Cracking fuzzy vaults and biometric encryption. In: 2007 Biometrics Symposium, pp. 1–6. IEEE (2007)

23. Poon, H.T., Miri, A.: A collusion attack on the fuzzy vault scheme. ISC Int. J. Inf. Secur. **1**(1), 27–34 (2009)

24. Bringer, J., Chabanne, H., Cohen, G., Kindarji, B., Zemor, G.: Theoretical and practical boundaries of binary secure sketches. IEEE Trans. Inf. Forensics Secur. **3**(4), 673–683 (2008)

25. Blanton, M., Aliasgari, M.: Analysis of reusability of secure sketches and fuzzy extractors. IEEE Trans. Inf. Forensics Secur. **8**(9), 1433–1445 (2013)

26. Hermans, J., Mennink, B., Peeters, R.: When a bloom filter is a doom filter: security assessment of a novel iris biometric template protection system. In: 2014 International Conference of the Biometrics Special Interest Group (BIOSIG), pp. 1–6. IEEE (2014)

27. Yang, W., Wang, S., Shahzad, M., Zhou, W.: A cancelable biometric authentication system based on feature-adaptive random projection. J. Inf. Secur. Appl. **58**, 102704 (2021)

28. Zuo, J., Ratha, N.K., Connell, J.H.: Cancelable iris biometric. In: 2008 19th International Conference on Pattern Recognition, pp. 1–4. IEEE (2008)

29. Lacharme, P.: Analysis of the iriscodes bioencoding scheme. Int. J. Comput. Sci. Softw. Eng. (IJCSSE 2012) **6**(5), 315–321 (2012)

30. Evans, D.L., Leemis, L.M., Drew, J.H.: The distribution of order statistics for discrete random variables with applications to bootstrapping. INFORMS J. Comput. **18**(1), 19–30 (2006)

31. Gomez-Barrero, M., Galbally, J., Rathgeb, C., Busch, C.: General framework to evaluate unlinkability in biometric template protection systems. IEEE Trans. Inf. Forensics Secur. **13**(6), 1406–1420 (2017)

32. Institute of Automation, C.A.o.S.: CASIA iris image database (2017)

33. Rathgeb, C., Uhl, A., Wild, P.: Iris Biometrics: From Segmentation to Template Security, vol. 59. Springer, Cham (2012). https://doi.org/10.1007/978-1-4614-5571-4

34. Ma, L., Tan, T., Wang, Y., Zhang, D.: Efficient iris recognition by characterizing key local variations. IEEE Trans. Image Process. **13**(6), 739–750 (2004)

35. Ajish, S., AnilKumar, K.: Iris template protection using double bloom filter based feature transformation. Comput. Secur. **97**, 101985 (2020)

EWMIGCN: Emotional Weighting Based Multimodal Interaction Graph Convolutional Networks for Personalized Prediction

Qing Liu[1,2,3] , Qian Gao[1,2,3(✉)] , and Jun Fan[4]

[1] Key Laboratory of Computing Power Network and Information Security, Ministry of Education, Shandong Computer Science Center, Qilu University of Technology (Shandong Academy of Sciences), Jinan, China
10431210776@stu.qlu.edu.cn, gq@qlu.edu.cn
[2] Shandong Engineering Research Center of Big Data Applied Technology, Faculty of Computer Science and Technology, Qilu University of Technology (Shandong Academy of Sciences), Jinan, China
[3] Shandong Provincial Key Laboratory of Computer Networks, Shandong Fundamental Research Center for Computer Science, Jinan, China
[4] China Telecom Digital Intelligence Technology Co., Ltd., No.1999, Shunhua Road, Jinan 250101, Shandong, China
fanjun.sd@chinatelecom.cn

Abstract. To address the challenges of information overload and cold start in personalized prediction systems, researchers have proposed graph neural network-based recommendation methods. However, existing studies have largely overlooked the shared or similar characteristics among different modal features. Moreover, there is a mismatch between the focuses of multimodal feature extraction (MFE) and user preference modeling (UPM). To tackle these issues, this paper establishes an interaction graph by extracting multimodal information and addresses the mismatch between MFE and UPM by constructing an emotion-weighted bisymmetric linear graph convolutional network (EW-BGCN). Specifically, this paper introduces a novel model called EWMIGCN, which combines multimodal information extraction using parallel CNNs to build an interaction graph, propagates the information on EW-BGCN, and predicts user preferences by summing the expressions of users and items through inner product calculations. Notably, this paper incorporates sentiment information from user comments to finely weigh the neighborhood aggregation in EW-BGCN, enhancing the overall quality of items. Experimental results demonstrate that the proposed model achieves superior performance compared to other baseline models on three datasets, as measured by HitsRatio with Normalized Discounted Cumulative Gain.

Keywords: Multimodal Feature Extraction · Emotional Weighting · Graph Convolutional Networks · Personalized Prediction

---

*Supported by the Natural Science Foundation of Shandong Province (ZR2022MF333).

B. Luo et al. (Eds.): ICONIP 2023, LNCS 14449, pp. 357–369, 2024.
https://doi.org/10.1007/978-981-99-8067-3_27

1 Introduction

Recent research has integrated user-item interactions into personalized recommendation systems [1] in e-commerce. Traditional methods are replaced by multilayer perceptron models [2] and bias terms [3] to enhance interaction functions. SVD++ [4] focuses on modeling user embeddings via historical item interactions. Graph convolutional networks (GCN) [5,6] gain traction for information interaction in personalized recommendations. MEGCN [7] uses graph representations and weighted sums for final recommendations. Sim-Graphrec [8] enriches data with neighbor similarity and attention mechanisms. However, graph neural network-based methods overlook cross-modality commonality, often introducing noise when integrating multimodal data.

Addressing these issues involves extracting shared features and noise reduction. DeepCoNN [9] uses a CNN module to extract features from images and text, capturing user behavior and item attributes. Similarly, MMGCN [10] extracts audio, video, and text modal features, while MGAT [11] constructs attention networks for user preference weights. GRCN [12] combines multimodal features via weighted summation, enhancing the graph structure. Yet, these approaches contend with noise during multimodal extraction, affecting information accuracy. This paper proposes advanced extraction methods for denoising, refining data for better user preference prediction. Additionally, it addresses the gap in capturing user sentiment, crucial for fusing multimodal information with preferences.

Another challenge lies in reconciling MFE [13] and UPM [14]. Recent studies incorporate user sentiment [15,16] into GCN for personalized prediction, emphasizing sentiment mining in reviews. However, these lack research on multimodal-item and user sentiment interaction, impacting preference prediction accuracy. This paper integrates advanced sentiment analysis models into multimodal graph neural networks, capturing latent sentiment in user comments, enhancing preference prediction accuracy, and refining MFE and UPM integration.

To tackle these challenges, this paper introduces interaction graphs by extracting multimodal information and incorporates them into the EW-BGCN module for information dissemination and user preference prediction. Contributions include:

(1) Proposing emotion-weighted EWMIGCN model for personalized prediction. It extracts emotional information and uses EW-BGCN module for semantic cross-modality associations and preference prediction.

(2) Utilizing advanced parallel CNN method to denoise pre-processed multimodal data, enhancing accuracy.

(3) Constructing EW-BGCN module enhancing associations via emotions. It propagates messages on interaction graphs for high-order collaborative signals and multimodal correlations. An advanced sentiment analysis model weighs correlated user sentiment features in reviews, embedding them in the interaction graph for improved prediction accuracy.

(4) Experiments on real datasets demonstrating EWMIGCN's superior performance over baseline models.

2   Methodology

This paper proposes an emotion-weighted multi-modal interaction graph convolutional network model for personalized prediction, referred to as EWMIGCN. The model framework is depicted in Fig. 1. The model framework is divided into three main parts: ① information extraction layer, ② information propagation layer, and ③ prediction optimization layer, where ④ is the embedding module. This model employs inner product summation of user-item expressions to predict user preferences for items. Specifically, the paper utilizes deep learning techniques to extract semantic knowledge from multiple modalities, and embeds emotion-weighted information into the interaction graph to capture higher-order information. The process involves extracting multimodal information through parallel CNNs from various modalities, initializing the information into the interaction graph. Subsequently, an EW-BGCN module is constructed, integrating a sentiment analysis model to extract emotion information from user comments and embedding it into the interaction graph with weighting to capture higher-order emotional information. Ultimately, the inner product is used to predict user preferences for items. This design and structure contribute to the EWMIGCN model's enhanced accuracy and interpretability in personalized prediction.

2.1   Information Extraction Layer

To tackle multimodal data noise, this paper suggests a parallel CNN method to extract emotions from each modality and link them to items. The following section elaborates on this approach.

Information Extraction. The user set is $U = \{u_1, u_2, u_3, ..., u_u\}$, the item set is $I = \{i_1, i_2, i_3, ..., i_i\}$, and the entity set is $E = \{e_1, e_2, e_3, ..., e_e\}$. Removing redundant or intentionally created text items and various types of noise from image items is a challenge in improving data accuracy. As the data is not time-dependent, this paper employs multimodal information as input and utilizes a difference operation to extract precise data from learned noisy information. This accurate data becomes the output of the parallel CNN model. Inspired by Deep-CoNN [9], this paper applies the parallel CNN approach, allowing simultaneous feature extraction from image and text data. The pre-processed data (V, T) is processed using parallel CNN, and the formula (1) is shown as follows:

$$D_M = f_{PCNN}(M), M = \{V, T\} \tag{1}$$

where D_M represents the vector processed by the parallel CNN, $f()$ is the model of the parallel CNN, and M represents the two modal data.

Enhanced Multimodal Interaction Diagram. First, in this paper, the interaction graph (user item interaction graph G_1, image feature node interaction graph G_v, text feature node interaction graph G_t) is transformed into a bipartite graph structure $G_1 = G_v = G_t = \{(u, r_{ui}, i)|u \in U, i \in I\}$, where U denotes the set of users, I denotes the set of items, and $r_{ui} = 1$ means there is interaction between U and I, otherwise $r_{ui} = 0$. Due to the differences between I and $E = E_V \cup E_T$, this paper constructs a bipartite graph

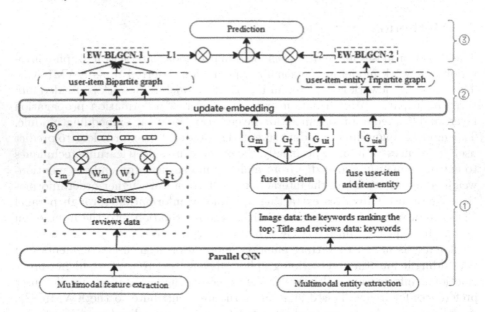

Fig. 1. Framework of the EWMIGCN model.

structure $G = \{(i, r_{ie}, e) | i \in I, e \in E\}$ of project and entity information, where $r_{ie} = 1$ indicates that E is extracted from multimodal data of i, otherwise $r_{ie} = 0$. Finally, this article uses the project node as a connection point and merges the above two parts to form a new user project entity tripartite graph $G_2 = \{(u, r_{ui}, i), (i, r_{ie}, e) | u \in U, i \in I, e \in E\}$, which is called a multimodal interaction graph in this paper.

2.2 Information Dissemination Layer

To capture CF signals and multimodal semantic correlations, this paper constructs an information propagation layer to facilitate graph convolution operations on interaction graphs and multimodal interaction graphs. At the same time, this paper embeds the neighbor nodes into the interaction graph with sentiment weighting to capture strong correlations, respectively.

Embed Initialization. In this paper, we mainly predict user preferences after incorporating GCN for neighbor aggregation by ID embedding interaction graph. The extracted user-item-entities, user-items and low-dimensional vectors of user-item-contents under two modalities are initialized in the following Eqs. (2) (3) (4), respectively, as shown below:

$$D = \left\{ d_{u_1}^0, ..., d_{u_{|U|}}^0, d_{i_1}^0, ..., d_{i_{|I|}}^0, d_{e_1}^0, ..., d_{e_{|E|}}^0 \right\} \tag{2}$$

$$D_{UI} = \left\{ d_{u_1}^0, ..., d_{u_{|U|}}^0, d_{i_1}^0, ..., d_{i_{|I|}}^0 \right\} \tag{3}$$

$$D_M = \left\{ d_{u_1}^0, ..., d_{u_{|U|}}^0, d_{i_1}^0, ..., d_{i_{|I|}}^0 \right\}, M = \{V, T\} \tag{4}$$

where D, D_{UI}, and D_M represent the vectors after initialization. The embedding matrices correspond to the nodes in $|U|, |I|$, and $|E|$, which represent the number of users, items, and semantic entities, respectively. It is worth noting that the user and item nodes in G_1, G_2, G are parameter-shared, ensuring consistency across the graphs.

Comment-Based Sentiment Extraction Model. This paper incorporates an advanced sentiment analysis model, SentiWSP [17], to extract sentiment information from user comments. Since reviews are highly subjective and reflect users' sentiments towards specific items, this paper utilizes the advanced sentiment analysis technique of SentiWSP to extract sentiment information from reviews, enabling fine-grained weighting of items. Drawing inspiration from MEGCF [7] and SentiWSP, this paper extracts emotional information from user comments and applies fine-grained weighting to items, ultimately enabling accurate user preference prediction. Equation (5) is as follows:

$$W_{s\_i} = \frac{\sum_{t \in T_i} f(t)}{|T_i|} \tag{5}$$

where W_{s_i} represents the emotional score, where T_i is the comment set of item i, $|T_i|$ is the number of T_i, $f()$ represents the pre-trained SentiWSP model, which outputs the emotional score of comments, and W_{s_i} is the average emotional score for item i.

Emotional Weighting. Emotions are diverse among individuals, and this paper categorizes emotions into two distinct types known as emotional polarity. The first type is referred to as power-added emotion, which primarily includes keyword information from positive product reviews and evaluations of user reviews expressing moderate to high emotions (determined by ratings of 3 stars and above on a 5-star scale). The second type is known as power-reduced emotion, which encompasses keyword information from negative product reviews and evaluations of user reviews with below moderate emotions (indicated by ratings of 3 stars and below on a 5-star scale). Additionally, negative keywords related to product evaluation and user reviews expressing low emotions (determined by ratings of 2 stars or less) fall into this category.

By analyzing the implied emotional polarity in user reviews, it helps e-commerce platforms understand users' feelings about using items, which provides a basis for item recommendation improvement and better personalized prediction. Inspired by LightGCN [5] and MEGCF [7], this paper constructs an EW-BGCN model that performs message propagation on user-item interaction graphs (G_u, G_v, G_t) respectively in order to better capture higher-order emotional information.

In this paper, message propagation is performed in three interaction graphs: the user-project interaction graph G_1, which is called EW-BGCN-1, the visual interaction graph G_v, which is called EW-BGCN-v, and the textual interaction graph G_t, which is called EW-BGCN-t. For the target user u_1 and project i_1, we define $q = u_1, i_1$, and generate embeddings for q in (G_1, G_v, G_t) using

Eqs. (6) (7) (8), respectively. Message propagation is then performed iteratively on the multimodal interaction graph G_2, called EW-BGCN-2. For the target user node u_1, this paper formulates the embedding output vector at layer l as shown in Eq. (9) below:

$$d_q^l = \sum_{u \in N_{i_1} \cup i_1} \frac{(W_{s\_i_1})^\gamma |I|}{\sum_{i \in I} (W_{s\_i})^\gamma} \cdot \frac{1}{|N_{i_1}|^{0.5}|N_u^{0.5-\alpha}|} \cdot d_u^{(l-1)} \tag{6}$$

$$d_q^{'l} = \sum_{u \in N_{i_1} \cup i_1} \frac{(W_{s\_i_1})^\gamma |I|}{\sum_{i \in I} (W_{s\_i})^\gamma} \cdot \frac{1}{|N_{i_1}|^{0.5}|N_u^{0.5-\alpha}|} \cdot d_u^{'(l-1)} \tag{7}$$

$$d_q^{''l} = \sum_{u \in N_{i_1} \cup i_1} \frac{(W_{s\_i_1})^\gamma |I|}{\sum_{i \in I} (W_{s\_i})^\gamma} \cdot \frac{1}{|N_{i_1}|^{0.5}|N_u^{0.5-\alpha}|} \cdot d_u^{''(l-1)} \tag{8}$$

$$d_{u_1}^{*l} = \sum_{u \in N_{u_1} \cup u_1} \frac{(W_{s\_i})^\gamma |I|}{\sum_{i \in I} (W_{s\_i})^\gamma} \cdot \frac{1}{|N_{u_1}|^{0.5}|N_i^{0.5-\alpha}|} \cdot d_i^{*(l-1)} \tag{9}$$

where l is the convolution level, $|I|$ represents the item set size. To further enhance the node representation, this paper adaptively adjusts the parameters α of the improved Laplace paradigm $\frac{1}{|N|^{0.5}|N|^{0.5-\alpha}}$, where α is used to adjust the model's response to the popularity features (popular vocabulary keywords in recent years such as key words such as dream catcher, retrograde, half-point fake, pan him, etc.), and $|N|$ represents the node size, $\gamma = 0.1$ is used to smooth emotional scores when $I = u_1$, set $s\_i = 1.0$. Next, for item i and entity e_1, this paper performs embedding to generate vectors, and Eq. (10) is shown below:

$$d_{e_1}^{*l} = \sum_{i \in N_{e_1} \cup e_1} \frac{(W_{s\_i})^\gamma |I|}{\sum_{i \in I} (W_{s\_i})^\gamma} \cdot \frac{d_i^{*(l-1)}}{|N_{e_1}|^{0.5}|N_i^{0.5-\alpha}|} \tag{10}$$

Considering the association relationship between users, projects and entities in G_2, where the neighbors of project nodes include both user nodes and entity nodes, this paper needs to separate neighbor nodes according to node types and generate project embedding representations by aggregating their neighbors. Finally this paper develops embedding generation for the target project i_1 in EW-BGCN-2, and Eq. (11) is shown as follows:

$$d_{i_1}^{*l} = \sum_{u \in N_{i_1}^{(u)} \cup i_1} \frac{(W_{s\_i_1})^\gamma |I|}{\sum_{i \in I} (W_{s\_i})^\gamma} \frac{d_u^{*(l-1)}}{|N_{i_1}|^{0.5}|N_u^{0.5-\alpha}|}$$
$$+ \sum_{e \in N_{i_1}^{(e)} \cup i_1} \frac{(W_{s\_i_1})^\gamma |I|}{\sum_{i \in I} (W_{s\_i})^\gamma} \frac{d_e^{*(l-1)}}{|N_{i_1}|^{0.5}|N_e^{0.5-\alpha}|} \tag{11}$$

2.3   Model Prediction and Optimization Layer

Prediction Function. Drawing inspiration from LightGCN [5], this paper selects only the output of the last GCN layer as the final representation for all nodes. Inspired by MGAT [14], the inner product operation is performed on the target user u and item i and the preference scores of users for items are calculated by summing different GCNs, and Eq. (12) is shown as follows:

$$\hat{y} = (d_u^L)^T \cdot d_i^L + (d_u^{'L})^T \cdot d_i^{'L} + (d_u^{''L})^T \cdot d_i^{''L} + (d_u^{*L})^T \cdot d_i^{*L} \tag{12}$$

where $\hat{y}$ is the predicted preference score.

Objective Function. In this paper, the BPR loss function [18] is applied to personalized prediction. Inspired by MEGCF [7], this paper constructs the corresponding BPR loss L for the embedding outputs of EW-BGCN-1, EW-BGCN-v, EW-BGCN-t, and EW-BGCN-2, respectively, and Eq.(13) is shown below:

$$
\begin{aligned}
L_1 &= -\sum_{(u,i,j)\in O} ln\sigma(([d_u^L]^T \cdot d_i^L - [d_u^L]^T \cdot d_j^L) \cdot ([d_u^{*L}]^T \cdot d_i^{*L} \\
&\quad -[d_u^{*L}]^T \cdot d_j^{*L})) + \lambda(||D_M||^2), M = \{V,T\} \\
L_2 &= -\sum_{(u,i,j)\in O} ln\sigma(([d_u'^L]^T \cdot d_i'^L - [d_u'^L]^T \cdot d_j'^L) \cdot ([d_u''^L]^T \cdot d_i''^L \\
&\quad -[d_u''^L]^T \cdot d_j''^L)) + \lambda(||D||^2 + ||D_{UI}||^2) \\
L &= L_1 + L_2
\end{aligned}
\tag{13}
$$

where σ is the sigmoid function, λ is L_2 regularization parameter and O is the complete training data. Finally, this paper uses joint losses to calculate the losses in this paper, using Adam [19] to minimize the losses in equation L.

3 Experiment

3.1 Experimental Setup

Datasets. In this paper, the real datasets Beauty and Arts_crafts_Sewing (referred to as Art) under the Amazon platform and Taobao, as well as the fashion matching dataset of Taobao platform, are selected respectively. The details of these three datasets are introduced as shown in Table 1.

Table 1. Data set introduction.

Dataset	User	Item	Interaction
Art	25165	9324	201427
Beauty	15576	8678	139318
Taobao	12539	8735	83648

Parameter Settings. To test the sequential as well as the accuracy of the predictions, two evaluation metrics, Hits Ratio (HR@k) and Normalized Discounted Cumulative Gain (NDCG@k), are used in this paper, where k denotes the average performance metric for each user in the computational test set. Similar to the adjustment of MEGCF [7], this paper uses the Xavier initializer [20] to initialize the embedding initialization of all models and the Adam optimizer to optimize the models by adaptively adjusting the learning rate and the L_2 regularization parameters λ to $\{0.0001, 0.001, 0.01, 0.1, 1\}$.

Baseline. To assess the validity of the EWMIGCN proposed in this paper, the present model was compared with the following baseline.

SVD++ [4]: The proposed model introduces a new adaptive learning rate (ALR) function to optimize the performance of the algorithm.

NGCF [6]: The model improves model performance through multi-layer iterative embedding propagation.

MMGCN [10]: The model uses fully connected networks with long and short-term memory networks to extract their features and improve data accuracy.

LightGCN [5]: The model enhances the experimental validity by aggregating the linear GCN neighborhoods.

GRCN [12]: The model uses linear GCN for weighted aggregation of features to improve experimental performance.

MEGCF [7]: The model incorporates features from sentiment analysis into user item interactions to improve data accuracy.

3.2   EWMIGCN Comparative Study

To evaluate the performance of the proposed EWMIGCN in this paper, comparative experiments are conducted against state-of-the-art baselines in Table 2.

First, all GCN-based CF methods (NGCF, LightGCN) consistently outperform the traditional CF methods (SVD++), which are heavily influenced by cold starts, leading to possible inaccuracies in the interaction of user items. In addition, for the issue of GCN linearity versus nonlinearity, the experimental results of LightGCN and MEGCF outperform NGCF because linear GCNs are more suitable for capturing CF signals than nonlinear GCNs, providing evidence for why linear GCNs were chosen for this study.

Second, all multimodal-based methods (MMGCN, GRCN, MEGCF) outperform GCN-based methods in most cases because multimodal data can enrich GCN data and enhance model accuracy. Among them, MMGCN and NGCF differ in NDCG performance because MMGCN extends multimodal features to multiple layers through user-item interaction graphs, which enriches preference-independent information, thus weakening the modeling of user preferences. In contrast, the EWMIGCN proposed in this paper combines CF signals and multimodal features to achieve high performance in NDCG.

Third, compared to the baseline model, the EWMIGCN proposed in this paper achieves the best performance on three datasets, which illustrates the importance of mining CF signals and multimodal semantic associations. In particular, compared to the more advanced baseline method MEGCF, this paper improves the performance of the EWMIGCN model by incorporating a new sentiment analysis model and multimodal information up to $1\% - 5\%$. Among them, the experimental results of this model in the Art dataset are lower than the other two datasets because the sentiment information in this dataset is weak and significant recommendation results can be obtained using only interaction data.

3.3   EWMIGCN Ablation Study

In this paper, we first introduce the model variants and then investigate the effects of different factors on the model. To better study the influence of different

Table 2. Comparative experiment.

Model	HR@k	SVD++	NGCF	MMGCN	LightGCN	GRCN	MEGCF	**EWMIGCN**	%IMP
Art	k = 5	0.6530	0.6742	0.6769	0.6814	0.6905	0.7116	**0.7196**	1.13%
	k = 10	0.7425	0.7541	0.7702	0.7639	0.7743	0.7902	**0.8016**	1.44%
	k = 20	0.8285	0.8287	0.8546	0.8329	0.8532	0.8651	**0.8763**	1.30%
Beauty	k = 5	0.4584	0.4853	0.4934	0.5002	0.5087	0.5439	**0.5711**	5.00%
	k = 10	0.5520	0.5820	0.6067	0.6063	0.6204	0.6464	**0.6730**	4.11%
	k = 20	0.6659	0.6810	0.7166	0.7178	0.7241	0.7448	**0.7750**	4.06%
Taobao	k = 5	0.3374	0.3575	0.3649	0.3848	0.3865	0.4045	**0.4169**	3.06%
	k = 10	0.4293	0.4593	0.4695	0.4893	0.4996	0.5212	**0.5329**	2.24%
	k = 20	0.5466	0.5841	0.5902	0.6237	0.6375	0.6516	**0.6680**	2.52%
Model	NDCG@k	SVD++	NGCF	MMGCN	LightGCN	GRCN	MEGCF	**EWMIGCN**	%IMP
Art	k = 5	0.5627	0.5882	0.5643	0.5886	0.5937	0.6144	**0.6279**	2.20%
	k = 10	0.5916	0.6141	0.5945	0.6153	0.6208	0.6398	**0.6531**	2.07%
	k = 20	0.6134	0.6330	0.6159	0.6340	0.6407	0.6588	**0.6705**	1.77%
Beauty	k = 5	0.3592	0.3776	0.3714	0.3807	0.3910	0.4257	**0.4484**	5.33%
	k = 10	0.3895	0.4089	0.4081	0.4152	0.4272	0.4590	**0.4798**	4.53%
	k = 20	0.4157	0.4339	0.4359	0.4435	0.4533	0.4838	**0.5048**	4.33%
Taobao	k = 5	0.2523	0.2658	0.2709	0.2840	0.2861	0.3020	**0.3156**	4.49%
	k = 10	0.2819	0.2986	0.3047	0.3176	0.3225	0.3397	**0.3525**	3.77%
	k = 20	0.3114	0.3301	0.3351	0.3515	0.3573	0.3726	**0.3865**	3.74%

factors on the model, the following EWMIGCN variants are established in this paper: out_V: remove the visual emotional information obtained; out_T: remove the text emotional information obtained; $out\_V\&T$: remove the text and visual emotional information; out_Gv: keep only G_t in symmetric linear GCN; out_Gt: keep only G_v in symmetric linear GCN; out_L1: loss function L_1 is removed, L_1 is the text-image level modal loss; out_L2: loss function L_2 is removed, L_2 is the Regularization loss; out_EW: emotional weighting strategy is removed.

Study Multiple Modalities. It is noteworthy that in Fig. 2, the $out\_V\&T$ model performs the worst, underscoring the importance of capturing multimodal semantic associations. Overall, out_V outperforms out_T, indicating that textual semantics are richer than visual semantics. After fusing visual and textual semantics, experimental results consistently demonstrate that EWMIGCN outperforms $out\_V\&T$, out_V, and out_T, confirming the correlation between visual and textual features in sparking user interest and the improved recommendation effect through their combined modeling.

Study the EW-BGCN Module. In most cases, out_Gv slightly outperforms out_Gt, indicating that Gv captures better content correlations. However, both variants inherently include preference-independent multimodal noise. Figure 3 experimental results demonstrate EWMIGCN's superior performance over the other two variants, affirming the effectiveness of combining multimodal content information.

Study the Loss Function. Figure 4 indicates that, in most cases, out_L2 outperforms out_L1, suggesting that optimizing multimodal semantic relevance is slightly more effective than optimizing CF signals. Specifically, under the Art dataset, the two variants show minimal differences, likely due to the dataset's content richness but lack of emotional information. EWMIGCN consistently outperforms both out_L1 and out_L2 across all three datasets, highlighting the joint loss function's effectiveness in model optimization.

Study Emotional Enhancement. Figure 5 reveals higher NDCG performance percentages on the right vertical axis, indicating this model stronger emphasis on sequential nature, particularly in lower k values where performance improvement is more pronounced. Notably, the Beauty dataset exhibits significantly higher percentage improvement than the Art dataset, consistent with Table 2 experiment results. For a detailed explanation, please refer to Sect. 3.2 Experiments, given space limitations.

3.4   Parameter Experiment

Figure 6 displays a histogram illustrating the impact of graph convolution layers, embedding size, and batch size on this model performance. Increasing graph convolution layers enhances performance across datasets, showcasing this model efficacy. However, performance declines beyond a critical layer count, while maintaining performance levels. Notably, Layer 3 exhibits relatively superior performance. For the Art dataset, k=10 was used for HR performance due to chapter length influence. The line graph in Fig. 6 indicates optimal results were achieved at embedding size 64 and batch size 2048.

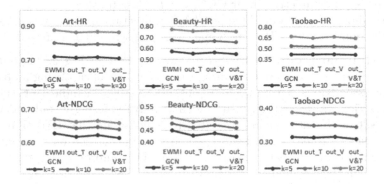

Fig. 2. The effect of semantic relevance of different modalities on model performance.

3.5   Complexity Analysis of EWMIGCN

Compared to the baseline model, this paper incorporates multimodal data (text and images), addressing mutual exclusion between multimodal data and user preferences through emotional feature integration. The complexity of Graph-Convolution and BPR Loss in EWMIGCN increases, yet emotional feature

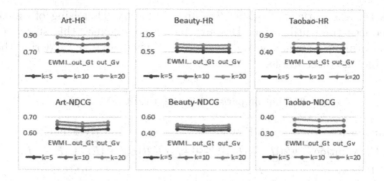

Fig. 3. Effect of bisymmetric linear GCN modules on model performance.

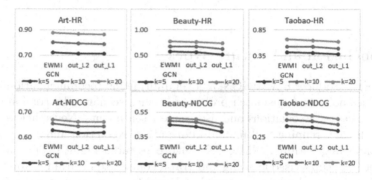

Fig. 4. Impact of the loss function on the performance of the model.

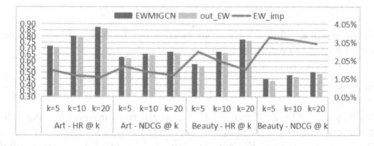

Fig. 5. Impact of neighbor aggregation utilizing emotion-enhanced associative relationships on model performance.

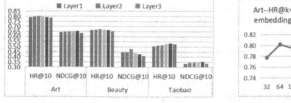

Fig. 6. Effects of layers (left) and size (right) on model performance

weights remain static, not adding training difficulty. Denoising of multimodal data reduces time utilization. Despite higher complexity, this study achieves good performance, efficiency, and resource savings (Table 3), rendering the complexity increase acceptable.

Table 3. The comparison of analytical time complexity.

Component	MEGCF	EWMIGCN
Graph Convolution	$O(2(|E| + |E_m|)\frac{Lds|E|}{B})$	$O(2(|E_v| + |E_t| + |E| + |E_m|)\frac{Lds|E|}{B})$
BPR Loss	$O(4|E|ds)$	$O(8|E|ds)$
Time	$O(T)$	$O(T/2)$

4   Conclusion and Future Work

This paper introduces EWMIGCN, a novel multimodal recommendation method that tackles noise and mismatch problems in such scenarios by combining multimodal semantic associations and user review sentiments. Nevertheless, there's room for further refining multimodal information extraction and semantic association modeling within EWMIGCN. Future research might explore advanced techniques like supervised learning and graph comparison learning. Additionally, integrating advanced preference emotional cues like knowledge graphs, information detection, and cross-domain sentiment preferences could heighten model interpretability.

Acknowledgements. This work was supported by the Natural Science Foundation of Shandong Province (ZR2022MF333).

References

1. Hu, L., Song, G., Xie, Z., Zhao, K.: Personalized recommendation algorithm based on preference features. Tsinghua Sci. Technol. **19**(3), 293–299 (2014)
2. He, X., Liao, L., Zhang, H., Nie, L., Hu, X., Chua, T.S.: Neural collaborative filtering. In: Proceedings of the 26th International Conference on World Wide Web, pp. 173–182 (2017)
3. Fan, J., Ji, R., Tian, Y.: Collaborative filtering recommendation algorithm based on improved BiasSVD. In: 2021 6th International Conference on Control, Robotics and Cybernetics (CRC), pp. 297–300. IEEE (2021)
4. Jiao, J., Zhang, X., Li, F., Wang, Y.: A novel learning rate function and its application on the SVD++ recommendation algorithm. IEEE Access **8**, 14112–14122 (2019)
5. He, X., Deng, K., Wang, X., Li, Y., Zhang, Y., Wang, M.: LightGCN: simplifying and powering graph convolution network for recommendation. In: Proceedings of the 43rd International ACM SIGIR Conference on Research and Development in Information Retrieval, pp. 639–648 (2020)

6. Wang, X., He, X., Wang, M., Feng, F., Chua, T.S.: Neural graph collaborative filtering. In: Proceedings of the 42nd International ACM SIGIR Conference on Research and Development in Information Retrieval, pp. 165–174 (2019)

7. Liu, K., Xue, F., Guo, D., Wu, L., Li, S., Hong, R.: MEGCF: multimodal entity graph collaborative filtering for personalized recommendation. ACM Trans. Inf. Syst. **41**(2), 1–27 (2023)

8. Jia, Z., Gao, Q., Fan, J.: Research on social recommendation model based on enhanced neighbor perception. In: 2022 IEEE International Conference on Systems, Man, and Cybernetics (SMC), pp. 3415–3420. IEEE (2022)

9. Zheng, L., Noroozi, V., Yu, P.S.: Joint deep modeling of users and items using reviews for recommendation. In: Proceedings of the tenth ACM International Conference on Web Search and Data Mining, pp. 425–434 (2017)

10. Wei, Y., Wang, X., Nie, L., He, X., Hong, R., Chua, T.S.: MMGCN: multi-modal graph convolution network for personalized recommendation of micro-video. In: Proceedings of the 27th ACM International Conference on Multimedia, pp. 1437–1445 (2019)

11. Tao, Z., Wei, Y., Wang, X., He, X., Huang, X., Chua, T.S.: MGAT: multimodal graph attention network for recommendation. Inf. Process. Manage. **57**(5), 102277 (2020)

12. Wei, Y., Wang, X., Nie, L., He, X., Chua, T.S.: Graph-refined convolutional network for multimedia recommendation with implicit feedback. In: Proceedings of the 28th ACM International Conference on Multimedia, pp. 3541–3549 (2020)

13. Ouaari, S., Tashu, T.M., Horváth, T.: Multimodal feature extraction for memes sentiment classification. In: 2022 IEEE 2nd Conference on Information Technology and Data Science (CITDS), pp. 285–290. IEEE (2022)

14. Wang, Y., Gao, Q., Fan, J.: KEAN: knowledge-enhanced and attention network for news recommendation. In: Memmi, G., Yang, B., Kong, L., Zhang, T., Qiu, M. (eds.) International Conference on Knowledge Science, Engineering and Management. KSEM 2022. LNCS, vol. 13368, pp. 437–449. Springer, Cham (2022). https://doi.org/10.1007/978-3-031-10983-6_34

15. Yuan, P., Chen, Q., Wang, Z., Yang, J.: Personalized tourism recommendation algorithm integrating tag and emotional polarity analysis. In: 2022 Tenth International Conference on Advanced Cloud and Big Data (CBD), pp. 163–168. IEEE (2022)

16. Wu, W., Wang, Y., Xu, S., Yan, K.: SFNN: semantic features fusion neural network for multimodal sentiment analysis. In: 2020 5th International Conference on Automation, Control and Robotics Engineering (CACRE), pp. 661–665. IEEE (2020)

17. Fan, S., et al.: Sentiment-aware word and sentence level pre-training for sentiment analysis. arXiv preprint arXiv:2210.09803 (2022)

18. Rendle, S., Freudenthaler, C., Gantner, Z., Schmidt-Thieme, L.: BPR: Bayesian personalized ranking from implicit feedback. arXiv preprint arXiv:1205.2618 (2012)

19. Kingma, D.P., Ba, J.: Adam: a method for stochastic optimization. arxiv preprint arxiv:1412.6980 (2014)

20. Glorot, X., Bengio, Y.: Understanding the difficulty of training deep feedforward neural networks. In: Proceedings of the Thirteenth International Conference on Artificial Intelligence and Statistics, pp. 249–256. JMLR Workshop and Conference Proceedings (2010)

Neighborhood Learning for Artificial Bee Colony Algorithm: A Mini-survey

Xinyu Zhou$^{(\boxtimes)}$, Guisen Tan, Yanlin Wu, and Shuixiu Wu

School of Computer and Information Engineering, Jiangxi Normal University,
Nanchang 330022, China
xyzhou@jxnu.edu.cn

Abstract. Artificial bee colony (ABC) algorithm is a representative
paradigm of swarm intelligence optimization (SIO) algorithms, which has
received much attention in the field of global optimization for its good
performance yet simple structure. However, there still exists a drawback for ABC that it owns strong exploration but weak exploitation,
resulting in slow convergence speed and low convergence accuracy. To
solve this drawback, in recent years, the neighborhood learning mechanism has emerged as an effective method, becoming a hot research topic
in the community of ABC. However, there has been no surveys on it,
even a short one. Considering the appeal of the neighborhood learning mechanism, we are motivated to provide a mini-survey to highlight
some key aspects about it, including 1) how to construct a neighborhood topology? 2) how to select the learning exemplar? and 3) what
are the advantages and disadvantages? In this mini-survey, some related
neighborhood-based ABC variants are reviewed to reveal the key aspects.
Furthermore, some interesting future research directions are also given
to encourage deeper related works.

Keywords: Artificial bee colony · Neighborhood learning ·
Neighborhood topology · Learning exemplar

1 Introduction

In recent years, in many academic and industrial fields, the optimization problems are becoming more difficult to solve due to some unfavorable features,
such as nonconvexity, strong nonlinearity, and multimodality [1]. In general,
these problems beyond the scope of many traditional global optimization methods, especially for the gradient information-based methods [33]. Fortunately, the
swarm intelligence optimization (SIO) algorithms can be an effective alternative way for solving these problems [8,39], which derive from mimicking the
collective behavior of social animals in nature. The SIO algorithms have some
attractive characteristics [2,32], such as simple structure, good performance, and
easy implementation. Among the SIO algorithms, some representative paradigms
include genetic algorithm (GA) [19,25], differential evolution (DE) [26,30], estimation of distribution algorithm (EDA) [23,41], particle swarm optimization

B. Luo et al. (Eds.): ICONIP 2023, LNCS 14449, pp. 370–381, 2024.
https://doi.org/10.1007/978-981-99-8067-3_28

(PSO) [20,29], firefly algorithm (FA) [28,38], and artificial bee colony (ABC) [16,45], etc.

The basic ABC has a strong exploration behavior in the entire search space due to its solution search equation [18,45], which might easily jump out of the local optimum region. Nevertheless, each coin has two sides, that its local search capability is insufficient leading to slow convergence and poor exploitation results. To alleviate this drawback, a number of methods have been proposed by researchers to enhance the exploitation of ABC, including probability distributions [31], logistic regression model [9], variable updating mechanisms [11], elite individuals [47], and neighborhood topologies [45], etc. Among them, it may be considered as the most direct way to improve the exploitation capacity by using elite individuals [47]. Unfortunately, the elite individuals may easily be overused to some extent, which makes the algorithm falls into a local optimum prematurely, i.e., evolutionary stagnation [45]. In contrast, the neighborhood topology is expected to be an effective technique to balance the exploration and exploitation capabilities due to some merits, such as, easy implementation, flexible structure, and good diversity [27,35,45].

In fact, the neighborhood learning technology mainly focus on how to select the valuable neighbors from a neighborhood topology and then learn useful information from the neighbors for other general individuals. Therefore, for this technology, two key aspects have been involved: 1) what kinds of neighborhood topologies can be used, and 2) how to learn the useful information. In fact, in the scenario of ABC, the two aspects can be considered as: 1) how to organize the food sources for constructing the neighborhood topology, and 2) how to design the solution search equation based on the neighborhood topology. For example, there exists a simple but effective neighborhood learning approach for ABC, i.e., the best neighbor-guided ABC (NABC) [27]. In the NABC, for each food source, a random neighborhood topology is first constructed by randomly selecting five other food sources from the population as the neighbors, and then the neighbor with the best fitness value, namely the so-called best neighbor, is utilized as the starting search point in the modified solution search equation. Due to that the neighborhood learning technology has been emerged as an appealing way to enhance the exploitation capability of ABC [24,27,35,45], it is necessary to survey this technology by reviewing some related representative work from the perspective of two involved key aspects, which may be helpful to the researchers to design more powerful ABC variants in the future.

The remaining sections are organized as follows. Section 2 describes some common neighborhood topologies and neighborhood learning approaches. The conclusion and further discussion are given in Sect. 3.

2 Neighborhood Learning for ABC

Since the basic ABC algorithm performs well in exploration yet poor in exploitation performance. To effectively address this shortcoming, neighborhood learning approach is introduced to ABC resulting in various ABC variants. These variants can be roughly divided into two types based on the number of different neighborhood topologies used in the single ABC variant, i.e., ABC variants with single neighborhood topology and ABC variants with multiple neighborhood topologies. There are also significant differences among neighborhood learning approaches due to the different connectivity among nodes (food sources) in various neighborhood topologies causing variations in abilities of disseminating search information. To further analyze the characteristics of various neighborhood learning approaches, it is necessary to be presented in the following section that commonly used neighborhood topology illustration and selection rules of the optimal exemplar (the best neighbor) for neighbors.

2.1 ABC with a Single Neighborhood Topology

Some ABC variants usually use single neighborhood topology to enhance the local search capability of ABC, where these topologies include ring topology, random topology, small-world topology and cellular topology. It is worth mentioning that there is also a class of topologies with custom metrics, which are flexibly tailored by researchers using information such as fitness values and position during the algorithm iteration. Moreover, some details about the above mentioned topologies are listed as follows.

Ring Topology: Wang et al. [35] proposed an ABC variant with a ring neighborhood topology (NSABC) with a fixed size of neighborhood radius k as Fig. 1. In addition, a novel solution search equation is formulated by combining the advantages of GABC [47] for the employed bee phase shown as Eq. (1). Among the determined neighbors of X_i, only the optimal neighbor is updated, serving to replace the probability selection during the onlooker bee phase, as expressed in Eq. (2).

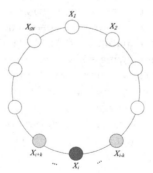

Fig. 1. Illustration of a ring topology, where X_i denotes the current food source while other food sources in the dashed ellipse are neighbors of X_i and k denotes the neighborhood radius.

$$v_{i,j}^* = x_{ib,j} + \phi_{i,j} \cdot (x_{ib,j} - x_{k,j}) + \varphi_{i,j} \cdot (x_{best,j} - x_{ib,j}) \tag{1}$$

$$v_{ib,j}^* = x_{ib,j} + \phi_{i,j} \cdot (x_{ib,j} - x_{k,j}) \tag{2}$$

In the neighborhood learning approach of NSABC, for employed bee phase, the optimal learning target (food source) comes from the neighbor with the highest fitness value (X_{ib}) from the neighbors of X_i, and randomly selecting one from the remaining food sources (X_k) in whole population as the basis for maintaining exploration ability. For the ring topology, food sources can interact with others situated on both sides based on their index, which implies that this topology exhibits a slower dissemination of search information, and to some extent, the best neighbor in the neighborhood can be used to enhance exploitation ability.

Xiao et al. [37] presented an improved version of ABC with an adaptive neighborhood radius size k in ring neighborhood topology (ABCNG), where the k is dynamically adjusted according to whether X_i is updated successfully or not as Eq. (3)

$$k = \begin{cases} k+1, & \text{if } f(V_i) < f(X_i) \\ k-1, & \text{otherwise} \end{cases} \tag{3}$$

One of $2k$ neighbors is selected by the neighbor learning approach as the learning object, and other food sources are selected in a manner and function similar to Eq. (1) in ABCNG.

Random Topology: It is also called panmictic neighborhood topology in paper [45]. ABC with a random neighborhood topology (NABC) was given by Peng et al. [27]. In NABC, the population is randomly divided into some subgroups of the same size in each iteration, so that after several iterations, each food source may has exchanged information with almost all other food sources indicated in Fig. 2.

$$v_{i,j} = x_{nbest,j} + \phi_{i,j} \cdot (x_{nbest,j} - x_{i,j}) \tag{4}$$

The neighbor with the highest fitness value (X_{nbest}) is used as a learning exemplar in each group and it has no additional food sources involved in the generation of candidate food sources except X_i, as shown in Eq. (4). Although this neighborhood learning approach has more greediness, it is an effective approach to maintain a balance between efficiency and effectiveness by carefully adjusting the size of subgroups and thus changing the exemplar of X_i. In addition, compared to the ring topology, the random topology has very fast information dissemination capabilities, since all the food sources in the random topology are connected to each other.

Cellular Topology: Zhang et al. [40] proposed the CGABC algorithm where two different types of cellular topologies were implemented, i.e., linear and compact structures. The figure containing eight neighbors is shown in Fig. 3. In the C9 case, neighbors are tightly clustered around X_i, and location information is

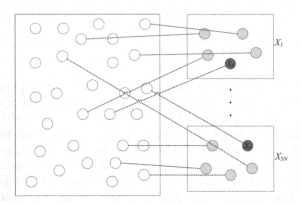

Fig. 2. Illustration of a random topology. The circles inside the large box on the left indicate all food sources while the circles on the right are the neighbors of X_i outside the box, and the solid red circle is the neighbor with the highest fitness value. (Color figure online)

propagated only through neighboring food sources. In the L9 case, the neighbors of X_i are selected horizontally or vertically in the arranged food source combinations, the information is only transmitted to some of the food sources immediately adjacent to X_i yet the spread is much deeper. Therefore, this topology shows a slower speed of information dissemination, which is more suitable for exploration.

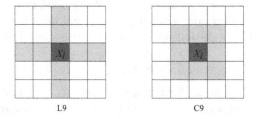

Fig. 3. Illustrations of two different ways of implementing cellular topology. The red and light red squares indicate the food source X_i and the neighbors of X_i, respectively. The abbreviated term L9 denotes a linear neighborhood structure containing eight neighbors while C9 means a compact neighborhood structure. (Color figure online)

$$v_{i,j} = N\left(s_{i,j}, \alpha * |s_{i,j} - x_{k,j}|\right) \tag{5}$$

$$s_{i,j} = \lambda * x_{lbest,j} + (1 - \lambda) * x_{gbest,j} \tag{6}$$

$$s_{i,j} = \lambda * x_{lbest,j} + (1 - \lambda) * x_{k,j} \tag{7}$$

Instead of using the solution search equation in the above paper to generate V_i, a Gaussian function as Eq. (5) is employed to perturb the position of X_i

by varying the mean and variance depending on the number of used function evaluations in this neighborhood learning approach. Further, the best neighbor (X_{lbest}) is used with both the employed and onlooker bee phases, but the global optimal food source (X_{gbest}) is utilized only in the latter phase as Eq. (6) and a random food source (X_k) for the former as Eq. (7) with the aim of allowing the population to fully coarse-grain search before fine-grained exploitation.

Note: Von Neumann Topology has also been employed by some researchers to enhance the exploitation capacity of ABC [48], but Von Neumann and cellular topology have highly similar structures. Therefore, for the sake of simplicity, they could be roughly classified as the same topology.

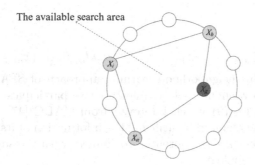

Fig. 4. Illustration of a global neighborhood search topology. The red solid circle X_g indicate the global optimal food source. The two light red solid circles X_a and X_b are the randomly selected food sources from the population and X_i indicates the exhausted food source. (Color figure online)

Global Neighborhood Search Operator: Zhou et al. [44] proposed the MABC-NS algorithm by combining MABC [13] and global neighborhood search operation. The new food source V_i is produced from a region surrounded by four different food sources as Fig. 4.

$$TX_i = r_1 \cdot X_i + r_2 \cdot X_g + r_3 \cdot (X_a - X_b) \tag{8}$$

In particular, it does not update each food source, only the probabilistically selected food source (X_i) can enter the global neighborhood search operation which employing the global optimal food source (X_g) and two random food sources (X_a and X_b) to reduce the waste of evaluation times using Eq. (8) in this neighborhood learning approach.

Scale-Free Network Topology: Ji et al. [15] introduced the scale-free network topology into ABC and proposed a novel SFABC algorithm as shown in Fig. 5. The neighbor relationship among food sources is no longer established by the index after initialization but using the BA algorithm [3] based on fitness value.

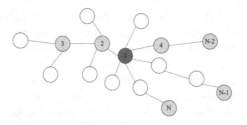

Fig. 5. Illustration of a scale-free network topology. Each of these circles represents a food source, and the circles connected to it indicate its neighbors. Numbers indicate index of food source in the population from the highest to the lowest according to the fitness value.

$$v_{i,j} = x_{i,j} + \phi_{i,j} \cdot (x_{i,j} - x_{r,j}) + rand_{i,j}[0,1] \cdot (x_{nr,j} - x_{i,j}) \tag{9}$$

Similarly, by this neighborhood learning approach of SFABC, in neighbors connected to X_i, a randomly selected neighbor to participate in the generation of V_i as shown in Eq (9) which improved from GABC [47]. The exploitation capacity of ABC is enhanced by utilizing the information of its neighbors, as the inferior food sources have the potential to connect with the superior ones in the scale-free network topology.

Custom Metrics: Zhou et al. [43] proposed a neighborhood construction method with custom metrics where two commonly used food source information, i.e., fitness value and distance, were used as the base for neighbor selection, as shown in Eq. (10).

$$X_{lb}^i = argmax_{X_k, k \in \{1,2,...,SN\} \wedge k \neq i} \frac{f(X_i) - f(X_k)}{\|X_i - X_k\|} \tag{10}$$

For this neighborhood learning approach, food sources in the population except X_i are treated as alternative neighbors and then the neighbor with the largest ratio (X_{lb}) is selected as the optimal neighbor in the exploitation operation. Similarly, the neighbor with the smallest value (by using $argmin(\cdot)$ function) is chosen for the exploration operation.

2.2   ABC with Multiple Neighborhood Topologies

It may have a ceiling situation in improving the performance of ABC using neighborhood learning approach based on single neighborhood topology due to its almost fixed structure and limited parameters. To explore more possibilities for ABC enhancement, those cases may increase gradually that multiple neighborhood topologies are applied to the same algorithm simultaneously, because a proper combination of topologies could almost fully exploit the advantages of each neighborhood topology, and configurations in their neighborhood learning approaches may be also diverse and flexible for different problems.

Different Topologies in Different Subpopulations: Three different types of topologies were utilized in the same ABC variant (ABC-MNT) proposed by Zhou et al. [45]. They are panmictic, small-world and ring topologies as shown in Fig. 6. Furthermore, the neighbor relationships are constructed by these topologies severally for three carefully divided subpopulations.

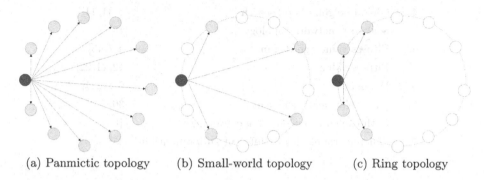

(a) Panmictic topology   (b) Small-world topology   (c) Ring topology

Fig. 6. The structure of the three neighborhood topologies, where the red solid circles represent food source X_i, and the light red and white solid circles denote the neighbors of X_i and other food sources in the population, respectively. (Color figure online)

One possible scheme is implemented in the neighborhood learning approach that three neighbor selection rules is formulated based on the connectivity among food sources in three topologies, i.e., all food sources are treated as neighbors of X_i in panmictic topology, the food sources selected by probability are the neighbors of X_i in small-world topology, and in ring topology, the neighbors of X_i come from the food sources close to X_i. It may be an appropriate way for (a), (b) and (c) in Fig. 6 in ABC-MNT to be used for exploitation, balance and exploration operations using their optimal neighbors as leaders, respectively. A more detailed description can be found in [27,35,45,46]. The advantage of using multi-neighborhood topology is that the performance of ABC can be better improved by complementing the strengths of different topologies. However, the effective utilization of the multi-neighborhood topology remains a challenging task.

2.3   A Short Summary

From the above presentation, it could be seen that different neighborhood learning approaches have significant differences in neighbor selection and optimal neighbor determination. Certainly, the specific neighborhood learning approach should be combined with detailed topology to be meaningful. To present the content more concisely, some common neighborhood construction methods, both mentioned and unmentioned above, are summarized below in Table 1 (6.x in the first column indicates the food source information used to construct neighborhood using custom metrics.).

Table 1. ABC with neighborhood technology

No	Neighborhood construction	Literature
1	Ring neighborhood topology	[14,35,37]
2	Random neighborhood topology	[27,42]
3	Cellular neighborhood topology	[40,48]
4	Global neighborhood search	[5,34,44]
5	Scale-free network topology	[15]
6.1	Fitness value and distance	[4,7,43]
6.2	Fitness value	[12,21,22]
6.3	Distance	[6,17,24]
6.4	Grey relational coefficient	[36]
6.5	Path exchange neighborhood topology	[10]
7	Different topologies in different subpopulations	[45]

In general, in the neighborhood learning approach, the neighbor selection rule is determined by the characteristics of the specific topology. For example, neighbors of X_i come from a fixed number of randomly selected food sources in the population in Table No. 2, and the optimal and the worst neighbor are identified in terms of the information of fitness values and distances among X_i and other food sources in Table No. 6.1. In addition, the neighbor with the highest fitness value has a higher probability of being the leader in solution search function.

3   Conclusion and Further Discussion

ABC is a new type of swarm intelligence algorithm with simple structure and few control parameters. However, its solution search equation emphasizes excessively on exploration which leads to slow convergence and low accuracy of the solution. To alleviate this disadvantage, the neighborhood learning approaches are incorporated into ABC by developing various ABC variants to enhance the exploitation capability. More specifically, the approaches based on single neighborhood topology have shown strong performance (verified by experiments in their paper). Neighbor selection rule is simple and efficient according to the specific topology. The neighbor with the highest fitness value is often used as the optimal neighbor which may lead selected neighbors to perform a strong local region search, and the random neighbor as the leader could keep a balance between exploitation and exploration.

Nevertheless, this is a limited improvement for ABC performance by using neighborhood learning approaches based on single neighborhood topology, which may be manifested by the inflexible structure of the topology and few adjustable parameters etc. Based on this, ABC variants with multiple neighborhood topologies are gradually developed which are derived from well-established or customized neighborhood topology schemes, and the neighborhood learning approaches may come from the original paper, semi- or fully customized neighborhood learning rules according to the characteristics of given variants or proposed motivations. Therefore these approaches are more flexible.

The neighborhood topology could effectively describe the relationship among food sources, while the neighborhood learning approach is useful for neighbor selection and optimal neighbor determination. Yet, after all, this is an improvement at the technical level with little consideration of the topography of the decision space. In fact, fitness landscape correlation has been used in many cases in ABC. In the future, it could be a potential research direction to improve ABC by adopting the hybrid scheme in advanced approach of multiple neighborhood topologies combined with fitness landscape correlation. Further, more effective parameter (in neighborhood topology) adaption schemes may also be worth exploring depending on the variation of fitness landscape for different problems.

Acknowledgments. This work is supported by the National Natural Science Foundation of China (Nos. 61966019 and 62366022), the Jiangxi Provincial Natural Science Foundation (No. 20232BAB202048), and the Science and Technology Plan Projects of Jiangxi Provincial Education Department (No. GJJ210324).

References

1. Akay, B., Karaboga, D.: Artificial bee colony algorithm for large-scale problems and engineering design optimization. J. Intell. Manuf. **23**(4), 1001–1014 (2012)
2. Bäck, T.: Evolutionary Algorithms in Theory and Practice: Evolution Strategies, Evolutionary Programming, Genetic Algorithms, vol. 996. Oxford University Press, Oxford (1996)
3. Barabási, A.L., Albert, R.: Emergence of scaling in random networks. Science **286**(5439), 509–512 (1999)
4. Biswas, S., Das, S., Kundu, S., Patra, G.R.: Utilizing time-linkage property in DOPs: an information sharing based artificial bee colony algorithm for tracking multiple optima in uncertain environments. Soft. Comput. **18**(6), 1199–1212 (2014)
5. Cai, Q., et al.: Enhancing artificial bee colony algorithm with dynamic best neighbor-guided search strategy. In: Proceedings of the IEEE Congress on Evolutionary Computation, pp. 1–8. IEEE (2020)
6. Cui, L., Li, G., Lin, Q., Chen, J., Lu, N., Zhang, G.: Artificial Bee colony algorithm based on neighboring information learning. In: Hirose, A., Ozawa, S., Doya, K., Ikeda, K., Lee, M., Liu, D. (eds.) ICONIP 2016. LNCS, vol. 9949, pp. 279–289. Springer, Cham (2016). https://doi.org/10.1007/978-3-319-46675-0_31
7. Cui, L., et al.: A smart artificial bee colony algorithm with distance-fitness-based neighbor search and its application. Futur. Gener. Comput. Syst. **89**, 478–493 (2018)
8. Dasgupta, D., Michalewicz, Z.: Evolutionary Algorithms in Engineering Applications. Springer, Heidelberg (2013). https://doi.org/10.1007/978-3-662-03423-1
9. Dedeturk, B.K., Akay, B.: Spam filtering using a logistic regression model trained by an artificial bee colony algorithm. Appl. Soft Comput. **91**, 106229 (2020)

10. Dongli, Z., Xinping, G., Yinggan, T., Yong, T.: An artificial bee colony optimization algorithm based on multi-exchange neighborhood. In: Proceedings OT the Fourth International Conference on Computational and Information Sciences, pp. 211–214. IEEE (2012)

11. Gao, H., Fu, Z., Pun, C.M., Zhang, J., Kwong, S.: An efficient artificial bee colony algorithm with an improved linkage identification method. IEEE Trans. Cybern. **52**, 4400–4414 (2020)

12. Gao, W., Chan, F.T., Huang, L., Liu, S.: Bare bones artificial bee colony algorithm with parameter adaptation and fitness-based neighborhood. Inf. Sci. **316**, 180–200 (2015)

13. Gao, W., Liu, S.: A modified artificial bee colony algorithm. Comput. Oper. Res. **39**(3), 687–697 (2012)

14. Jadon, S.S., Bansal, J.C., Tiwari, R., Sharma, H.: Artificial bee colony algorithm with global and local neighborhoods. Int. J. Syst. Assur. Eng. Manage. **9**(3), 589–601 (2018)

15. Ji, J., Song, S., Tang, C., Gao, S., Tang, Z., Todo, Y.: An artificial bee colony algorithm search guided by scale-free networks. Inf. Sci. **473**, 142–165 (2019)

16. Karaboga, D., Basturk, B.: On the performance of artificial bee colony (ABC) algorithm. Appl. Soft Comput. **8**(1), 687–697 (2008)

17. Karaboga, D., Gorkemli, B.: A quick artificial bee colony (QABC) algorithm and its performance on optimization problems. Appl. Soft Comput. **23**, 227–238 (2014)

18. Karaboga, D., Gorkemli, B., Ozturk, C., Karaboga, N.: A comprehensive survey: artificial bee colony (ABC) algorithm and applications. Artif. Intell. Rev. **42**(1), 21–57 (2014)

19. Katoch, S., Chauhan, S.S., Kumar, V.: A review on genetic algorithm: past, present, and future. Multimedia Tools Appl. **80**(5), 8091–8126 (2021)

20. Kennedy, J., Eberhart, R.: Particle swarm optimization. In: Proceedings of the IEEE International Conference on Neural Networks, vol. 4, pp. 1942–1948 (1995)

21. Kiran, M.S., et al.: Improved artificial bee colony algorithm for continuous optimization problems. J. Comput. Commun. **2**(04), 108 (2014)

22. Kong, D., Chang, T., Dai, W., Wang, Q., Sun, H.: An improved artificial bee colony algorithm based on elite group guidance and combined breadth-depth search strategy. Inf. Sci. **442**, 54–71 (2018)

23. Krejca, M.S., Witt, C.: Theory of estimation-of-distribution algorithms. In: Theory of Evolutionary Computation. NCS, pp. 405–442. Springer, Cham (2020). https://doi.org/10.1007/978-3-030-29414-4_9

24. Liao, Z., Gong, W., Wang, L.: A hybrid swarm intelligence with improved ring topology for nonlinear equations. Sci. Sinica Informationis **50**(3), 396–407 (2020)

25. Mirjalili, S.: Genetic algorithm. In: Evolutionary Algorithms and Neural Networks. SCI, vol. 780, pp. 43–55. Springer, Cham (2019). https://doi.org/10.1007/978-3-319-93025-1_4

26. Pant, M., Zaheer, H., Garcia-Hernandez, L., Abraham, A., et al.: Differential evolution: a review of more than two decades of research. Eng. Appl. Artif. Intell. **90**, 103479 (2020)

27. Peng, H., Deng, C., Wu, Z.: Best neighbor-guided artificial bee colony algorithm for continuous optimization problems. Soft. Comput. **23**(18), 8723–8740 (2019)

28. Peng, H., Zhu, W., Deng, C., Wu, Z.: Enhancing firefly algorithm with courtship learning. Inf. Sci. **543**, 18–42 (2021)

29. Piotrowski, A.P., Napiorkowski, J.J., Piotrowska, A.E.: Population size in particle swarm optimization. Swarm Evol. Comput. **58**, 100718 (2020)

30. Price, K.V.: Differential evolution. In: Zelinka, I., Snášel, V., Abraham, A. (eds.) Handbook of Optimization. Intelligent Systems Reference Library, vol. 38, pp. 187–214. Springer, Heidelberg (2013). https://doi.org/10.1007/978-3-642-30504-7_8

31. Rajasekhar, A., Abraham, A., Pant, M.: Levy mutated artificial bee colony algorithm for global optimization. In: Proceedings of the IEEE International Conference on Systems, Man, and Cybernetics, pp. 655–662. IEEE (2011)

32. Slowik, A., Kwasnicka, H.: Evolutionary algorithms and their applications to engineering problems. Neural Comput. Appl. **32**, 12363–12379 (2020)

33. Snyman, J.A.: Practical Mathematical Optimization. Springer, New York (2005). https://doi.org/10.1007/b105200

34. Sun, C., Zhou, X., Wang, M.: A multi-strategy artificial bee colony algorithm with neighborhood search. In: Tan, Y., Shi, Y., Niu, B. (eds.) ICSI 2019. LNCS, vol. 11655, pp. 310–319. Springer, Cham (2019). https://doi.org/10.1007/978-3-030-26369-0_29

35. Wang, H., Wang, W., Xiao, S., Cui, Z., Xu, M., Zhou, X.: Improving artificial bee colony algorithm using a new neighborhood selection mechanism. Inf. Sci. **527**, 227–240 (2020)

36. Xiang, W.L., Li, Y.Z., Meng, X.L., Zhang, C.M., An, M.Q.: A grey artificial bee colony algorithm. Appl. Soft Comput. **60**, 1–17 (2017)

37. Xiao, S., Wang, H., Wang, W., Huang, Z., Zhou, X., Xu, M.: Artificial bee colony algorithm based on adaptive neighborhood search and gaussian perturbation. Appl. Soft Comput. **100**, 106955 (2021)

38. Yang, X.S.: Firefly algorithm, stochastic test functions and design optimisation. Int. J. Bio-Inspired Comput. **2**(2), 78–84 (2010)

39. Yu, X., Gen, M.: Introduction to Evolutionary Algorithms. Springer, London (2010). https://doi.org/10.1007/978-1-84996-129-5

40. Zhang, M., Tian, N., Palade, V., Ji, Z., Wang, Y.: Cellular artificial bee colony algorithm with gaussian distribution. Inf. Sci. **462**, 374–401 (2018)

41. Zhang, Q., Zhou, A., Jin, Y.: RM-MEDA: a regularity model-based multiobjective estimation of distribution algorithm. IEEE Trans. Evol. Comput. **12**(1), 41–63 (2008)

42. Zhong, F., Li, H., Zhong, S.: An improved artificial bee colony algorithm with modified-neighborhood-based update operator and independent-inheriting-search strategy for global optimization. Eng. Appl. Artif. Intell. **58**, 134–156 (2017)

43. Zhou, J., et al.: An individual dependent multi-colony artificial bee colony algorithm. Inf. Sci. **485**, 114–140 (2019)

44. Zhou, X., Wang, H., Wang, M., Wan, J.: Enhancing the modified artificial bee colony algorithm with neighborhood search. Soft. Comput. **21**(10), 2733–2743 (2017)

45. Zhou, X., Wu, Y., Zhong, M., Wang, M.: Artificial bee colony algorithm based on multiple neighborhood topologies. Appl. Soft Comput. **111**, 107697 (2021)

46. Zhou, X., Wu, Z., Wang, H., Rahnamayan, S.: Gaussian bare-bones artificial bee colony algorithm. Soft. Comput. **20**(3), 907–924 (2016)

47. Zhu, G., Kwong, S.: Gbest-guided artificial bee colony algorithm for numerical function optimization. Appl. Math. Comput. **217**(7), 3166–3173 (2010)

48. Zou, W., Zhu, Y., Chen, H., Shen, H.: Artificial bee colony algorithm based on von Neumann topology structure. In: Proceeding of the IEEE International Conference on Computer and Electrical Engineering. IEEE (2012)

Human Centred Computing

Channel Attention Separable Convolution Network for Skin Lesion Segmentation

Changlu Guo[1](✉), Jiangyan Dai[2], Márton Szemenyei[1], and Yugen Yi[3](✉)

[1] Budapest University of Technology and Economics, Budapest, Hungary
clguo.ai@gmail.com
[2] Weifang University, Weifang, China
[3] Jiangxi Normal University, Nanchang, China
yiyg510@jxnu.edu.cn

Abstract. Skin cancer is a frequently occurring cancer in the human population, and it is very important to be able to diagnose malignant tumors in the body early. Lesion segmentation is crucial for monitoring the morphological changes of skin lesions, extracting features to localize and identify diseases to assist doctors in early diagnosis. Manual de-segmentation of dermoscopic images is error-prone and time-consuming, thus there is a pressing demand for precise and automated segmentation algorithms. Inspired by advanced mechanisms such as U-Net, DenseNet, Separable Convolution, Channel Attention, and Atrous Spatial Pyramid Pooling (ASPP), we propose a novel network called Channel Attention Separable Convolution Network (CASCN) for skin lesions segmentation. The proposed CASCN is evaluated on the PH2 dataset with limited images. Without excessive pre-/post-processing of images, CASCN achieves state-of-the-art performance on the PH2 dataset with Dice similarity coefficient of 0.9461 and accuracy of 0.9645.

Keywords: Skin lesion segmentation · Dermoscopic · U-Net · Channel Attention · Separable Convolution

1 Introduction

Melanoma is a highly lethal and rapidly spreading cancer that is prevalent throughout the world. It is estimated that in the United States alone, there will be approximately 97,610 new cases and 7,990 deaths from melanoma in 2023 [1]. Cancer staging at diagnosis refers to the severity of the cancer in the body, which greatly affects a patient's chances of survival and determines the doctor's treatment options. Generally, cancer is localized (sometimes called stage 1) if it is only found in the part of the body where it started. If cancer cells have metastasized

---

*This work is supported by the National Natural Science Foundation of China under Grants 62062040 and 61672150, the Outstanding Youth Project of Jiangxi Natural Science Foundation (No. 20212ACB212003), the Jiangxi Province Key Subject Academic and Technical Leader Funding Project (No. 20212BCJ23017).

to other regions of the body, the stage is regional or distant. The 5-year relative survival rate was 99.6% when diagnosed with local-stage melanoma of the skin, whereas it was only 35.1% when diagnosed with distant-stage melanoma. Therefore, the earlier skin melanoma is detected, the better a person's chance of surviving five years after being diagnosed. Most commonly, experienced ophthalmologists can diagnose malignant melanoma by looking at the images produced by dermoscopy, but this work is time-consuming, tedious, and subjective. Therefore, it is very necessary to help dermatologists diagnose malignant melanoma with the help of computer-aided diagnosis system (CAD), and it can also improve the accuracy of diagnosis. However, patient-specific attributes may vary in skin texture, color, location, size, and the existence of a large number of artifacts. (e.g., reflective bubbles, body hair, markings, shadows, and non-uniform lighting) leading to automatic differentiation of lesion area and healthy skin becomes a challenging task.

In the past few years, the widespread adoption of deep learning in the domain of computer vision has inspired many innovative solutions to various problems. Among them, deep learning-based semantic segmentation, as an end-to-end method, is favored by various related fields because it does not need to design cumbersome preprocessing and postprocessing steps. Long et al. introduced the Fully Convolutional Network (FCN) [2] in which training is performed end-to-end and pixel-to-pixel, and the fully connected layers are all replaced with convolution layers and deconvolution layers to preserve the initial spatial resolution. However, FCN obviously ignores the correlation between pixels, which leads to the loss of certain spatial information, and the segmentation effect is not impressive. To this end, Ronneberger et al. [3] proposed a fully convolutional network structure called U-Net, which is currently one of the most commonly used FCNs in medical image segmentation. In U-Net, skip connections were incorporated to enable the decoder to retrieve important features learned during each stage of the encoder, which might have been lost due to pooling. In particular, U-Net has demonstrated outstanding performance in different medical image segmentation tasks with various imaging modalities, such as Residual U-Net [4], U-Net++ [5], MultiResUNet [6], CAR-UNet [7], by extracting contextual features based on the encoder-decoder architecture. In order to tackle the challenge of skin lesion segmentation, Bi et al. [8] employed a multi-stage fully convolutional network (mFCN) with a parallel integration strategy to achieve accurate segmentation of skin lesions. Similarly, Tang et al. [9] introduced a multi-stage U-Net for the segmentation of skin lesions. However, this model does not consider the crucial global contextual information that is necessary for precisely identifying the location of skin lesions. In addition, Al-masni et al. introduced the full-resolution convolutional network (FrCN) [10], which removes all encoder subsampling layers to retain the complete resolution of the input image and prevent any loss of spatial information. Nevertheless, the absence of subsampling, CNN models are prone to overfitting due to the presence of redundant features and limited feature map coverage [11].

In this paper, we introduce the Channel Attention Separable Convolution Network (CASCN), which is a semantic segmentation network designed for precise and robust dermoscopic skin lesion segmentation. To eliminate the need for learning redundant features in the encoder, we utilize dense blocks and transition blocks inspired by DenseNet [12]. To create a more general and lightweight network, we adopt depthwise separable convolutions in the decoder, similar to Xception [13] and MobileNet [14]. To enhance the discriminative ability of the network and recover the spatial information lost during pooling in each stage of the encoder, we introduce a novel Modified Efficient Channel Attention (MECA) [7] applied to traditional "skip connections" rather than just copying the encoder's feature map to the corresponding decoder. Furthermore, we incorporate Atrous Spatial Pyramid Pooling (ASPP) [15] between the encoder and decoder to capture multi-scale information utilizing dilated parallel convolutions at different sampling rates. The novelty of this work is that CASCN possesses a distinct learning ability among existing CNN networks for skin lesion segmentation. The experimental results demonstrate that CASCN achieves state-of-the-art performance on the PH2 dataset.

2   Methods

2.1   DenseNet

In 2017, Huang et al. [12] proposed a convolutional neural network (CNN) architecture known as DenseNet. It is known for its efficient use of parameters and its ability to combat the vanishing gradient problem, which can occur in very deep neural networks. The architecture of DenseNet is based on the idea of dense connectivity, which means that each layer is densely connected to all previous layers in a feedforward manner. In other words, the output of each layer is concatenated with the input of every subsequent layer in the network. This allows for feature reuse and encourages the network to learn more compact representations of the input, making it more efficient and less prone to overfitting.

DenseNet also introduces the concept of dense blocks, which are groups of layers that are densely connected to each other. In each dense block, the feature maps of all previous layers are concatenated before passing to the next layer. This concatenation preserves the spatial information in the feature maps and enables the network to learn more intricate features.

2.2   Depthwise Separable Convolution

Depthwise separable convolution, which decomposes standard convolution into depthwise convolution and pointwise convolution, was first proposed in [16] and is widely known due to the application of MobileNet [14]. In the convolution kernel of standard convolution, all channels in the corresponding image area are considered at the same time, this method greatly increases the calculation amount of model parameters, running time and memory capacity, making the model complex and cumbersome. The depthwise convolution considers spatial regions and channels separately, processing the spatial and depth dimensions, respectively. First, the depthwise convolution uses different convolution kernels for different input channels for convolution, and then the pointwise convolution uses 1×1 convolution for the previous output and merges the final output, which effectively reduces the amount of computation and model parameter amount. For an input of dimensions $H \times W \times N$ convolved with stride 1 with a kernel of size $D_k \times D_k$ and M output channels, the cost of a standard convolution is $H \times W \times D_K^2 \times N \times M$ while the cost of a depthwise separable convolution is $H \times W \times N \times (D_K^2 + M)$. This means that the total computation cost of a depthwise separable convolution is $\frac{D_K^2 \times M}{D_K^2 + M}$ times lower than a standard convolution, while still being able to achieve a similar level of performance.

2.3 Channel Attention

Channel Attention was initially introduced in the Squeeze-and-Excitation Networks (SENet) [17]. The key idea of Channel Attention is to capture feature channel dependencies in Convolutional Neural Networks (CNNs) by selectively weighting the importance of different channels in each layer. In SENet, Channel Attention is implemented through a Squeeze-and-Excitation block (SEB) which is a simple and efficient approach that can be easily integrated into existing network models. The SEB block computes a weighting factor for each channel by performing global pooling and applying a set of fully connected layers, and then scales the channel activations by the computed weights. The SEB has demonstrated its effectiveness in improving the performance of various tasks, such as object detection and image classification, by reducing the impact of irrelevant channels and increasing the discriminative power of important channels. To further enhance the capability of capturing meaningful features in both spatial and channel dimensions, the Convolutional Block Attention Module (CBAM) was proposed by Woo et al. [18]. The CBAM sequentially applies channel and spatial attention modules to learn where and what to pay attention to in channel and spatial dimensions, respectively. The channel attention module in CBAM is similar to that in SENet, but it employs both average pooling and max pooling features and feeds them into a shared Multi-layer Perceptron (MLP). However, many methods that aim to improve performance tend to increase the complexity of the model. To address this issue, Wang et al. [19] proposed an Efficient Channel Attention (ECA) module, which uses $1D$ convolutions to avoid dimensionality reduction operations in the Squeeze-and-Excitation block, resulting in significantly reduced model complexity while maintaining superior performance compared to previous attention mechanisms. In other words, ECA replaces the MLP in the channel attention module with a $1D$ convolution. However, in ECA, only average-pooling is utilized for gathering spatial information, while max-pooling can also provide valuable information about distinctive object features to infer a more detailed channel-wise attention, as pointed out in [18]. Therefore, to address the limitations of using only average pooling for aggregating spatial information in ECA, Modified Efficient Channel Attention (MECA) was proposed in [7]. MECA employs both average pooling and max pooling to obtain finer channel-wise attention, which can capture more distinctive object features and improve the network's ability to learn discriminative features. Figure 1 illustrates the operation of MECA.

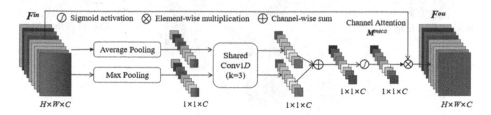

Fig. 1. Diagram of Modified Efficient Channel Attention (MECA)

2.4 Atrous Spatial Pyramid Pooling

Before introducing Atrous Spatial Pyramid Pooling (ASPP) [15], we briefly introduce atrous convolution. Atrous convolution, also known as dilated convolution, is a type of convolutional operation that allows for the expansion of a filter's receptive field without increasing the number of parameters or computational cost. The operation is achieved by inserting spaces between the kernel elements and filling the gaps with zeros. The amount of space, or dilation rate, can be adjusted to control the size of the receptive field. The resulting sparse similarity filter is then convolved with the input feature map. Atrous convolution has the advantage of increasing the effective receptive field without increasing the number of parameters or computation, making it an efficient way to capture multi-scale contextual information.

ASPP involves performing parallel convolutional operations with different dilation rates (or "atrous rates") on the same input feature map, followed by pooling and concatenation of the resulting feature maps, as shown in Fig. 2. The use of multiple dilation rates allows the network to capture multi-scale contextual information, which is useful for semantic segmentation tasks where objects of different sizes need to be identified and segmented accurately.

2.5 Proposed CASCN Architecture

A convolutional neural network (CNN) for semantic segmentation typically comprises an encoder and a decoder. The encoder includes convolutional and subsampling layers responsible for automatic feature extraction. Convolutional layers generate feature maps by applying a set of learnable filters to the input image, while subsampling layers reduce the spatial resolution of the feature maps to

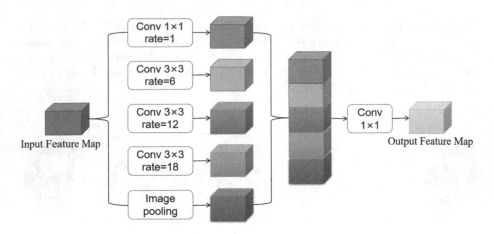

Fig. 2. Diagram of Atrous Spatial Pyramid Pooling

achieve spatial invariance and larger receptive fields. This reduction in resolution allows the network to capture more abstract and high-level features while also reducing computational complexity. In our proposed CASCN, we utilize an encoder that is based on the DenseNet architecture with 121 layers to avoid learning redundant features. The encoder is capable of learning abstract features of skin lesions, mitigating the issue of gradient disappearance, and improving feature propagation. The CASCN is designed such that each encoding layer can directly access the gradient of the loss function of all previous encoding layers, as shown in Fig. 3. The CASCN encoder comprises feature layers, dense blocks, transformation blocks, and auxiliary operations. As illustrated in Fig. 3, the encoder is positioned in the first half of the CASCN architecture before the ASSP boundary and includes DenseNet121 and a separable convolution module.

In CASCN, the decoder takes the low-resolution features produced by the encoder and projects them onto the high-resolution pixel space to perform dense pixel-wise classification. However, the downsampling process in the encoder often leads to a reduction in spatial resolution of the feature maps, which can result in roughness, loss of edge information, checkerboard artifacts, and over-segmentation in the resulting semantic segmentation masks. To address these issues, Ronneberger et al. [3] proposed the use of skip connections in U-Net, which enable the decoder to recover important features that were lost during pooling at each stage of the encoder. Inspired by U-Net, CASCN also employs skip connections to overcome subsampling limitations and deconvolution overlaps. To enhance the discriminative power of the network, we have incorporated Modified Efficient Channel Attention (MECA) into the skip connection, which allows for more sophisticated feature selection and weighting. This ensures that the feature maps obtained from the encoder are not simply copied to the decoder, but are instead refined and optimized for improved segmentation accuracy. The first four pooling layers of CASCN are all connected to a deconvoluted feature map with the same dimensionality via MECA, as shown in Fig. 4, which acts as a compensation connection for the lost spatial information due to subsampling.

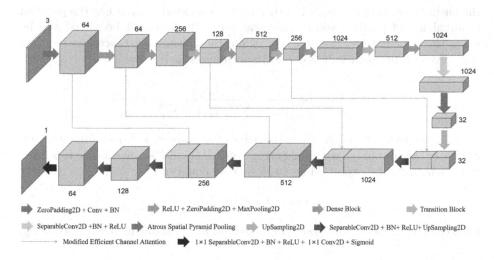

Fig. 3. Diagram of Channel Attention Separable Convolution Network (CASCN)

3 Designing Experiments and Reporting Results

3.1 Dataset

To verify the effectiveness of CASCN we evaluated using an independent dataset named PH2 [20] collected from the Dermatology Service of the Hospital Pedro Hispano, Portugal. PH2 is a dermoscopy dataset containing 200 dermoscopy images including 40 melanomas, 80 atypical moles and 80 common moles. Each image of PH2 is an RGB image of size 560 × 768 and provides a corresponding human-annotated mask. This dataset is publicly available and is mainly used for experimental and research purposes to facilitate research on dermoscopy image classification and segmentation algorithms.

3.2 Loss Function

In the input image, for any pixel x_i, the corresponding CASCN output p_i represents the estimated posterior probability that the pixel belongs to a skin lesion. Since CASCN essentially performs pixel-wise classification, cross-entropy can be used as the loss function:

$$L_{CE} = -\frac{1}{N} \sum_{i=1}^{N} [x_i \ln p_i + (1 - x_i)(1 - \ln p_i)] \tag{1}$$

where N represents all pixel quantities, and $x_i \in \{0, 1\}$ is the truth class of x_i with for background and for lesion.

In dermoscopy, the area of skin lesions often occupies only a small part of the entire dermoscopic image. In other words, this kind of data has the problem of imbalance of positive and negative samples, which cannot be solved well by using the cross-entropy loss function, while the Jaccard Distance loss function can solve this problem well [35], and its formula is as follows:

$$L_{JD} = 1 - \frac{\sum_{i=1}^{N}(x_i p_i)}{\sum_{i=1}^{N}(x_i) + \sum_{i=1}^{N}(p_i) - \sum_{i=1}^{N}(x_i p_i)} \tag{2}$$

In order to combine the characteristics of the above two loss functions, we minimize the sum of binary cross-entropy and Jaccard Distance as the loss function , which can be expressed by the following equation:

$$L_{SEG} = L_{CE} + L_{JD} \tag{3}$$

which can be efficiently integrated into backpropagation during network training.

3.3   Key Settings Validation

First, we analyze the impact of some key settings of CASCN on skin lesion segmentation performance. These key settings include the input image size, optimization method, loss function and image augmentation. In each experiment, we replace one setting with another setup, while keeping the others unchanged. In order to quantitatively evaluate the performance of CASCN, some widely used evaluation indicators are employed for comparison. Specifically, we choose Sensitivity (SE), Specificity (SP), Accuracy (AC), Dice Similarity Coefficient (DI) and Jaccard index (JA) as evaluation metrics.

1)Image Size: To investigate how various input image sizes impact segmentation performance, we select three commonly used input sizes: 96×128, 192×256 and 384×512. As shown in Table 1, when the size is enlarged from 96×128 to 192×256, the DI increases from 93.87% to 94.57% and the JA increases from 88.89% to 90.13%. However, when the image size is further expanded to 384×512, the JA does not improve further but decreases to 87.01% compared to 192×256. This drop in performance is due to the dramatic increase in the number of pixels making the model more difficult to train. On the other hand, training 100 epochs for 384×512 size takes 7608 s, while 192×256 is 1700 s and 96×128 is 1333 s. The above discussion shows that an image size of 192×256 achieves a good balance between computational cost and segmentation performance, so we resize all images to this size before putting them into the CASCN model.

2)Optimization Method and Loss Function: We compare two different optimization methods (Adam vs. SGD with Nesterov momentum) and three different loss functions (Cross-entropy, Jaccard Distance and their sum) in combination. The learning rate for both optimization methods is set to 0.003, and the

Table 1. Comparative Experiments on Different Image Sizes

Model	Image Size	$SE(\%)$	$SP(\%)$	$AC(\%)$	$DI(\%)$	$JA(\%)$	Training Time
CASCN	96 × 128	94.77	95.84	96.12	93.87	88.89	**1333 s**
CASCN	192 × 256	95.79	**96.21**	**96.45**	**94.61**	**90.18**	1700 s
CASCN	384 × 512	**96.00**	93.97	95.07	92.59	87.01	7608 s

SGD momentum is set to 0.9. As shown in Table 2, we use the tick to determine the different combinations. It is obvious from the results that no matter which loss function is chosen, Adam outperforms SGD. Under the premise of using the Adam optimization method, using the sum of Cross-entropy and Jaccard Distance as the loss function yields highest SP of 96.16%, highest AC of 96.41%, highest DI of 94.57% and highest JA of 90.13%, only the SE is slightly lower than using Jaccard Distance (95.70% vs. 95.91%). The above results demonstrate that the combination of choosing Adam as the optimization method and the sum of Cross-entropy and Jaccard Distance as the loss function is effective.

Table 2. Comparative Experiments on Different Augmentation Methods

Loss Functions	SGD	Adam	$SE(\%)$	$SP(\%)$	$AC(\%)$	$DI(\%)$	$JA(\%)$
Binary Cross-entropy	✓		92.88	95.40	95.16	92.44	86.62
Binary Cross-entropy		✓	95.87	95.03	95.88	94.02	89.46
Jaccard Distance	✓		91.28	95.08	94.12	90.63	83.44
Jaccard Distance		✓	**95.91**	95.20	96.00	94.12	89.47
Sum	✓		93.70	95.29	95.24	92.24	86.27
Sum		✓	95.79	**96.21**	**96.45**	**94.61**	**90.18**

3)Image Augmentation: We employ 4 image augmentation methods in this work including rotation, horizontal flip, vertical rotation and diagonal flip. To examine the effect of each image augmentation method on segmentation performance, we conduct five experiments. In the first experiment, the CASCN was trained using images without any image augmentation applied, resulting in a test result of JA of 86.01%. Then we artificially augment the training set by applying one of the four augmentation strategies in the third to fifth experiments, respectively. As can be seen from Table 3, the four image augmentation methods all improved the segmentation performance, increasing JA by 1.83%, 2.60%, 1.58% and 1.87% respectively. Finally, the combination of the four augmentation methods makes JA improve by 4.12%. From the above experimental results, it can be shown that data augmentation is necessary for small sample datasets like PH2.

Table 3. Comparative Experiments on Different Augmentation Methods

Model	Data Augmentation	$SE(\%)$	$SP(\%)$	$AC(\%)$	$DI(\%)$	$JA(\%)$
CASCN	W/O	94.30	93.83	94.57	91.90	86.01
CASCN	Rotation only	95.24	95.12	95.63	93.22	87.84
CASCN	Horizontal flip only	95.24	95.48	95.80	93.65	88.61
CASCN	Vertical flip only	93.75	96.21	95.45	93.04	87.59
CASCN	Diagonal flip only	93.98	95.68	95.79	93.21	87.88
CASCN	full	**95.79**	**96.21**	**96.45**	**94.61**	**90.18**

3.4 Ablation Experiments

Through the previous experiments, we verified the effectiveness of some key settings of CASCN, and to further verify that the key components of CASCN can improve the performance of skin lesion segmentation, we conduct ablation experiments. In our ablation study, all methods are under the same setting and the same computing environment to guarantee fair comparison. As shown in Table 4, in order to verify the effectiveness of separable convolution in this work, we conduct comparative experiments on "DenseNet121 + stConv" and "DenseNet121 + seConv", where stConv refers to standard convolution and seConv refers to separable convolution. From the results of the above two experiments, compared to using standard convolution, the network using separable convolution achieves higher DI (93.56% vs. 94.06%) and higher JA (88.70% vs. 89.09%), which indicates that the use of separable convolution to replace standard convolution is meaningful. Next, on the basis of "DenseNet121 + seConv", we added ASPP and MECA modules and conducted experiments respectively. From the experimental results in Table 4, compared with "DenseNet121 + seConv", adding ASPP or MECA alone improves the performance of the network to a certain extent, which shows that the ASPP and MECA modules are effective in improving the performance of skin lesion segmentation. Moreover, by comparing CASCN (i.e. "DenseNet121 + seConv + ASPP + MECA") with "DenseNet121 + seConv + MECA" and "DenseNet121 + seConv + ASPP" respectively, the experimental performance can also illustrate the effectiveness of ASPP and MECA. Finally, we compare CASCN with "DenseNet121 + seConv", and from the experimental results, SE, DI and JA are improved by 2.27%, 0.17%, 0.51% and 1.04%, respectively, despite the a slight decrease for SP. It is demostrated that the combination of ASPP and MECA is successful in this work. In addition, compared with "DenseNet121 + stConv", CASCN delivers the most favorable segmentation results due to the effective fusion of separation convolution, ASPP, and MECA.

Table 4. Ablation Experiments on PH2 Dataset

Models	$SE(\%)$	$SP(\%)$	$AC(\%)$	$DI(\%)$	$JA(\%)$
DensNet121 + stConv	95.20	94.47	95.44	93.56	88.70
DensNet121 + seConv	93.53	**96.74**	96.24	94.06	89.09
DensNet121 + seConv + ASPP	95.10	95.73	96.22	93.99	89.17
DensNet121 + seConv + MECA	95.62	96.07	96.27	94.12	89.37
CASCN (ours)	**95.79**	96.21	**96.45**	**94.61**	**90.18**

4 Results

4.1 Results on the PH2 Dataset

In this section, we compare and contrast the performance of CASCN with several recent networks that are commonly used for medical image segmentation, such as U-Net, Residual U-Net, U-Net++, MultiResUNet, and CAR-UNet. We used the same dataset and parameter settings for training, validation, and testing in all experiments. It is evident from Table 5 that the segmentation performance of UNet as a baseline model for skin lesion segmentation can barely reach the average level, while Residual UNet, which introduces the Residual mechanism, has not significantly improved the accuracy of skin lesion segmentation. UNet++ relies on the improvement of skip connection to obtain better accuracy performance than traditional U-Net. MultiResUNet proposed MultiRes blocks to replace traditional convolutional blocks and ResPath to replace UNet's skip connection, further improving segmentation performance. CAR-UNet proposed Modified Efficient Channel Attention (MECA) and applies it to residual blocks and skip connections to achieve a relatively lightweight network. In this task, it achieves comparable performance to the UNet++ method. As for the proposed CASCN, it outperforms other competitors in general and performs the best in all indicators. It can be concluded from Table 5 that SE, SP, AC, DI, and JA of CASCN achieved the highest values of 95.79%, 96.21%, 96.45%, 94.61%, and 90.18% respectively compared with other methods.

Table 5. Experiments on PH2 Dataset

Models	$SE(\%)$	$SP(\%)$	$AC(\%)$	$DI(\%)$	$JA(\%)$
U-Net [3]	94.67	93.61	94.89	92.56	86.84
Residual U-Net [4]	94.14	94.50	95.07	92.53	86.84
U-Net++ [5]	94.84	94.01	95.35	92.81	87.11
MultiResUNet [6]	94.88	94.87	95.92	93.56	88.48
CAR-UNet [7]	93.77	94.95	95.24	92.85	87.30
CASCN (ours)	**95.79**	**96.21**	**96.45**	**94.61**	**90.18**

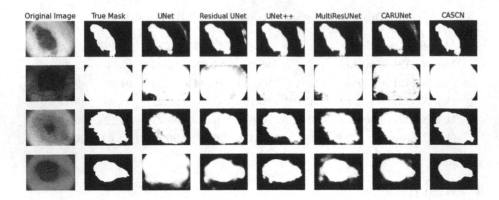

Fig. 4. Visual comparison of challenging cases in the PH2 dataset

Furthermore, we conduct a visual comparison of typical challenging cases in the PH2 dataset using several commonly used networks for medical image segmentation, namely U-Net, Residual U-Net, U-Net++, MultiResUNet, and CAR-UNet. As shown in Fig. 5, for most challenging situations with blurred boundaries and complex brightness distribution, Our method is still able to achieve the most favorable segmentation results, especially in other methods where it is difficult to avoid mis-segmented areas, our CASCN can still distinguishing them as non-lesion regions clearly showcases the effectiveness of the proposed CASCN.

4.2   Comparison with Other State-of-the-art Methods

Lastly, we compare the performance of CASCN with other current state-of-the-art methods utilized for the task of skin cancer segmentation. Table 6 presents a summary of the release years of various methods and their corresponding performance on the PH2 dataset. From the results, it can be concluded that CASCN achieves the best performance on PH2. When compared to other methods, CASCN achieves the highest specificity of 96.21%, the highest accuracy of

Table 6. Comparison with other state-of-the-art methods

Models	Year	$SE(\%)$	$SP(\%)$	$AC(\%)$	$DI(\%)$	$JA(\%)$
FCN [2]	2015	–	–	–	93.80	–
FrCN [10]	2018	93.72	95.65	95.08	91.77	84.79
DCL-PSI [21]	2019	96.23	94.52	95.30	92.10	85.90
DSNet [22]	2020	93.70	**96.90**	–	–	87.00
ASCU-Net [23]	2021	96.00	93.70	94.30	90.90	84.20
AS-Net [24]	2022	**96.24**	94.31	95.20	93.05	87.60
CASCN (ours)	–	95.79	96.21	**96.45**	**94.61**	**90.18**

96.45%, the highest Dice Similarity Coefficient of 94.61%, the highest Jaccard Distance of 90.18%, and the sensitivity is comparable to other methods. From the above results, it is shown that our proposed CASCN achieves the state-of-the-art performance.

5   Conclusions

In this study, we present a novel fully automated architecture, called the Channel Attention Separable Convolution Network (CASCN), which is designed for accurate skin lesion segmentation. Unlike traditional UNet-based encoders, we adopt a 121-layer DenseNet encoder structure to remove the necessity of learning redundant features. To reduce the number of parameters and computational costs in the network, we use lightweight depthwise separable convolutions instead of standard convolutions. In addition, to increase the network's discriminative power, we employ ASPP to learn the characteristics of different receptive fields of images and use MECA to replace the traditional "skip connection" to overcome the limitations of deconvolution overlap and subsampling. We carry out comprehensive experiments on the PH2 dataset, which is publicly accessible, to assess the performance of our proposed skin lesion segmentation method. The comparison with state-of-the-art methods demonstrates that CASCN achieves superior accuracy in this segmentation task.

References

1. National Cancer Institute, Melanoma of the Skin - Cancer Stat Facts (2023). www.seer.cancer.gov/statfacts/html/melan.html
2. Long, J., Shelhamer, E., Darrell, T.: Fully convolutional networks for semantic segmentation. In: Proceedings of the IEEE Conference on Computer Vision and Pattern Recognition (2015)
3. Ronneberger, O., Fischer, P., Brox, T.: U-Net: convolutional networks for biomedical image segmentation. In: Navab, N., Hornegger, J., Wells, W.M., Frangi, A.F. (eds.) MICCAI 2015. LNCS, vol. 9351, pp. 234–241. Springer, Cham (2015). https://doi.org/10.1007/978-3-319-24574-4_28
4. Zhang, Z., Liu, Q., Wang, Y.: Road extraction by deep residual U-Net. IEEE Geosci. Remote Sens. Lett. **15**(5), 749–753 (2018)
5. Zhou, Z., Rahman Siddiquee, M.M., Tajbakhsh, N., Liang, J.: UNet++: a nested U-Net architecture for medical image segmentation. In: Stoyanov, D., et al. (eds.) DLMIA/ML-CDS -2018. LNCS, vol. 11045, pp. 3–11. Springer, Cham (2018). https://doi.org/10.1007/978-3-030-00889-5_1
6. Ibtehaz, N., Rahman, M.S.: MultiResUNet: rethinking the U-Net architecture for multimodal biomedical image segmentation. Neural Netw. **121**, 74–87 (2020)
7. Guo, C., et al.: Channel attention residual u-net for retinal vessel segmentation. In: ICASSP 2021–2021 IEEE International Conference on Acoustics, Speech and Signal Processing (ICASSP) (2021)
8. Bi, L., Kim, J., Ahn, E., Kumar, A., Fulham, M., Feng, D.: Dermoscopic image segmentation via multistage fully convolutional networks. IEEE Trans. Biomed. Eng. **64**(9), 2065–2074 (2017)

9. Tang, Y., Yang, F., Yuan, S.: A multi-stage framework with context information fusion structure for skin lesion segmentation. In 2019 IEEE 16th International Symposium on Biomedical Imaging (2019)

10. Al-Masni, M.A., et al.: Skin lesion segmentation in dermoscopy images via deep full resolution convolutional networks. Comput. Meth. Program. Biomed. **16**(2), 221–231 (2018)

11. Long, J., Shelhamer, E., Darrell, T.: Fully convolutional networks for semantic segmentation. In: Proceedings of the IEEE Conference on Computer Vision and Pattern Recognition, pp. 3431–3440 (2015)

12. Huang, G., et al.: Densely connected convolutional networks. In: Proceedings of the IEEE Conference on Computer Vision and Pattern Recognition (2017)

13. Chollet, F.: Xception: deep learning with depthwise separable convolutions. In: Proceedings of the IEEE Conference on Computer Vision and Pattern Recognition (2017)

14. Howard, A.G., et al.: MobileNets: efficient convolutional neural networks for mobile vision applications. arXiv preprint arXiv:1704.04861 (2017)

15. Chen, L.C., Papandreou, G., Kokkinos, I., Murphy, K., Yuille, A.L.: DeepLab: semantic image segmentation with deep convolutional nets, Atrous convolution, and fully connected CRFs. IEEE Trans. Pattern Anal. Mach. Intell. **40**(4), 834–848 (2017)

16. Sifre, L., Mallat, S.: Rigid-motion scattering for texture classification. arXiv preprint arXiv:1403.1687 (2014)

17. Hu, J., Shen, L., Sun, G.: Squeeze-and-excitation networks. In: Proceedings of the IEEE Conference on Computer Vision and Pattern Recognition, pp. 7132–7141 (2018)

18. Woo, S., Park, J., Lee, J.Y., Kweon, I.S.: CBAM: convolutional block attention module. In: Proceedings of the European Conference on Computer Vision (ECCV), pp. 3–19 (2018)

19. Wang, Q., Wu, B., Zhu, P., Li, P., Zuo, W., Hu, Q.: ECA-Net: efficient channel attention for deep convolutional neural networks. In: 2020 IEEE/CVF Conference on Computer Vision and Pattern Recognition (CVPR), 2020, pp. 11531–11539 (2020). https://doi.org/10.1109/CVPR42600.2020.01155

20. Mendonça, T., Ferreira, P.M., Marques, J.S., Marcal, A.R., Rozeira, J.: PH 2-A dermoscopic image database for research and benchmarking. In: 2013 35th Annual International Conference of the IEEE Engineering in Medicine and Biology Society (EMBC), pp. 5437–5440. IEEE (2013)

21. Bi, L., Kim, J., Ahn, E., Kumar, A., Feng, D., Fulham, M.: Step-wise integration of deep class-specific learning for dermoscopic image segmentation. Pattern Recogn. **85**, 78–89 (2019)

22. Hasan, M.K., Dahal, L., Samarakoon, P.N., Tushar, F.I., Martí, R.: DSNet: automatic dermoscopic skin lesion segmentation. Comput. Biol. Med. **120**, 103738 (2020)

23. Tong, X., Wei, J., Sun, B., Su, S., Zuo, Z., Wu, P.: ASCU-Net: attention gate, spatial and channel attention u-net for skin lesion segmentation. Diagnostics **11**(3), 501 (2021)

24. Hu, K., Lu, J., Lee, D., Xiong, D., Chen, Z.: AS-Net: attention synergy network for skin lesion segmentation. Expert Syst. Appl. **201**, 117112 (2022)

A DNN-Based Learning Framework for Continuous Movements Segmentation

Tian-yu Xiang[1,2], Xiao-Hu Zhou[1,2(✉)], Xiao-Liang Xie[1,2], Shi-Qi Liu[1,2], Zhen-Qiu Feng[1,2], Mei-Jiang Gui[1,2], Hao Li[1,2], and Zeng-Guang Hou[1,2,3,4(✉)]

[1] State Key Laboratory of Multimodal Artificial Intelligence Systems, Institute of Automation, Chinese Academy of Science, Beijing 100190, China
[2] School of Artificial Intelligence, University of Chinese Academy of Sciences, Beijing 100049, China
{xiaohu.zhou,zengguang.hou}@ia.ac.cn
[3] CAS Center for Excellence in Brain Science and Intelligence Technology, Beijing 100190, China
[4] Joint Laboratory of Intelligence Science and Technology, Institute of Systems Engineering, Macau University of Science and Technology, Taipa, Macao, China

Abstract. This study presents a novel experimental paradigm for collecting Electromyography (EMG) data from continuous movement sequences and a Deep Neural Network (DNN) learning framework for segmenting movements from these signals. Unlike prior research focusing on individual movements, this approach characterizes human motion as continuous sequences. The DNN framework comprises a segmentation module for time point level labeling of EMG data and a transfer module predicting movement transition time points. These outputs are integrated based on defined rules. Experimental results reveal an impressive capacity to accurately segment movements, evidenced by segmentation metrics (accuracy: 88.3%; Dice coefficient: 82.9%; mIoU: 72.7%). This innovative approach to time point level analysis of continuous movement sequences via EMG signals offers promising implications for future studies of human motor functions and the advancement of human-machine interaction systems.

Keywords: Human motor analysis · Electromyography · Deep neural networks

1 Introduction

Bio-signal-based analysis of human motor activity is a field of significant interest, given its capacity to provide direct insights into an individual's manipulation characteristics [24]. Among the myriad types of bio-signals, Electromyography (EMG) has emerged as one of the most salient and has been deployed across a broad spectrum of research areas, including motor recognition [10], human-machine interaction [22], skill level evaluation [23], and motor function rehabilitation [5]. Despite this, these studies have centered their focus primarily on

© The Author(s), under exclusive license to Springer Nature Singapore Pte Ltd. 2024
B. Luo et al. (Eds.): ICONIP 2023, LNCS 14449, pp. 399–410, 2024.
https://doi.org/10.1007/978-981-99-8067-3_30

isolated movements, thereby overlooking the sequences of movements. Evidence rooted in neuroscience underscores the fact that human cognition is more attuned to deciphering sequence information compared to discrete pieces of information [9]. This lends support to the existence of a hierarchical control strategy in human motor control, whereby commands from the upper cortex systems are followed by a series of movements executed by lower-level systems [20]. This research proposes that a shift in focus to sequence analysis could enhance the understanding of human motion and aid in the development of intuitive human-machine interfaces and effective rehabilitation therapies.

The challenge of acquiring bio-signal data during sequential movements requires a well-designed experimental paradigm. Most existing bio-signal datasets used for analyzing human motor characteristics focus on specific movements [3,7,13]. In these datasets, an experimental paradigm is commonly used, where subjects are asked to perform or imagine a particular movement within a set time interval. During this time, bio-signals like EMG are recorded for later analysis. While this approach facilitates the generation of structured bio-signal data conducive to later studies, it may inadvertently impose an unnatural rhythm on the subjects' movements, deviating from the natural flow of human motion. To address these limitations, this research introduces a novel experimental paradigm, aiming to capture bio-signal data as subjects engage in sequential manipulations at their natural rhythm.

Research efforts have increasingly utilized video data to understand human movement sequences, characterizing movements based on video frames [2,21]. Although valuable, this approach offers an observational perspective and may miss more nuanced aspects of human behavior. To capture these nuances, bio-signals, which directly reflect muscle and nerve activity, can be employed. However, despite their potential, the use of bio-signals in the analysis of human movement sequences remains largely uncharted territory. A significant obstacle is the inherent complex temporal-spatial relationship in data, which can make it difficult to identify specific types of movements within a sequence. Consequently, the development of a segmentation algorithm tailored for bio-signals is of crucial importance.

Segmentation tasks, which aim at classifying images at the pixel level, have been thoroughly investigated within the domain of image segmentation [15,16]. Analogously, the objective of this study is to classify Electromyography (EMG) signals at the time point level. Consequently, certain concepts from image segmentation methods can be effectively repurposed. Particularly, the U-net structure [16], recognized for its outstanding performance in medical image segmentation, is leveraged to extract multi-scale features from EMG signals. However, it is important to acknowledge that bio-signals, such as EMG, are typically inherent complex temporal-spatial information compared to image data [8], implying that a direct application of image processing algorithms may fail to analyze these data effectively.

To address the aforementioned challenges, this study presents a novel experimental paradigm for the acquisition of bio-signals during natural rhythm movement sequences and a deep-neural-network (DNN)-based learning framework for the segmentation of continuous movements from EMG signals. The proposed DNN-based learning framework consists of two primary modules: a segmentation module and a transfer module. The segmentation module aims to predict preliminary labels for the EMG data at the time-point level, while the transfer module estimates the transition points between movements. Both modules employ the U-structure for multi-scale temporal feature extraction and mixed-convolution for spatial information extraction. The outputs from both modules are integrated using a set of statistical rules to produce the final predictions.

The major contributions of this study are as follows:

- The design and implementation of an experimental paradigm for the collection of structured bio-signals of continuous movements, executed at the natural rhythm of the subject. Following this paradigm, a dataset comprising 4 distinct subjects and 240 movement sequences has been assembled.
- The formulation of a deep-neural-network (DNN)-based learning framework for segmenting continuous movement from EMG signals. To the best of our knowledge, this is the first work that can classify the EMG signals at the time point level.
- Evaluation results on the collected dataset demonstrate the effectiveness of the proposed learning framework in accurately segmenting movement (accuracy: 88.3%; Dice coefficient: 82.9%; mIoU: 72.7%).

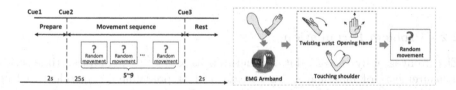

Fig. 1. General steps for data collection.

2 Method

This section will first describe the experimental paradigm, then formulate the problem, and lastly discuss the algorithm's implementation.

2.1 Experimental Paradigm

Traditional bio-signal collection paradigms often lead to unnatural, isolated movements as they require participants to execute specific actions within a fixed interval. To capture bio-signals during naturally paced, continuous movement sequences, a novel experimental paradigm was developed. This approach allows for the collection of bio-signals as participants perform a sequence of movements at their natural rhythm, facilitating a more authentic reflection of motor characteristics and aligning with the natural human motor control mechanism.

Data were collected from four right-hand dominant, healthy participants. As shown in Fig. 1, each participant performed a randomly selected sequence, composed of twisting wrist (TW), opening hand (OH), and touching shoulder (TS), varying in length from 5 to 9 movements. Each sequence was completed within a 25-s window at the participant's natural pace, followed by a rest and preparation period of 2 s each. Participants rested for about 5 min after completing 6 sequences. In total, 60 sequences per participant were recorded.

An EMG armband (Oymotion gForcePro+) collected EMG signals from the participants' forearms at 500 Hz, while a MSHIWI SUA33GC camera captured video data and timestamps at 60 Hz for EMG data labeling.

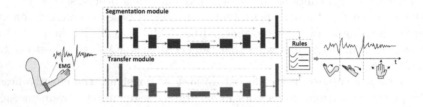

Fig. 2. Overall structure of the proposed method.

2.2   Problem Formulation

Electromyography (EMG) data, structured as $\mathbf{x} \in \mathbb{R}^{C \times T}$, comprise time-series measurements of muscular voltages captured at predefined electrodes. In this context, an EMG acquisition system utilizes C channels and T sampling points per trial. During the experimental procedure, subjects are required to execute n distinct movements randomly selected from the task set ($\mathcal{M}$).

This study aims to formulate an algorithm (f) characterized by a learnable parameter θ and a set of rules $\mathcal{R}$ that facilitate the segmentation of EMG signals corresponding to various movements:

$$f(x, \theta, \mathcal{R}) : \mathbf{x} \rightarrow [(m_0, t_0, t_1), ..., (m_{n-1}, t_{n-1}, t_n)] \qquad (1)$$

where m_i is a specific movement, t_i is start time for m_i, and t_{i+1} is end time for m_i ($i=0,1,2...,n-1$, $m_0, ..., m_{n-1} \in \mathcal{M}$, $0=t_0 < ... < t_n=T$).

The proposed methodology's general structure is illustrated in Fig. 2. The EMG data procured from muscular activity undergo processing by both a segmentation and transfer module. The former module generates movement labels for EMG signals, while the latter predicts the transition points between movements. The outputs of these modules undergo optimization via statistical rules, leading to the final movement predictions.

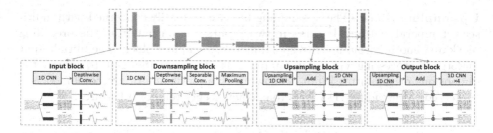

Fig. 3. Network structure of segmentation module and transfer module.

2.3   Networks Structure

The segmentation and transfer modules, as depicted in Fig. 3, exhibit a shared architecture, comprised of four distinct blocks: input, downsampling, upsampling, and output. Drawing inspiration from U-net [16] and EEGnet [12], this architecture is designed to effectively capture both multi-scale temporal and spatial features inherent in EMG signals. The U-shaped structure, adopted from U-net, facilitates the extraction of multi-scale temporal features, crucial for the interpretation of EMG signals which exhibit temporal fluctuations. On the other hand, the mixed convolutions, inspired by EEGnet, focus on the spatial features across the EMG channels, catering to the inherent spatial relationships prevalent in these signals.

Input Block. The segmentation and transfer modules share a similar input block configuration. Initially, the preprocessed EMG data undergoes a transformation via one-dimensional convolutional neural networks (1D CNNs). These networks function as dynamic filters, adept at extracting temporal information inherent in the EMG data. Subsequently, depthwise convolution [4] is employed to uncover spatial relationships across EMG channels.

Downsampling Block. Downsampling blocks, integral to both the segmentation and transfer modules, facilitate the extraction of multi-scale features inherent in EMG signals. Each downsampling block commences with a 1D CNN to extract the temporal information from the EMG signal. This is followed by depthwise convolution to derive the spatial relationships across the EMG channels. Subsequently, separable convolution is applied to generate a summary of each representation. Unlike images, the scale of EMG signals manifests temporally. Therefore, a hybrid CNN structure inspired by EEGnet [12], rather than the 2D CNNs employed in U-net [16], is utilized to imitate the process of common space filter.

These blocks produce two outputs: the first proceeds to the next block, while the second is directed to the corresponding upsampling block, as illustrated in Fig. 3. The first output emanates from the separable convolution, while the second is down-sampled by a maximum pooling operation applied post-separable convolution to extract the salient features.

Upsampling Block. The upsampling block is primarily concerned with multi-scale temporal features. It takes in two inputs: the output from the preceding block and another from its corresponding downsampling block. The initial input undergoes processing via an upsampling 1D CNN, enabling it to match the dimensions of the latter input. Following this, both inputs are combined and subjected to cascading 1D CNNs to extract deeper temporal features.

Output Block. The output block, akin to the upsampling block, incorporates an additional 1D CNN as the final output layer. This distinction provides a resolution to the binary classification problem in the transfer module, determining the transition point to a different movement, as well as addressing the multi-class problem in the segmentation module, differentiating among movement types. The output of this block corresponds to the probability distribution across the respective classes.

Loss Function. Distinct loss functions are employed in the training of the segmentation and transfer modules, tailored to the unique requirements of each. In the segmentation module, the cross-entropy loss is utilized, as defined by:

$$L_s(y_s, \hat{y}_s) = \sum_{c=0}^{|\mathcal{M}|-1} y_s^c \log \hat{y}_s \tag{2}$$

here, y_s^c refers to the cth element of the segmentation label y_s for a given EMG sequence, $\hat{y}_s$ represents the prediction of the segmentation module, and $|\mathcal{M}|$ denotes the total number of movements within the movement set $\mathcal{M}$.

The transfer module is trained with a focal loss function [14], due to the rarity of the transfer time in comparison to the movement execution time:

$$L_t(y_t, \hat{y}_t) = \begin{cases} -\alpha(1 - \hat{y}_t)^\lambda \log \hat{y}_t, & y_t = 1 \\ -(1 - \alpha)\hat{y}_t^\lambda \log(1 - \hat{y}_t), & y_t = 0 \end{cases} \tag{3}$$

where y_t is the transfer label for the given EMG data, $\hat{y}_t$ represents the prediction of the transfer module, α and λ are hyper-parameters.

2.4   Statistical Rules

Upon obtaining the predictions $\hat{y}_t$ and $\hat{y}_s$, they are integrated using a predefined set of rules to generate the final prediction. The primary motivation of these rules is to smooth out the noise in $\hat{y}_t$ and $\hat{y}_s$, and then combine both predictions. Initially, a dynamic threshold for $\hat{y}_t$ is established, taking into account the rarity of the transfer time:

$$\hat{y}_t = [\hat{y}_t > \text{mean}(\hat{y}_t) + \text{std}(\hat{y}_t)] \tag{4}$$

where if $\hat{y}_t(t) = 1$, the time t is considered an activation time. Subsequently, y_t is suppressed if activation does not occur for f_1 consecutive frames. Furthermore, any activations of y_t within a span of f_2 frames are combined, with the entire continuous activation period treated as a single activation. Here, f_1 and f_2 are hyperparameters. The mean time of each continuous activation interval is designated as the transfer time.

The next step identifies neighboring transfer times as the start and end times for each movement. Finally, non-maximum suppression is employed to determine the type of movement within each time interval:

$$m_i = \text{argmax}\left[\sum_{t_i}^{t_{i+1}} \hat{y}_s\right] \tag{5}$$

where $i = 0, 1, ..., n-1$, n is the number of movements in the sequence, $m_i \in \mathcal{M}$, and t_i and t_{n-1} are neighboring transfer times.

3   Experiments

3.1   Training Details

This work was implemented by the TensorFlow framework [1], with RMSprop [19] optimizing the segmentation and transfer modules. The initial learning rate (η_0) was 0.005 and was decayed throughout 500 epochs of training, as described by:

$$\eta = \eta_0 \text{DR}^{-\frac{1+E}{\text{EP}}} \tag{6}$$

Here, DR denotes the drop rate (0.8), EP is the epoch drop (50), and E represents the current epoch number.

Convolution operations were succeeded by batch normalization [11], elu [6], and dropout [18]. The hyper-parameters in the focal loss were $\alpha = 0.25$ and $\gamma = 3$. Both modules underwent L_2 regularization (0.05) and maintained a dropout rate of 0.5. The statistical rules parameters were $f_1 = 20$ and $f_2 = 300$.

Data was subjected to five-fold cross-validation per subject. Each fold comprised a randomly assigned 20% testing set, 10% validation set, and 70% training set. The best-performing model on the validation set was selected, and its testing set results were reported.

3.2   Evaluation Metrics

The proposed model's performance was evaluated at both the time point and movement levels using several metrics. At the time point level, popular image segmentation metrics were employed: accuracy (Acc.): $\frac{|\hat{y}=y|}{|y|}$, Dice coefficient (Dice): $2\frac{|\hat{y}\cap y|}{|\hat{y}|+|y|}$, and mIoU: $\frac{|\hat{y}\cap y|}{|\hat{y}\cup y|}$. Here, y represents the ground truth, $\hat{y}$ is the model's prediction, and $|\cdot|$ is the number of elements in a set. Additionally, time point level accuracy for each movement was reported.

Table 1. Comparison with other methods.

Model	Type	TW	OH	TS	R	Acc.	Dice	mIoU	Flops(B)
EEGnet [12]	0.2 s	52.3	59.7	70.3	96.1	76.8	67.4	54.1	5.8
	0.3 s	57.8	64.7	71.1	95.2	78.5	69.8	56.7	16.4
	0.4 s	60.2	68.6	72.1	93.0	78.8	70.4	57.6	21.8
Shawllownet [17]	0.2 s	58.7	66.6	71.4	93.4	78.2	69.6	56.4	37.8
	0.3 s	62.9	70.7	71.8	91.3	78.9	70.5	57.8	59.4
	0.4 s	64.1	72.9	73.3	91.7	79.9	71.3	58.8	81.0
Ours	w/o Trans.	54.8	68.3	76.3	97.9	81.0	71.0	59.2	**0.8**
	Trans.	**64.9**	**86.9**	**82.5**	**99.1**	**88.3**	**82.9**	**72.7**	1.6

At the movement level, the model's performance was assessed from three perspectives: accuracy in predicting movements within a specified time interval, precision in predicting transfer times, and the statistical correlation between the predicted time interval and the ground truth. The evaluation employed four metrics: Accuracy in predicting movements for a given interval; deviation in the predicted transfer time; accuracy in predicting the number of transfer times; p-value from an analysis of variance (ANOVA) test, comparing the length of a given movement with the predicted length.

Table 2. Time for different subjects finishing the given movement (s) and p-value between the predicted time interval and ground truth.

S	TW		OH		TS		R	
	time	p	time	p	time	p	time	p
s1	2.30 ± 0.47	0.336	2.17 ± 0.35	0.406	2.52 ± 0.34	0.327	2.24 ± 0.68	0.802
s2	2.29 ± 0.82	0.387	2.19 ± 0.80	0.746	2.52 ± 0.70	0.521	2.73 ± 0.93	0.231
s3	1.86 ± 0.35	0.537	1.79 ± 0.23	0.397	2.30 ± 0.40	0.791	2.99 ± 0.59	0.787
s4	1.61 ± 0.91	0.170	1.54 ± 0.67	0.548	2.01 ± 1.12	0.243	3.37 ± 0.79	0.131

3.3 Experimental Results

The dataset under investigation requires subjects to complete a designated movement sequence within 25 s, with any remaining time classified as rest. This presents a four-class segmentation problem, which includes three types of movement and rest. To the best of our knowledge, no movement segmentation algorithms specifically designed for EMG signals could be found in the literature. Existing bio-signal classification methods typically generate predictions over specified time windows and apply them successively across the entire sequence.

The proposed method is compared with two well-established and computationally efficient bio-signal classification algorithms [12,17], as shown in Table 1. These algorithms demonstrate a slight performance increase when longer time windows are utilized, but at the cost of escalating computational complexity. The proposed method, however, offers superior performance in terms of accuracy and efficiency. Moreover, the inclusion of a transfer module significantly enhances the performance, leading to improvements of 9.0%, 16.8%, and 22.8% for accuracy, Dice coefficient, and mIoU, respectively.

Further analysis of the proposed model's performance at the movement level is conducted. However, such a level of examination is not feasible for classification methods due to their inability to precisely predict the start and end times of a specific movement. The proposed model achieves an accuracy of 87.4% in predicting movements and correctly identifies 98.1% of transfer times with a deviation of 0.285 s. Additionally, the predicted movement duration exhibits a distribution similar to the subjects', with a p-value exceeding 0.1 between the time taken to complete a designated movement and the predictions generated by the model (Table 2), which shows no significant difference between the predictions and ground truth. These results validate the model's capability to accurately segment movements while encapsulating unique manipulation characteristics such as different subjects' rhythms.

4   Discussion

4.1   Individual Differences

Table 2 illustrates the time taken by subjects to complete a designated movement, denoted as mean±std in seconds. This data reveals distinct interindividual variations. Notably, subjects 1 and 3 exhibit a lower standard deviation in movement completion times relative to subjects 2 and 4, implying a potential heightened level of experimental engagement.

Furthermore, variations are evident in muscle activities across subjects, even within the same channel. These disparities underscore the existence of unique control mechanisms in individuals, which might be responsible for differences in the results.

Despite these variations in manipulation rhythm and style among subjects, the robustness of the proposed method is demonstrated by its consistently high performance.

4.2   Contribution of the Transfer Module

Beyond the segmentation module, which is typically employed in image segmentation tasks, the proposed model includes a transfer module. The incorporation of this module markedly enhances the time point level evaluation outcomes, as shown in Table 1.

The significance of the transfer module to the overall performance of the proposed model becomes apparent when examining the direct outputs of the

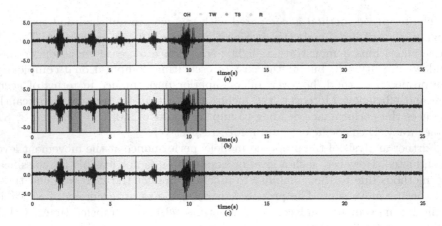

Fig. 4. Example of the output of the segmentation module and combined output with the transfer module (ours). (a) ground truth; (b) output of segmentation module; (c) output of the proposed method.

segmentation module alongside the combined outputs of the transfer module (Fig. 4). Upon comparison, the combined outputs of the transfer module align closely with the ground truth, indicating the substantial role of the transfer module in enhancing the model's performance during movement transfer.

As illustrated in Fig. 4, the EMG signal maintains relative stability during movement transfer, absent of notable patterns. The transfer module is designed specifically to detect periods of stable EMG rather than differentiate specific movements. Thus, the proposed model heightens the accuracy of EMG signal segmentation by identifying periods of stable EMG as indicative of movement transfer times.

5   Conclusion

This study introduces a novel experimental paradigm for the collection of bio-signals that accurately reflect individual rhythms of movement and presents a Deep Neural Network (DNN)-based learning framework for segmenting contin-uous movements from these signals. The assembled dataset, which includes 240 EMG sequences collected from 4 subjects, was used for empirical evaluation of the proposed method, demonstrating its high accuracy in classifying EMG sig-nals at the time point level. These encouraging results underline the potential of the proposed experimental paradigm and learning framework in providing novel perspectives for studying human manipulation. Furthermore, this research paves the way for potential advancements in human-machine interaction sys-tems, offering a trajectory towards the development of more intuitive system designs. Future research efforts will involve the incorporation of additional bio-signals, such as EEG, with the aim of expanding our understanding of the control mechanisms underlying human motor functions.

Acknowledgements. This work was supported in part by the National Natural Science Foundation of China under Grant 62003343, Grant 62222316, Grant U1913601, Grant 62073325, Grant U20A20224, and Grant U1913210; in part by the Beijing Natural Science Foundation under Grant M22008; in part by the Youth Innovation Promotion Association of Chinese Academy of Sciences (CAS) under Grant 2020140; in part by the CIE-Tencent Robotics X Rhino-Bird Focused Research Program.

References

1. Abadi, M., et al.: Tensorflow: a system for large-scale machine learning. In: Proceedings of the 12th USENIX Symposium on Operating Systems Design and Implementation (OSDI 2016), pp. 265–283 (2016)
2. van Amsterdam, B., et al.: Gesture recognition in robotic surgery with multimodal attention. IEEE Trans. Med. Imaging **41**(7), 1677–1687 (2022)
3. Blankertz, B., et al.: The BCI competition 2003: progress and perspectives in detection and discrimination of EEG single trials. IEEE Trans. Biomed. Eng. **51**(6), 1044–1051 (2004)
4. Chollet, F.: Xception: deep learning with depthwise separable convolutions. In: Proceedings of the IEEE Conference on Computer Vision and Pattern Recognition, pp. 1251–1258 (2017)
5. Cisnal, A., Pérez-Turiel, J., Fraile, J.C., Sierra, D., de la Fuente, E.: Robhand: a hand exoskeleton with real-time EMG-driven embedded control quantifying hand gesture recognition delays for bilateral rehabilitation. IEEE Access **9**, 137809–137823 (2021)
6. Clevert, D.A., Unterthiner, T., Hochreiter, S.: Fast and accurate deep network learning by exponential linear units (elus). arXiv preprint arXiv:1511.07289 (2015)
7. Côté-Allard, U., et al.: Deep learning for electromyographic hand gesture signal classification using transfer learning. IEEE Trans. Neural Syst. Rehabil. Eng. **27**(4), 760–771 (2019)
8. De Luca, C.J., Gilmore, L.D., Kuznetsov, M., Roy, S.H.: Filtering the surface EMG signal: movement artifact and baseline noise contamination. J. Biomech. **43**(8), 1573–1579 (2010)
9. Henin, S., et al.: Learning hierarchical sequence representations across human cortex and hippocampus. Sci. Adv. **7**(8), eabc4530 (2021)
10. Inam, S., et al.: A brief review of strategies used for EMG signal classification. In: Proceedings of 2021 International Conference on Artificial Intelligence (ICAI), pp. 140–145. IEEE (2021)
11. Ioffe, S., Szegedy, C.: Batch normalization: accelerating deep network training by reducing internal covariate shift. In: Proceedings of the International Conference on Machine Learning, pp. 448–456 (2015)
12. Lawhern, V.J., Solon, A.J., Waytowich, N.R., Gordon, S.M., Hung, C.P., Lance, B.J.: EEGNet: a compact convolutional neural network for EEG-based brain-computer interfaces. J. Neural Eng. **15**(5), 056013 (2018)
13. Lee, M.H., et al.: EEG dataset and OpenBMI toolbox for three BCI paradigms: an investigation into BCI illiteracy. GigaScience **8**(5), giz002 (2019)
14. Lin, T.Y., Goyal, P., Girshick, R., He, K., Dollár, P.: Focal loss for dense object detection. In: Proceedings of the IEEE International Conference on Computer Vision, pp. 2980–2988 (2017)

15. Minaee, S., Boykov, Y., Porikli, F., Plaza, A., Kehtarnavaz, N., Terzopoulos, D.: Image segmentation using deep learning: a survey. IEEE Trans. Pattern Anal. Mach. Intell. **44**(7), 3523–3542 (2022). https://doi.org/10.1109/TPAMI.2021.3059968

16. Ronneberger, O., Fischer, P., Brox, T.: U-Net: convolutional networks for biomedical image segmentation. In: Navab, N., Hornegger, J., Wells, W.M., Frangi, A.F. (eds.) MICCAI 2015. LNCS, vol. 9351, pp. 234–241. Springer, Cham (2015). https://doi.org/10.1007/978-3-319-24574-4_28

17. Schirrmeister, R.T., et al.: Deep learning with convolutional neural networks for EEG decoding and visualization. Hum. Brain Mapp. **38**(11), 5391–5420 (2017)

18. Srivastava, N., Hinton, G., Krizhevsky, A., Sutskever, I., Salakhutdinov, R.: Dropout: a simple way to prevent neural networks from overfitting. J. Mach. Learn. Res. **15**(1), 1929–1958 (2014)

19. Tieleman, T., Hinton, G., et al.: Lecture 6.5-rmsprop: divide the gradient by a running average of its recent magnitude. COURSERA: Neural Netw. Mach. Learn. **4**(2), 26–31 (2012)

20. Voytek, B., et al.: Oscillatory dynamics coordinating human frontal networks in support of goal maintenance. Nat. Neurosci. **18**(9), 1318–1324 (2015)

21. Xiang, T.Y., et al.: Quantitative movement analysis using scaled information implied in monocular videos. IEEE Trans. Med. Rob. Bionics **5**, 88–99 (2023). https://doi.org/10.1109/TMRB.2023.3240285

22. Xiong, D., Zhang, D., Zhao, X., Zhao, Y.: Deep learning for EMG-based human-machine interaction: a review. IEEE/CAA J. Automatica Sinica **8**(3), 512–533 (2021)

23. Zhou, X.H., et al.: Surgical skill assessment based on dynamic warping manipulations. IEEE Trans. Med. Rob. Bionics **4**(1), 50–61 (2022)

24. Zhou, X.H., et al.: Learning skill characteristics from manipulations. IEEE Trans. Neural Netw. Learn. Syst. (2022). https://doi.org/10.1109/TNNLS.2022.3160159

Neural-Symbolic Recommendation with Graph-Enhanced Information

Bang Chen[1], Wei Peng[2], Maonian Wu[1(✉)], Bo Zheng[1], and Shaojun Zhu[1]

[1] School of Information Engineering, Huzhou University, Huzhou, China
wmn@zjhu.edu.cn
[2] College of Computer Science, Guizhou University, Guiyang, China

Abstract. The recommendation task is not only a problem of inductive statistics from data but also a cognitive task that requires reasoning ability. The most advanced graph neural networks have been widely used in recommendation systems because they can capture implicit structured information from graph-structured data. However, like most neural network algorithms, they only learn matching patterns from a perception perspective. Some researchers use user behavior for logic reasoning to achieve recommendation prediction from the perspective of cognitive reasoning, but this kind of reasoning is a local one and ignores implicit information on a global scale. In this work, we combine the advantages of graph neural networks and propositional logic operations to construct a neuro-symbolic recommendation model with both global implicit reasoning ability and local explicit logic reasoning ability. We first build an item-item graph based on the principle of adjacent interaction and use graph neural networks to capture implicit information in global data. Then we transform user behavior into propositional logic expressions to achieve recommendations from the perspective of cognitive reasoning. Extensive experiments on five public datasets show that our proposed model outperforms several state-of-the-art methods, source code is avaliable at [https://github.com/hanzo2020/GNNLR].

Keywords: Recommendation Systems · Neuro-Symbolic · Graph Neural Network

1 Introduction

The explosive growth in internet information has made recommendation systems increasingly valuable as auxiliary decision-making tools for online users in various areas, including e-commerce [16], video [13], and social networks [11]. Classic recommendation methods mainly include matrix factorization-based approaches [14], neural network methods [7], time-series-based methods [10], and others that leverage richer external heterogeneous information sources, such

B. Chen, W. Peng and M. Wu—Contributed equally to the work.

as sentiment space context [20] and knowledge graphs [19]. Graph neural networks have recently gained attention for their success in structured knowledge tasks. They have been widely used in recommendation systems, including Wang et al.'s graph collaborative filtering approach [18], He et al.'s lightweight graph collaborative filtering method [6], and Wang's recommendation algorithm based on a graph attention model [17]. The advantage of graph neural networks is that they can aggregate information from neighbor nodes through the data structure of graphs in a global view, which allows them to better capture implicit high-order information compared to other types of neural network methods.

Although the above methods have their advantages, they all have one obvious drawback: they only learn matching patterns in the data from a perception perspective, not reasoning [15]. Although graph neural networks can utilize structured knowledge from graphs, such aggregation learning is essentially a weak reasoning mode within the scope of perceptual learning and does not consider explicit logic reasoning relationships between entities [2]. As a task that requires logic reasoning ability, recommendation problems are more like decision-making processes based on past known information. For example, a user who has recently purchased a computer does not need recommendations for similar products but needs peripheral products such as keyboards and mice. However, in current recommendation system applications, users are often recommended similar products immediately after purchasing an item, even if their demand has already disappeared.

Some researchers have attempted to incorporate logic reasoning into recommendation algorithms to address the above issue. For example, Shi et al. [15] proposed a neural logic reasoning algorithm that uses propositional logic to achieve recommendations. Subsequently, Chen et al. [3] proposed neural collaborative reasoning and added user information to improve the model's personalized reasoning ability for users. However, these advanced methods only perform logic reasoning based on the current user's historical interaction behavior, which is just a local range of reasoning and lacks implicit high-order information from the global perspective. Especially when there is an enormous amount of recommendation data available, the number of items interacted with by a single user compared to all items is usually very small; therefore, it is evident that large amounts of implicit global information will be ignored.

To address the above challenges, this paper proposes a neural-symbolic recommendation model based on graph neural networks and Proposition Logic. We use logic modules to compensate for the lack of reasoning ability of neural networks and graph neural networks to compensate for the logic module's weakness in focusing only on local information. Our model can use both implicit messages from the global perspective and explicit reasoning from the local perspective to make recommendations. In addition, we also designed a more suitable knowledge graph construction method for the model to construct an item-item graph from existing data. Our main contributions are as follows:

1. We propose a neural-symbolic recommendation model, which combines graph neural networks with logic reasoning, called GNNLR. The model can not only

obtain information aggregation gain from the graph but also use propositional logic to reason about users' historical behaviors.

2. We design a new method of constructing graphs for the proposed model, building item-item graphs based on the adjacency principle.

3. We experimented with the proposed model on several real public datasets and compared it with state-of-the-art models. We have also explored different GNN architectures.

2   Methodology

Figure 1 illustrates the overall architecture of the proposed model, called GNNLR, which mainly consists of five parts: 1. item-item graph construction; 2. node information fusion; 3. propositional logic convert; 4. neural logic computing; and 5. prediction and training. We will describe these five parts in detail as follows.

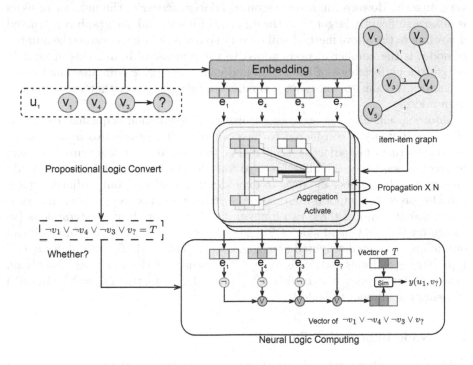

Fig. 1. GNNLR framework

2.1   Item-Item Graph Construction

We first describe how the graph required for the model are constructed. We constructs the graph differently from the previous method, as shown in Fig. 2

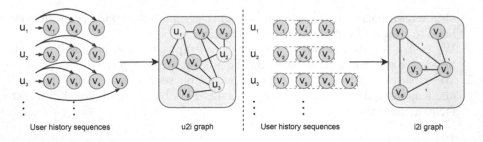

Fig. 2. Traditional graph construction method (left) and our method (right).

Previous method (e.g., NGCF [18]) typically construct a graph based on known user-item interaction relationships (as shown on the left side of Fig. 2) and use it for subsequent graph neural network calculations. The strength of this approach is that the interaction between the user and the item is retained very directly. However, in a real recommendation scenario, the number of users is often significantly larger than the number of items, and the graph constructed according to the above method will be very sparse and large, because the number of nodes is the number of users added to the number of items, this ultimately affects the performance of the model and causes excessive computational costs. Therefore, we aim to construct a smaller and denser graph that only contains item nodes.

More specifically, for each user's historical interaction sequence, each pair of adjacent items is considered to have an edge. There are two main reasons for considering only adjacent items rather than all items in the same historical interaction sequence: 1. if all items were considered without restrictions, it would be easy for the number of edges to explode. 2. considering only adjacent items can also preserve temporal information. Furthermore, the weight of each edge is the adjacent count of these two items in the history of all user interactions (as shown on the right-hand side of Fig. 2). Through the above procedure, we can obtain an undirected weighted homogeneous graph $\mathcal{G} = (\mathcal{V}, \mathcal{E})$ with nodes $v_i \in \mathcal{V}$ representing all items and edges $(v_i, v_j) \in \mathcal{E}$ connecting them. We can also obtain a weighted adjacency matrix $A \in \mathbb{R}^{N \times N}$ and degree matrix $D_{ii} = \sum_j A_{ij}$ with N being the number of nodes.

2.2 Node Information Fusion

After the item-item graph is constructed, we embed all items as vectors and propagate and aggregate these vectors based on the item-item graph. GNNLR uses a graph convolutional network [9] to perform this operation.

Assume that X represents the features of all nodes and Θ represents all trainable parameters. The formula for obtaining new features in each information aggregation is:

$$X' = \widehat{D}^{-\frac{1}{2}} \widehat{A} \widehat{D}^{-\frac{1}{2}} X \Theta \tag{1}$$

where $\widehat{A} = A + I$ indicates that the information of the node itself is kept, $\widehat{D}$ is the degree matrix. For each node x_i, its information aggregation formula is:

$$x_i' = \Theta^\top \sum_{j \in \mathcal{N}(i) \cup \{i\}} \frac{e_{j,i}}{\sqrt{\widehat{d_j}\widehat{d_i}}} x_j \tag{2}$$

where $e_{j,i}$ represents the weight of the edge from x_j to x_i, and $\widehat{d_i} = 1 + \sum_{j \in \mathcal{N}(i)} e_{j,i}$. In short, each node's features are influenced by its neighbors' features, and the degree of influence is related to the weights of the edges. This aggregation process can be repeated several times, which is beneficial because the information of neighbor nodes' neighbor nodes is also aggregated. In addition, we use the ReLU activation function at the end of each aggregation.

2.3 Propositional Logic Convert

After obtaining the aggregated item features, we transform the existing user history behavior into propositional logic expressions to train the model and implement recommendation prediction.

The main symbols of classical propositional logic include $\wedge, \vee, \neg, \rightarrow, \leftrightarrow$, representing 'and', 'or', 'not', 'if...then', and 'equivalent to'. Among them, the operator $\neg$ has the highest calculation priority when there are no parenthesis.

In recommendation tasks, we can naturally convert the historical behavior of each user into propositional logic expressions. For example, if a user has a history of interaction $h = (v_1, v_4, v_3)$ and v_3 is considered as the target item while v_1, v_4 is viewed as their historical interaction item, then we can derive the following logic rule:

$$(v_1 \rightarrow v_3) \vee (v_4 \rightarrow v_3) \vee (v_1 \wedge v_4 \rightarrow v_3) = T \tag{3}$$

Formula 3 represents that a user's preference for v_3 may be due to their previous liking of v_1, or v_4, or because they have liked both v_1 and v_4 at the same time.

Subsequently, according to the logic implication equivalence $p \rightarrow q = \neg p \vee q$, the implied formula can be transformed into a disjunctive normal form consisting of Horn clauses. The above formula can be transformed as:

$$(\neg v_1 \vee v_3) \vee (\neg v_4 \vee v_3) \vee (\neg(v_1 \wedge v_4) \vee v_3) = T \tag{4}$$

At this point, the transformed logic expressions still have the problem of computational complexity. When the number of items in the history interaction increases, the length of the expressions and the number of logic variables will explode, especially the number of variables in the higher-order Horn clauses. Based on our calculation, when there are n items in the history interaction, the number of Horn clause terms in this disjunctive normal form is $\sum_{r=1}^{n} \frac{n!}{r!(n-r)!} = 2^n - 1$ and its computational complexity is $O(2^n)$. Therefore, we further simplify those high-order Horn clauses using De Morgan's law. According to De Morgan's law $\neg(p \wedge q) \leftrightarrow \neg p \vee \neg q$, we can further transform Eq. 4 into:

$$(\neg v_1 \vee v_3) \vee (\neg v_4 \vee v_3) \vee (\neg v_1 \vee \neg v_4 \vee v_3) = T \tag{5}$$

Finally, according to the propositional logic associative law, we can remove the parentheses and eliminate the duplicate variables. The simplified horn clause is obtained as follows:

$$\neg v_1 \lor \neg v_4 \lor v_3 = T \tag{6}$$

At this point, the computational complexity is reduced to $O(n)$. Similarly, we can transform any length of user historical interaction behavior into simplified propositional logic expression to train the model. When we make a prediction, we only need to construct a propositional logic expression consisting of the user's history interaction and the target item and let the model determine whether the expression is true. For example, based on the above historical interaction, we can let the model determine whether the following logic expression is true:

$$\neg v_1 \lor \neg v_4 \lor \neg v_3 \lor v_? \tag{7}$$

where $v_?$ is the predicted item, the closer the logic expression becomes to true, the more likely the user is to like item $v_?$.

2.4 Neural Logic Computing

In this section, we describe how the model computes the converted logic expressions. We adopt the method from paper NLR [15] and use neural networks to perform logic operations.

Specifically, in GNNLR, the propositional logic expression contains two operators $\neg$ and $\lor$. We train two independent neural network modules $NOT(\cdot)$ and $OR(\cdot, \cdot)$ to perform their corresponding logic operations, and the neural network modules use a multilayer perceptron structure. If the dimension of the item vector after graph neural network aggregation is d, then for module $NOT(\cdot)$, its input is a d-dimension vector, and its output is also a d-dimension vector representing the logic negation of that vector.

$$\neg e_i = NOT(e_i) = W_2^{not}\sigma(W_1^{not}e_i + b_1^{not}) + b_2^{not} \tag{8}$$

For module $OR(\cdot, \cdot)$, its input is a vector of dimension 2d concatenated from two vectors, and the output is a vector of dimension d representing the result of logic disjunction operation on the two input vectors.

$$e_i \lor e_j = OR(e_i, e_j) = W_2^{or}\sigma(W_1^{or}(e_i \oplus e_j) + b_1^{or}) + b_2^{or} \tag{9}$$

where $W_2^{not}, W_1^{not}, W_2^{or}, W_1^{or}, b_2^{not}, b_1^{not}, b_2^{or}, b_1^{or}$ are all learnable model parameters and σ is the activation function. The neural logic computing model will compute propositional logic expression according to its order of operation and ultimately outputs a vector e_l of dimension d representing the expression. As shown at the bottom of Fig. 1.

2.5   Prediction and Training

In this section, we describe how to use the computed vectors of logic expressions for prediction and training. When the GNNLR model is initialised, a benchmark vector T of dimension d is generated. We determine whether a logic expression is true by computing the similarity of vector e_l with vector T and compute the similarity by using the following formula:

$$Sim(e_l, T) = sigmoid(\varphi \frac{e_l \cdot T}{\|e_l\| \times \|T\|}) \tag{10}$$

where φ is an optional parameter that can be combined with the Sigmoid function to make the model more flexible when dealing with different datasets, the similarity result ranges from 0 to 1. A result closer to 1 indicates that the logic expression is closer to true, and the target item $v_?$ is more likely to be preferred by users. During training, we adopt a pair-wise learning strategy that for each item v^+ liked by a user, we randomly sample an item v^- that has not been interacted with or disliked by the user. We then calculate the loss function based on the following formula:

$$L = -\sum_{v^+} \log(sigmoid(p(v^+) - p(v^-))) \tag{11}$$

where $p(v^+)$ and $p(v^-)$ are the predicted results of model on item v^+ and v^-. We also use the logic rule loss $\mathcal{L}_{logic}$ defined in [15] to constrain the training of logic operator modules, and use regularization terms $\sum_{e \in E} \|e\|_F^2$ for constraining vector lengths and $\|\Theta\|_F^2$ for constraining parameter lengths. The final loss function is shown in Eq. 12, where $\lambda_{\mathcal{L}}$, λ_l, λ_Θ are the weights of three constraint losses.

$$L = -\sum_{v^+} \log(sigmoid(p(v^+) - p(v^-))) + \lambda_{\mathcal{L}}\mathcal{L}_{logic} + \lambda_l \sum_{e \in E} \|e\|_F^2 + \lambda_\Theta \|\Theta\|_F^2 \tag{12}$$

3   Experiments

3.1   Datasets and Evaluation Metrics

We conducted experiments on five real datasets with different categories and data volumes, including GiftCard, Luxury, Software, Industry from the Amazon review website and MovieLens-100k from the MovieLens website. Table 1 shows the specific information of these five datasets, where edge_num represents the number of item-item edges generated by the method in Sect. 2.1.

Considering that some baseline models are based on sequence algorithms, according to the suggestion in [3], we adopt a leave-one-out strategy to process and divide the dataset: we sort each user's historical interactions by time and use each user's last two positive interactions as validation set and test set.

We use two metrics, H@K (Hit Rate) and N@K (Normalized Discounted Cumulative Gain), to evaluate the performance of our model. A higher value

Table 1. General statistical information about the five real-world datasets.

Dataset	User	Item	Interaction	Edge_num	Density
GiftCard	459	149	2972	2580	4.35%
Software	1826	802	12805	21408	0.874%
Luxury	3820	1582	34278	7906	0.567%
Industry	11042	5335	77071	75758	0.131%
ML100k	943	1682	1000000	71066	6.3%

for H@K indicates that the target item appears more frequently in the top K predicted items, while a higher value for N@K indicates that the target item has a more advanced ranking. We randomly sample 50 negative items for the first three datasets as interference for each correct answer during testing and randomly sample 100 negative items as interference for the last two larger datasets.

3.2 Comparison Methods

We will compare the proposed GNNLR with the following baseline models, which cover different recommendation approaches including shallow models, deep models, sequence models, graph neural networks and reasoning models:

- **BMF** [14]: A matrix factorization model based on Bayesian personalized ranking, which is a very classic recommendation algorithm.
- **NCF** [7]: Neural Collaborative Filtering is an improved collaborative filtering algorithm that replaces vector dot products with neural networks and integrates traditional matrix factorization.
- **STAMP** [12]: A popular model that takes into account both short-term attention and long-term user behavior memory.
- **NARM** [10]: A powerful sequence recommendation model that combines attention mechanism and gated recurrent networks.
- **GRU** [8]: A powerful sequence recommendation model that applies gated recurrent networks to recommendation algorithms.
- **NGCF** [18]: This is a state-of-the-art recommendation model based on GNN, which utilizes graph neural networks for collaborative filtering algorithms. It models user-item interactions as a graph structure and performs information aggregation.
- **NLR** [15]: Neural logic reasoning, a neural-symbolic model based on modular propositional logic operation of neural networks. This is a state-of-the-art reasoning-based recommendation framework.

3.3 Parameter Settings

All models were trained with 200 epochs using the Adam optimizer and a batch-size of 128. The learning rate was 0.001 and early-stopping was conducted according to the performance on the validation set. Both λ_l and λ_Θ were set to 1×10^{-4}

and applied to the baseline models equally; λ_r was set to 1×10^{-5}. The vector embedding dimension was set to 64 for all baseline models. The maximum history interaction length was set to 5 for sequence-based models. More details can be obtained from the code link provided in the abstract.

3.4 Recommendation Performance

Table 2 shows the recommendation performance of our model and baseline models on five datasets. The best results are highlighted in bold, while the second-best results are underlined.

Table 2. Performance comparison of all models on five datasets.

Dataset	Metric	BMF	NCF	SMP	NAM	GRU	NGCF	NLR	Ours
GiftCard	N@10	0.3028	0.3032	0.2926	0.3409	0.3582	0.3169	0.3308	**0.3646**
	N@20	0.3442	0.3381	0.3344	0.3727	0.3988	0.3656	0.3697	**0.4134**
	H@10	0.5772	0.5894	0.5306	0.5918	0.5967	0.5732	0.5813	**0.6057**
	H@20	0.7398	0.7276	0.6939	0.7492	0.7492	0.7520	0.7358	**0.7642**
Software	N@10	0.2903	0.2937	0.3524	0.3831	0.3682	0.3339	0.3794	**0.4487**
	N@20	0.3386	0.3424	0.3971	0.4305	0.4125	0.3767	0.4384	**0.4842**
	H@10	0.4582	0.4757	0.5919	0.6554	0.6597	0.5894	0.6433	**0.7513**
	H@20	0.6487	0.6894	0.7681	0.8370	0.8326	0.7587	0.8354	**0.8809**
Luxury	N@10	0.5075	0.4707	0.5090	0.5205	0.5135	0.5021	0.5189	**0.5541**
	N@20	0.5505	0.5133	0.5459	0.5568	0.5466	0.5306	0.5522	**0.5841**
	H@10	0.6236	0.5951	0.7196	0.7308	0.7340	0.6317	0.7428	**0.7727**
	H@20	0.7969	0.7749	0.8644	0.8740	0.8636	0.8149	0.8684	**0.8907**
Industry	N@10	0.2553	0.2213	0.2383	0.2611	0.2600	0.2526	0.2612	**0.3163**
	N@20	0.2935	0.2492	0.2697	0.2953	0.2944	0.2873	0.2966	**0.3409**
	H@10	0.4138	0.3401	0.3791	0.4147	0.4232	0.3915	0.4253	**0.4934**
	H@20	0.5425	0.4715	0.5037	0.5558	0.5593	0.5293	0.5603	**0.6217**
ML100k	N@10	0.3578	0.3595	0.3907	0.4084	0.4094	0.3841	0.4151	**0.4239**
	N@20	0.4085	0.4066	0.4303	0.4435	0.4424	0.4259	0.4458	**0.4581**
	H@10	0.6281	0.6338	0.6602	0.6795	0.6752	0.6488	0.6833	**0.6956**
	H@20	0.8184	0.8081	0.8137	0.8210	0.8092	0.8124	0.8215	**0.8296**

The experimental results show that the GNNLR model exhibits the best performance on all four metrics of the five data sets due to its ability to utilize global implicit information from graph neural networks and local explicit reasoning from propositional logic. Sequence-based models (e.g., NARM and GRU) and reasoning-based models (NLR) achieve most of the second-best performance, probably because these models are good at utilizing the temporal information in the data, which we retain during the data processing. In addition, GNNLR outperforms both NGCF, which relies solely on graph neural networks for recommendations, and NLR, which relies solely on neural logic for recommendations, and verifies that our contributions are meaningful from the perspective of ablation experiments. Although NGCF performs not very well, the significant improvement of GNNLR over NLR of recommendation results proves the usefulness of graph neural networks. In conclusion, the experimental results show that our proposed GNNLR model and item-item graph construction method can efficiently combine the advantages of neural and symbolic methods and significantly enhance the recommendation results.

3.5   Research on Different GNN Model

For GNNLR, the GNN module is a plug-and-play component. Therefore, we further explored the impact of different GNN architectures on the performance of GNNLR models. In addition to GCN (the GNN architecture used by GNNLR), we selected five other different GNN models for testing:

– **GAT** [17]: The Graph Attention Network.

Table 3. Comparison of GNNLR with different GNN architectures.

Dataset	Metric	GCN	GAT	LGNN	ChebGC	GC2N	FAC
GiftCard	N@10	**0.3646**	0.3352	0.3284	0.3593	0.3547	0.3437
	N@20	**0.4134**	0.3846	0.3751	0.4023	0.4098	0.3822
	H@10	**0.6057**	0.5913	0.5688	0.5991	0.5913	0.5994
	H@20	0.7642	0.7558	0.7517	**0.7683**	0.7682	0.7539
Software	N@10	**0.4487**	0.4318	0.4325	0.4342	0.4291	0.3915
	N@20	**0.4842**	0.4708	0.4746	0.4776	0.4693	0.4345
	H@10	**0.7513**	0.7379	0.7208	0.7204	0.7236	0.6824
	H@20	0.8809	0.8873	0.8805	**0.8879**	0.8826	0.8526
Luxury	N@10	**0.5541**	0.5486	0.5511	0.5405	0.4908	0.4692
	N@20	**0.5841**	0.5782	0.5714	0.5729	0.5219	0.5104
	H@10	0.7727	0.7723	**0.7814**	0.7611	0.7152	0.6793
	H@20	0.8907	0.8788	**0.8947**	0.8879	0.8573	0.8410
Industry	N@10	**0.3163**	0.2881	0.3003	0.2871	0.3009	0.2635
	N@20	**0.3409**	0.3242	0.3349	0.3206	0.3335	0.2970
	H@10	**0.4934**	0.4832	0.4818	0.4747	0.4770	0.4266
	H@20	0.6217	**0.6260**	0.6195	0.6070	0.6062	0.5595
ML100k	N@10	**0.4239**	0.3939	0.4127	0.4178	0.3840	0.3864
	N@20	**0.4581**	0.4359	0.4524	0.4566	0.4215	0.4285
	H@10	**0.6956**	0.6602	0.6763	0.6849	0.6624	0.6517
	H@20	0.8296	0.8253	**0.8328**	0.8382	0.8103	0.8189

- **Light-GNN** [6]: The Light Graph Convolution (LGC) operator.
- **ChebGCN** [5]: The Chebyshev spectral Graph Convolutional operator.
- **GCN2Conv** [4]: The Graph Convolutional operator with initial residual connections and identity mapping.
- **FAConv** [1]: The Frequency Adaptive Graph Convolution operator.

The experimental results using different GNN modules are shown in Table 3. The traditional GCN architecture achieves the best results in most metrics, Light-GNN, ChebGCN and GAT also showed the best performance on some metrics. This indicates that different GNN architectures have their own advantages when facing different types of data or recommendation metrics. As for the worse performance of the GCN2Conv and FAConv models, we thought it might be due to the complex structure that takes more time to converge. We used a uniform number of epochs in our experiments, and more epochs may further improve the performance of these two GNN architectures.

4   Conclusion

In this work, we propose a Neural-Symbol recommendation model that combines the advantages of graph neural networks and logic reasoning, named GNNLR. GNNLR uses both global implicit information from graphs and local explicit reasoning from propositional logic for recommendation prediction. We also design a method for constructing item-item graphs for GNNLR better to integrate graph neural networks with propositional logic reasoning. We conduct extensive experiments on five real-world datasets and explore the effects of different graph neural network architectures on GNNLR performance. Extensive experiments verified the effectiveness of the GNNLR model; and showed that different graph neural network architectures have their advantages when facing datasets with different characteristics.

In future work, we will explore and construct more graphs with different perspectives and combine them to enable graph neural networks to further extract rich global implicit information from multiple perspectives. Meanwhile, we will incorporate user information in the logic reasoning module and utilize first-order logic to enhance its flexibility and scalability.

Acknowledgements. This work was supported by the National Natural Science Foundation of China (No. 61906066), Natural Science Foundation of Zhejiang Province (No. LQ18F020002), Zhejiang Provincial Education Department Scientific Research Project(No. Y202044192), Postgraduate Research and Innovation Project of Huzhou University (No. 2022KYCX43).

References

1. Bo, D., Wang, X., Shi, C., Shen, H.: Beyond low-frequency information in graph convolutional networks. In: Proceedings of the AAAI Conference on Artificial Intelligence, vol. 35, pp. 3950–3957 (2021)
2. Chen, H., Li, Y., Shi, S., Liu, S., Zhu, H., Zhang, Y.: Graph collaborative reasoning. In: Proceedings of the Fifteenth ACM International Conference on Web Search and Data Mining, pp. 75–84 (2022)
3. Chen, H., Shi, S., Li, Y., Zhang, Y.: Neural collaborative reasoning. In: Proceedings of the Web Conference 2021, pp. 1516–1527 (2021)
4. Chen, M., Wei, Z., Huang, Z., Ding, B., Li, Y.: Simple and deep graph convolutional networks. In: International Conference on Machine Learning, pp. 1725–1735. PMLR (2020)
5. Defferrard, M., Bresson, X., Vandergheynst, P.: Convolutional neural networks on graphs with fast localized spectral filtering. In: Advances in Neural Information Processing Systems, vol. 29 (2016)
6. He, X., Deng, K., Wang, X., Li, Y., Zhang, Y., Wang, M.: LightGCN: simplifying and powering graph convolution network for recommendation. In: Proceedings of the 43rd International ACM SIGIR Conference on Research and Development in Information Retrieval, pp. 639–648 (2020)
7. He, X., Liao, L., Zhang, H., Nie, L., Hu, X., Chua, T.S.: Neural collaborative filtering. In: Proceedings of the 26th International Conference on World Wide Web, pp. 173–182 (2017)
8. Hidasi, B., Karatzoglou, A.: Recurrent neural networks with top-k gains for session-based recommendations. In: Proceedings of the 27th ACM International Conference on Information and Knowledge Management, pp. 843–852 (2018)
9. Kipf, T.N., Welling, M.: Semi-supervised classification with graph convolutional networks. In: International Conference on Learning Representations (2017). www.openreview.net/forum?id=SJU4ayYgl
10. Li, J., Ren, P., Chen, Z., Ren, Z., Lian, T., Ma, J.: Neural attentive session-based recommendation. In: Proceedings of the 2017 ACM on Conference on Information and Knowledge Management, pp. 1419–1428 (2017)
11. Liao, J., et al.: SocialLGN: light graph convolution network for social recommendation. Inf. Sci. **589**, 595–607 (2022)
12. Liu, Q., Zeng, Y., Mokhosi, R., Zhang, H.: Stamp: short-term attention/memory priority model for session-based recommendation. In: Proceedings of the 24th ACM SIGKDD International Conference on Knowledge Discovery & Data Mining, pp. 1831–1839 (2018)
13. Liu, S., Chen, Z., Liu, H., Hu, X.: User-video co-attention network for personalized micro-video recommendation. In: The World Wide Web Conference, pp. 3020–3026 (2019)
14. Rendle, S., Freudenthaler, C., Gantner, Z., Schmidt-Thieme, L.: BPR: Bayesian personalized ranking from implicit feedback. In: Proceedings of the Twenty-Fifth Conference on Uncertainty in Artificial Intelligence, pp. 452–461 (2009)
15. Shi, S., Chen, H., Ma, W., Mao, J., Zhang, M., Zhang, Y.: Neural logic reasoning. In: Proceedings of the 29th ACM International Conference on Information & Knowledge Management, pp. 1365–1374 (2020)
16. Wang, J., Louca, R., Hu, D., Cellier, C., Caverlee, J., Hong, L.: Time to shop for valentine's day: shopping occasions and sequential recommendation in e-commerce. In: Proceedings of the 13th International Conference on Web Search and Data Mining, pp. 645–653 (2020)

17. Wang, X., He, X., Cao, Y., Liu, M., Chua, T.S.: Kgat: knowledge graph attention network for recommendation. In: Proceedings of the 25th ACM SIGKDD International Conference on Knowledge Discovery & Data Mining, pp. 950–958 (2019)
18. Wang, X., He, X., Wang, M., Feng, F., Chua, T.S.: Neural graph collaborative filtering. In: Proceedings of the 42nd International ACM SIGIR Conference on Research and Development in Information Retrieval, pp. 165–174 (2019)
19. Yang, Y., Huang, C., Xia, L., Li, C.: Knowledge graph contrastive learning for recommendation. In: Proceedings of the 45th International ACM SIGIR Conference on Research and Development in Information Retrieval, pp. 1434–1443 (2022)
20. Zhou, Y., Yang, G., Yan, B., Cai, Y., Zhu, Z.: Point-of-interest recommendation model considering strength of user relationship for location-based social networks. Expert Syst. Appl. **199**, 117147 (2022)

Contrastive Hierarchical Gating Networks for Rating Prediction

Jingwei Ma[1], Jiahui Wen[2(✉)], Chenglong Huang[2], Mingyang Zhong[3], Lu Wang[2], and Guangda Zhang[2]

[1] Shandong Normal University, Shandong, China
[2] Defense Innovation Institute, Beijing, China
`wen_jiahui@outlook.com`, `huangchenglong16@alumni.nudt.edu.cn`
[3] Southwest University, Chongqing, China

Abstract. Review-based recommendations suffer from text noises and the absence of supervised signals. To address those challenges, we propose a novel hierarchical gated sentiment-aware model for rating prediction in this paper. Specifically, to automatically suppress the influence of noisy reviews, we propose a hierarchical gating network to select informative textual signals at different levels of granularity. Specifically, a local gating module is proposed to select reviews with personalized end-to-end differential thresholds. The aim is to gate reviews in a relatively "hard" way to minimize the information flow from noisy reviews while facilitating the model training. A global gating module is employed to evaluate the overall usefulness of the review signals by estimating the uncertainties encoded in the historical reviews. In addition, a discriminative learning module is proposed to supervise the learning of the hierarchical gating network. The essential intuition is to exploit the sentiment consistencies between the target reviews and the target ratings for developing self-supervision signals so that the hierarchical gating network can select relevant reviews related to the target ratings for better prediction. Finally, extensive experiments on public datasets and comparison studies with state-of-the-art baselines have demonstrated the effectiveness of the proposed model, additional investigations also provide a deep insight into the rationale underlying the superiority of the proposed model.

Keywords: rating prediction · hierarchical gating network · discriminative learning module

1 Introduction

Recommendations based on user-item interactions usually face the data-sparseness [1,2] and cold-start [3] problems. Existing works resort to historical reviews [3–5] for boosting performance. The intuitive idea is to exploit rich semantics from the reviews and learn comprehensive user/item representations for better recommendations. Therefore, review-based recommendations [1,6,7] have drawn much more attention in the past decade.

© The Author(s), under exclusive license to Springer Nature Singapore Pte Ltd. 2024
B. Luo et al. (Eds.): ICONIP 2023, LNCS 14449, pp. 424–446, 2024.
https://doi.org/10.1007/978-981-99-8067-3_32

Recent works propose to extract textual features from reviews with deep learning techniques due to their successful applications in text-related tasks [8,9]. Early review-based recommendations [3,10] employ neural networks to independently derive representations over the reviews for the users and items, followed by user-item interactions modeling [11,12] for predicting ratings. Advanced review-based methods [5,7] bring forward the interaction modeling to the feature extraction phase, so that pair-specific text semantics can be captured to boost performance. For example, DeepCoNN [3] employs two convolutional neural networks to infer user/item embeddings, and then interacts the embeddings to predict the final ratings. DAML [7] proposes to learn local and mutual textual features at the review-level. The aim is to bring the user-item interactions modeling forward to the earlier stage, and pay close attention to informative text signals for each specific user-item pair. To address the coarse-grained challenge of the historical reviews, some other works utilize the target reviews to facilitate user-item interaction modeling. Representative works include [13–15].

Recently, graph learning methods [16] have demonstrated their effectiveness in graph representation learning [17]. Since user-item relations in review-based recommendations can be formulated as a user-item bipartite graph, advanced works employ graph learning techniques to sufficiently explore graph structure for better recommendation. For instance, RMG [18] proposes to utilize useful information from both user-item graph and review content, where a hierarchical attention network and a graph neural network are proposed to model reviews and user-item interactions, respectively. The former is to select informative hierarchical review features, while the latter models trilateral relatedness among the users and items. RGCL [19] proposes a graph-based contrastive learning model that explores review information for rating prediction. The reviews are treated as edges during the information propagation process to exploit the unique structure of the user-item graph. In addition, contrastive learning is incorporated to develop supervised signals and enhance the graph learning process based on the reviews.

Despite great progress, the aforementioned works are still facing the following challenges. First, due to the randomness and diversity of user behaviors, the historical reviews always contain texts that are not sentimental consistent with the target rating. Those irrelevant reviews tend to overwhelm the informative signals from a few relevant ones. Even though many works try to approach this challenge in a "soft" way, which learns to pay more attention to some important reviews for better representation, noisy texts still exist in the historical reviews and can potentially jeopardize prediction performance. Second, review-based recommendations aim to make accurate pair-specific prediction by learning to select relevant reviews from history. However, this is not always guaranteed, since no supervised signals are available to tell the relevancies of the reviews. Therefore, selecting informative reviews without review-level supervision signals is difficult for better discriminative user understanding.

To address the aforementioned challenges, in this paper, we proposed a hierarchical gated sentiment-aware rating prediction model (HGS for short), which can effectively uncover useful reviews with self-supervised information for better rating prediction. The proposed HGS model consists of a local gating module, a global gating module, and a discriminative learning module. The local gating module aims to select informative reviews with personalized trainable thresholds, while the global module evaluates the overall usefulness of the textual signals. As for the discriminative learning module, it leverages the sentiment consistencies between the target reviews and the target ratings, and develops self-supervised signals for learning discriminative gating modules. The advantages of HGS are two-fold. First, the hierarchical gating modules learn to select reviews in a relatively "hard" way within an end-to-end framework. Therefore, HGS can learn to suppress noisy reviews for accurate prediction. Second, the discriminative learning module develops supervised signals, which guide the learning of gating functions for accurate review selection. Therefore, our HGS can automatically select relevant reviews for each specific pair, and outperform the existing competitive models based on the mutual enhancement of the three modules.

To conclude, the contributions of this paper are as follows:

- We propose a hierarchical gating network that exploits informative review features at both local and global levels. To the best of our knowledge, this is the first work of exploring reviews denoising in an end-to-end framework for rating prediction.
- We propose a novel discriminative learning process to supervise the learning of the hierarchical gating network. The discriminative learning module can develop self-supervised signals, and improve the effectiveness and efficiency of relevant reviews discovery.
- Extensive comparisons on four real-world datasets demonstrate the effectiveness of our HGS over the state-of-the-art baselines. Additional investigations provide a deep insight of the rationale underlying the superiority of the proposed model.

The remainder of this paper is organized as follows: Section 2 reviews advanced in the related area of review-based recommendation, signals denoising and contrastive learning. Section 3 introduces the problem definition of rating prediction. Section 4 provides a detailed description of the proposed model. Section 5 presents the comparison results of the proposed model with the baselines and also the evaluations of the proposed model in different respects. Finally, Sect. 6 concludes this work.

2   Related Work

In this section, we first review recent works in the research area of review-based recommendations. Then, we review the works involving signals denoising, since we employ a hierarchical gated network to suppress noisy reviews. Finally, as we propose to develop self-supervision signals, we introduce the related works about contrastive learning.

2.1   Review-Based Recommendation

Traditional recommender systems [20] that depend on user-item interactions [21] have shortcomings in data sparseness and code start. To alleviate this problem, recent works propose to exploit textual features from reviews for better user understanding. Primitive works mainly extract textual features with deep learning techniques [22] due to their wide applications in text-related tasks [23]. For example, ConvMF [10] employs Convolutional Neural Networks (CNNs) to explore reviews for learning better item representations, and feeds them into a probabilistic matrix factorization framework for recommendation. DeepCoCNN [3] concatenates the historical reviews into a long document for each user-item pair, and applies two parallel CNNs to model their respective representations. Then, a Factorization Machine [11] is utilized to capture the high-order pairwise interactions for rating prediction. D-attn [24] proposes a dual local and global attention mechanism [25] for modeling latent factors from reviews, and then employs neural networks to extract high-level abstract features. Finally, D-attn interacts the pair-wise features for the final rating prediction. NARRE [4] also proposes an attention mechanism to estimate the usefulness of the reviews, and then formulate the latent features as a weighted sum over the historical reviews for interaction modeling.

In recent years, graph learning techniques [26,27] have enabled much research to discover high-order connectivities among the users/items, and model deep user-item interrelations for recommendations. For instance, GC-MC [28] uses relational graph convolution [29] for learning user/item embeddings, and treats rating prediction as link prediction based on the user-item bipartite graph. RMG [18] combines reviews with user-item graph through a double-view framework to enhance user/item representation learning. In the review-view, a hierarchical framework is used to learn user/item representations from their respective reviews. In the graph-view, a hierarchical graph neural network is used to mine the user-item correlations from the graph. Afterward, representations from different views are concatenated to comprise the final representations, and the inter-product is calculated as the prediction. RGCL [19] proposes to integrate reviews into graph learning. The goal is to exploit unique structure of the bipartite graph with review features. In addition, it incorporates contrastive learning for better node embedding and interaction modeling.

Although effective, for review-concatenate methods such as DeepCoNN and graph-propagation methods such as RGCL, irrelevant reviews can easily introduce noises and affect the model performance. While for attention-based methods such as NARRE, the resultant performance can be inferior without the supervised signals regarding the review usefulness.

2.2   Signals Denoising

Due to the randomness and diversity of user behaviors, not all historical reviews are relevant to predict a user's rating given a target item. Therefore, it is necessary to perform reviews denoising and select credibly relevant reviews to enhance the prediction performance. One line of works proposes to address this challenge in a "soft" way, which tries to implicitly reduce the influence of noisy reviews by learning from training data. For example, NARRE [4] estimates the usefulness of the reviews with an attention mechanism, which learns to pay close attention to the important reviews and assigns low weights to some less important ones. DMAL [7] proposes a hierarchical soft-denoising method, where word importance weights are learned via the self-attention method, while review-level relevancies are learned through a mutual-attention method. Although the soft-attention methods are proven to be effective in distilling useful textual information, the noisy signals still exist and may potentially jeopardise model performance. Related works in the sequence recommendation area [30] proposed to denoise irrelevant items in a "binary" way for a better recommendation. The binary-item-selection method is based on Gumbel Softmax [31–33], making it end-to-end differentiable and facilitate the model training. For instance, CLEA [34] develops a denoising generator that adaptively identifies the relevancies of the items in the basket with respect to the target item, and uses contrastive learning process to supervise the relevance learning. HSD [30] designs a hierarchical sequence denoising, which generates noiseless subsequences by learning two levels of inconsistency signals from input sequences. Although effective, the aforementioned binary-selection methods may mistakenly filter out reviews that turn out to be important for the target prediction.

2.3   Contrastive Learning

Contrastive learning has shown great advancement in many areas, such as computer vision [35] and natural language processing [36]. The essential idea of contrastive learning is to pull an anchor close to a "positive" sample, and push it away from many "negative" samples. As contrastive learning is first explored in graph representation learning [37,38], its early application in the area of recommendation is observed in graph-based recommendations. For instance, SGL [39] generates multiple views of a node, and reinforces self-discriminative node learning by maximizing the node agreement of different views. The self-supervised task is then integrated into the classical supervised task for recommendation. RGCL [19] devises node discrimination and edge discrimination tasks for boosting embedding learning and interaction modeling. The former achieves better node embedding by maximizing node similarities of different sub-graphs, and the latter models better interaction by reinforcing the consistency between pairwise interaction and the target review.

While most of the works focus on developing self-supervised signals in graph-based collaborative filtering, we leverage self-supervised signals to uncover relevant texts in review-based recommendations. Specifically, in this work, we exploit the sentiment consistencies between the target reviews and the target ratings to develop supervised signals. Even though the target reviews are not accessible at the test stage, they are exploited during the training phase to supervise the learning of the hierarchical gating network, so that we can effectively discover the most relevant reviews to facilitate the prediction task.

3   Problem Definition

In a typical rating prediction task, we have a user set $\mathcal{U} = \{u_1, ..., u_M\}$ and an item set $\mathcal{V} = \{v_1, ..., v_N\}$, where $u_i \in \mathcal{U}$ and $v_j \in \mathcal{V}$ indicate the i-th user and j-th item. $\|\mathcal{U}\| = M$ and $\|\mathcal{V}\| = N$ denote the total number of users and items, respectively. The explicit rating records can be represented as a rating matrix $\mathbf{R} = \mathcal{R}^{M \times N}$, where each element r_{ij} is the rating score of u_i to v_j, representing the user's preference over the item. In this work, we focus on explicit rating prediction, hence the rating scores are real values in a specific range (e.g., $\{1, 2, 3, 4, 5\}$). A rating r_{ij} is formally associated with a targeted review d_{ij} that u_i writes about v_j when he/she rates the item. In addition, we denote the user u_i's historical reviews as a set $\mathcal{D}_i = \{d_{i,1}, ..., d_{i,m}\}$, and the item v_j's reviews as $\mathcal{D}_j = \{d_{j,1}, ..., d_{j,n}\}$, where m and n are the respective review number. For simplicity, we leverage a pre-trained model to transform each review into a fix-size feature vector, denoted as $\overrightarrow{\mathcal{D}}_i = \{\mathbf{d}_{i,1}, ..., \mathbf{d}_{i,m}\}$, $\overrightarrow{\mathcal{D}}_j = \{\mathbf{d}_{j,1}, ..., \mathbf{d}_{j,n}\}$.

Given a user u_i and an item v_j, the prediction task is to explore the historical ratings and reviews, and produce the estimated rating $\hat{r}_{ij}$ of u_i over v_j.

4   The Proposed Model

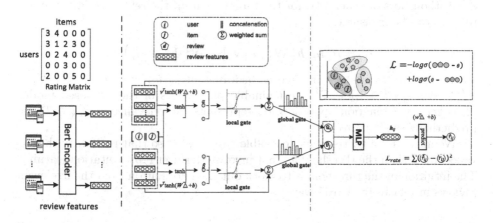

Fig. 1. Overall architecture of the proposed model.

As shown in Fig. 1, the proposed model takes observed ratings and historical reviews as input. For each user-item pair, the respective historical reviews are pre-processed into review vectors. The proposed model consists of two components. In the first component, the review vectors are further processed by a local and a global gating module. Those two gating modules are designed to filter noisy reviews and exploit the most relevant texts for boosting rating prediction. Specifically, the local gating module aims to uncover the most relevant reviews in an end-to-end differential fashion, while the global gating module estimates the overall usefulness of the textual features, and incorporates them for user-item interaction modeling.

In the second component, the self-supervised information in the target reviews is exploited, and the margin between the target reviews and the negative reviews is explicitly enforced. The rationale is to leverage the sentiment consistency between the target reviews and the ratings. The discriminative signals of the target reviews are developed to supervise the learning of the hierarchical gating modules, so that they can effectively discover relevant reviews for the prediction task. Compared with traditional review-based models, the proposed model can discriminatively suppress noisy textual features for better prediction. In the following subsection, each module of the proposed model is introduced.

4.1   Local Gating Module

For a given user-item pair (u_i, v_j), the local gating module is to explore their historical reviews $\mathcal{D}_i = \{d_{i,1}, ..., d_{i,m}\}$, $\mathcal{D}_j = \{d_{j,1}, ..., d_{j,n}\}$, and select the most relevant ones for better user/item representations. Similar to [19,40], we employ BERT-whitening [41] to transform the reviews, $d_{i,s} \in \mathcal{D}_i, d_{j,t} \in \mathcal{D}_j$ into fix-size vectors $\mathbf{d}_{i,s} \in \mathbb{R}^d, \mathbf{d}_{j,t} \in \mathbb{R}^d$. The review vectors are fixed during the training process, as it is shown to be time- and memory-saving without performance degradation.

Without loss of generality, for the target item v_j, the relevance over the u_i's reviews can be measured as:

$$\alpha_s = \boldsymbol{\nu}_1^T tanh\Big(\mathbf{W}_1^T[\mathbf{u}_i||\mathbf{v}_j] + \mathbf{W}_2^T\mathbf{d}_{i,s} + \mathbf{b}_1\Big) \tag{1}$$

where $\mathbf{u}_i \in \mathbb{R}^d, \mathbf{v}_j \in \mathbb{R}^d$ are free embeddings of u_i, v_j, and $\boldsymbol{\nu}_1 \in \mathbb{R}^k, \mathbf{W}_1 \in \mathbb{R}^{d \times k}, \mathbf{W}_2 \in \mathbb{R}^{d \times k}, \mathbf{b}_1 \in \mathbb{R}^k$ are trainable parameters of the proposed model, $[||]$ is the concatenation operation. α_s represents the relevance of $d_{i,s} \in \mathcal{D}_i$ for predicting u_i's rating to v_j.

We introduce a user-specific learnable threshold θ_i so that reviews with relevance lower than the threshold are not involved for user representation learning. The intuition of the threshold is to block out the information from the irrelevant reviews in relatively "hard" way.

$$I(\alpha_s, \theta_i) = \frac{1}{1 + e^{-\lambda(\alpha_s - \theta_i)}} \tag{2}$$

where λ is a scalable hyper-parameter, $I(\alpha_s, \theta_i)$ acts as a gating function when λ is adequately large, as $I(\alpha_s, \theta_i) \approx 1$ if $\alpha_s > \theta_i$, and 0 otherwise. With the gating function, the relevance scores $\{\alpha_s\}_{s=1}^n$ can be normalized with the softmax function, and used for summarizing the review vectors:

$$\alpha_s' = \frac{I(\alpha_s, \theta_i)exp(\alpha_s)}{\sum_{d_{i,s} \in \mathcal{D}_i} I(\alpha_s, \theta_i)exp(\alpha_s)}$$

$$\mathbf{d}_i = \sum_{d_{i,s} \in \mathcal{D}_i} \alpha_s'\mathbf{d}_{i,s} \tag{3}$$

from the equations, we can see that the summarized review vector d_i mainly consists of the vectors of the reviews having relevance scores greater than the threshold. The advantage of the gating function is twofold. First, it is personalized, and can be fit into the data for each individual to obtain better performance. Second, it makes the downstream parameters (e.g., $\mathbf{W}_1, \mathbf{W}_2$) differentiable with respect to the objective function, which can result in an end-to-end learnable model and facilitate the training process. Similarly, the summarized review vector of item v_j can be obtained as follows:

$$\beta_t = \nu_2^T tanh\left(\mathbf{W}_3^T[\mathbf{u}_i||\mathbf{v}_j] + \mathbf{W}_4^T d_{j,t} + \mathbf{b}_2\right)$$

$$I(\beta_t, \theta_j) = \frac{1}{1 + e^{-\lambda(\beta_t - \theta_j)}}$$

$$\beta_t' = \frac{I(\beta_t, \theta_j)exp(\beta_t)}{\sum_{d_{j,t} \in \mathcal{D}_j} I(\beta_t, \theta_j)exp(\beta_t)} \tag{4}$$

$$d_j = \sum_{d_{j,t} \in \mathcal{D}_j} \beta_t' d_{j,t}$$

where $\nu_2, \mathbf{W}_3, \mathbf{W}_4, \mathbf{b}_2$ are model parameters, θ_j is a trainable item-specific parameter for item v_j.

4.2   Global Gating Module

The local gating module evaluates the relevance of each individual review, while the global gating module measures the overall usefulness of the text features for the prediction task. The overall usefulness of the reviews information can be measured by the degree of separation between the relevant reviews (e.g., $\alpha_s > \theta_i$) and irrelevant ones (e.g., $\alpha_s < \theta_i$) [42]. The basic idea is that if it is hard to distinguish between relevant and irrelevant reviews, meaning there are high uncertainties in the historical reviews and they are not reliable for learning user preferences, then the information flow from the historical reviews should be blocked. On the contrary, clear boundary between the relevant and irrelevant reviews denotes that the review features convey small uncertainties, hence the information flow from the historical reviews is encouraged.

To do this, we first calculate the mean relevance scores for relevant and irrelevant reviews, respectively. Again, to make the model differential, the gating function is employed in the calculation:

$$f(\theta_i) = \frac{\sum_{d_{i,s} \in \mathcal{D}_i} I(\alpha_s, \theta_i)\alpha_s}{\sum_{d_{i,s} \in \mathcal{D}_i} I(\alpha_s, \theta_i)}$$

$$g(\theta_i) = \frac{\sum_{d_{i,s} \in \mathcal{D}_i} I(\theta_i, \alpha_s)\alpha_s}{\sum_{d_{i,s} \in \mathcal{D}_i} I(\theta_i, \alpha_s)} \tag{5}$$

where $f(\theta_i)$ is the mean relevant score of the relevant reviews, while $g(\theta_i)$ is that of the irrelevant ones. Notice that, the gating function has the property that $I(\alpha_s, \theta_i) \approx 1$ if $\alpha_s > \theta_i$, and 0 otherwise. Therefore, $\sum_{s=1}^{m} I(\alpha_s, \theta_i)\alpha_s$ is the summation score over the relevant reviews, and $\sum_{s=1}^{m} I(\alpha_s, \theta_i)$ indicates the number of relevant reviews. As for the irrelevant reviews, $I(\theta_i, \alpha_s) \approx 1$ if $\alpha_s < \theta_i$, then $\sum_{s=1}^{m} I(\theta_i, \alpha_s)\alpha_s$ is the summation score over the irrelevant reviews, and $\sum_{s=1}^{m} I(\theta_i, \alpha_s)$ denotes the number of irrelevant reviews.

A non-linear transformation function is employed as the global gating function that controls the information transferred from the historical reviews.

$$\phi(\theta_i) = \sigma\Big(\big(f(\theta_i) - \theta_i\big) * \big(\theta_i - g(\theta_i)\big)\Big) - 0.5$$
$$\hat{\mathbf{u}}_i = \big(1 - \phi(\theta_i)\big) * \mathbf{u}_i + \phi(\theta_i) * \mathbf{d}_i \tag{6}$$

where $\sigma(x) = \frac{1}{1+e^{-x}}$ is the sigmoid function, $\hat{\mathbf{u}}_i$ is the learned user embedding for u_i, and $\phi(x)$ acts as the global gating function exploiting the overall usefulness of the review features, and controls the information flowing from the summarized review vector $\mathbf{d}_i$. The intuition of the global gating function is that if it is hard to differentiate relevant reviews from irrelevant ones, then $\phi(\theta_i)$ is close to 0, $\hat{\mathbf{u}}_i$ is mainly comprised of $\mathbf{u}_i$. On the other extreme, $\mathbf{u}_i$ and $\mathbf{d}_i$ contribute equally to $\hat{\mathbf{u}}_i$. Similarly, the learned item embeddings can be obtained as follows:

$$f(\theta_j) = \frac{\sum_{d_{j,t} \in \mathcal{D}_j} I(\beta_t, \theta_j)\beta_t}{\sum_{d_{j,t} \in \mathcal{D}_j} I(\beta_t, \theta_j)}$$
$$g(\theta_j) = \frac{\sum_{d_{j,t} \in \mathcal{D}_j} I(\theta_j, \beta_t)\beta_t}{\sum_{d_{j,t} \in \mathcal{D}_j} I(\theta_j, \beta_t)} \tag{7}$$
$$\phi(\theta_j) = \sigma\Big(\big(f(\theta_j) - \theta_j\big) * \big(\theta_j - g(\theta_j)\big)\Big) - 0.5$$
$$\hat{\mathbf{v}}_j = \big(1 - \phi(\theta_j)\big) * \mathbf{v}_j + \phi(\theta_j) * \mathbf{d}_j$$

4.3 Discriminative Learning Module

The hierarchical local and global gating modules aim to discover informative reviews to better characterize users and items, and it is expected to provide accurate rating predictions by drawing a more precise boundary between the relevant and irrelevant reviews. However, the margin is not guaranteed due to the absence of supervised signals (i.e., labels denoting relevant reviews) for learning discriminative local/global gating modules. In Eq. (6) and Eq. (7), we only represent users/item with relevant reviews, while the relevancies of the reviews are not explicitly given. Existing works [34] show that the lack of supervised signals can easily result in unstable model performance, unexpectedly causing the model to update other parameters such as free embeddings (e.g., $\mathbf{u}_i, \mathbf{v}_j$).

To address this challenge, we resort to the contrastive learning paradigm [30,43,44]. The goal is to exploit self-supervised signals to complement the local/global gating modules. Our intuition is that the target review d_{ij} that u_i writes about v_j has strong indications of u_i's preference over the item. Even though the historical reviews do not have labels of relevancy, however, the target review d_{ij} usually presents high sentiment alignment with the target rating r_{ij}. Therefore, the target review can be treated as the relevant review and is expected to achieve a higher relevance score than the user-specific threshold. On the contrary, randomly sampled reviews from the training set are regarded as irrelevant reviews, and should result in low relevance scores. This process can be formulated as follows:

$$\alpha_j = \boldsymbol{\nu}_1^T tanh\left(\mathbf{W}_1^T [\mathbf{u}_i || \mathbf{v}_j] + \mathbf{W}_2^T \mathbf{d}_{i,j} + \mathbf{b}_1\right)$$

$$\alpha_j' = \boldsymbol{\nu}_1^T tanh\left(\mathbf{W}_1^T [\mathbf{u}_i || \mathbf{v}_j] + \mathbf{W}_2^T \mathbf{d}_{i,j}' + \mathbf{b}_1\right) \tag{8}$$

$$\mathcal{L}_i = -\left\{ log\sigma(\alpha_j - \theta_i) + log\sigma(\theta_i - \alpha_j') \right\}$$

where $\boldsymbol{\nu}_1, \mathbf{W}_1, \mathbf{W}_2, \mathbf{b}_1, \theta_i$ are the same parameters as those in Eq. (1) and Eq. (2), $\sigma(x) = \frac{1}{1+e^{-x}}$ is the sigmoid function, and $d_{i,j}'$ indicates a randomly sampled review from the training set. Similarly, to learn discriminative item relevance threshold θ_j, we have the following objective function:

$$\beta_i = \boldsymbol{\nu}_2^T tanh\left(\mathbf{W}_3^T [\mathbf{u}_i || \mathbf{v}_j] + \mathbf{W}_4^T \mathbf{d}_{i,j} + \mathbf{b}_2\right)$$

$$\beta_i' = \boldsymbol{\nu}_2^T tanh\left(\mathbf{W}_3^T [\mathbf{u}_i || \mathbf{v}_j] + \mathbf{W}_4^T \mathbf{d}_{i,j}' + \mathbf{b}_2\right) \tag{9}$$

$$\mathcal{L}_j = -\left\{ log\sigma(\beta_i - \theta_j) + log\sigma(\theta_j - \beta_i') \right\}$$

where $\boldsymbol{\nu}_2, \mathbf{W}_3, \mathbf{W}_4, \mathbf{b}_2$ are the same parameters as those in Eq. (4). Note that even though the target review is not accessible in the testing stage, it can be leveraged in the training stage to provide discriminative signals. The rationale underlying this module is to propagate the self-supervised signals from the target reviews, and guide the discriminative learning of the local/global gating modules, so that they can identify informative reviews for the prediction task.

4.4   Model Learning

The defined objective function consists of two parts: rating prediction loss and discriminative learning loss. These two objective functions are calculated dependently, and are combined for join optimization.

Rating Prediction Loss. As shown in Fig. 1, when the user/item embeddings (i.e., $\hat{\mathbf{u}}_i, \hat{\mathbf{v}}_j$) containing information flow from respective historical reviews are obtained, they are fed into a Multi-Layer Perceptron (MLP) to model the user-item interactions. The MLP uses a softmax function as the last layer, where the dimensionality is set to the number of rating class:

$$\mathbf{e}_{ij} = MLP\left([\hat{\mathbf{u}}_i || \hat{\mathbf{v}}_j]\right)$$

$$\hat{r}_{ij} = \mathbf{w}^T \mathbf{e}_{ij} + b \tag{10}$$

where $\mathbf{w}, b$ are trainable parameters, and $\mathbf{e}_{ij}$ is the learned user-item interaction features. $MLP(\cdot)$ denotes two-layer MLP with GELU activation function. After getting predicted rating $\hat{r}_{ij}$, we employ square error between r_{ij} and $\hat{r}_{ij}$ to define the the rating prediction loss:

$$\mathcal{L}_{rate} = \sum_{<u_i,v_j,r_{ij},d_{ij}>\in\mathcal{T}} (r_{ij} - \hat{r}_{ij})^2 \tag{11}$$

where $\mathcal{T}$ is the set of training records.

Discriminative Learning Loss. The discriminative learning loss is defined as the sum of $\mathcal{L}_i$ and $\mathcal{L}_j$ in Eq. (8) and Eq. (9) for each user-item pair:

$$\mathcal{L}_{dis} = \sum_{<u_i,v_j,r_{ij},d_{ij}>\in\mathcal{T}} \left\{ \mathcal{L}_i + \mathcal{L}_j \right\} \tag{12}$$

Finally, these two objective functions are combined for joint optimization:

$$\mathcal{L} = \mathcal{L}_{rate} + \eta\mathcal{L}_{dis} \tag{13}$$

where η is a trade-off hyper-parameter to balance the two objective functions.

5   Experiments

5.1   Experiment Settings

Table 1. Statistics of the four datasets.

Statistics	Musical Instruments	Grocery and Gourmet Food	Video Games	Sports and Outdoors
#users	1,429	14,681	24,303	35,598
#items	900	8,731	10,672	18,357
#records	10,261	151,254	231,577	296,337
#reviews/user[a]	10	10	9	8
#reviews/item[a]	17	22	21	14
density	0.798%	0.118%	0.089%	0.045%

[a] #review/user and #review/item indicate the average number of reviews per user and item, respectively.

Datasets. We evaluate the proposed model on four publicly available datasets, which are from different domains of Amazon 5-core [45, 46][1] : "Musical Instruments", "Grocery and Gourmet Food", "Video Games" and "Sports and Outdoors". The 5-core indicates each user/item has at least 5 interaction records, and hence 5 reviews. Following to [19, 40], we utilize BERT-whitening [41] to generate representation vectors $\mathbf{d}_{ij}$ for each review d_{ij} that u_i commented on v_j, and fixed them during the back-propagation process, which is shown to be time- and memory-saving without performance degradation. Similar to previous works [7, 13, 19], we randomly split each dataset into training, validation and test sets with 8:1:1 ratio. We present the statistics of the four datasets in Table 1.

Setup. Table 2 presents configuration of the hyper-parameters of the proposed model. Specifically, the scalar λ in Eq. (2) is varied amongst [0,2,4,6,8,10] to study the impact of the local gating module, while the tradeoff η is varied in the range of [0,1] to study the influence of the discriminative learning module. Our model is trained over shuffled mini-batches, where the batch size and epoch are set to 16 and 10 for small dataset "Musical Instruments", and 128 and 5 for other large datasets. The objective loss is minimized using Adam [47] with the initial learning rate of 1e-4. The embedding size and latent size are set as $d = 64, k = 64$. The user-item interactions are captured with a 2-layer MLP, where the latent dimensions are set to 64 and 32, respectively. We implement the proposed model

Table 2. Configuration of hyper-parameters.

Configuration	Value(s)	Description
λ	$[0,2,...,10]^a$	scalar λ in Eq. (2)
η	$[0,0.2,...,1.0]^a$	tradeoff η in Eq. (13)
lr	1e-4	learning rate
epoch	$5/10^b$	training epoches
batch size	$16/128^c$	training batch size
embedding size	64	dimensions of $\mathbf{u}_i, \mathbf{v}_j$
latent size	64	k in Eq. (1)
MLP	$[64,32]^d$	mlp in Eq. (10)

^{a} The model performance is studied w.r.t those values.
^{b} The epoch is set to 10 for small dataset "Musical Instruments", and 5 for other large datasets.
^{c} The batch size is set to 16 for small dataset "Musical Instruments", and 128 for other large datasets.
^{d} 64 and 32 indicate the hidden size of the first and second layer of the MLP, respectively.

---

[1] http://snap.stanford.edu/data/amazon/productGraph/categoryFiles/.

with TensorFlow[2] and conduct the experiments on an NVIDIA TITAN graphics card with 24 G memory.

Metric. Following [7], we evaluate the competitive models with Mean Absolute Error (MAE), which is commonly used for rating prediction in recommendation area. To avoid random bias, we repeat each experiment for 10 runs, and report means over the 10 runs for model comparison. However, very minimal differences in performance can be noticed over different runs.

Baselines. To demonstrate the advantage of the proposed model, we select recent-published review-based baselines for comparison. These competitive models are listed as follows:

- PMF [48] is a classical matrix factorization model that predicts ratings with user-item inner product.
- NeuMF [49] combines generalized matrix factorization and multi-layer neural network for estimating ratings.
- DeepCoNN [3] represents users and items with their respective reviews, and interacts them with a factorization machine for the prediction task.
- NARRE [4] boosts DeepCoNN by employing an attention mechanism to exploit the usefulness of the reviews.
- DAML [7] proposes to jointly learn review features by exploiting local and mutual features of the Convolution Neural Network (CNN).
- GC-MC [28] leverages relational graph convolution to represent users/items and formulate recommendations as link prediction tasks on the user-item bipartite graph.
- RMG [18] integrates graph and review information for rating estimation.
- RGCL [19] incorporates graph and contrastive learning for better uncovering graph structure and alleviating user behavior limitations.

5.2 Evaluation

Overall Comparison. Table 3 reports the overall comparison results across the four datasets. From the table, we can observe the following findings.

First, PMF achieves the worst performance on all the datasets. Since it is solely based on user-item inner product to capture the rating behaviors, it suffers from the data sparseness of the user-item interactions in recommendation problems. NeuMF shows a significant improvement over PMF in terms of MAE. The reason can be attributed to the multi-layer neural networks for capturing complex user/item interrelations.

Second, the review-based models (Table 3 (3)–(5)) generally obtain better performance than the latent factor models (i.e., PMF, NeuMF), demonstrating the effectiveness of exploiting review features for modeling user behaviors

---

[2] https://tensorflow.google.cn/api_docs.

Table 3. Model comparison in terms of MAE across the datasets of different methods.

Models	Musical Instruments	Grocery and Gourmet Food	Video Games	Sports and Outdoors
(1)PMF	1.137	1.397	1.395	1.203
(2)NeuMF	0.7198	0.9434	0.8693	0.7516
(3)DeepCoNN	0.7590	0.8016	0.8752	0.7192
(4)NARRE	0.6949	0.7467	0.7991	0.6897
(5)DAML	0.6510	0.7354	0.7881	0.6676
(6)GC-MC	0.6394	0.7400	0.7892	0.6649
(7)RMG	0.6381	0.7322	0.7880	0.6613
(8)RGCL	0.6113	0.7201	0.7734	0.6438[a]
(9)HGS	0.6044	0.7013	0.7437	0.6223
Δ^b	1.1%	2.7%	3.8%	3.3%

[a] The second-best MAE is marked as underlined.
[b] Δ indicates the performance improvement of HGS over the best baseline, a relative improvement over 1% is considered as significant [50,51].

and item attributes. NARRE shows improved MAE over DeepCoNN across the datasets, indicating the benefit of discovering textual information at different levels of granularity. DeepCoNN concatenates all historical reviews into a long document, which can easily introduce text noises and irrelevant information. On the contrary, NARRE employs an attention mechanism to select informative reviews for accurate prediction. DAML yields the best performance among the review-based models, suggesting the superiority of capturing pair-wise correlations between each user-item pair, as it can dynamically reveal complex interactions for rating prediction for each specific user-item pair.

Third, the graph-based models (Table 3 (6)–(8)) outperform the review-based models significantly. One possible explanation is that graph learning can discover high-order connectivities among the user-item bipartite graph, and learn comprehensive representation for boosting the prediction task. RGCL obtains the best performance among the competitive models, supporting the importance of enhancing user-item graph with reviews features and utilizing contrastive learning to provide self-supervised signals.

Finally, the proposed HGS model yields the best performance across the four datasets. Specifically, compared with the best baseline (i.e. RGCL), the relative MAE improvements on the four datasets are 1.1% ("Musical Instruments"), 2.7% ("Grocery and Gourmet Food"), 3.8% ("Video Games") and 3.3% ("Sports and Outdoors"), respectively. The reasons underlying the superiority of our HGS model are two-fold. First of all, the hierarchical gating modules can gate the irrelevant review noises and discover the most useful reviews, so that the noise problem in reviews can be properly addressed and the relevant reviews can be fully exploited for the prediction task. Moreover, we propose a discriminative learning module that can transfer knowledge from the target reviews and provide self-supervised signals, which can help to learn discriminative local gating function and guide the HGS to pay close attention to relevant textual signals.

Module Effectiveness. In this subsection, we investigate the effectiveness of each sub-module of the proposed HGS model. To do this, HGS is compared with the following variants:

- HGS-local is a HGS variant that replaces local gating function with a traditional attention mechanism.
- HGS-global is a HGS variant that replaces global gating function with a native mean pooling operation.
- HGS-dis is a HGS variant that excludes the discriminative learning module.

Table 4. Model comparison of different variants.

Variants	Musical Instruments	Grocery and Gourmet Food	Video Games	Sports and Outdoors
HGS-local	0.6127	0.7165	0.7635	0.6238
HGS-global	0.6080	0.7127	0.7469	0.6258
HGS-dis	0.6082	0.7444	0.7983	0.6672
HGS	0.6044	0.7013	0.7437	0.6223

The comparison results are presented in Table 4. From the table, we have the following conclusions. First, HGS has better performance than all its variants, showing the effectiveness of the sub-modules for filtering noises and exploring supervised signals. Second, HGS-local experiences more performance loss than HGS-global, indicating the importance of performing denoising at the local review level. Third, HGS-dis ranks last in terms of prediction accuracy, validating the effectiveness of transferring knowledge from the target reviews to provide supervised signals and learn discriminative local gating parameters. We can speculate that without supervising signals to tell the review relevancies, the local gating module is not guaranteed to draw informative reviews for prediction tasks, which could easily result in inferior learning of corresponding parameters.

To provide a deep insight into the efficacy of HGS in relevant review selection, we compare it with traditional attention mechanism (ATT for short) set select reviews without self-supervision signals. For comparison, we develop two metrics to measure the performance, namely Mean Relevant Ratio (MRR) and Weighted Mean Relevant Ratio (MRR-W). The former calculates the average ratio of selected relevant reviews to the reviews yielding a relevance score higher than the threshold. MRR-W is MRR weighted by the relevance score. Higher MRR and MRR-W mean more relevant reviews are selected. The metrics MRR and MRR-W are defined as follows:

$$MRR = \underset{<u_i,v_j>\in T'}{mean} \left(\frac{\|\{d_s | d_s \in \mathcal{D}_i \ \& \ \alpha_s > \theta_i \ \& \ r_{d_s} = r_{d_{ij}}\}\|}{\|\{d_s | d_s \in \mathcal{D}_i \ \& \ \alpha_s > \theta_i\}\|} \right)$$

$$MRR - W = \underset{<u_i,v_j>\in T'}{mean} \left(\frac{\sum\{\theta_s | d_s \in \mathcal{D}_i \ \& \ \alpha_s > \theta_i \ \& \ r_{d_s} = r_{d_{ij}}\}}{\sum\{\theta_s | d_s \in \mathcal{D}_i \ \& \ \alpha_s > \theta_i\}} \right) \tag{14}$$

where $\mathcal{T}'$ denotes the test set. $r_{d_s} = r_{d_{ij}}$ indicates that historical review d_s shares the same rating class as the target review d_{ij}. The results are presented in Fig. 2. From the figures, we can observe that HGS consistently achieves better performance than ATT in terms of MRR and MRR-W, demonstrating that HGS can effectively select more relevant reviews than the classical attention mechanisms, underlying the superiority of HGS in exploiting historical reviews for providing accurate rating predictions.

Hyper-parameter Study

Tradeoff η. This hyper-parameter controls the balance between the rating prediction loss and the discriminative learning loss. The experiment results are shown in Fig. 3, where η is varied amongst [0.0, 0.2, 0.4, 0.6, 0.8, 1.0]. From the figure, we can observe that η needs to be set to somewhere between [0,1] to obtain the best performance, namely 0.8 ("Musical Instruments"), 1.0 ("Grocery and Gourmet Food"), 0.8 ("Video Games") and 0.6 ("Sports and Outdoors"), respectively. The inferior performance with $\eta = 0.0$ verifies that the discriminative learning module is essential for discovering the most relevant reviews for the rating prediction. In addition, we can learn from the table that the performance degrades with large η, as in that case, the optimizer allocates more effort to minimize the discriminative learning loss rather than the rating prediction loss, which results in sub-optimal performance.

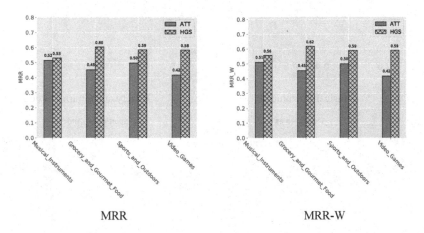

MRR   MRR-W

Fig. 2. MRR and MRR-W comparisons between HGS and ATT over the datasets.

Scalar λ. λ controls the sharpness of the local gating function in Eq. (2), large λ indicates selecting the historical reviews in a "binary" way, while small λ makes the selection in a "soft" way. Figure 4 shows the model performance over the datasets when λ is varied in the range [0, 10]. From the table, we can see that the setting of λ for obtaining the best performance varies for different datasets. Specifically, the optimal configurations are $\lambda = 2.0$ ("Musical Instruments"),

$\lambda = 4.0$ ("Grocery and Gourmet Food"), $\lambda = 10.0$ ("Video Games") and $\lambda = 4.0$ ("Sports and Outdoors"), respectively. Also, we can conclude from the figure that extremely small or large λ incurs performance loss, and one possible reason is that small λ can introduce textual noises and negatively affect the prediction performance, while large λ may mistakenly gate otherwise informative reviews.

Case Study. To provide a deep insight into the efficacy of HGS in discovering relevant reviews for the prediction task, we visualize the relevance scores of the historical reviews when estimating the ratings. To do this, we randomly sample some user-item records from the test set of the "Sports and Outdoors" dataset, and plot the α_s (i.e., Eq. (1)) and θ_i (i.e., Eq. (2)) for the corresponding user and historical reviews, respectively. As shown in Fig. 5, the figure presents the visualizations of three user-item pairs, where the x-axis indicates the historical

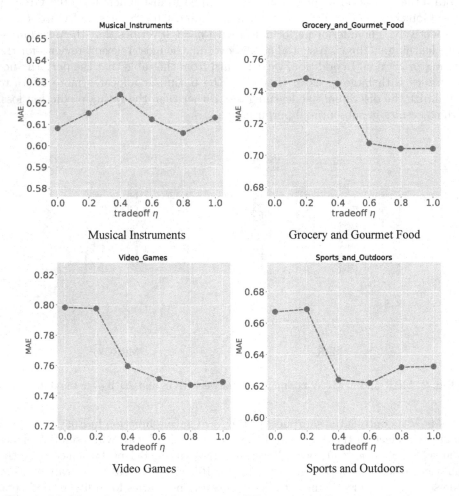

Fig. 3. Model performance with respect to different tradeoff η across the datasets.

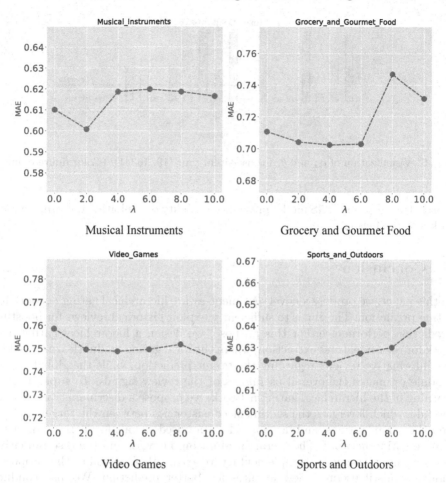

Fig. 4. Model performance with respect to different scalar λ in Eq. (2) across the datasets.

reviews, while the y-axis is the corresponding α_s. We also draw the personalized threshold θ_i to facilitate analysis. The "grey" bars indicate that the corresponding reviews are relevant (i.e., have the sentiment as the target review), while the "white" ones mean irrelevant reviews.

From the figures, we can observe that the relevant reviews generally obtain a relevance score higher than the irrelevant ones. In addition, the relevant reviews tend to be ranked above the threshold, while the irrelevant ones present the opposite trends. The results reveal the rationale underlying the effectiveness of the proposed discriminative learning module, which can transfer supervised signals from the target reviews and learn to draw a precise boundary between the relevant and irrelevant reviews. Together with the hierarchical gating modules that can automatically incorporate informative semantics and suppress irrelevant

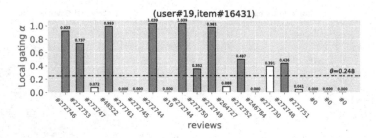

Fig. 5. Visualization of α_s and θ_i for user-item pair (19, 16341). (Color figure online)

noises, the proposed HGS model presents capability of exploiting relevant reviews to boost prediction.

6   Conclusion

In this paper, we propose a novel sentiment-aware hierarchical gating network for rating prediction. The aim is to sufficiently exploit historical reviews for boosting prediction performance. For this purpose, we design a hierarchical gating network for review features extraction, where the local gating module learns to select the informative reviews regarding the target prediction, while the global gating module estimates the overall usefulness of the review signals. To supervise the learning of the hierarchical gating network, we propose a discriminative learning module, which levergaes the sentiment consistencies between the target reviews and the target ratings, and develop self-supervised signals for learning discriminative gating network. The hierarchical gating network and the discriminative learning module can be complementary to each other, enabling the proposed model to distil useful review features for better prediction. We also conduct extensive experiments on public datasets to demonstrate the superiority of HGS over state-of-the-art baselines. Further investigation also provides a deep insight of the rationale underlying the effectiveness of the proposed model.

Acknowledgements. Funding: This work was partially supported by Shandong Provincial Natural Science Foundation (ZR2021QF014), the National Natural Science Foundation of China (62102437), Fundamental Research Funds for the Central Universities (SWU021001), Beijing Nova Program (Z211100002121116, 2021108), Oversea Study and Innovation Foundation of Chongqing (CX2021105). Conflict of Interest: The authors declare that they have no known competing financial interests or personal relationships that could have appeared to influence the work reported in this paper.

References

1. McAuley, J., Leskovec, J.: Hidden factors and hidden topics: understanding rating dimensions with review text. In: Seventh ACM Conference on Recommender Systems, RecSys 2013, Hong Kong, China, 12–16 October 2013, pp. 165–172 (2013)

2. Shi, C., et al.: Deep collaborative filtering with multi-aspect information in heterogeneous networks. IEEE Trans. Knowl. Data Eng. **33**(4), 1413–1425 (2021)
3. Zheng, L., Noroozi, V., Yu, P.S.: Joint deep modeling of users and items using reviews for recommendation. In: Proceedings of the Tenth ACM International Conference on Web Search and Data Mining, WSDM 2017, Cambridge, United Kingdom, 6–10 February 2017, pp. 425–434 (2017)
4. Chen, C., Zhang, M., Liu, Y., Ma, S.: Neural attentional rating regression with review-level explanations. In: Proceedings of the 2018 World Wide Web Conference on World Wide Web, WWW 2018, Lyon, France, 23–27 April 2018, pp. 1583–1592 (2018)
5. Wu, L., Quan, C., Li, C., Wang, Q., Zheng, B., Luo, X.: A context-aware user-item representation learning for item recommendation. ACM Trans. Inf. Syst. **37**(2), 22:1–22:29 (2019)
6. Dong, X., et al.: Asymmetrical hierarchical networks with attentive interactions for interpretable review-based recommendation. In: The Thirty-Fourth AAAI Conference on Artificial Intelligence, AAAI 2020, The Thirty-Second Innovative Applications of Artificial Intelligence Conference, IAAI 2020, The Tenth AAAI Symposium on Educational Advances in Artificial Intelligence, EAAI 2020, New York, NY, USA, 7–12 February 2020, pp. 7667–7674 (2020)
7. Liu, D., Li, J., Du, B., Chang, J., Gao, R.: DAML: dual attention mutual learning between ratings and reviews for item recommendation. In: Proceedings of the 25th ACM SIGKDD International Conference on Knowledge Discovery & Data Mining, KDD 2019, Anchorage, AK, USA, 4–8 August 2019, pp. 344–352 (2019)
8. Le, Q., Mikolov, T.: Distributed representations of sentences and documents. In: Proceedings of the 31th International Conference on Machine Learning, ICML 2014, Beijing, China, 21–26 June 2014, volume 32 of JMLR Workshop and Conference Proceedings, pp. 1188–1196 (2014)
9. Mikolov, T., Sutskever, I., Chen, K., Corrado, G.S., Dean, J.: Distributed representations of words and phrases and their compositionality. In: Advances in Neural Information Processing Systems 26: 27th Annual Conference on Neural Information Processing Systems 2013. Proceedings of a Meeting Held 5–8 December 2013, Lake Tahoe, Nevada, United States, pp. 3111–3119 (2013)
10. Kim, D., Park, C., Oh, J., Lee, S., Yu, H.: Convolutional matrix factorization for document context-aware recommendation. In: Proceedings of the 10th ACM Conference on Recommender Systems, pp. 233–240. ACM (2016)
11. Rendle, S.: Factorization machines. In: ICDM 2010, the 10th IEEE International Conference on Data Mining, Sydney, Australia, 14–17 December 2010, pp. 995–1000 (2010)
12. Rendle, S., Gantner, Z., Freudenthaler, C., Schmidt-Thieme, L.: Fast context-aware recommendations with factorization machines. In: Proceeding of the 34th International ACM SIGIR Conference on Research and Development in Information Retrieval, SIGIR 2011, Beijing, China, 25–29 July 2011, pp. 635–644 (2011)
13. Catherine, R., Cohen, W.: Transnets: learning to transform for recommendation. In: Proceedings of the Eleventh ACM Conference on Recommender Systems, RecSys 2017, Como, Italy, 27–31 August 2017, pp. 288–296 (2017)
14. Chen, X., Zhang, Y., Xu, H., Qin, Z., Zha, H.: Adversarial distillation for efficient recommendation with external knowledge. ACM Trans. Inf. Syst. **37**(1), 12:1-12:28 (2019)
15. Sun, P., Wu, L., Zhang, K., Fu, Y., Hong, R., Wang, M.: Dual learning for explainable recommendation: towards unifying user preference prediction and review gen-

eration. In: Huang, Y., King, I., Liu, T.Y., van Steen, M., (eds.) WWW 2020: The Web Conference 2020, Taipei, Taiwan, 20–24 April 2020, pp. 837–847 (2020)

16. Kipf, T.N., Welling, M.: Semi-supervised classification with graph convolutional networks. In: ICLR (2017)

17. Hamilton, W., Ying, Z., Leskovec, J.: Inductive representation learning on large graphs. In: Proceedings of the 31st International Conference on Neural Information Processing Systems, pp. 1025–1035 (2017)

18. Wu, C., Wu, F., Qi, T., Ge, S., Huang, Y., Xie, X.: Reviews meet graphs: enhancing user and item representations for recommendation with hierarchical attentive graph neural network. In: Proceedings of the 2019 Conference on Empirical Methods in Natural Language Processing and the 9th International Joint Conference on Natural Language Processing, EMNLP-IJCNLP 2019, Hong Kong, China, 3–7 November 2019, pp. 4883–4892 (2019)

19. Shuai, J., et al.: A review-aware graph contrastive learning framework for recommendation. In: The 45th International ACM SIGIR Conference on Research and Development in Information Retrieval, SIGIR 2022, Madrid, Spain, 11–15 July 2022, pp. 1283–1293 (2022)

20. Parvin, H., Moradi, P., Esmaeili, S., Qader, N.N.: A scalable and robust trust-based nonnegative matrix factorization recommender using the alternating direction method. Knowl. Based Syst. **166**, 92–107 (2019)

21. Wang, X., He, X., Wang, M., Feng, F., Chua, T.S.: Neural graph collaborative filtering. In: Proceedings of the 42nd International ACM SIGIR Conference on Research and Development in Information Retrieval, pp. 165–174 (2019)

22. Kim, Y.: Convolutional neural networks for sentence classification. In: Proceedings of the 2014 Conference on Empirical Methods in Natural Language Processing (EMNLP), pp. 1746–1751 (2014)

23. Kalchbrenner, N., Grefenstette, E., Blunsom, P.: A convolutional neural network for modelling sentences. In: Proceedings of the 52nd Annual Meeting of the Association for Computational Linguistics (Volume 1: Long Papers), pp. 655–665 (2014)

24. Seo, S., Huang, J., Yang, H., Liu, Y.: Interpretable convolutional neural networks with dual local and global attention for review rating prediction. In: Proceedings of the Eleventh ACM Conference on Recommender Systems, pp. 297–305. ACM (2017)

25. Yang, Z., Yang, D., Dyer, C., He, X., Smola, A., Hovy, E.: Hierarchical attention networks for document classification. In: Proceedings of the 2016 Conference of the North American Chapter of the Association for Computational Linguistics: Human Language Technologies, pp. 1480–1489 (2016)

26. Wu, L., Sun, P., Fu, Y., Hong, R., Wang, X., Wang, M.: A neural influence diffusion model for social recommendation. In: Proceedings of the 42nd International ACM SIGIR Conference on Research and Development in Information Retrieval, pp. 235–244 (2019)

27. Ying, R., He, R., Chen, K., Eksombatchai, P., Hamilton, W.L., Leskovec, J.: Graph convolutional neural networks for web-scale recommender systems. In: Proceedings of the 24th ACM SIGKDD International Conference on Knowledge Discovery and Data Mining, pp. 974–983 (2018)

28. Berg, R.V.D., Kipf, T.N., Welling, M.; Graph convolutional matrix completion. CoRR (2017)

29. Schlichtkrull, M., Kipf, T.N., Bloem, P., van den Berg, R., Titov, I., Welling, M.: Modeling relational data with graph convolutional networks. In: Gangemi, A., et al. (eds.) ESWC 2018. LNCS, vol. 10843, pp. 593–607. Springer, Cham (2018). https://doi.org/10.1007/978-3-319-93417-4_38

30. Zhang, C., Du, Y., Zhao, X., Han, Q., Chen, R., Li, L.: Hierarchical item inconsistency signal learning for sequence denoising in sequential recommendation. In: Proceedings of the 31st ACM International Conference on Information & Knowledge Management, Atlanta, GA, USA, 17–21 October 2022, pp. 2508–2518 (2022)

31. Wang, Q., Yin, H., Wang, H., Nguyen, Q.V.H., Huang, Z., Cui, L.: Enhancing collaborative filtering with generative augmentation. In: Proceedings of the 25th ACM SIGKDD International Conference on Knowledge Discovery & Data Mining, KDD 2019, Anchorage, AK, USA, 4–8 August 2019, pp. 548–556 (2019)

32. Sixing, W., Li, Y., Zhang, D., Zhou, Y., Zhonghai, W.: Topicka: Generating commonsense knowledge-aware dialogue responses towards the recommended topic fact. In: Proceedings of the Twenty-Ninth International Joint Conference on Artificial Intelligence, IJCAI 2020, pp. 3766–3772 (2020)

33. Yu, J., Gao, M., Yin, H., Li, J., Gao, C., Wang, Q.: Generating reliable friends via adversarial training to improve social recommendation. In: 2019 IEEE International Conference on Data Mining, ICDM 2019, Beijing, China, 8–11 November 2019, pp. 768–777 (2019)

34. Qin, Y., Wang, P., Li, C.: The world is binary: contrastive learning for denoising next basket recommendation. In: The 44th International ACM SIGIR Conference on Research and Development in Information Retrieval, SIGIR 2021, Virtual Event, Canada, 11–15 July 2021, pp. 859–868 (2021)

35. Hjelm, R.D., et al.: Learning deep representations by mutual information estimation and maximization. In: 7th International Conference on Learning Representations, ICLR 2019, New Orleans, LA, USA, 6–9 May 2019 (2019)

36. Alexandridis, G., Tagaris, T., Siolas, G., Stafylopatis, A.: From free-text user reviews to product recommendation using paragraph vectors and matrix factorization. In: Companion of the 2019 World Wide Web Conference, WWW 2019, San Francisco, CA, USA, 13–17 May 2019, pp. 335–343 (2019)

37. You, Y., Chen, T., Sui, Y., Chen, T., Wang, Z., Shen, Y.: Graph contrastive learning with augmentations. In: Advances in Neural Information Processing Systems 33: Annual Conference on Neural Information Processing Systems 2020, NeurIPS 2020, 6–12 December 2020 (2020)

38. You, Y., Chen, T., Wang, Z., Shen, Y.: Bringing your own view: graph contrastive learning without prefabricated data augmentations. In: The Fifteenth ACM International Conference on Web Search and Data Mining, Virtual Event/Tempe, WSDM 2022, AZ, USA, 21–25 February 2022, pp. 1300–1309 (2022)

39. Wu, J., et al.: Self-supervised graph learning for recommendation. In: The 44th International ACM SIGIR Conference on Research and Development in Information Retrieval, SIGIR 2021, Virtual Event, Canada, 11–15 July 2021, pp. 726–735 (2021)

40. Hyun, D., Park, C., Yang, M.C., Song, I., Lee, J.T., Yu, H.: Review sentiment-guided scalable deep recommender system. In: The 41st International ACM SIGIR Conference on Research & Development in Information Retrieval, SIGIR 2018, Ann Arbor, MI, USA, 08–12 July 2018, pp. 965–968 (2018)

41. Su, J., Cao, J., Liu, W., Ou, Y.: Whitening sentence representations for better semantics and faster retrieval. CoRR, abs/2103.15316 (2021)

42. Ma, J., Wen, J., Zhang, P., Zhong, M., Zhang, G., Li, X.: A unified model for recommendation with selective neighborhood modeling. Inf. Process. Manag. **57**(6), 102363 (2020)

43. Chen, Y., Liu, Z., Li, J., McAuley, J., Xiong, C.: Intent contrastive learning for sequential recommendation. In: The ACM Web Conference 2022, WWW 2022, Virtual Event, Lyon, France, 25–29 April 2022, pp. 2172–2182 (2022)

44. Zheng, Y., et al.: Disentangling long and short-term interests for recommendation. In: The ACM Web Conference 2022, WWW 2022, Virtual Event, Lyon, France, 25–29 April 2022, pp. 2256–2267 (2022)

45. Ni, J., Li, J., McAuley, J.: Justifying recommendations using distantly-labeled reviews and fine-grained aspects. In: Proceedings of the 2019 Conference on Empirical Methods in Natural Language Processing and the 9th International Joint Conference on Natural Language Processing, EMNLP-IJCNLP 2019, Hong Kong, China, 3–7 November 2019, pp. 188–197 (2019)

46. He, R., McAuley, J.: Ups and downs: modeling the visual evolution of fashion trends with one-class collaborative filtering. In: Proceedings of the 25th International Conference on World Wide Web, WWW 2016, Montreal, Canada, 11–15 April 2016, pp. 507–517 (2016)

47. Kingma, D.P., Ba, J.: Adam: a method for stochastic optimization. In: 3rd International Conference on Learning Representations, ICLR 2015, San Diego, CA, USA, 7–9 May 2015, Conference Track Proceedings (2015)

48. Mnih, A., Salakhutdinov, R.R.: Probabilistic matrix factorization. In: Advances in Neural Information Processing Systems, vol. 20, Proceedings of the Twenty-First Annual Conference on Neural Information Processing Systems, Vancouver, British Columbia, Canada, 3–6 December 2007, pp. 1257–1264 (2007)

49. He, X., Liao, L., Zhang, H., Nie, L., Hu, X., Chua, T.S.: Neural collaborative filtering. In: Proceedings of the 26th International Conference on World Wide Web, WWW 2017, Perth, Australia, 3–7 April 2017, pp. 173–182 (2017)

50. Li, Z., Cheng, W., Kshetramade, R., Houser, J., Chen, H., Wang, W.: Recommend for a reason: unlocking the power of unsupervised aspect-sentiment co-extraction. In: Findings of the Association for Computational Linguistics: EMNLP 2021, Virtual Event/Punta Cana, Dominican Republic, 16–20 November 2021, pp. 763–778 (2021)

51. Tay, Y., Luu, A.T., Hui, S.C.: Multi-pointer co-attention networks for recommendation. In: Proceedings of the 24th ACM SIGKDD International Conference on Knowledge Discovery & Data Mining, KDD 2018, London, UK, 19–23 August 2018, pp. 2309–2318 (2018)

Interactive Selection Recommendation Based on the Multi-head Attention Graph Neural Network

Shuxi Zhang$^{(\boxtimes)}$, Jianxia Chen, Meihan Yao, Xinyun Wu, Yvfan Ge, and Shu Li

School of Computer Science, Hubei University of Technology, Wuhan, China
1375018832@qq.com

Abstract. The click-through rate prediction of users is a critical task in the recommendation system. As a powerful machine learning method, graph neural networks have been favored by scholars to solve the task recently. However, most graph neural network-based click-through rate prediction models ignore the effectiveness of feature interaction and generally model all feature combinations, even if some are meaningless. Therefore, this paper proposes a Multi-head attention Graph Neural Network with Interactive Selection, named MGNN_IS in short, to capture the complex feature interactions via graph structures. In particular, there are three sub-graphs to be constructed to capture internal information of users and items respectively, and interactive information between users and items, namely the user internal graph, item internal graph, and user-item interaction graph correspondingly. Moreover, the proposed model designs a multi-head attention propagation module for the aggregation with an interactive selection strategy. This module can select the constructed graph and increase diversity with multiple heads to achieve the high-order interaction from the multiple layers. Finally, the proposed model fuses the features, and predicts. Experiments on three public datasets demonstrate that the proposed model outperformed other advanced models.

Keywords: Click-through Rate · Feature Interaction · Internal Graph · Interaction Graph · Multi-head Attention

1 Introduction

The recommendation system aims to cope with the information [16] overload that users may face. Usually, the click behavior has been regarded as a behavior that expresses the users preferences, thus the click-through rate (CTR) prediction is a crucial task of the recommendation system. Traditional recommendation

This work is supported by National Natural Science Foundation of China (Grant No.61902116).

B. Luo et al. (Eds.): ICONIP 2023, LNCS 14449, pp. 447–458, 2024.
https://doi.org/10.1007/978-981-99-8067-3_33

methods mainly include the content-based and collaborative filtering (CF)-based approaches [7,17]. However, there are still some limits of them because of the sparse interaction between users and items [18].

As an effective model to capture the information for the graph data, graph neural networks(GNNs) have achieved a state-of-the-art performance in various tasks such as the semantic segmentation [10], machine translation [1], and recommendation systems [6,17].In particular, GNNs show its great potential in modeling the high-order feature interaction to predict the CTR as well. For example, Fi-GNN [9] utilizes a complete graph to interact with each pair of features. However, there are not beneficial for all feature interactions in a complete graph.

Inspired by the Fi-GNN model, the proposed model constructs a feature responding to a graph node and enables different features to interact each other via edges. However, since not all edge interactions are beneficial, Fi-GNN is not a good choice for modeling the interactions. To overcome this limitation, this paper not only enriches the graph construction with the attribute information but also filters out the helpful feature interactions via a special selection step for the interactions.

CF models are good at obtaining more detailed collaborative information to reveal the similarity between attributes via the feature embedding [13]. Usually, if considering the interactions between different features, the feature embedding can utilizes more useful information to improve the performance of the prediction [11,12]. Recently, the features interactions are proposed in an interpret-able way with attention mechanisms. For example, HoAFM [15] updates the feature representations by aggregating the representations of the co-occurring features. AutoInt [12] first attempts to utilize a multi-head self-attention mechanism to explicitly model feature interactions. GMCF [14] designs a cross-interaction module before the feature interaction on both users and items sides.

Inspired by the GMCF [14], the proposed model interacts within the user side and the item side at the same time. Different from the GMCF, however, the proposed model revises the propagation aggregation of the GNN structure with a multi-head attention, and added multi-layer structure. Thus, the proposed model can learn the higher-order interaction information.

From the above analysis, this paper proposes a Multi-head attention Graph Neural Network with Interactive Selection, named MGNN_IS in short. In particular the proposed MGNN_IS model explicitly aggregates the internal-interactions and the cross-interactions in various ways in the graph structure. In addition, it also proposes a novel multi-layer network, in which each layer generates the higher-order interaction on the existing basis. The main contributions of the paper are described as follows:

(1) Designs a feature interaction model with the multi-head attention mechanism that incorporated the idea of the residual connection.
(2) Calculates an attention score via the feature interaction and the multi-layer perceptron (MLP), in order to select the edges with the highest score in the graph.
(3) Demonstrates the effectiveness and interpretability of proposed model through the experimental results.

2 MGNN_IS Model

The MGNN_IS model mainly consists of four sub-modules: graph construction & feature embedding, interaction selection & propagation aggregation, feature fusion, and prediction. The model architecture is shown in Fig. 1. Symbol definition and each sub-module are described as follows.

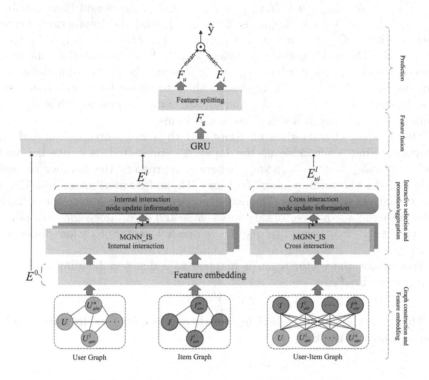

Fig. 1. Overall architecture of MGNN_IS model

2.1 Symbol Definition

The set of users and their attributes is defined as $\mathcal{U} = \{u_1, u_2, \cdots, u_a, u_{attr1}, u_{attr2}, \cdots, u_{attrb}\}$, the set of items and their attributes is presented as $\mathcal{I} = \{i_1, i_2, \cdots, i_c, i_{attr1}, i_{attr2}, \cdots, i_{attrd}\}$, and the set of all nodes is $\mathcal{V} = \mathcal{U} \cup \mathcal{I}$, and $\mathcal{E}$ is presented as the set of relations generated by users, items and their attributes.

Each different user u has multiple attributes u_{attr}, and each different item i has multiple attributes i_{attr}. Since the training data of a recommendation system usually consists of historical interactions between users and items, each pair of (u, i) is utilized to represent them, where $u \in \mathcal{U}$ and $i \in \mathcal{I}$.

The input of the task that this paper deals with is a graph $\mathcal{G}$, which includes the users and its attributes, items and its attributes, and structural-semantic information. The final output result includes a class label $\hat{y}$, which is the label of (u, i), indicating whether u and i interact.

2.2 Graph Construction and Feature Embedding Sub-module

This paper constructs three sub-graphs, including users and their attributes $\mathcal{G}_{uu} = \{(u, u_{attr}, e_{uu}) \mid u \in \mathcal{U}, u_{attr} \in \mathcal{U}, e_{uu} \in \mathcal{E}\}$, items and their attributes $\mathcal{G}_{ii} = \{(i, i_{attr}, e_{ii}) \mid i \in \mathcal{I}, i_{attr} \in \mathcal{I}, e_{ii} \in \mathcal{E}\}$, and the interactions between users and items $\mathcal{G}_{ui} = \{(u, i, u_{attr}, i_{attr}, e_{ui}) \mid u \in \mathcal{U}, i \in \mathcal{I}, u_{attr} \in \mathcal{U}, i_{attr} \in \mathcal{I}, e_{ui} \in \mathcal{E}\}$. Defined respectively, where $e_{uu} \in \mathcal{E}$ represents the relationship between a user and its attribute, $e_{ii} \in \mathcal{E}$ represents the relationship between an item and its attribute, $e_{ui} \in \mathcal{E}$ represents the interaction between a user and an item. It should be noted that $\mathcal{G}_{uu}$ and $\mathcal{G}_{ii}$ are complete graphs, while $\mathcal{G}_{ui}$ is the interconnection between nodes of users and items.

This module characterizes all users and their attributes, items and their attributes. First, each node as the input is represented as a one-hot vector $Node = [node_1, node_2, \cdots, node_z]$, where z represents the number of nodes, Node includes the total number of all user IDs and its attributes, item IDs and its attributes in the datasets. The $node_i$ represents the one-hot vector of the i-th node. Since the one-hot vectors are very sparse and high-dimensional, a trainable matrix $V \in \mathbb{R}^{z \times d}$ is needed to map these one-hot vectors to a low-dimensional latent space.

Specifically, the vector $node_i$ is mapped to a dense embedding $e_i \in \mathbb{R}^d$, as shown in the Eq. (1):

$$e_i = V node_i \tag{1}$$

Therefore, the feature embedding matrix can be composed by feature embedding as shown in the Eq. (2):

$$E^0 = [e_1, e_2, \cdots, e_z] \tag{2}$$

2.3 Interaction Selection and Propagation Aggregation Sub-module

This sub-module adopts a multi-head attention mechanism to perform the message propagation and aggregation. As shown in Fig. 2, this module consists of multiple layers, in which each layer includes a GNNs and an Add & Norm part. Meanwhile, the left side shows the multi-layer structure of the model, and the right side shows the specific calculation of each layer in the multi-layer. In particular, the output $H^{(l)}$ of the GNNs results from the updating of the node features in each layer. The output $H^{(l)'}$ of the Add & Norm is the input of the next layer. The result of feature embedding E^0 is the input in the first layer represented by $\{H^{(0)}, H^{(0)}_{attr1}, \cdots, H^{(0)}_{attrn}\}$. Finally, the results of node feature updates at each layer is concatenated to be the final output $E^l \in \mathbb{R}^{n \times l * d}$.

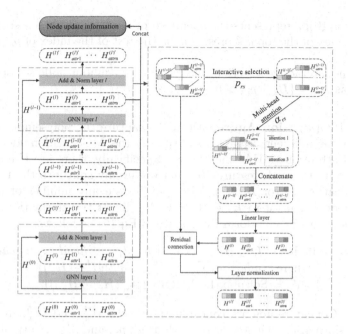

Fig. 2. Structure diagram of the interaction selection and propagation aggregation sub-module.

Interaction Selection Mechanism. Since not all node interactions are beneficial, the MGNN_IS model designs an interaction selection mechanism, in which a MLP with a hidden layer is designed to calculate the weight of the edge between two nodes by the dot product of the node pair, as shown in the Eq. (3):

$$p_{rs} = \sigma(W_2\delta(W_1(H_r \odot H_s) + b_1) + b_2) \tag{3}$$

Where, (H_r, H_s) are the feature vectors of a pair of neighboring nodes; $\odot$ represents the dot product; $W_1 \in \mathbb{R}^{e \times d \times hidden}$ represents the weight of the first linear layer of MLP; $b_1 \in \mathbb{R}^{e \times 1}$ represents the bias of the first linear layer of MLP; δ is the activation function ReLU of the first layer of MLP; $W_2 \in \mathbb{R}^{e \times hidden \times 1}$ represents the weight of the second linear layer of MLP; $b_2 \in \mathbb{R}^{e \times 1}$ represents the bias of the second linear layer of MLP; σ is the activation function Sigmoid; $p_{rs} \in \mathbb{R}^{e \times 1}$ is the result obtained by calculating equation.

After obtaining the attention score p_{rs}, the top k edges are selected and the weights of the other edges are set to 0. The number of k is set as a fixed proportion multiplied by the number of edges in the graph. The calculation process is shown in the Eq. (4):

$$id_k = \mathrm{argtop}_k p_{rs}$$
$$p_{rs}[-id_k] = 0 \tag{4}$$

Where, argtop_k represents the operation of selecting the top k scores of p_{rs}; id_k is the index of the top k scores, $-id_k$ is the rest of the index of p_{rs} excluding id_k.

After the interactive selection, the remaining neighbor node set of nodes H_r is defined as $\mathcal{N}_r = \{H_s \mid p_{rs} > 0, s = 1, 2, \cdots, n_r\}$.

Message Propagation Aggregation. To capture the polysemous of feature interactions in different semantic sub-spaces, the MGNN_IS model adopts a multi-head attention (MHA) mechanism. Specifically, there are H-independent attentions, and the node features H_r that are evenly split into the H parts. To make the feature vector $H_r \in \mathbb{R}^d$ be split by the any number of heads, the proposed model maps it to $H_r \in \mathbb{R}^{H*d}$ with a linear transformation. The split features independently perform the update of Eq. (5) as follows:

$$H_r^o = \text{MultiHead}(H_r \odot H_s) = \text{Concat}[head_1; \ldots; head_h; \ldots; head_H]$$
$$\text{where } head_h = \sigma(\textstyle\sum_{s \in \mathcal{N}_r} \alpha_{rs}^h p_{rs} W_b^h (H_r^h \odot H_s^h)) \qquad (5)$$
$$\alpha_{rs}^h = \text{Softmax}(\text{LeakyReLU}(W_a^h (H_r^h \odot H_s^h)))$$

Where, Concat represents concatenation; both W_a^h and W_b^h are trainable linear transformation matrices at the h-th head; p_{rs} and α_{rs}^h are attention scores calculated by different functions; σ and LeakyReLU are activation functions; H_r^o is the updated node feature.

Moreover, the proposed model links the above features together to obtain the updated feature $H_r^o \in \mathbb{R}^{H*d}$. Afterward, it utilizes another linear transformation to make $H_r^o \in \mathbb{R}^d$ to facilitate subsequent calculations. In the case of multiple layers, the paper performs the addition operation for the output of the current GNN layer and the output of the previous GNN layer, followed by layer normalization, to obtain the result of the Add & Norm sub-module as $H_r^{O'} \in \mathbb{R}^d$. The purpose of Add & Norm is to improve the performance and stability of the network.

2.4 Feature Fusion Sub-module

As shown in Fig. 1, the MGNN_IS module utilizes the RNN to integrate three kinds of the node information. In particular, through the feature embedding module, the MGNN_IS model can obtain the set of all node features E^0 in the graph $\mathcal{G}$. Meanwhile, it can obtain the updated set of all node features E^l of the internal-interaction graphs $\mathcal{G}_{uu}$ and $\mathcal{G}_{ii}$, and the updated set of all node features E_{ui}^l of the cross-interaction graph $\mathcal{G}_{ui}$. Afterward, the node features in E^l and E_{ui}^l are the concatenation of the outputs of each layer of the GNN module.

To make the concatenated features be able to perform subsequent calculations, the MGNN_IS model utilizes a linear layer to map the concatenated dimension to the original dimension size. Moreover, it utilizes a gated recurrent unit (GRU) [3] model to combine the three sets of node features E^0, E^l, and E_{ui}^l, to result in the final set of node features $\mathcal{F}_g$, in which $\mathcal{F}_g = \text{GRU}(E^0, E^l, E_{ui}^l) = \{e_g^* | g \in V\}$.

2.5   Prediction Sub-module

The prediction module divides the nodes into two parts such as the users and items nodes, and corresponding average values of nodes are the feature representation for the users and items respectively. Afterward, the dot product is utilized to calculate whether the user and item interact, that is, to predict $\hat{y}$.

In particular, the MGNN_IS model divides $\mathcal{F}_g$ into the user feature set $\mathcal{F}_u$ and the item feature set $\mathcal{F}_i$. Moreover, to calculate the whole-graph attributes of both the user-graph and the item-graph, the MGNN_IS model utilizes the average values of their respective node sets $\mathcal{F}_u$ and $\mathcal{F}_i$ to capture the user-graph attributes E_u^F and the item-graph attribute E_i^F.

Finally, MGNN_IS model predicts the final value $\hat{y}$ with the sum of the dot products based on the two graph of the user and the item attributes, as shown in the Eq. (6):

$$\hat{y} = \sigma(sum(E_u^F \odot E_i^F)) \tag{6}$$

Where $E_u^F, E_i^F \in \mathbb{R}^{b \times l*d}$, b is the batch size, σ represents the Sigmoid function, and the values in the result $\hat{y}$ ranging from 0 to 1.

Since the task of this paper is the binary classification whether the user is interested in the item or not, the proposed model utilizes the binary cross-entropy loss function (BCELoss) shown in the Eq. (7):

$$\mathcal{L} = -(y \cdot \log \hat{y} + (1 - y) \cdot \log (1 - \hat{y})) \tag{7}$$

Where y is the true label, $\hat{y}$ is the predicted value, and the optimizer utilizes the Adam [8] algorithm.

3   Experiment

3.1   Datasets

The MGNN_IS model was tested on the following three benchmark datasets. Table 1 summarizes the statistical details of these datasets.

- MovieLens 1M [5]: Contains user-movie ratings, and the user attributes and movie attributes.
- Bookcrossing [20]: Contains user-book ratings, and both users and books have attributes.
- AliEC [19]: Displays advertising click-through rate prediction datasets from Taobao.com.

3.2   Parameter Settings

This paper randomly splits each dataset into the training, validation, and test sets at a ratio of 6:2:2. It utilizes three evaluation metrics, namely Area Under the ROC Curve (AUC), Normalized Discounted Cumulative Gain top 5 (NDCG@5), and Normalized Discounted Cumulative Gain top 10 (NDCG@10). The specific hyper-parameter settings are shown in Table 2.

Table 1. Statistical information of the datasets.

Datasets	Interaction	User_ID	Item_ID	User_features	Item_features
MovieLens 1M	1 144 739	6060	3952	30	6049
Bookcrossing	1 050 834	4873	53 168	87	43 157
AliEC	2 599 463	4532	371 760	36	4 344 254

Table 2. Hyper-parameter description.

Symbol	Size	Meaning
batch_size	128	Batch size
d	64	Feature embedding dimension size
epochs	50	Number of training iterations
hidden	64	Number of hidden units in the interaction selection function
lr	1×10^{-3}	Learning rate

3.3   Baseline Model

This paper compares the following baseline models with the MGNN_IS model.

- FM [11]: Computes relevance in a low-dimensional dense space, rather than directly computing the relevance of the input vectors themselves.
- NFM [7]: Combines FM with neural networks to capture multi-order interactions between features.
- W&D [2]: A hybrid model composed of a single-layer Wide part and a multi-layer Deep part with a strong "memory ability" and "generalization ability".
- DeepFM [4]: Utilizes FM to replace the Wide side of W&D to simultaneously learn low-order explicit feature combinations and high-order implicit feature combinations.
- AutoInt [12]: Proposes a multi-head attention mechanism to implement the high-order explicit interactions between features.
- Fi-GNNs [9]: Models features as a complete graph and utilizes gated graph neural networks to model feature interactions.
- GMCF [14]: A graph-based CF method that utilizes both internal and cross interactions.

3.4   Experimental Results and Analysis

Comparison with Baselines. As shown in Table 3, the best-performing model is shown in bold, the second-best model is shown with an underline, and the last row is the relative improvement of the proposed MGNN_IS model compared to the best baseline.

Compared with the best performance of the baseline models, the proposed MGNN_IS model improved the AUC score by 4.86%, the NDCG@5 score by 3.08%, and the NDCG@10 score by 2.77% on the Book-Crossing datasets; the

Table 3. Model performance comparison.

	MovieLens 1M			Book-Crossing			AliEC		
	AUC	*NDCG @5*	*NDCG @10*	*AUC*	*NDCG @5*	*NDCG @10*	*AUC*	*NDCG @5*	*NDCG @10*
FM	0.8761	0.8761	0.8761	0.7417	0.7616	0.8029	0.6171	0.0812	0.1120
NFM	0.8985	0.8486	0.8832	0.7988	0.7989	0.8326	0.6550	0.0997	0.1251
W&D	0.9043	0.8538	0.8538	0.8105	0.8048	0.8381	0.6531	0.0959	0.1242
DeepFM	0.9049	0.8510	0.8848	0.8127	0.8088	0.8400	0.6550	0.0974	0.1243
AutoInt	0.9034	0.8619	0.8931	0.8130	0.8127	0.8472	0.6434	0.0924	0.1206
Fi-GNN	0.9063	0.8705	0.9029	0.8136	0.8094	0.8522	0.6462	0.0986	0.1241
GMCF	0.8998	0.9412	0.9413	0.8255	0.8843	0.8989	0.6566	0.0995	0.1347
MGNN_IS	**0.9091**	**0.9458**	**0.9460**	**0.8656**	**0.9115**	**0.9238**	**0.6635**	**0.1013**	**0.1353**
Improve(%)	0.31	0.49	0.50	4.86	3.076	2.77	1.05	1.81	0.45

AUC score by 0.31%, the NDCG@5 score by 0.49%, and the NDCG@10 score by 0.50% on the MovieLens 1M datasets; and the AUC score by 1.05%, the NDCG@5 score by 1.81%, and the NDCG@10 score by 0.45% on the AliEC datasets. Therefore, it is obvious that the proposed MGNN_IS model improved the performance on all three datasets and achieved the best improvement.

Comparison of Different Numbers of Heads and Layers. As shown in Fig. 3, the comparison with different numbers of heads and layers on datasets, it should be noted that the best performance of the model is not obtained with the same number of heads and layers for different datasets.

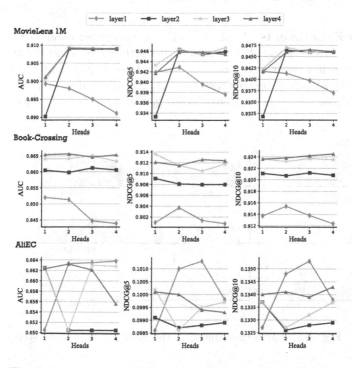

Fig. 3. Comparison of different numbers of heads and layers.

In Fig. 3, the paper conducted experiments on three datasets and utilized line charts to visualize the model performance when the number of heads was 1, 2, 3, and 4 and the number of layers was 1, 2, 3, and 4 respectively. The horizontal axis identifies the number of heads, the vertical axis identifies the scores of three different evaluation indicators, and different line colors indicate different numbers of layers. The legend shows the line colors and their corresponding numbers of layers. It is easy to find that for the MovieLens 1M and Book-Crossing datasets, the model has the best performance when the number of heads is 2 and the number of layers is 4, and after three layers, the number of layers has very little impact on the model performance. However, this situation is not the same for the AliEC datasets. This is because AliEC has a large amount of data that has more interactions between users and items and their attributes, in which the number of layers increasing will make the final features smoother.

In addition, it can be found that the proposed model has best performance while the number of heads is defined as two or three. With the number of heads increasing, however, it does not necessarily improve the performance of the model. The features need to be split evenly before sending them into the multiple heads. This means that each head gets less information as the number of heads increases. The purpose of the multi head attention mechanism is to learn information in multiple semantic sub-spaces. This can increase diversity and make the model more generalizable. By adjusting the number of heads, the model can balance the amount of information obtained by each head and the variation of the generalization performance.

Ablation Experiments. As shown in Fig. 4, this paper conducted an ablation experiment to verify the effectiveness of the interactive selection of the model. The experiments remove the interactive selection step and utilize the optimal number of heads and layers for each dataset: 2 heads and 4 layers for MovieLens 1M, 2 heads and 4 layers for Book-Crossing, and 3 heads and 1 layer for AliEC.

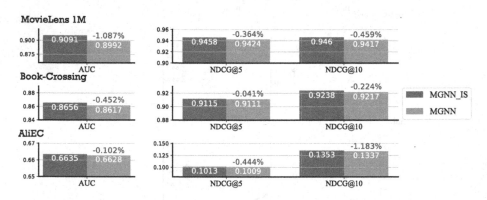

Fig. 4. Ablation experiment on the interactive selection step.

In the Fig. 4, the horizontal axis identifies the evaluation indicators, the green bars are the results of MGNN_IS, and the orange bars are the results of MGNN_IS without the interactive selection step. The white font is the specific value, and the green font on the orange bar is the decrease in the evaluation indicator score after ablating the interactive selection step.

The ablation experimental results demonstrate that interactive selection improves the performance of MGNN_IS model. In addition, it is best for the interactive selection sub-module to be combined with the multi-layer and multi-head attention sub-module together. If only interactive selection modules or multi-layer multi-head attention sub-modules are utilized, the performance is not as good as the separate multi-layer attention sub-module.

4   Conclusion

This paper proposes a novel interactive selection recommendation model named MGNN_IS, which solves the click-through rate prediction problem and improves performance of the generalization and interpretability. In particular, the MGNN_IS model constructs three sub-graphs including the user internal-interaction, item internal-interaction, and user-item cross-interaction. After feature encoding, it utilizes the MHA-based GNN with the proposed interactive selection to propagate and aggregate messages for the internal-interaction and the cross-interaction separately. Moreover, it utilizes the GRU to fuse all features of above interactions. Afterward, the MGNN_IS model divides the nodes into user's and item's nodes and combines their respective information to calculate the features of the user and item graph separately. Finally, the MGNN_IS model utilizes the dot product to predict the final click-through rate.

Compared with the baselines, the experimental results demonstrate that the MGNN_IS model improves the recommendation performance greatly. In addition, the paper also explores the function of multi-head and multi-layer, and verifies the effectiveness of the interactive selection step by the ablation study. In the future work, the paper would like to propose the cross features while reduce the noise information and achieve the personalized cross features at the sample granularity.

References

1. Beck, D., Haffari, G., Cohn, T.: Graph-to-sequence learning using gated graph neural networks. arXiv preprint arXiv:1806.09835 (2018)
2. Cheng, H.T., et al.: Wide & deep learning for recommender systems. In: Proceedings of the 1st Workshop on Deep Learning for Recommender Systems, pp. 7–10 (2016)
3. Cho, K., et al.: Learning phrase representations using RNN encoder-decoder for statistical machine translation. arXiv preprint arXiv:1406.1078 (2014)
4. Guo, H., Tang, R., Ye, Y., Li, Z., He, X.: DeepFM: a factorization-machine based neural network for CTR prediction. arXiv preprint arXiv:1703.04247 (2017)

5. Harper, F.M., Konstan, J.A.: The movielens datasets: history and context. ACM Trans. Interact. Intell. Syst. (TIIS) **5**(4), 1–19 (2015)
6. He, X., Deng, K., Wang, X., Li, Y., Zhang, Y., Wang, M.: LightGCN: simplifying and powering graph convolution network for recommendation. In: Proceedings of the 43rd International ACM SIGIR conference on research and development in Information Retrieval, pp. 639–648 (2020)
7. He, X., Liao, L., Zhang, H., Nie, L., Hu, X., Chua, T.S.: Neural collaborative filtering. In: Proceedings of the 26th International Conference on World Wide Web, pp. 173–182 (2017)
8. Kingma, D.P., Ba, J.: Adam: a method for stochastic optimization. arXiv preprint arXiv:1412.6980 (2014)
9. Li, Z., Cui, Z., Wu, S., Zhang, X., Wang, L.: FI-GNN: modeling feature interactions via graph neural networks for CTR prediction. In: Proceedings of the 28th ACM International Conference on Information and Knowledge Management, pp. 539–548 (2019)
10. Qi, X., Liao, R., Jia, J., Fidler, S., Urtasun, R.: 3D graph neural networks for RGBD semantic segmentation. In: Proceedings of the IEEE International Conference on Computer Vision, pp. 5199–5208 (2017)
11. Rendle, S.: Factorization machines. In: 2010 IEEE International Conference on Data Mining, pp. 995–1000. IEEE (2010)
12. Song, W., et al.: AutoInt: automatic feature interaction learning via self-attentive neural networks. In: Proceedings of the 28th ACM International Conference on Information and Knowledge Management, pp. 1161–1170 (2019)
13. Su, Y., Erfani, S.M., Zhang, R.: MMF: attribute interpretable collaborative filtering. In: 2019 International Joint Conference on Neural Networks (IJCNN), pp. 1–8. IEEE (2019)
14. Su, Y., Zhang, R.M., Erfani, S., Gan, J.: Neural graph matching based collaborative filtering. In: Proceedings of the 44th International ACM SIGIR Conference on Research and Development in Information Retrieval, pp. 849–858 (2021)
15. Tao, Z., Wang, X., He, X., Huang, X., Chua, T.S.: HoAFM: a high-order attentive factorization machine for CTR prediction. Inf. Process. Manag. **57**(6), 102076 (2020)
16. Wang, H., Zhao, M., Xie, X., Li, W., Guo, M.: Knowledge graph convolutional networks for recommender systems. In: The World Wide Web Conference, pp. 3307–3313 (2019)
17. Wang, X., He, X., Wang, M., Feng, F., Chua, T.S.: Neural graph collaborative filtering. In: Proceedings of the 42nd International ACM SIGIR Conference on Research and Development in Information Retrieval, pp. 165–174 (2019)
18. Wang, X., Wang, C.: Recommendation system of e-commerce based on improved collaborative filtering algorithm. In: 2017 8th IEEE International Conference on Software Engineering and Service Science (ICSESS), pp. 332–335. IEEE (2017)
19. Zhou, G., et al.: Deep interest network for click-through rate prediction. In: Proceedings of the 24th ACM SIGKDD International Conference on Knowledge Discovery & Data Mining, pp. 1059–1068 (2018)
20. Ziegler, C.N., McNee, S.M., Konstan, J.A., Lausen, G.: Improving recommendation lists through topic diversification. In: Proceedings of the 14th International Conference on World Wide Web, pp. 22–32 (2005)

CM-TCN: Channel-Aware Multi-scale Temporal Convolutional Networks for Speech Emotion Recognition

Tianqi Wu, Liejun Wang[✉], and Jiang Zhang

Xinjiang Key Laboratory of Signal Detection and Processing, College of Information Science and Engineering, Xinjiang University, Urumqi 830046, China
wljxju@xju.edu.cn

Abstract. Speech emotion recognition (SER) plays a crucial role in understanding user intent and improving human-computer interaction (HCI). Currently, the most widely used and effective methods are based on deep learning. In the existing research, the temporal information becomes more and more important in SER. Although some advanced deep learning methods can achieve good results, such as convolutional neural networks (CNN) and attention module, they often ignore the temporal information in speech, which can lead to insufficient representation and low classification accuracy. In order to make full use of temporal features, we proposed channel-aware multi-scale temporal convolutional networks (CM-TCN). Firstly, channel-aware temporal convolutional networks (CATCN) is used as the basic structure to extract multi-scale temporal features combining channel information. Then, global feature attention (GFA) captures the global information at different time scales and enhances the important information. Finally, we use the adaptive fusion module (AFM) to establish the overall dependency of different network layers and fuse features. We conduct extensive experiments on six dataset, and the experimental results demonstrate the superior performance of CM-TCN.

Keywords: Speech emotion recognition · Channel-aware · Temporal convolutional networks · Global attention

1 Introduction

With the development of human-computer interaction (HCI) technology in recent years, speech emotion recognition (SER) has become an increasingly important research area. In the process of HCI, recognizing emotion in speech signal can better understand the user's intention and can enhance the user interaction experience. Usually, the model in SER consists of two parts: feature extractor and classifier. The most challenging part is how to extract the

---

This research was funded by the Scientific and technological innovation 2030 major project under Grant 2022ZD0115800, Xinjiang Uygur Autonomous Region Tianshan Excellence Project under Grant 2022TSYCLJ0036.

key emotion features from the speech. Before deep learning was widely used, most researchers used hand-extracted features and traditional machine learning methods in SER [1,2]. However, hand-extracted features often contain information that is not related to emotion, so it is immature to pass hand-extracted features directly to classifiers. Deep learning is using an end-to-end approach for emotion classification, and commonly used model architectures include convolutional neural networks (CNN), long short-term memory (LSTM), and attention modules. It is well known that temporal information is one of the important features in SER. Some existing studies have enhanced the temporal emotion representation by using Gate recurrent unit (GRU) and LSTM, but there is the problem that it cannot be computed in parallel and the results are not good compared with the latest networks. Another part of TCN-based SER methods ignore the correlation of temporal features and the global information, so it leads to low classification accuracy. In addition, the existing methods treat feature channels equally and do not consider the correlation between channels, which also limits the feature representation. In general, the existing methods have the following problems: (1) The model has room for improvement in extracting time-dependent features. (2) CNN performs the same processing for different channels, which hinders the expression of the emotion information of the different channels. (3) The existing methods do not consider the correlation of features among layers. To solve these problems, we proposed channel-aware multi-scale temporal convolutional networks (CM-TCN) that the channel-aware temporal convolutional network (CATCN) is chosen as the basic architecture, and using global feature attention (GFA) to establish the dependence of multi-scale temporal features. Finally, the adaptive fusion module (AFM) is used to fuse the features of different network layers for emotion classification. In summary, our main contributions are as follows:

(1) CM-TCN, a speech emotion recognition network fusing the multi-scale temporal features of different network layers, is proposed to enhance the emotion representation and improve the classification accuracy.
(2) Channel-aware temporal convolutional block (CATCB) is designed. Deep features are learned by connecting multiple blocks to form a network and retaining multi-scale temporal information during feature extraction. In addition, channel-aware mechanism can interact with the emotion information in different channels and adaptively adjust weights by information importance.
(3) Global feature attention (GFA) is introduced to obtain global temporal information by modeling the correlation of multi-scale features.
(4) Extensive experiments on six dataset in SER task show that our model yields the best performance compared to other networks that have performed temporal modeling.

The rest of the paper is structured as follows, with Sect. 2 describing the related work. Section 3 describes the network structure of the proposed model. Section 4 evaluates the performance of the model through experiments and analysis. Finally, Sect. 5 summarizes the conclusions.

2   Related Work

In recent years, the widespread use of deep learning has advanced the research in SER. In 2014, Han et al. proposed the first end-to-end SER model based on deep learning [3]. In the same year, Huang et al. proposed the use of CNN for extracting emotion features [4]. Lorenzo Tarantino et al. combined the self-attention mechanism with CNN [5]. Xu et al. proposed a multi-head fusion of attention and area attention to improve SER accuracy by improving existing attention mechanisms [6,7]. Zhang et al. designed a model combining multiple deep network consisting of 1-D, 2-D, and 3-D CNN, which provides more abundant emotion representation for the classifier by combining multi-scale features in speech [8]. Due to the strong scalability of CNN, researchers often design models using CNN structures in combination with other network modules. There has been a lot of work in the SER field to improve CNN. These techniques combine different network layers to reduce the reliance on manual features and to improve the classification accuracy in SER.

However, CNN cannot effectively model the temporal dependence in speech [9]. Therefore, many studies have designed deep learning models in SER with the ability to handle sequential data. The main popular methods based on temporal information are recurrent neural network(RNN), LSTM, GRU and temporal convolutional network(TCN). For example, Murugan proposed the use of RNN to maintain the temporality of speech signals [10]. Xie et al. proposed LSTM based on attention to extract the feature by dot product to capture dependencies from each frame [11]. Su et al. proposed a graph attention-gated recurrent unit (GA-GRU) to handle SER task [12], and Lin et al. combined gated mechanism and LSTM in a flexible way to preserve temporal information of sentences [13]. Wang et al. proposed a two-stage LSTM to exploit temporal information from different time frequency resolutions [14]. Zhong et al. used CNN with bi-GRU and focal loss to learn integrated spatio-temporal representation [15]. Rajamani et al. proposed an attention-based ReLU in GRU to capture remote interactions between features [16]. Zhao et al. made full use of CNN and bidirectional LSTM to learn spatio-temporal representation [17]. Kwon et al. proposed a model based on 1-D CNN with a multi-learning strategy for learning spatial and temporal features in parallel [18].

In addition, TCN is a new CNN-based model for sequence analysis, which is also used to maintain the temporal information in speech. It offers the ability of massively parallel processing with low training cost because it does not process data sequentially like RNN [19], thus avoiding high training costs [20]. In contrast to other CNN structures,TCN structures can be extended with arbitrary lengths of receptive domains and processing sequences through the use of dilated causal convolution. Although TCN have been widely used for time-series tasks such as speech separation and other domains [21], few studies have explored the effectiveness of TCN in SER task. Moreover, most studies have demonstrated that TCN adding attention modules can establish internal dependence information of the input [22,23]. Meanwhile, few studies have focused on the dependencies at different time scales and global information between different network layers, which leads to the poor feature representation ability of existing research methods. To address the above problems, we design CM-TCN. Our proposed networks architecture will be described in detail in the next section.

3   Methods

In this section, we introduce the proposed CM-TCN. As shown in Fig. 1, the model framework consists of four components: multiple sequential channel-aware temporal convolution blocks (CATCB) form channel-aware temporal convolution network (CATCN) to extract depth features, global feature attention (GFA) for global information aggregation, adaptive fusion module (AFM) to fuse features at network layers, and classifier consisting of a fully connected layer and softmax function for emotion classification. **Algorithm Description** illustrates the process.

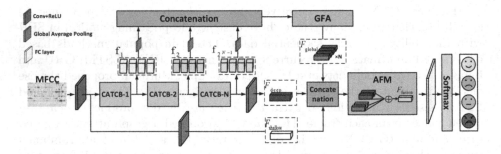

Fig. 1. Model overview

3.1   Input Pipeline

The current methods of multi-modal features in SER are becoming increasingly popular, but we believe that emotion information can be extracted from a single type of feature using efficient feature extraction methods to achieve higher recognition accuracy, and that for some scenarios, unimodal emotion recognition is still the simplest and most effective method. Therefore, only the most commonly used Mel frequency cepstrum coefficients (MFCC) are used as the input to the model in our study [24]. The MFCC extraction process is shown in Fig. 2.

We first set the sampling rate according to different dataset, and perform the frame-splitting operation and Hamming window on the raw speech data with each frame length of 50 ms and displacement of 12.5 ms. Subsequently, the spectrum is obtained by performing a 2048-point fast Fourier transform on the speech signal of each frame. Then it passes through a set of Meier-scale triangular filter banks. Finally, a discrete cosine transform is applied to each frame to obtain the 39-dimensional MFCC features.

Algorithm Description
Input:MFCC
Output:Predictions for the emotion classifications
/*Feature extraction steps*/
1. For the input MFCC, shallow features $F_{shallow}$ are obtained by a 1-D dilated causal convolution operation.
2. Pass through N CATCBs in turn, each block outputs middle layer features f_j (j=1,2,4 ...2^{N-1})and j is the expansion rate in the CATCB.
3. Global average pooling and concatenate the middle layer features f_j, followed by GFA to establish multi-timescale correlations to obtain global temporal features F_{global}.
4. Get the output deep features F_{deep} of the Nth CATCB.
5. Global average pooling for $F_{shallow}$ and F_{deep}.
/*Fusion step*/
6. Concatenate $F_{shallow}$, F_{global} and F_{deep}, AFM assigns weights according to feature importance, and finally performs F_{fusion} based on the weighting results.
/*Classifier step*/
7. Pass F_{fusion} through the fully connected layer and output predictions for emotion classification through softmax classifier.

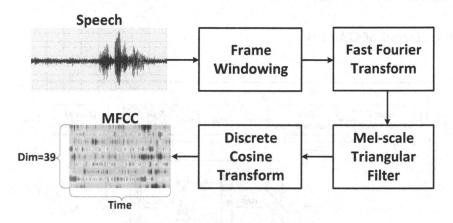

Fig. 2. MFCC feature extraction

3.2   Deep Feature Extraction

Compared with CNN, TCN mainly consist of two components: causal convolution and dilated convolution. Causal convolution is a network structure with strict temporal constraints and thus no access to future information. The dilation convolution allows the network to have a larger perceptual field to build stronger dependencies. Our proposed channel-aware temporal convolutional network (CATCN) is the backbone part. It consists of a 1-D dilation causal convolutional layer and a stack of N dilation CATCBs growing exponentially. The CATCN accepts as input two-dimensional features of size $T \times F$, where T denotes time dimension and F denotes feature dimension(i.e., channels). The first layer of the network performs 1-D dilated causal convolution with a kernel size of 1 and a number of channels of 39. The output of the first layer is then fed to successive CATCBs. The output $F^{(j)}$ of each CATCB is concatenated for global feature attention(GFA), and the obtained F_{global} are fed to the fusion module as global features. In the feature extraction stage, a nonlinear mapping relationship between the sequence of inputs and the hidden states of recorded historical information is established by overlaying multiple CATCBs, and then the learned features are spliced to obtain a representation of multi-scale temporal features. CM-TCN not only captures the variations between speech through long time memory, but also extracts more complementary emotion information from different time scales.

3.3   Detail of CATCB

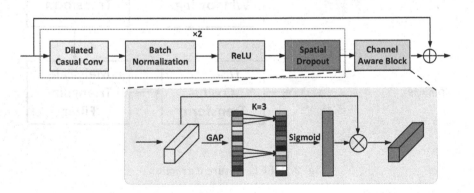

Fig. 3. Channel-Aware Temporal Convolution Block (CATCB)

We designed the CATCB as a residual block based on the TCN structure. As shown in Fig. 3, each CATCB contains two temporal feature extraction units and one channel-aware module, and is connected by a residual structure. The feature extraction units consist of dilated causal convolution, batch normalization, ReLU function, and spatial dropput. It is known that different speech segments and features contain different emotion information, and irrelevant feature information introduces unnecessary redundant noise. Current models using TCN to assume that different channels contribute equally. Therefore, we propose to introduce a channel-aware module in the temporal convolution block for adaptively calculating contribution of different channels. Influenced by the successful application of channel attention mechanism in computer vision and considering the channel size of features in the network [25,26], we use channel-aware module for obtaining the importance of each feature channel, increasing the weight of useful information contribution and suppressing features that are not useful for emotion classification. Features are extracted from the units by temporal features with dimensionality $T{\times}C$.

First, the features χ are compressed in the time dimension using global average pooling, which transforms the time dimensional feature channels into a $1{\times}C$ vector. This vector has a global perceptual field. It represents the global distribution of responses on the feature channels. To adjust the weights of each channel feature, 1-D convolution is used in the network for the interaction of cross-channel features. According to the channel dimension size we set the convolution kernel size as 3, where the size of the convolution kernel indicates the range of cross-channel interaction. After that, the result of convolution is activated by sigmoid function σ to obtain the weights of each channel ω.The specific operation is shown:

$$\omega = \sigma(C1D_3(Gap(\chi))) \qquad (1)$$

Then, we multiply the channel weights with the feature vector and output the high-level features η :

$$\eta = \chi \otimes \omega \qquad (2)$$

The channel-aware module has less complexity and comparable performance compared to SE-net [25]. This is because channel features perform cross-channel interaction not by dimensionality reduction, which reduces the loss of information. The channel attention allows the model to learn the weight coefficients of each channel, enabling the model to better distinguish the features of each channel. Finally, we integrate the features of the previous time scale and the output of the current time scale through the residual structure to obtain high-level features of the current time scale. Overall, we use the residual structure stacking CATCBs not only from being able to learn the deep representation of features, but also to automatically enhance the feature representation using different time scales and channel correlations.

3.4   Global Feature Extraction by GFA

However, CATCN utilizes time-dependent information and extracts features at different time scales, but it cannot establish the dependencies between different time scales and loses some global information. For this reason, we designed GFA to extract the relationships between features at different time scales. To prevent overfitting, the output f_j of CATCB is converted into a $1{\times}C$ vector using global

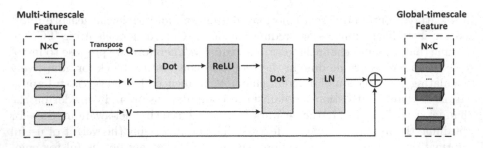

Fig. 4. Global Feature Attention (GFA)

average pooling (GAP) before inputting the GFA, and subsequently cacatenate with N f_j and shape into a two-dimensional vector F. The structure of the GFA is shown in Fig. 4, where the input is a multi-scale temporal feature of $N \times C$, where N is the number of CATCBs and C is the number of channels. The correlation matrix $W \in R^{N \times N}$ between the different layers is calculated as follows:

$$W_{m,n} = ReLU(F \times F^T), (i, j = 1, 2 ... N) \tag{3}$$

F^T denotes the transpose matrix, and (m, n) in the matrix denotes the correlation weights between the mth and n^{th} CATCBs. The global feature F_{global} are subsequently obtained by multiplying F with the weight matrix $W_{m,n}$ and performing residual concatenation:

$$F_{global} = LN(W_{m,n} \otimes F) + F \tag{4}$$

It is worth noting in Eq. (3) that the activation function for common attention mechanism is softmax, but the function produces a dense distribution and gives little attention to irrelevant features, which complicates the analysis of emotion information. GFA uses ReLU as the activation function for attention based on the fact that ReLU removes all query contextually less relevant negative scores, automatically ensuring the sparsity of attention weights $W_{m,n}$. Its output can be any non-negative value and more flexible. The application of the ReLU function automatically learns sparse attention in a flexible and efficient way. Using GFA, the network can assign attention weights according to the importance of different time scales, thus improving the representation of emotion features. In addition, we use layer normalization(LN) to eliminate the problem of gradient instability due to different time scales and variances of the ReLU output.

3.5   Feature Fusion with AFM

In order to address the different pronunciation habits of different speakers, so the model has to achieve the dynamic fusion of multi-scale temporal features, so we proposed AFM fuses the features. The fused features consist of three main components: shallow features $F_{shallow}$ obtained by extracting the original features, deep features F_{deep} extracted by CATCBs stacking, and global features F_{global}

obtained by GFA. First, the features of the three branches are concatenated by pooling them through GAP, followed by feature fusion in AFM, which assigns learnable weights w_i to different features during training, and the final add operation, which weighted fusion of the features output from multiple branches.

$$F_{\text{fusion}} = \sum_{i=1}^{n} w_i F_i \tag{5}$$

The fused vectors have abundant emotion representation and can be fed directly into the fully connected layer classification with softmax function.

4 Experiment

4.1 Experimental Setup

Dataset: Our experiments were conducted on six SER dataset. The details of the dataset are shown in Table 1. The **Speakers** indicates the number of males and females. The **Number** is the number of sentences in the dataset. The **Emotion** shows the number of emotion type in the dataset. The **Frequency** indicates the sampling rate of the dataset.

Table 1. The detailed information of speech emotion dataset.

Copus	Language	Speakers	Number	Emotion	Frequency
CASIA [27]	Chinese	2/2	1200	6	22.1
EMODB [28]	German	5/5	535	7	16.0
EMOVO [29]	Italian	3/3	588	7	48.0
IEMOCAP [30]	English	5/5	5531	4	48.0
RAVDESS [31]	English	12/12	1440	8	48.0
SAVEE [32]	English	4/0	480	7	44.1

Implementation Details: We performed a 10-fold cross-validation in line with previous studies. In our experiments, we extracted 39-dimensional MFCC using the Librosa toolkit. The training process used cross-entropy loss function and an Adam optimizer with batchsize of 64 and an initial learning rate of 0.001. To avoid overfitting during the training phase, we used label smoothing and set the spatial dropout to 0.1.

Evaluation Metrics: Due to class imbalance, we evaluate the performance of the SER using two widely used metrics, weighted average recall (WAR) (i.e., accuracy) and unweighted average recall (UAR). WAR uses class probabilities to balance the recall metrics across classes, while UAR treats each class equally. These two metrics are defined as follows:

$$WAR = \sum_{k=1}^{K} \frac{M_k}{N} \times \frac{\sum_{i=1}^{M_k} TP_{ki}}{\sum_{i=1}^{M_k}(TP_{ki} + FN_{ki})} \tag{6}$$

$$UAR = \frac{1}{K} \sum_{k=1}^{K} \frac{\sum_{i=1}^{M_k} TP_{ki}}{\sum_{i=1}^{M_k}(TP_{ki} + FN_{ki})} \tag{7}$$

where K, M_k and N represent the number of categories, the number of voices in category k and the number of all voices in the dataset, respectively. TP_{ki}, TN_{ki} and FN_{ki} represent the correct positive, correct negative, and incorrect negative for category k of speech i, respectively.

4.2   Experimental Results

In order to verify the effectiveness of the proposed model, we conducted relevant experiments. We verified the validity of the proposed model on the Chinese dataset CASIA, the German dataset EMODB, the Italian dataset EMOVO, the English dataset IEMOCAP, RAVDESS, and SAVEE, respectively. In addition, we compared with recent state-of-the-art models. The experimental results are shown in Table 2, where our proposed model performs better in most cases, obtaining the highest accuracy on six dataset.

More recent studies have paid attention to temporal emotion information in MFCC, such as CNN+Bi-GRU to model temporal information with bi-GRU, respectively. GM-TCN used a TCN-based approach on CASIA, EMODB, RAVDESS, and SAVEE dataset to obtain an average WAR of 88.20% and an average UAR of 88.58% [35]. Our work uses the same MFCC extraction method and experimental setup as theirs. The average WAR and UAR of our model on the four dataset improved by 3.29% and 2.07%, respectively, to 91.64% and 90.78%. Meanwhile, the WAR and UAR of our model are higher than the Convolution-Pooling Attention CapsNet (CPAC) on the four SER dataset [36]. This is because our model not only retains temporal information that facilitates emotion classification, but also establishes dependencies on temporal information at different time scales. Figure 6 shows the confusion matrix on six dataset, and the results demonstrate that CM-TCN can achieve the best overall performance and better balanced results (Fig. 5).

Table 2. WAR and UAR comparison against state-of-the-art SER models.

Copus	Model	Year	WAR	UAR
CASIA	DT-SVM [33]	2019	85.08	85.08
	TLFMRF [34]	2020	85.83	85.83
	GM-TCN [35]	2022	90.17	90.17
	CPAC [36]	2022	92.75	92.75
	Ours	2023	93.75	93.59
EMODB	TSP+INCA [37]	2021	90.09	89.47
	GM-TCN [35]	2022	91.39	90.48
	Light-SERNet [38]	2022	94.21	94.15
	CPAC [36]	2022	94.95	94.32
	Ours	2023	95.15	94.68
EMOVO	RM+CNN [39]	2021	68.93	68.93
	SVM [40]	2021	73.30	73.30
	TSP+INCA [37]	2021	79.08	79.08
	CPAC [36]	2022	85.40	85.40
	Ours	2023	91.15	90.91
IEMOCAP	CNN+Bi-GRU [15]	2020	70.39	71.75
	TF-GCN [41]	2021	71.88	69.60
	Hierarchical-Net [42]	2021	70.50	72.50
	Light-SERNet [38]	2022	70.23	70.76
	Ours	2023	71.49	71.79
RAVDESS	INCA+TS-CNN [43]	2021	85.00	–
	TSP+INCA [37]	2022	87.43	87.43
	GM-TCN [35]	2022	87.35	87.64
	CPAC [36]	2022	90.83	88.41
	Ours	2023	90.76	89.99
SAVEE	3D CNN [44]	2019	81.05	–
	TSP+INCA [37]	2021	84.79	83.38
	GM-TCN [35]	2022	83.88	86.02
	CPAC [36]	2022	85.63	83.69
	Ours	2023	86.88	84.84

4.3 Ablation Experiments

We conducted ablation experiments on six dataset. In this section, we conduct ablation experiments from three different components, including the effectiveness of CATCB, GFA, and AFM. Also, we do ablation experiments on the activation function in GFA. In this section, we use the weighted average recall (WAR) (i.e., accuracy) as a measure of model performance.

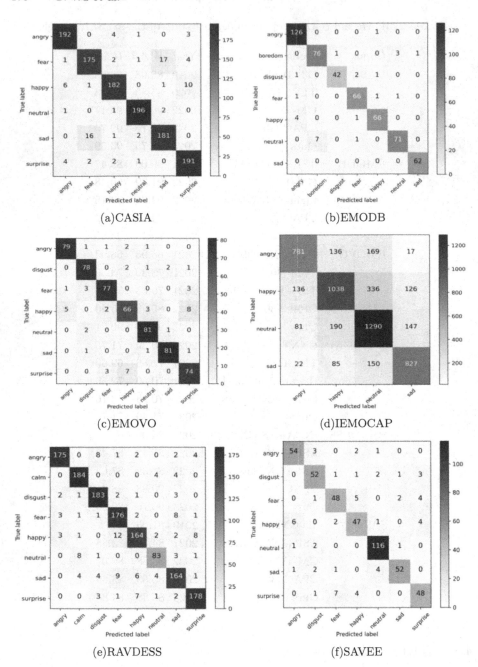

Fig. 5. The 10-fold CV confusion matrix obtained on various dataset.

Table 3. The performance of ablation studies and our model under 10-fold CV on all six dataset. The 'w/o' means removing the component from model.

Copus	TCN	CATCN	w/o GFA	w/o AFM	CM-TCN
CASIA	85.07	87.82	92.17	91.33	93.75
EMODB	88.41	91.26	92.33	90.46	95.15
EMOVO	81.33	83.37	89.29	85.54	91.15
IEMOCAP	60.29	62.89	70.42	70.11	71.18
RAVDESS	81.36	84.74	89.10	86.74	90.76
SAVEE	77.16	80.62	85.00	84.58	86.88
Average	78.94	81.78	86.39	84.79	88.15

Effectiveness of CATCN: In the experiments, our model learns to extract deep features by stacking CATCBs. Channel-aware module can focus important channel information in features, increase the weight of useful information contribution, and reduce useless redundant information. The experimental results demonstrate that the powerful feature extraction capability of the network can lead to better performance. We first used TCN as a baseline to achieve an average accuracy of 78.94% on six dataset. After that, we used the CATCN for experiments on six dataset and achieved an average accuracy of 81.78%. Compared to TCN without channel attention, our method improves the average accuracy on six dataset by 2.84%.

Effectiveness of GFA: Enhancement of multi-scale temporal feature representation by connecting multiple CATCBs, but inability to model different time-scale dependent relationships. The GFA adaptively extracts global feature information and adjusts the correlations at different time scales. Observation of the experimental results yields an average accuracy improvement of 1.71% for GFA on the six dataset. To visualize the impact of the module on emotion classification, we visualized the classification results by the t-sne visualizations. It is observed from Fig. 7(a) that there are some problems of sample overlap and unclear classification boundaries between emotion categories before GFA is introduced. In contrast, Fig. 7(b) can observe that there are clear classification boundaries between each category. The results prove that GFA has better ability to classify emotion.

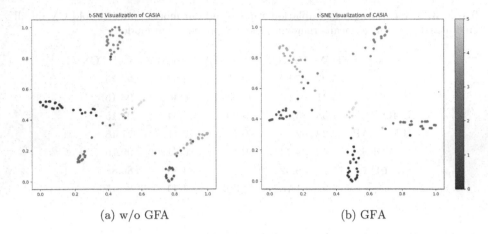

(a) w/o GFA   (b) GFA

Fig. 6. t-SNE visualizations of features learned from w/o GFA.

Effectiveness of AFM: AFM enables information from different time scales to interact and complement each other by fusing multi-level features and taking advantage of adaptive weight adjustment. The abundant emotion information in the fused features has a significant improvement for SER. Observing the results in Table 3 we can see that AFM brings a 3.31% improvement to the model performance compared to using the join operation.

Effectiveness of ReLU in GFA: Our proposed GFA is a sparse attention module without softmax. ReLU is used as the activation function in the attention module to remove all negative attention scores and generate sparse attention. We did ablation experiments in order to verify the performance impact brought by softmax and Relu functions in the GFA module. In the Table 4, we see that the model classification accuracy is improved on five dataset after the replacement of softmax by ReLU, and one dataset with the same classification accuracy. We analyze this because the ReLU function removes negative attention scores for

Table 4. The ablation studie from 'softmax' and 'ReLU'

Copus	softmax	ReLU
CASIA	93.08	93.75
EMODB	94.40	95.15
EMOVO	91.15	91.15
IEMOCAP	71.0	71.18
RAVDESS	89.96	90.76
SAVEE	86.46	86.88
Average	87.68	88.15

useless information. Thus, it reduces the invalid attention and makes the network pay attention to more information that is beneficial for emotion classification. We believe that ReLU is more effective in GFA compared to softmax and can directly replace the softmax activation function.

5   Conclusion

In this paper, we propose channel-aware multi-scale temporal convolutional networks (CM-TCN) for SER task. CM-TCN can fully extract emotion features in time-domain sequences and model correlations at different time scales. We designed CATCB to enhance the ability of network to extract channel information. At the same time, in order to fully extract the global multi-scale temporal information, we propose GFA to establish the dependencies between multi-scale temporal features. Finally, the proposed model fuses the features at different levels to perform emotion classification. The experimental results validate the effectiveness of CM-TCN and demonstrate the importance of modeling the dependencies between different time scales in SER. Compared with other SER methods for modeling time series, we achieve the best performance on six commonly used SER dataset. However, the dataset we used in our experiments are short time speech files, and their performance in real applications needs further testing, so our future research will focus on improving the generalization capability of long speech.

References

1. Schuller, B., Rigoll, G., Lang, M.: Hidden Markov model-based speech emotion recognition. In: 2003–2003 IEEE International Conference on Multimedia and Expo (ICME), pp. 1–401. IEEE, Baltimore (2003)
2. Mower, E., Mataric, M.J., Narayanan, S.: A framework for automatic human emotion classification using emotion profiles. IEEE Trans. Audio Speech Lang. Process. **19**(5), 1057–1070 (2011). https://doi.org/10.1109/TASL.2010.2076804
3. Han, K., Yu, D., Tashev, I.: Speech emotion recognition using deep neural network and extreme learning machine. In: Interspeech, pp. 223–227. ISCA, Singapore (2014)
4. Huang, Z., Dong, M., Mao,Q., Zhan, Y.: Speech emotion recognition using CNN. In: Proceedings of the 22nd ACM International Conference on Multimedia, pp. 801–804 (2014)
5. Tarantino, L., Garner, P.N., Lazaridis, A., et al.: Self-attention for speech emotion recognition, In: Interspeech, pp. 2578–2582. ISCA, Graz (2019)
6. Xu, M., Zhang, F., Khan, S.U.: Head fusion: improving the accuracy and robustness of speech emotion recognition on the IEMOCAP and RAVDESS dataset. IEEE Access **9**, 1058–1064 (2020)
7. Xu, M., Zhang, F., Cui, X., Zhang, W.: Speech emotion recognition with multiscale area attention and data augmentation. In: ICASSP 2021–2021 IEEE International Conference on Acoustics, Speech and Signal Processing (ICASSP), pp. 6319–6323. IEEE, Toronto (2021)

8. Jahangir, R., Teh, Y.W., Hanif, F., Mujtaba, G.: Deep learning approaches for speech emotion recognition: state of the art and research challenges. Speech Commun. **127**, 73–81 (2021)
9. Zhang, S., Tao, X., Chuang, Y., Zhao, X.: Learning the sequential temporal information with recurrent neural networks. Multimedia Tools Appl. **80**(16), 23745–23812 (2021)
10. Murugan, P.: Learning deep multimodal affective features for spontaneous speech emotion recognition. R abs 1807.02857 (2018). https://doi.org/10.48550/arXiv.1807.02857
11. Xie, Y., Liang, R., Liang, Z., Zhao, L.: Attention-based dense LSTM for speech emotion recognition. IEICE Trans. Inf. Syst. **102**(7), 1426–1429 (2019)
12. Su, B., Chang, C., Lin, Y., Lee, C.: Improving speech emotion recognition using graph attentive bi-directional gated recurrent unit network. In: Interspeech, pp. 506–510. ISCA, Shanghai (2020)
13. Lin, W., Busso, C.: An efficient temporal modeling approach for speech emotion recognition by mapping varied duration sentences into fixed number of chunks. In: Interspeech, pp. 2322–2326. ISCA, Shanghai (2020)
14. Wang, J., Xue, M., Culhane, R., et al.: Speech emotion recognition with dual-sequence LSTM architecture. In: ICASSP 2020–2020 International Conference on Acoustics, Speech and Signal Processing (ICASSP), pp. 6474–6478. IEEE, Barcelona (2020)
15. Zhong, Y., Hu, Y., Huang, H., Silamu, W.: A lightweight model based on separable convolution for speech emotion recognition. In: Interspeech, pp. 3331–3335. ISCA, Shanghai (2020)
16. Rajamani, S.T., Rajamani, K.T., Mallol-Ragolta, A., et al.: A novel attention-based gated recurrent unit and its efficacy in speech emotion recognition. In: ICASSP 2021–2021 IEEE International Conference on Acoustics, Speech and Signal Processing (ICASSP), pp. 6294–6298. IEEE, Toronto (2021)
17. Zhao, Z., Zheng, Y., Zhang, Z., et al.: Exploring spatio-temporal representations by integrating attention-based bidirectional-LSTM-RNNs and FCNs for speech emotion recognition. In: Interspeech, pp. 272–276.ISCA , Hyderabad (2018)
18. Mustaqeem, S.K.: MLT-DNet: speech emotion recognition using 1D dilated CNN based on multi-learning trick approach. Expert Syst. Appl. **167**, 114177 (2021)
19. Bai, S., Kolter, J.Z., Koltun, V.: An empirical evaluation of generic convolutional and recurrent networks for sequence modeling, CoRR abs/1803.01271 (2018). https://doi.org/10.48550/arXiv.1803.01271
20. Salehinejad, H., Baarbe, J., Sankar, S., Barfett, J., Colak, E., Valaee, S.: Recent advances in recurrent neural networks. CoRR abs/1801.01078 (2018). https://doi.org/10.48550/arXiv.1801.01078
21. Farha, Y.A., Gall, J.: MS-TCN: multi-stage temporal convolutional network for action segmentation. IEEE Trans. Pattern Anal. Mach. Intell. **45**(6), 6647–6658 (2018)
22. Zhao, Y., Wang, D., Xu, B., Zhang, T.: Monaural speech dereverberation using temporal convolutional networks with self attention. IEEE Trans. Audio Speech Lang. Process. **28**, 1057–1070 (2020)
23. Luo, Y., Mesgarani, N.: Conv-TasNet: surpassing ideal time-frequency magnitude masking for speech separation. IEEE Trans. Audio Speech Lang. Process. **27**(8), 1256–1266 (2019)

24. Peng, Z., Lu, Y., Pan, S., et al.: Efficient speech emotion recognition using multi-scale CNN and attention. In: ICASSP 2021–2021 IEEE International Conference on Acoustics, Speech and Signal Processing (ICASSP), pp. 3020–3024. IEEE, Toronto (2021)
25. Hu, J., Shen, L., Sun, G.: Squeeze-and-excitation networks. In: 2018 IEEE/CVF Conference on Computer Vision and Pattern Recognition (CVPR), pp. 7132–7141.IEEE, Salt Lake City (2018)
26. Wang, Q., Wu, B., Zhu, P., Li, P., Zuo, W., Hu, Q.: ECA-Net: efficient channel attention for deep convolutional neural networks. In:2020 IEEE/CVF Conference on Computer Vision and Pattern Recognition (CVPR), pp. 11531–11539. IEEE, Seattle (2020)
27. Tao, J., Liu, F., Zhang, M., Jia, H.: Design of speech dataset for mandarin text to speech. In: Blizzard Challenge 2008 Workshop (2008)
28. Burkhardt, F., Paeschke, A., Rolfes, M., et al.: A database of German emotional speech. In: Interspeech, pp. 1517–1520. ISCA, Lisbon (2005)
29. Costantini, G., Iaderola, I., Paoloni, A., Todisco, M.: Emovo dataset: an Italian emotional speech database. In: Proceedings of the Ninth International Conference on Language Resources and Evaluation (LREC 2014), pp. 3501–3504. European Language Resources Association (ELRA), Reykjavik (2014)
30. Busso, C., Bulut, M., Lee, C.C., et al.: IEMOCAP: interactive emotional dyadic motion capture database. Lang. Res. Eval. **42**(4), 335–359 (2008)
31. Livingstone, S.R, Russo, F.A.: The ryerson audio-visual database of emotional speech and song (ravdess): a dynamic, multimodal set of facial and vocal expressions in North American English. PLOS ONE **13**(5), e0196391 (2018)
32. Philip J., Haq, S.: Surrey Audio-Visual Expressed Emotion (savee) Database. University of Surrey, Guildford (2014)
33. Sun, L., Fu, S., Wang, F.: Decision tree SVM model with fisher feature selection for speech emotion recognition. EURASIP J. Audio Speech Music. Process. **2019**, 2 (2019)
34. Chen, L., Su, W., Feng, Y., et al.: Two-layer fuzzy multiple random forest for speech emotion recognition in human-robot interaction. Inf. Sci. **509**, 150–163 (2020)
35. Ye, J., Wen, X., Wang, X., et al.: GM-TCNet: gated multi-scale temporal convolutional network using emotion causality for speech emotion recognition. Speech Commun. **145**, 21–35 (2022)
36. Wen, X., Ye, J., Luo, Y., et al. CTL-MTNet: a novel capsnet and transfer learning-based mixed task net for single-dataset and cross-dataset speech emotion recognition. In: International Joint Conferences on Artificial Intelligence (IJCAI) 2022, Vienna, Austria, pp. 2305–2311 (2022)
37. Tuncer, T., Dogan, S., Acharya, U.R.: Automated accurate speech emotion recognition system using twine shuffle pattern and iterative neighborhood component analysis techniques. Knowl.-Based Syst. **211**, 106547 (2021)
38. Aftab, A., Morsali, A., Ghaemmaghami, S., et al.: LIGHT-SERNET: a lightweight fully convolutional neural network for speech emotion recognition. In: ICASSP 2022–2022 International Conference on Acoustics, Speech and Signal Processing (ICASSP), pp. 6912–6916. IEEE, Virtual and Singapore (2022)
39. Ozer, I.: Pseudo-colored rate map representation for speech emotion recognition. Biomed. Signal Process. Control **66**, 102502 (2021)
40. Ancilin, J., Milton, A.: Improved speech emotion recognition with mel frequency magnitude coefficient. Appl. Acoust. **179**, 108046 (2021)

41. Liu, J., Song, Y., Wang, L., Dang, J., Yu, R.: Time-frequency representation learning with graph convolutional network for dialogue-level speech emotion recognition. In: Interspeech, pp. 4523–4527. ISCA, Brno (2020)
42. Cao, Q., Hou, M., Chen, B., Zhang, Z., Lu, G.: Hierarchical network based on the fusion of static and dynamic features for speech emotion recognition. In: ICASSP 2021–2021 IEEE International Conference on Acoustics. Speech and Signal Processing (ICASSP), pp. 6334–6338. IEEE, Toronto (2021)
43. Mustaqeem, Kwon, S.: Optimal feature selection based speech emotion recognition using two-stream deep convolutional neural network. Int. J. Intell. Syst. **36**(9), 5116–5135 (2021)
44. Hajarolasvadi, N., Demirel, H.: 3D CNN-based speech emotion recognition using k-means clustering and spectrograms. Entropy **21**(5), 479 (2019)

FLDNet: A Foreground-Aware Network for Polyp Segmentation Leveraging Long-Distance Dependencies

Xuefeng Wei[✉] and Xuan Zhou[✉]

Institut Polytechnique de Paris, Rte de Saclay, 91120 Palaiseau, France
{xuefeng.wei,xuan.zhou}@ip-paris.fr

Abstract. Given the close association between colorectal cancer and polyps, the diagnosis and identification of colorectal polyps play a critical role in the detection and surgical intervention of colorectal cancer. In this context, the automatic detection and segmentation of polyps from various colonoscopy images has emerged as a significant problem that has attracted broad attention. Current polyp segmentation techniques face several challenges: firstly, polyps vary in size, texture, color, and pattern; secondly, the boundaries between polyps and mucosa are usually blurred, existing studies have focused on learning the local features of polyps while ignoring the long-range dependencies of the features, and also ignoring the local context and global contextual information of the combined features. To address these challenges, we propose **FLD-Net** (**F**oreground-**L**ong-**D**istance Network), a Transformer-based neural network that captures long-distance dependencies for accurate polyp segmentation. Specifically, the proposed model consists of three main modules: a pyramid-based Transformer encoder, a local context module, and a foreground-Aware module. Multilevel features with long-distance dependency information are first captured by the pyramid-based transformer encoder. On the high-level features, the local context module obtains the local characteristics related to the polyps by constructing different local context information. The coarse map obtained by decoding the reconstructed highest-level features guides the feature fusion process in the foreground-Aware module of the high-level features to achieve foreground enhancement of the polyps. Our proposed method, FLDNet, was evaluated using seven metrics on common datasets and demonstrated superiority over state-of-the-art methods on widely-used evaluation measures.

Keywords: Deep Learning · Polyp Segmentation · Colorectal Cancer

1 Introduction

Colorectal cancer (CRC) is the third most common type of cancer worldwide, with polyps on the intestinal mucosa considered as precursors of CRC that can

---

X. Wu and Z. Zhou—Equal contribution.

© The Author(s), under exclusive license to Springer Nature Singapore Pte Ltd. 2024
B. Luo et al. (Eds.): ICONIP 2023, LNCS 14449, pp. 477–487, 2024.
https://doi.org/10.1007/978-981-99-8067-3_35

easily become malignant. Therefore, the early detection of polyps has significant clinical implications for the treatment of CRC. Numerous studies have shown that early colonoscopy reduces the incidence of CRC by 30% [7]. Fortunately, due to advancements in computing, several automatic polyp segmentation techniques [1,2] have been proposed, demonstrating promising performance [3,6]. These techniques, however, still face challenges in accurately segmenting polyps due to their varied sizes and shapes, and the indistinct boundaries between polyps and the mucosa [9].

To tackle these challenges, the application of conventional Convolutional Neural Networks (CNNs) has been explored, which possess a local receptive field. However, CNNs may neglect the global information of polyps, which could lead to reduced performance in polyp segmentation [15,20]. The Transformer model [11], with its self-attention mechanism, is able to focus on the global context of input data. This property is particularly important when dealing with the complexity and diversity of polyp images.

Moreover, some studies have proposed methods that either leverage attention mechanisms [8] or utilize individual edge supervision [12] to handle the varying polyp sizes and shapes. But these methods tend to focus more on local information, lacking the consideration of the global context. Therefore, in this study, we adopt a Transformer-based model, which we expect can overcome the limitations of CNNs in handling polyp images and achieve higher segmentation performance.

To further address these challenges and enhance the accuracy and generalization of polyp segmentation techniques, we propose the PVT-based FANet polyp segmentation network, designed to capture long-distance dependencies between image patches. To compensate for the shortcomings of PVT [5], we introduce a local context module and a foreground perception module. The local context module builds different context information through a multi-branch structure to obtain polyp local information, thus resolving the issue of attention dispersion and indistinct features of local small objects. The foreground perception module addresses the problem of low contrast and blurred boundaries between polyps and surrounding mucosa. By highlighting foreground features in a chaotic background, we aim to achieve more accurate segmentation.

The main contributions of this paper are:

1. We introduce a pyramidal Transformer, enhancing the model's ability to capture long-distance dependencies and boosting the model's capacity to seize global information and generalization capability.
2. We introduce a local context module and a foreground-Aware module, addressing the deficiency of local features in polyps and the blurring of the boundaries between polyps and surrounding mucosa, thus enhancing the network's capability to capture local information.
3. Extensive experiments demonstrate that our proposed FLDNet surpasses most state-of-the-art models on challenging datasets.

2   Related Work

The field of polyp segmentation has witnessed significant developments, transitioning from relying on handcrafted features to adopting deep learning methods. Early studies heavily relied on handcrafted features such as color and texture [1,2]. However, these features often struggle to capture global information and exhibit poor stability in complex scenes.

In recent years, deep learning methods have gained dominance in the field of polyp segmentation. Approaches such as Fully Convolutional Networks (FCN) [3], U-Net [14], U-Net++ [15], ACS-Net [19], and PraNet [8] have been applied to polyp segmentation. The U-Net model [14] utilizes a well-known structure for medical image segmentation, consisting of a contracting path that captures context and an expanding path that restores precise details. U-Net++ [15] and ResUNet++ [9] further improve the original U-Net by incorporating dense connections and better pre-trained backbone networks, achieving satisfactory segmentation performance. Although these methods yield good results for the main body of polyps, the boundary regions are often neglected. To enhance boundary segmentation, Psi-Net [24] proposes a method that combines both body and boundary features. Similarly, SFA [18] explicitly applies region-boundary constraints to supervise the learning of polyhedron regions and boundaries, while PraNet [8] introduces a reverse attention mechanism that first localizes the polyp region and then implicitly refines the object boundary.

Recently, with the advent of Transformer architectures in computer vision tasks [11], researchers have started to explore their potential in medical image analysis, including polyp segmentation. Unlike traditional convolutional layers, the Transformer model captures long-range dependencies in the data, which could be particularly beneficial in handling polyp images' complexity and diversity. However, directly applying Transformer models in medical image segmentation tasks, such as polyp segmentation, is not straightforward [25]. These models were originally designed for natural language processing tasks, and their suitability for the specific challenges of medical image segmentation is still an open question. For instance, the self-attention mechanism of Transformer models, while being powerful, is also computationally expensive and could be less effective in localizing small-sized polyps due to its global nature.

Moreover, it's worth mentioning that while deep learning techniques have significantly improved polyp segmentation, issues still persist. One key challenge is the foreground-background imbalance present in most polyp images, which could affect the segmentation performance. Another challenge is the large variance in polyps' appearance, including their size, shape, texture, and color. These variances make it hard for models to generalize across different patients and imaging conditions.

To overcome these challenges, there is an increasing trend towards developing more sophisticated models that combine the strengths of different architectures. For example, hybrid models combining the local feature extraction capability of CNNs and the long-range dependencies modeling of Transformer models have been proposed.

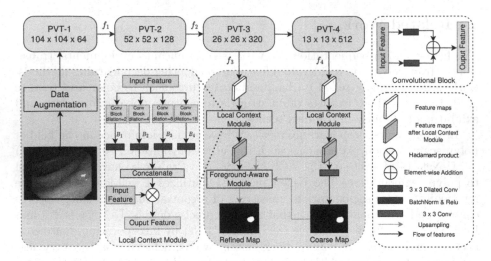

Fig. 1. An overview of the proposed FLDNet. The model consists of a Pyramid Vision Transformer (PVT)-based encoder, a Local Context Module, and a Foreground-Aware Module. The PVT encoder captures long-distance dependencies and forms the backbone of the model. It is divided into four sections, each corresponding to different resolution feature maps. Features from the last two blocks $f_{i_{i=3}}^4$ are further utilized considering their higher semantic information. The Local Context Module and the Foreground-Aware Module further refine the features for precise polyp segmentation. The specific operations and interactions between these modules are described in detail in Sect. 3.

3   Method

In the proposed FLDNet, we employ a PVT-based encoder, Local Context Module, and Foreground-Aware Module to achieve precise polyp segmentation (Fig. 1). The training procedure and loss function used for network optimization are also discussed in this section.

The PVT-based encoder forms the backbone of the model, capturing long-distance dependencies. Deeper features from the PVT encoder, bearing more semantic information, are specifically employed in our approach. Further details on each component of FLDNet are provided in the ensuing discussion.

3.1   Local Feature Capture via Local Context Module

CNN-based models [9,14,15] commonly used in medical image segmentation can struggle to extract critical local details such as textures and contours. Our Local Context Module mitigates this by using standard and dilated convolutions to build diverse context information.

The Local Context Module, visualized in Fig. 1, consists of four branches. Each branch involves both standard and dilated convolutions, processes the input feature map X, and combines the convolution outputs using element-wise addition. The outputs from all branches are then concatenated and element-wise

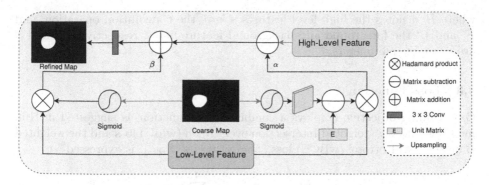

Fig. 2. An overview of the proposed Foreground-Aware Module, See Sect. 3.2 for details.

multiplied with the original input feature map X to yield the final module output Y. This process can be encapsulated in a single equation:

$$Y = X \otimes Concat(Conv_i(X) \oplus ConvDilated_{2^i}(X))_{\forall i} \qquad (1)$$

where $\oplus$ represents the element-wise addition operation, $\otimes$ denotes the element-wise multiplication, and the subscript $\forall i$ signifies that the operation applies for all branches i. This approach allows effective capture of both global and local context information, which is pivotal for accurate polyp segmentation.

3.2 Foreground-Aware Module

In contrast to strategies in [4,7,8], which use features from all stages for segmentation, we adaptively fuse high-level features in two parallel streams, thereby improving foreground depiction. Our method, inspired by [8], leverages features from PVT, known for their rich contextual information, instead of conventional deepest-layer CNN features.

Our Foreground-Aware Module, illustrated in Fig. 2, employs a two-step approach: First, high-level features are decoded to generate a coarse segmentation map, guiding the fusion of deep-layer features to accentuate the foreground. Second, we derive a foreground probability map via upsampling the segmentation map and applying the Sigmoid function. A corresponding background map is computed by subtracting the foreground map from 1, enabling the separation of foreground and background features.

Next, two learnable parameters, α and β, are used to balance the impact of foreground and background features on the final refined segmentation map. The adjusted feature map, generated by reducing α times the foreground map and increasing β times the background map, is then convolved to yield the final segmentation map. This is described by the equation:

$$R = Conv(Upsample(H) - \alpha \cdot F' + \beta \cdot B') \qquad (2)$$

where H denotes the high-level features, $Conv$ the convolution operation, and F' and B' the foreground and background feature maps, respectively. α and β are learnable parameters.

3.3 Loss Function

Our training objective employs a combined loss function as suggested in [16], incorporating the weighted Intersection over Union (wIoU) loss and the weighted Binary Cross Entropy (wBCE) loss. The total loss, L_{total}, is expressed as:

$$L_{\text{total}} = L_{\text{wIoU}} + L_{\text{wBCE}} + \sum_{i=3}^{4} L(G, S_i^{up}) \tag{3}$$

In these equations, L_{wIoU} and L_{wBCE} represent the wIoU and wBCE losses respectively. G indicates the ground truth, S_i^{up} refers to the upscaled prediction at scale i, and L symbolizes the loss function, identical to L_{total} without the summation term. The variables w_i, p_i, and y_i correspond to the weights, predictions, and ground truth at the ith pixel respectively.

This formulation facilitates our model in capturing the complex task of polyp segmentation, ensuring the retention of detailed information across various scales.

4 Experiment

4.1 Datasets and Baselines

We assess the FLDNet using three polyp segmentation datasets: CVC-ClinicDB [10], CVC-ColonDB [17], and ETIS [13]. CVC-ClinicDB and CVC-ColonDB offer a broad spectrum of polyps in different imaging conditions, while ETIS is distinctive for its low contrast, complex images.

UNet [14] and **UNet++** [15] are known for their effective feature capture. **SFA** [18] has a shared encoder and two constrained decoders for accurate predictions. **ACSNet** [19] addresses polyp size, shape, and spatial context, while **PraNet** [8] uses a reverse attention mechanism for incremental object refinement. **DCRNet** [20] leverages contextual relations to enhance accuracy, **MSEG** [23] incorporates the HarDNet68 backbone. **EU-Net** [22] combines the SFEM and AGCM to prioritize important features. Lastly, **SANet** [21] uses a shallow attention module and a probability correction strategy to ensure accuracy, particularly for small polyps.

4.2 Evaluation Metrics

To evaluate the performance of our model, we employ several widely used metrics, including the Dice Coefficient (Dice), Intersection over Union (IoU), weighted F-measure (F_β^w), S-measure (S_α), mean absolute error (MAE), and

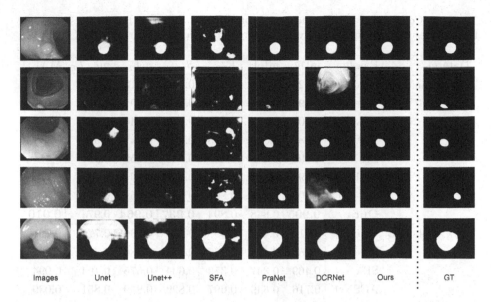

Fig. 3. Visual comparison between proposed FLDNet and other methods. Example polyp images sourced from the CVC-ClinicDB and ETIS datasets. The first row presents the original polyp images, followed by the corresponding segmentation results, where the polyp regions are denoted in white and the background in black. To facilitate comparison, the Ground Truth (GT) and the segmentation predictions generated by our model are separated by dashed lines. (Color figure online)

E-measure. For Dice and IoU, we report the mean values, denoted as mDice and mIoU, respectively. For the E-measure, we provide both the mean and max values, represented as mE_ξ and $maxE_\xi$. These metrics are extensively utilized in the literature for segmentation and object detection tasks [8].

4.3   Experiment Settings

We normalized the input images to a size of 416×416 and employed data augmentation techniques such as random flipping and brightness variations. The framework used was PyTorch 1.8.1 with CUDA 11.1, running on a hardware configuration consisting of a GeForce RTX 3090 with 24 GB of compute memory. The initial learning rate was set to 10^{-4}, and the Adam optimizer was utilized. The learning rate was reduced by a factor of 10 every 50 epochs, and a total of 200 epochs were trained.

4.4   Quantitative Comparison

In order to thoroughly evaluate the effectiveness of our proposed FLDNet for polyp segmentation, we conducted a comparative analysis with several state-of-the-art models. The results of this comparison on the CVC-ClinicDB dataset [10], CVC-ColonDB dataset [17], and ETIS dataset [13] are detailed in Table 1.

Table 1. Comparison of different models on different datasets. Bold values represent the best results, and underlined values indicate the second best results.

Dataset	Baseline	mDic	mIoU	F_β^ω	S_α	mE_ξ	$maxE_\xi$	MAE $\downarrow$
CVC-ClinicDB	UNet	0.823	0.755	0.811	0.889	0.913	0.954	0.019
	UNet++	0.794	0.729	0.785	0.873	0.891	0.931	0.022
	SFA	0.700	0.607	0.647	0.793	0.840	0.885	0.042
	ACSNet	0.882	0.826	0.873	0.927	0.947	0.959	0.011
	PraNet	0.899	**0.849**	**0.896**	0.936	0.959	0.965	<u>0.011</u>
	DCRNet	0.896	0.844	0.890	0.933	0.964	**0.978**	0.010
	EU-Net	<u>0.902</u>	0.846	0.891	<u>0.936</u>	<u>0.959</u>	0.965	0.011
	Ours	**0.905**	<u>0.848</u>	<u>0.894</u>	**0.937**	**0.964**	<u>0.974</u>	**0.010**
CVC-ColonDB	UNet	0.512	0.444	0.498	0.712	0.696	0.776	0.061
	UNet++	0.483	0.410	0.467	0.691	0.680	0.760	0.064
	SFA	0.469	0.347	0.379	0.634	0.675	0.764	0.094
	ACSNet	0.716	0.649	0.697	0.829	0.839	0.851	0.039
	PraNet	0.712	0.640	0.699	0.820	0.847	0.872	0.043
	MSEG	0.735	0.666	0.724	0.834	0.859	0.875	<u>0.038</u>
	DCRNet	0.704	0.631	0.684	0.821	0.840	0.848	0.052
	EU-Net	<u>0.756</u>	<u>0.681</u>	<u>0.730</u>	0.831	0.863	0.872	0.045
	SANet	0.753	0.670	0.726	<u>0.837</u>	<u>0.869</u>	<u>0.878</u>	0.043
	Ours	**0.801**	**0.717**	**0.770**	**0.869**	**0.902**	**0.916**	**0.030**
ETIS	UNet	0.398	0.335	0.366	0.684	0.643	0.740	0.036
	UNet++	0.401	0.344	0.390	0.683	0.629	0.776	0.035
	SFA	0.297	0.217	0.231	0.557	0.531	0.632	0.109
	ACSNet	0.578	0.509	0.530	0.754	0.737	0.764	0.059
	PraNet	0.628	0.567	0.600	0.794	0.808	0.841	0.031
	MSEG	0.700	0.630	0.671	0.828	0.854	0.890	0.015
	DCRNet	0.556	0.496	0.506	0.736	0.742	0.773	0.096
	EU-Net	0.687	0.609	0.636	0.793	0.807	0.841	0.067
	SANet	<u>0.750</u>	<u>0.654</u>	<u>0.685</u>	<u>0.849</u>	**0.881**	<u>0.897</u>	**0.015**
	Ours	**0.758**	**0.670**	**0.698**	**0.857**	<u>0.868</u>	**0.906**	<u>0.023</u>

Our FLDNet consistently shows an impressive performance across different datasets, in several cases outperforming the other models by a significant margin. Particularly on the CVC-ColonDB dataset [17], FLDNet surpasses all other models, exhibiting a remarkable improvement in mean DICE (over 6%)

Apart from our own results, it is also notable that different models perform differently across various datasets. For example, the ACSNet model, which generally shows strong performance, struggles on the CVC-ColonDB dataset

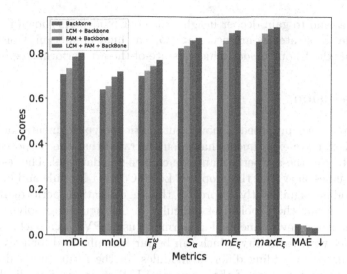

Fig. 4. Ablation Study Results of the different components of our model on different metrics.

compared to other datasets. This could be due to the model's difficulty in handling the specific complexities presented in the CVC-ColonDB dataset. On the other hand, the MSEG model performs remarkably well on the ETIS dataset, where it provides highly competitive results, especially in terms of the mE_ξ and $maxE_\xi$ metrics.

These observations demonstrate the uniqueness and challenges of each dataset and underline the importance of model versatility when tackling different types of data. It is precisely in this aspect that our FLDNet excels, exhibiting robust performance regardless of dataset complexity, thereby confirming its effectiveness for polyp segmentation.

4.5   Visual Comparison

Figure 3 illustrates the prediction maps for polyp segmentation by different models. We have compared the prediction results of our proposed model with those of other models. The results reveal that our model significantly excels in emphasizing the polyp region and suppressing background noise, compared to other approaches. Especially when faced with challenging scenarios, our model can handle them adeptly and generate more accurate segmentation masks. These outcomes comprehensively demonstrate the significant superiority of our proposed model in polyp segmentation over other models.

4.6   Ablation Study

To investigate the importance of each proposed component in the FLDNet, the ColonDB dataset was employed for ablation studies to validate the efficacy of the modules and to gain deeper insights into our model. As shown in Fig. 4, all the modules or strategies are necessary for the final prediction. Combining all proposed methods, our model achieves state-of-the-art performance.

5   Conclusion

In this paper, we proposed a novel automatic polyp segmentation network, FLDNet. Extensive experiments have demonstrated that the proposed FLDNet achieved state-of-the-art performance on challenging datasets. The results of the ablation studies show that the proposed Local Context Module and Foreground-Aware Module significantly improved the segmentation performance of the model, alleviating the problem of difficulty in distinguishing polyps from surrounding mucosal environments. Furthermore, using PVT as a backbone endows our FLDNet with versatility, which can be further enhanced for polyp segmentation accuracy by adding different modules. In the future we will study the segmentation performance of the proposed FLDNet on more medical datasets, such as lung and MRI images to explore the generality of FLDNet.

References

1. Mamonov, A.V., Figueiredo, I.N., Figueiredo, P.N., Tsai, Y.H.R.: Automated polyp detection in colon capsule endoscopy. IEEE Trans. Med. Imaging **33**(7), 1488–1502 (2014)
2. Tajbakhsh, N., Gurudu, S.R., Liang, J.: Automated polyp detection in colonoscopy videos using shape and context information. IEEE Trans. Med. Imaging **35**(2), 630–644 (2015)
3. Brandao, P., et al.: Fully convolutional neural networks for polyp segmentation in colonoscopy. In: Medical Imaging 2017: Computer-Aided Diagnosis, vol. 10134, pp. 101–107 (2017)
4. Naseer, M.M., Ranasinghe, K., Khan, S.H., Hayat, M., Shahbaz Khan, F., Yang, M.H.: Intriguing properties of vision transformers. Adv. Neural. Inf. Process. Syst. **34**, 23296–23308 (2021)
5. Wang, W., et al.: Pyramid vision transformer: a versatile backbone for dense prediction without convolutions. In: Proceedings of the IEEE/CVF International Conference on Computer Vision, pp. 568–578 (2021)
6. Kim, T., Lee, H., Kim, D.: Uacanet: uncertainty augmented context attention for polyp segmentation. In: Proceedings of the 29th ACM International Conference on Multimedia, pp. 2167–2175 (2021)
7. Chen, S., Tan, X., Wang, B., Hu, X.: Reverse attention for salient object detection. In: ECCV, pp. 234–250 (2018)
8. Fan, D.-P., et al.: PraNet: parallel reverse attention network for polyp segmentation. In: Martel, A.L., et al. (eds.) MICCAI 2020. LNCS, vol. 12266, pp. 263–273. Springer, Cham (2020). https://doi.org/10.1007/978-3-030-59725-2_26
9. Jha, D., et al.: Resunet++: an advanced architecture for medical image segmentation. In: IEEE ISM, pp. 225–2255 (2019)

10. Bernal, J., ánchez F.J., Fernández-Esparrach G., Gil D., Rodríguez de Miguel C., Vilariño F: WM-DOVA maps for accurate polyp highlighting in colonoscopy: validation vs. saliency maps from physicians. Comput. Med. Imaging Graph. **43**, 99–111 (2015)
11. Vaswani, A., et al.: Attention is all you need. Adv. Neural. Inf. Process. Syst. **30**, 1–11 (2017)
12. Raghu, M., Unterthiner, T., Kornblith, S., Zhang, C., Dosovitskiy, A.: Do vision transformers see like convolutional neural networks? Adv. Neural. Inf. Process. Syst. **34**, 12116–12128 (2021)
13. Silva, J., Histace, A., Romain, O., Dray, X., Granado, B.: Toward embedded detection of polyps in WCE images for early diagnosis of colorectal cancer. Int. J. Comput. Assist. Radiol. Surg. **9**(2), 283–293 (2014)
14. Ronneberger, O., Fischer, P., Brox, T.: U-Net: convolutional networks for biomedical image segmentation. In: Navab, N., Hornegger, J., Wells, W.M., Frangi, A.F. (eds.) MICCAI 2015. LNCS, vol. 9351, pp. 234–241. Springer, Cham (2015). https://doi.org/10.1007/978-3-319-24574-4_28
15. Zhou, Z., Rahman Siddiquee, M.M., Tajbakhsh, N., Liang, J.: UNet++: a nested U-net architecture for medical image segmentation. In: Stoyanov, D., et al. (eds.) DLMIA/ML-CDS -2018. LNCS, vol. 11045, pp. 3–11. Springer, Cham (2018). https://doi.org/10.1007/978-3-030-00889-5_1
16. Wei, J., Wang, S., Huang, Q.: F3Net: fusion, feedback and focus for salient object detection. In: AAAI (2020)
17. Bernal, J., Sánchez, J., Vilarino, F.: Towards automatic polyp detection with a polyp appearance model. Pattern Recogn. **45**(9), 3166–3182 (2012)
18. Fang, Y., Chen, C., Yuan, Y., Tong, K.Y.: Selective feature aggregation network with area-boundary constraints for polyp segmentation. In: International Conference on Medical Image Computing and Computer-Assisted Intervention, pp. 302–310 (2019)
19. Zhang, R., Li, G., Li, Z., Cui, S., Qian, D., Yu, Y.: Adaptive context selection for polyp segmentation. In: Marte, A.L., et al. (eds.) MICCAI 2020. LNCS, vol. 12266, pp. 253–262. Springer, Cham (2020). https://doi.org/10.1007/978-3-030-59725-2_25
20. Yin, Z., Liang, K., Ma, Z., Guo, J.: Duplex contextual relation network for polyp segmentation. In: IEEE International Symposium on Biomedical Imaging (2022)
21. Wei, J., Hu, Y., Zhang, R., Li, Z., Zhou, S.K., Cui, S.: Shallow attention network for polyp segmentation. In: de Bruijne, M., et al. (eds.) MICCAI 2021. LNCS, vol. 12901, pp. 699–708. Springer, Cham (2021). https://doi.org/10.1007/978-3-030-87193-2_66
22. Patel, K., Bur, A.M., Wang, G.: Enhanced u-net: a feature enhancement network for polyp segmentation. In: Proceedings of International Robot and Vision Conference, pp. 181–188 (2021)
23. Huang, C.-H., Wu, H.-Y., Lin, Y.-L.S.: HarDNet-MSEG: a simple encoder-decoder polyp segmentation neural network that achieves over 0.9 mean dice and 86 FPS. arXiv:2101.07172 (2021)
24. Murugesan, B., Sarveswaran, K., Shankaranarayana, S.M., Ram, K., Joseph, J., Sivaprakasam, M.: Psi-Net: shape and boundary aware joint multi-task deep network for medical image segmentation. In: 2019 41st Annual International Conference of the IEEE Engineering in Medicine and Biology Society (EMBC), pp. 7223–7226 (2019)
25. Shamshad, F., et al.: Transformers in medical imaging: a survey. arXiv preprint arXiv:2201.09873 (2022)

Domain-Invariant Task Optimization for Cross-domain Recommendation

Dou Liu[1,2,3], Qingbo Hao[1,2,3], Yingyuan Xiao[1,2,3](✉), Wenguang Zheng[1,2,3], and Jinsong Wang[1,2,3]

[1] School of Computer Science and Engineering, Tianjin University of Technology, Tianjin 300384, China
yyxiao@tjut.edu.cn
[2] Engineering Research Center of Learning-Based Intelligent System, Ministry of Education, Tianjin University of Technology, Tianjin 300384, China
[3] Tianjin Key Laboratory of Intelligence Computing and Novel Software Technology, Tianjin University of Technology, Tianjin 300384, China

Abstract. The challenge of cold start has long been a persistent issue in recommender systems. However, Cross-domain Recommendation (CDR) provides a promising solution by utilizing the abundant information available in the auxiliary source domain to facilitate cold-start recommendations for the target domain. Many existing popular CDR methods only use overlapping user data but ignore non-overlapping user data when training the model to establish a mapping function, which reduces the model's generalization ability. Furthermore, these CDR methods often directly learn the target embedding during training, because the target embedding itself may be unreasonable, resulting in an unreasonable transformed embedding, exacerbating the difficulty of model generalization. To address these issues, we propose a novel framework named Domain-Invariant Task Optimization for Cross-domain Recommendation (DITOCDR). To effectively utilize non-overlapping user information, we employ source and target domain autoencoders to learn overlapping and non-overlapping user embeddings and extract domain-invariant factors. Additionally, we use a task-optimized strategy for target embedding learning to optimize the embedding and implicitly transform the source domain user embedding to the target feature space. We evaluate our proposed DITOCDR on three real-world datasets collected by Amazon, and the experimental results demonstrate its excellent performance and effectiveness.

Keywords: Cross-domain recommendation · Cold-start recommendations · Task optimization

1 Introduction

The development of the network world creates a challenge in extracting useful information from vast data, making recommender systems crucial for information filtering. However, collaborative filtering and deep learning models often struggle with the cold start problem in real-world scenarios due to data sparsity. Cross-domain recommendation (CDR) [1–5] addresses this issue by leveraging user interactions across multiple fields, such as ratings, articles, and videos. CDR aims to transfer information from the auxiliary domain to the target domain, alleviating the cold start problem.

Most existing CDR approaches employ embedding and mapping methods. Initially, user preferences in the source and target domains are encoded separately as embeddings. An explicit mapping function is then learned based on overlapping users to minimize the distance between the target embedding and the mapped embeddings.

Previous research indicates a significant challenge faced by most embedding and mapping models: the low proportion of overlapping users between the source and target domains, typically less than 10% in real-world scenarios. Our experiments involve two tasks, with overlap rates of 9.95% for Task 1 and only 5.41% for Task 2, as shown in Table 1. This limitation hinders model generalization and impairs the performance of cold-start users in the target domain. Additionally, the mapping method learns to map user embeddings from the source domain to the target domain, aiming to minimize the distance. However, the target domain user embeddings are obtained through encoding learning, potentially leading to unreasonable target embeddings and further complicating the challenge of model generalization.

Table 1. Statistics of the tasks datasets (*#Rate* denotes the overlap ratio)

	#Domain	#Users	#Items	#Ratings	#Overlap	#Rate
Task1	Movie	123,960	50,052	1,697,533	18,031	9.95%
	Music	75,258	64,443	1,097,592		
Task2	Book	603,668	367,982	8,898,041	37,388	5.41%
	Movie	123,960	50,052	1,697,533		

To address these challenges, we propose a novel method called Domain-Invariant Task Optimization for Cross-domain Recommendation (DITOCDR). DITOCDR effectively tackles the cold-start problem caused by data sparsity by utilizing non-overlapping user data and training task-oriented methods to optimize the learning process of target embeddings, resulting in improved model performance. Experimental results demonstrate the practicality and generalization ability of DITOCDR.

The main contributions of our work can be summarized in three aspects:

- We introduce DITOCDR, a domain-invariant task optimization framework, to tackle the cold-start problem in CDR. DITOCDR employs autoencoder networks in both the source and target domains to learn embedding information for all users, overcoming the generalization challenge caused by a low number of overlapping users.
- We utilize a task-oriented optimization approach that focuses on rating rather than mapping. This technique mitigates the negative impact of unreasonable user embeddings, leading to improved recommendation accuracy.
- We adopt a joint training approach that simultaneously trains multiple tasks mentioned above. This approach involves shared user embedding parameters, mutually

regularized tasks, and implicit data augmentation. Experimental results demonstrate the effectiveness of our method in addressing cold-start scenarios.

2 Related Work

2.1 cross-domain Recommendation

According to CMF [6] assumes that user embeddings in different domains can be shared to establish a global user embedding matrix. In recent years, several models utilizing deep learning methods have been proposed for Cross-domain Recommendation. Prior work relevant to ours is based on cross-domain embedding mapping methods [2–5, 7–11]. EMCDR [3] is a commonly used CDR model that learns a mapping function as a bridge for user embeddings to map from the source domain to the target domain. SSCDR [2] uses a semi-supervised method and encodes distance information according to unlabeled data when learning the mapping function. LACDR [5] optimizes user embedding mapping by utilizing non-overlapping user data. DCDCSR [4] considers rating sparsity of individual users and items in different domains when learning the mapping function. PTUPCDR [11] learns user feature embeddings through a meta-network to generate personalized bridge functions for each user, achieving personalized preference transfer. However, PTUPCDR, like existing embedding mapping-based methods, requires training the mapping function using overlapping users and does not leverage non-overlapping user data.

The main difference between our work and previous literature is that we not only effectively utilized non-overlapping user data to learn cross-domain mapping but also optimized the task of the mapped user embeddings.

3 Approach

3.1 Problem Preliminary

In the CDR setting, two distinct domains are present: the source domain, also known as the auxiliary domain, and the target domain. Each domain comprises a set of users $\mathcal{U} = \{u_1, u_2, ...\}$, a set of items $\mathcal{V} = \{v_1, v_2, ...\}$, and a rating matrix, denoted by $r_{ij} \in \mathcal{R}$ which represents the rating assigned by a user to an item, signifying the interaction between them, and typically expressed as an integer, such as a scale from 1 to 5. Specifically, $\mathcal{U}^s, \mathcal{V}^s, \mathcal{R}^s$ represent the user set, item set, and rating matrix of the source domain, respectively, while $\mathcal{U}^t, \mathcal{V}^t, \mathcal{R}^t$ represent the user set, item set, and rating matrix of the target domain. An overlapping user set between the two domains exists, which is defined as $\mathcal{U}^o = \mathcal{U}^s \bigcap \mathcal{U}^t$. In this context, a cold start user refers to a user who has not interacted in the target domain, but has interacted in the auxiliary domain. Our main objective is to provide effective recommendations to cold start users in the target domain. The detailed description of the overall algorithm of DITOCDR can be found in Algorithm 1.

Algorithm 1 Domain-Invariant Task Optimization for CDR (DITOCDR)

Input: $\mathcal{U}^s$, $\mathcal{U}^t$, $\mathcal{V}^s$, $\mathcal{V}^t$, $\mathcal{U}^o$, $\mathcal{R}^s$, $\mathcal{R}^t$.
Input: task optimization network g_θ.
Pre-training Stage:
   1. A source model contains u^s, v^s.
   2. A target model contains u^t, v^t.
Extracting Domain Invariant Factors Stage:
   3. Learning source autoencoder and target autoencoder and extracting domain invariant factors by minimizing Equation (6), (7) and (8).
Task-based Optimization Stage:
   4. Learning a task optimization network g_θ by minimizing Equation (10).
Test Stage:
   5. for a cold-start user u, we use $g_\theta(u^s)$ as the user embedding for prediction.

We will explain in detail below. The overall structure of DITOCDR is shown in Fig. 1.

3.2 Pre-training

Pre-training involves training a latent factor model for both the source and target domains. This model converts users and items into high-dimensional vectors, where each user and item has a distributed representation in this high-dimensional space. These vectors are commonly known as word embeddings or embedding vectors, and are considered as trainable parameters. The two most common methods for training latent factor models are matrix factorization (MF) and deep learning-based methods. MF decomposes the rating matrix into two low-dimensional matrices, while deep learning-based methods use a multi-layer perceptron for training. In this paper, R denotes the rating matrix, d represents the dimensionality of latent factors, $u_i^s \in \mathbb{R}^d$ and $v_j^s \in \mathbb{R}^d$ represent the user embedding and item embedding in the source domain, respectively, and $u_i^t \in \mathbb{R}^d$ and $v_j^t \in \mathbb{R}^d$ represent the corresponding embeddings in the target domain. Through Gaussian observation noise, the probability of the observed rating r_{ij} for a user's evaluation of an item can be modeled.

$$P\left(r_{ij}|u_i, v_j; \sigma^2\right) = N\left(r_{ij}|u_i^T v_j, \sigma^2\right). \tag{1}$$

Subsequently, the user and item embedding parameters were estimated using maximum likelihood estimation, which requires minimizing the following loss function:

$$\left(u_i v_j, r_{ij}\right) = \frac{1}{|\mathcal{R}|}\sum\nolimits_{r_{ij}\in\mathcal{R}}\left(r_{ij} - u_i v_j\right)^2. \tag{2}$$

where r_{ij} denotes the ground truth label and $|\mathcal{R}|$ denotes the number of ratings.

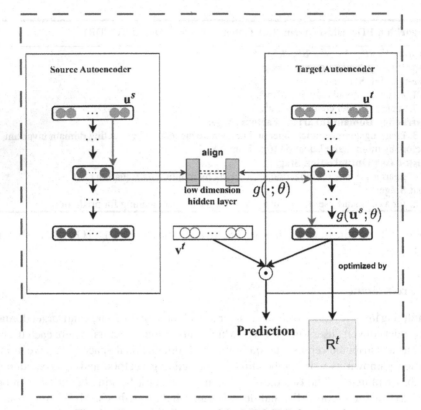

Fig. 1. Illustrative diagram of the DITOCDR framework.

3.3 Autoencoder Network

The embedding-based approach requires the user embeddings of the source domain and a mapping function that maps the user embeddings from the source domain to the target domain. Typically, the mapping function is trained by minimizing the following loss function:

$$\mathcal{L} = \sum_{u_i \in \mathcal{U}^o} \left\| f\left(\mathbf{u}_i^s; \phi\right) - \mathbf{u}_i^t \right\|_2. \tag{3}$$

In real-world CDR scenarios, the overlap of users between the two domains is typically minimal, usually no more than 10%, which poses a challenge for traditional embedding-based methods to train a mapping function that performs well for cross-domain recommendation of non-overlapping users. To this end, we adopt an encoder-decoder framework in both the source and target domains, using autoencoders to extract information from non-overlapping users. They can be formulated as follows:

$$\mathbf{h} = Encoder_s(\mathbf{u}). \tag{4}$$

$$\hat{\mathbf{u}} = Decoder_s(\mathbf{h}). \tag{5}$$

The variable **u** denotes the input, which is the user embedding of the source domain. The variable **h** denotes our intermediate layer, which exists in a low-dimensional space. The output **û** represents the reconstruction of the input, i.e., the source domain user embedding. Our goal in this part is to make the output **û** as close as possible to the input **u** and obtain reasonable domain-invariant factors. Therefore, for the source domain, we need to minimize the following reconstruction loss function:

$$\mathcal{L}_{re}^s = \sum\nolimits_{u \in \mathcal{U}^s} \|\hat{\mathbf{u}} - \mathbf{u}\|_2. \tag{6}$$

Similarly, we adopt the same framework for the autoencoder in the target domain as in the source domain. Therefore, for the target domain, we also need to minimize the following reconstruction loss function:

$$\mathcal{L}_{re}^t = \sum\nolimits_{u \in \mathcal{U}^t} \|\hat{\mathbf{u}} - \mathbf{u}\|_2. \tag{7}$$

And we need to minimize the following domain-invariant loss function in this part:

$$\mathcal{L}_{domain-invariant} = \sum\nolimits_{u_i \in \mathcal{U}^o} \left\| Encoder_s(\mathbf{u}_i^s) - Encoder_t(\mathbf{u}_i^t) \right\|_2. \tag{8}$$

3.4 Task-Based Optimization

The objective of task-based optimization is to optimize the transformed user embedding based on the task. Traditional embedding mapping methods mainly use matrix factorization or fully connected neural network models to map the source domain user embedding to the transformed target domain user embedding.

This method has demonstrated good performance in most models, but several is-sues exist. Unreasonable user embeddings and losing a large amount of rating information lead to the decrease in model accuracy. To tackle these issues, we propose a task-oriented optimization method to train the autoencoder network for the source and target domains. This optimization process directly utilizes the actual target domain users' item ratings as the optimization objective during training. Therefore, in this part, we aim to minimize the following task-oriented loss function:

$$\mathcal{L}_{task-oriented} = \frac{1}{|\mathcal{R}_o^t|} \sum\nolimits_{r_{ij} \in \mathcal{R}_o^t} \left(r_{ij} - g(\mathbf{u}_i^s; \theta)\mathbf{v}_j^t \right)^2. \tag{9}$$

where $\mathcal{R}_o^t = \{r_{ij} | u_i \in \mathcal{U}^o, v_j \in \mathcal{V}^t\}$ denotes the interactions of overlapping users in the target domain, g denotes the function process that transforms the source domain user embedding using the autoencoder network framework in both the source and target domains, as previously mentioned. g is parameterized by θ. $|\mathcal{R}_o^t|$ is the total number of user ratings for items in the target domain.

3.5 Joint Training

Initially, it is essential to delineate the joint training tasks required in this study, which include the four tasks. The four tasks are respectively source domain autoencoder training, target domain autoencoder training, domain-invariant factor extraction, and optimization based on real rating data. To execute multitasking concurrently, joint training for the aforementioned four tasks is conducted. More specifically, the source and target domain autoencoder models are trained using all users from both domains. Additionally, the third and fourth tasks are trained with overlapping users from both domains, while the remaining overlapping users serve as the testing set. Consequently, according to formula 6, 7, 8 and 9, the following joint training loss should be minimized:

$$\mathcal{L} = \mathcal{L}_{re}^s + \mathcal{L}_{re}^t + \gamma * \mathcal{L}_{domain-invariant} + \delta * \mathcal{L}_{task-oriented}. \tag{10}$$

where γ and δ represent hyperparameters employed to adjust relative weights.

Contrasting with most existing mainstream methods, which adhere to the embedding mapping scheme and train a single model for each task sequentially, our approach jointly trains four tasks. These tasks consist of two autoencoder tasks, a dimensionality reduction task for extracting domain-invariant factors, and a task optimized based on real rating data. These four tasks mutually regulate, share parameters, and utilize implicit data augmentation, thereby enhancing the model's performance.

4 Experiments

We conduct experiments to answer the following research questions: (Q1) How does our proposed CDR model perform in the cold-start cross-domain recommendation task compared to other baseline models that address this issue? (Q2) Does the use of non-overlapping user information and task-based optimization strategies help our model? (Q3) Why does our model achieve better performance?

4.1 Experiment Setup

Datasets. In order to conduct experiments directly in real CDR scenarios, we used the Amazon review dataset, which follows the approach of most existing methods [1, 3, 4, 9–11]. The specific statistics are shown in Table 1. Specifically, we proposed two real CDR tasks using the Amazon-5cores dataset, in which each user and item has at least five ratings. To better align with real-world scenarios, we used all the data in the dataset for CDR experiments.

Evaluation Metrics. The ground truth ratings in the Amazon review dataset are integers from 0 to 5. Following [3, 9, 11], we used the mean absolute error (MAE) and root mean square error (RMSE) as standards, which are commonly used metrics in regression tasks.

Baselines
TGT: Only applies the MF model in the target domain and trains it with the target domain data.

CMF [6]: An enhanced model that shares the embedding of overlapping users between the source and target domains, without considering the differences between the domains.

EMCDR [3]: A commonly used CDR model that mainly learns a mapping function as a bridge between the source and target domains for user embeddings.

SSCDR [2]: Uses a semi-supervised approach and adds distance information encoded according to unlabeled data in learning the mapping function.

DCDCSR [4]: Also a bridge-based method that considers the sparsity of individual user and item ratings in different domains in learning the mapping function.

PTUPCDR [11]: Considers the differences in user interests and adds an attention mechanism to the learning of the source domain user embeddings, learning a personalized bridge function.

Implementation Details. For optimization, we utilized the Adam optimizer [12] for all tasks and models. Following [3], we randomly selected a portion of overlapping users to construct the test set, with the proportion set to β, while the remaining users were allocated to the training set for the mapping function. To achieve reliable results, we ran the code randomly five times and obtained the average performance, as shown in Fig. 2 and Table 2.

Table 2. Cold-start results (RMSE) of two cross-domain tasks. We report the mean results over five runs. The best and the second best results are highlighted by boldface and underlined respectively. $\Delta\%$ denotes relative improvement over the best baseline.

Task	Task1			Task2		
β						
Method	10%	50%	90%	10%	50%	90%
TGT	5.097	5.158	5.177	4.798	4.769	4.824
CMF	2.029	2.212	2.929	1.883	1.987	3.338
DCDCSR	1.913	2.343	3.206	1.734	2.055	2.770
SSCDR	1.632	1.921	2.432	1.652	1.560	1.702
EMCDR	1.566	1.846	2.319	1.408	1.503	1.664
PTUPCDR	<u>1.424</u>	<u>1.648</u>	2.059	<u>1.364</u>	<u>1.443</u>	<u>1.587</u>
DITOCDR	**1.320**	**1.527**	**1.947**	**1.212**	**1.275**	**1.429**
DITOCDR_a	1.501	1.734	2.211	1.345	1.432	1.588
DITOCDR_b	1.438	1.653	2.142	1.311	1.416	1.546
DITOCDR_c	1.572	1.853	2.302	1.397	1.514	1.652
$\Delta\%$	7.30	7.34	5.43	11.14	11.64	9.95

4.2 Overall Performance of DITOCDR (Q1)

In this section, we discuss the experimental results of DITOCDR model in cold-start CDR scenarios. Based on existing bridge-based methods [1–4, 9–11], the performance of our DITOCDR model in cold-start CDR is presented in Table 2 and Fig. 2 for two real-world scenarios. The best performance in the experimental results is highlighted in bold. Δ% denotes the relative improvement of our model compared to the previous best baseline. Further analysis reveals the following two findings: (1) CMF is a data augmentation model that merges data from the source and target domains, improving the cold-start problem by employing auxiliary data for recommendations in the target domain. However, the performance of most baseline models surpasses that of CMF. The crux of the issue is that CMF disregards the differences between data from the source and target domains, potentially resulting in domain shift. Conversely, other baseline models are bridge-based methods that learn mapping functions and effectively alleviate this issue, which impedes CDR. (2) Lastly, we observe that DITOCDR's performance consistently outperforms the best baseline across all scenarios, indicating the efficacy of our proposed DITOCDR model in addressing the cold-start recommendation problem.

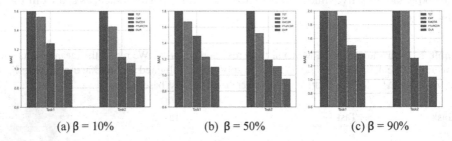

(a) β = 10%   (b) β = 50%   (c) β = 90%

Fig. 2. Cold-start experiments on TGT, CMF, EMCDR, PTUPCDR and DITOCDR for different proportions of test (cold-start) users: (a) β = 10%, (b) β = 50%, and (c) β = 90%.

4.3   Ablation Study (Q2)

Table 2 shows the results of DITOCDR_a, an ablation study we conducted to investigate the impact of non-overlapping user information on our model. In this experiment, we only used overlapping user data to train the autoencoder between the two domains, instead of using all user information from the source and target domains to train their respective autoencoder frameworks. The results indicate that using only the overlapping user information and adopting a task-based optimization strategy to implement the entire model significantly decreases the recommended performance of the model. Although its performance is weaker than the PTUPCDR model, it is still slightly stronger than the EMCDCR model. These findings suggest that introducing more auxiliary information, such as non-overlapping user information, can improve the recommendation accuracy of cross-domain recommendation. This is particularly important because it can more effectively alleviate the cold start problem that is commonly encountered in CDR.

We also conducted DITOCDR_b, an ablation study to investigate the impact of task-based optimization strategy on our model. In this experiment, we directly learned the target domain user embeddings learned in the latent factor model to predict user interests in target domain items for CDR, instead of learning the target domain user embeddings through a task-based optimization strategy. The data in Table 2 show that when we use non-overlapping user information for CDR without adopting a task-based optimization strategy, the recommended performance of the model still significantly decreases. Although its performance is weaker than the PTUPCDR model, it is still stronger than the EMCDR model. These findings suggest that task-based optimization strategy is also helpful for our model, as it directly learns the ground truth rating labels, which can lead to more reasonable embeddings for users compared to previous models that only learned user embeddings.

Finally, we conducted DITOCDR_c, an ablation study without using non-overlapping user information and a task-based optimization strategy. The performance of this experiment decreased the most compared to DITOCDR, and is very close to the EMCDR model's performance. This further illustrates the critical importance of non-overlapping user information for training the autoencoder framework of our model in the source and target domains, and the usefulness of task-based optimization strategy for better learning the target embeddings, i.e., the process of transforming user embeddings from the source domain to the target domain. Overall, both non-overlapping user information and task-based optimization strategy are crucial for improving the performance of our model for CDR.

4.4 Explanation of the Improvement (Q3)

In this section, we conducted experiments to visualize potential factors and illustrate why our DITOCDR model has further improvements. Specifically, we analyzed the embeddings in the feature space of the target domain to investigate the superiority of DITOCDR over EMCDR.

Following the default settings of t-SNE [13], We used it in Scikit-learn to visualize the user embeddings learned by EMCDR and DITOCDR on Task 2, with $\beta = 0.1$. Figure 3(a) shows the user embeddings learned by EMCDR, while Fig. 3(b) shows the visualized embeddings learned by DITOCDR. The blue dots denote the target embeddings of the target domain, which are considered the ground truth, and the red cross points denote the transformed embeddings learned by our model. To ensure clarity, we randomly selected 150 users for plotting.

Ideally, the distribution of the user embeddings we learned should be roughly similar to the distribution of the target embeddings. From Fig. 3(a), we observe that the distribution of the actual user embeddings is scattered, but the embeddings learned by EMCDR are relatively concentrated. This is due to the fact that EMCDR learns from overlapping user data, while in practice, the majority of users in most CDR tasks do not overlap. Therefore, ignoring the information of most non-overlapping users leads to a significant accuracy loss. Additionally, EMCDR directly learns the mapping function from the source domain user embeddings to the target domain user embeddings, while the target domain embeddings are learned through latent factor models, which leads to inaccurate target domain embeddings and a further performance decline. In contrast, DITOCDR achieves better results, as shown in Fig. 3(b). The transformed user embeddings learned by DITOCDR are not clustered together like those learned by EMCDR. Instead, they are scattered across the feature space of the target domain, indicating the superiority of using non-overlapping user data and task-based optimization strategies. Moreover, the distribution of the transformed embeddings by DITOCDR fits the distribution of the target embeddings better, which may be the fundamental reason why DITOCDR can achieve better overall performance.

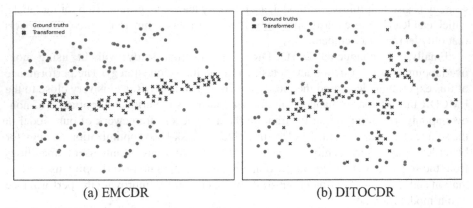

(a) EMCDR   (b) DITOCDR

Fig. 3. t-SNE visualization of randomly sampled user ground truths embeddings in target-domain feature space and transformed user embeddings. (a) and (b) denote the visualization results of EMCDR and DITOCDR, respectively.

5 Conclusion

In this paper, we studied the cold-start problem in cross-domain recommendation (CDR). Many popular CDR methods assume overlap between users and directly learn target embeddings. However, disregarding information from non-overlapping users and learning unreasonable target embeddings can lead to challenges in model generalization. To overcome these issues, we propose a new framework called Domain-Invariant Task Optimization for Cross-domain Recommendation (DITOCDR). Briefly, two autoencoder networks learn all user data from source and target domains and extract domain-invariant factors. The training of target embeddings is optimized using task-based optimization strategies. We conducted extensive experiments on three real-world datasets from Amazon to evaluate DITOCDR. Experimental results demonstrate the excellent performance and effectiveness of our proposed DITOCDR.

References

1. Fu, W., Peng, Z., Wang, S., Xu, Y., Li, J.: Deeply fusing reviews and contents for cold start users in cross-domain recommendation systems. In: AAAI (2019)
2. Kang, S., Hwang, J., Lee, D., Yu, H.: Semi-supervised learning for cross-domain recommendation to cold-start users. In: CIKM (2019)
3. Man, T., Shen, H., Jin, X., Cheng, X.: Cross-domain recommendation: an embedding and mapping approach. In: IJCAI (2017)
4. Zhu, F., Wang, Y., Chen, C., Liu, G., Orgun, M., Wu, J.: A deep framework for cross-domain and cross-system recommendations. arXiv preprint arXiv:2009.06215 (2020)

5. Wang, T., Zhuang, F., Zhang, Z., Wang, D., Zhou, J., He, Q.: Low-dimensional alignment for cross-domain recommendation. In: CIKM (2021)
6. Singh, A.P., Gordon, G.J.: Relational learning via collective matrix factorization. In: SIGKDD (2008)
7. Pan, W., Xiang, E., Liu, N., Yang, Q.: Transfer learning in collaborative filtering for sparsity reduction. In: AAAI (2010)
8. Zhang, Y., Liu, Y., Han, P., Miao, C., Cui, L., Li, B., Tang, H.: Learning personalized itemset mapping for cross-domain recommendation. In: IJCAI (2020)
9. Zhao, C., Li, C., Xiao, R., Deng, H., Sun, A.: CATN: cross-domain recommendation for cold-start users via aspect transfer network. In: SIGIR (2020)
10. Zhu, Y., et al.: Transfer-meta framework for cross-domain recommendation to cold-start users. In: SIGIR (2021)
11. Zhu, Y., et al.: Personalized transfer of user preferences for cross-domain recommendation. In: WSDM (2022)
12. Kingma, D.P., Ba, J.: Adam: a method for stochastic optimization. arXiv preprint arXiv:1412.6980 (2014)
13. Donahue, J., Jia, Y., Vinyals, J., Hoffman, J., Zhang, N.: DeCAF: a deep convolutional activation feature for generic visual recognition. In: ICML (2014)

Ensemble of Randomized Neural Network and Boosted Trees for Eye-Tracking-Based Driver Situation Awareness Recognition and Interpretation

Ruilin Li[1,2], Minghui Hu[1], Jian Cui[3]([✉]), Lipo Wang[1]([✉]), and Olga Sourina[4]

[1] School of Electrical and Electronic Engineering, Nanyang Technological University, Singapore 639798, Singapore
{ruilin001,e200008}@e.ntu.edu.sg, elpwang@ntu.edu.sg
[2] Yong Loo Lin School of Medicine, National University of Singapore, Singapore 117599, Singapore
[3] Research Center for Augmented Intelligence, Research Institute of Artificial Intelligence, Zhejiang Lab, Hangzhou, Zhejiang, China
cuijian@zhejianglab.com
[4] Fraunhofer, Nanyang Technological University, 50 Nanyang Avenue, Singapore 639798, Singapore
eosourina@ntu.edu.sg

Abstract. Ensuring traffic safety is crucial in the pursuit of sustainable transportation. Across diverse traffic systems, maintaining good situation awareness (SA) is important in promoting and upholding traffic safety. This work focuses on a regression problem of using eye-tracking features to perform situation awareness (SA) recognition in the context of conditionally automated driving. As a type of tabular dataset, recent advances have shown that both neural networks (NNs) and gradient-boosted decision trees (GBDTs) are potential solutions to achieve better performance. To avoid the complex analysis to select the suitable model for the task, this work proposed to combine the NNs and tree-based models to achieve better performance on the task of SA assessment generally. Considering the necessity of the real-time measure for practical applications, the ensemble deep random vector functional link (edRVFL) and light gradient boosting machine (lightGBM) were used as the representative models of NNs and GBDTs in the investigation, respectively. Furthermore, this work exploited Shapley additive explanations (SHAP) to interpret the contributions of the input features, upon which we further developed two ensemble modes. Experimental results demonstrated that the proposed model outperformed the baseline models, highlighting its effectiveness. In addition, the interpretation results can also provide practitioners with references regarding the eye-tracking features that are more relevant to SA recognition.

Keywords: Situation awareness (SA) · Eye-tracking · Ensemble deep random vector functional link (edRVFL) · Light gradient boosting machine (lightGBM) · Shapley additive explanations (SHAP)

B. Luo et al. (Eds.): ICONIP 2023, LNCS 14449, pp. 500–511, 2024.
https://doi.org/10.1007/978-981-99-8067-3_37

1   Introduction

Sustainable transportation aims to minimize the negative impacts of transportation systems on the environment and resources while meeting people's mobility needs. A crucial element in achieving sustainable transportation, particularly in the context of conditionally automated driving, is the enhancement of traffic safety. This entails the need for a well-developed situational awareness (SA) [1] to effectively take over control of the automated vehicle when it reaches its operational limitations.

Measuring SA efficiently and effectively is one of the most pressing issues in SA studies. Conventional approaches have been widely used as objective measures, including the Situation Awareness Global Assessment Technique (SAGAT) [1]. However, these methods have certain limitations, such as intrusiveness and limited applicability in real-world settings. Alternatively, researchers have discovered that SA states can be inferred by measuring physiological signals. One of the most representative bio-signals is the eye-tracking signals [2]. Physiological signal-based measures are gaining popularity in SA recognition due to their advantages of objectivity and real-time operation. Moreover, with the availability of data labeled using conventional methods, automated SA recognition and improved performance have been achieved using machine learning algorithms [3,4].

In this study, we specifically addressed the issue of eye-tracking-based SA recognition. Traditionally, eye-tracking-based recognition research revolves around the utilization of exported eye movement features [5], treating the eye-tracking feature-SA states data as a tabular dataset. Recent advancements in regression tasks pertaining to tabular datasets have yielded diverse outcomes. Both neural networks (NNs) [6] and tree-based models, especially gradient-boosted decision trees (GBDTs) [7], have exhibited superior performance and recent works have shown that either NNs or GBDTs could outperform the other model. Within the context of eye-tracking-based SA recognition, identifying the most appropriate models for achieving optimal recognition results still remains a research gap. While preliminary studies for tabular datasets have explored the influence of meta-features on model selection [8], the existing analysis remains intricate and lacks automation, particularly when considering the inclusion of additional features, datasets, and various practical scenarios. Consequently, it is desired to develop a comprehensive model capable of generally enhancing the performance of eye-tracking-based SA recognition.

Moreover, in most previous comparisons between NNs and GBDTs, NNs based on back-propagation algorithms have predominantly been employed [9]. Although they have demonstrated exceptional performance, the time-consuming intrinsic property poses a challenge for real-time requirements in driver's SA recognition. In contrast, randomized neural networks (RNNs) offer the advantages of fast computation and have global optimum solutions. Recent studies have also revealed the outstanding performance of edRVFL [10,11] and its variants [12,13] on tabular datasets. Hence, to achieve competitive performance while maintaining computational efficiency, the edRVFL was selected as the representative NN in our investigation.

In terms of GBDTs, it is worth noting that the light gradient boosting machine (lightGBM) [14] was recognized for its faster computation compared to other GBDT implementations, making it an ideal choice as the representative GBDT in our investigation. While lightGBM has been explored for eye-tracking-based sentiment analysis (SA) recognition [4], the default parameters were employed. It is important to highlight that using baseline models without proper parameter tuning may result in an unfair comparison and limit the understanding of lightGBM's full potential in SA recognition.

In the case of tabular datasets, understanding the contribution of input features to the final output and interpreting the model's decisions are crucial for practical applications. However, the interpretation of the edRVFL network remains an area that requires further research. In this study, we addressed this gap by utilizing SHAP additive explanations (SHAP) [15] to recognize the concentration of the edRVFL on the input features. Furthermore, leveraging this interpretation, we identified the importance of input features for individual and basic ensemble models, upon which two additional ensemble approaches were proposed to enhance SA recognition performance further.

Overall, to address the aforementioned challenges in eye-tracking-based SA recognition, this work made the following contributions:

- This work proposed a general ensemble framework to combine RNNs and GBDTs to boost the SA recognition performance while keeping the computational efficiency. The Bayesian optimization (BO) [16] was employed to optimize the hyper-parameters of the models and achieve a fair comparison.
- SHAP was used to interpret the importance of the input features.
  - By identifying the high-importance input features of the basic ensemble model, feature selection was performed to achieve superior performance.
  - We found that edRVFL and lightGBM had different focuses, upon which we developed an enhanced ensemble based on separate feature selections for each individual model.
- The proposed ensemble framework and the corresponding interpretation could be used as a general baseline/reference for practitioners to select suitable eye-tracking-based SA recognition models and suitable input features to achieve outstanding performance.

In the following part of the paper, the methodology is introduced in Sect. 2. The experiment results are discussed in Sect. 3. Finally, the conclusion and the future work are presented in Sect. 4.

2   Methods

Ensemble learning can usually lead to higher performance by capturing different aspects of the data and leveraging the strengths of each base learner, resulting in improved overall performance. Therefore, ensemble learning was employed in this work and we proposed a general ensemble framework to combine edRVFL [10] and lightGBM [14]. Three different ensemble models were developed under the framework.

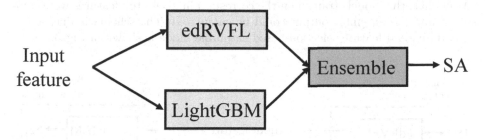

Fig. 1. Schematic diagram of basic ensemble mode.

Basic Ensemble Mode. At first, we proposed to perform a simple ensemble of two models to combine the outputs of two models directly. We followed the process of Shi *et al.* [10] to train the edRVFL. The hyper-parameter setting is presented in Sect. 3.2. The schematic diagram of the basic ensemble mode is shown in Fig. 1.

Feature Selection-Based Ensemble Mode. Then, SHAP [15] was employed to investigate the contribution of the input features (as shown in Sect. 3.4). Based on the obtained SHAP values, the features with higher importance were selected and further used to retrain the ensemble model. The schematic diagram of this feature selection-based ensemble mode is shown in Fig. 2.

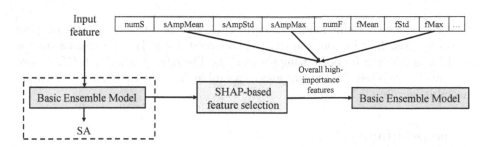

Fig. 2. Schematic diagram of feature selection-based ensemble mode.

Individual Feature-Selection-Based Esnemble Mode. In addition, using SHAP [15] to individually interpret the contribution of input features for edRVFL and lightGBM was performed. For each model, the high-importance features were selected separately based on the SHAP values (as shown in Sect. 3.4). After that, the models trained on the corresponding selected features were combined and the ensemble output could be calculated. The schematic diagram of this Individual feature selection-based ensemble mode is shown in Fig. 3.

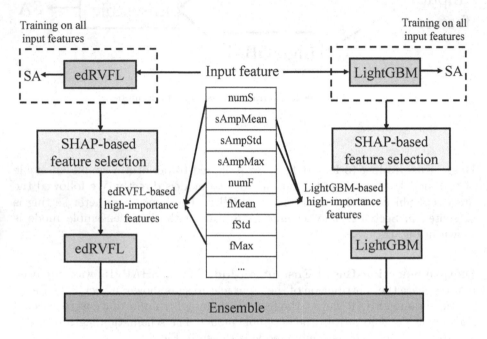

Fig. 3. Schematic diagram of individual feature selection-based ensemble mode.

For all three ensemble modes, the averaging was used to combine the outputs of two models. In addition, the BO implemented using Tree Parzen Estimator (TPE) was adopted for optimizing the models. The edRVFL and lightGBM were optimized separately in the proposed ensemble framework before performing ensemble learning.

3　Experiment

3.1　Dataset

This work investigated the SA recognition based on a public dataset [17] which involved 32 participants who watched 33 videos featuring a conditionally automated vehicle from a driver's perspective. The videos depicted a three-lane driving scenario with 5 or 6 vehicles. The video lengths varied from 1 to 20 s. Each lane contained one to two vehicles, with two to four vehicles in front of the ego lane. The videos did not include any lane changes. In 16 scenarios, participants had to take control of the automated vehicle. After watching each video, participants had to select the appropriate maneuver decision to avoid a collision: 'Evade left', 'Brake only', or 'No need to take over'. The SA score ground truth for the corresponding participants was created using SAGAT.

This study utilized the pre-processed dataset from Zhou *et al.* [4]. Specifically, the SA scores were normalized to a range of 0 to 1, and SA prediction was treated as a regression problem. Eye-tracking data was collected using the EyeLink 1000 Plus eye tracker at a sampling rate of 2000 Hz. Sixteen eye-tracking features were extracted for SA prediction. The eye-tracking features are listed in Table 1.

Table 1. Explantions of the eye-tracking features in the dataset [4].

Features	Explanantion
numS	number of the saccades of the participant in one video
sAmpMean	average saccade amplitude of the participant in one video
sAmpStd	standard deviation of saccade amplitudes of the participant in one video
sAmpMax	maximum value of saccade amplitudes of the participant in one video
numF	number of fixations of the participant in one video
fMean	average fixation duration of the participant in one video
fStd	standard deviation of fixation duration of the participant in one video
fMax	maximum value of fixation duration of the participant in one video
backMirror	number of fixations of the participant on the rear
leftMirror	number of fixations of the participant on the left mirror in one video
rightMirror	number of fixations of the participant on the right mirror in one video
road	number of fixations of the participant on the road in one video
sky	number of fixations of the participant on the sky in one video
pupilChange	pupil size change between the end of and the beginning of the video
pupilMean	average value of the pupil diameter of the participant in one video
pupilStd	standard deviation of the pupil diameter of the participant in one video

3.2 Experiment Setting

To investigate the effectiveness of the proposed ensemble framework, we compared the ensemble modes with 8 relevant models, which consisted of 3 baseline models: linear regression (LR) [18], support vector regression (SVR) [19] with RBF kernel and ridge regression [20], 2 GBDT models: extreme gradient boosting (XGBoost) [21] and lightGBM [14], as well as 3 RNNs: RVFL [22], extreme learning machine [23] and edRVFL [10].

To ensure a fair comparison, we conducted a 10-fold cross-validation using 10 different random seeds to assess the performance of the compared models, excluding the baseline models. Within each fold, 25% of the training set was utilized as the validation set. Additionally, we employed BO to optimize the hyper-parameters of the compared models. The search space utilized for hyper-parameter optimization is outlined in Table 2. The hyper-parameter names specified in the respective packages were listed in the table. Specifically, SVR, linear regression, and ridge regression employed the sklearn package [21], while scikit-learn interface of XGBoost [21] and lightGBM packages [14] was used for XGboost and lightGBM, respectively. For the hyper-parameter settings of RNNs, we refer the readers to [10,13]

Three evaluation metrics were used in this work, namely root mean square error (RMSE), mean absolute error (MAE), and Correlation.

Table 2. Regression models' hyper-parameter search space.

Model	Hyper-parameter	Search space
SVR	C	[1, 16]
	epsilon	[0.1, 1]
Ridge	alpha	$[10^{-5}, 0.5]$
XGBoost& LightGBM	learning_rate	[0.05, 0.5]
	n_estimators	[10, 100]
	max_depth	[2, 10]
	min_child_weight	[1, 6]
	reg_alpha	$[10^{-5}, 1]$
	reg_lambda	$[10^{-5}, 1]$
ELM & RVFL& edRVFL	Number of hidden nodes	[2, 256]
	Regularization parameter	$[10^{-5}, 0.5]$
	Weight scaling	[0.1, 1]

3.3 Basic Ensemble Mode

The comparison results of the proposed basic ensemble model and the relevant models are shown in Table 3. It is observed that the proposed basic ensemble model showed better RMSE and Correlation than the compared models while presented a comparable MAE with lightGBM.

Table 3. Comparison results on SA recognition

	LR	SVR	Ridge	ELM	RVFL	edRVFL	XGBoost	LightGBM	Ensemble
RMSE	0.1269	0.1272	0.1269	0.1308	0.1277	0.1261	0.1246	0.1241	**0.1236**
MAE	0.1025	0.1001	0.1025	0.1045	0.1024	0.1003	0.0991	**0.0986**	0.0987
Correlation	0.3571	0.3678	0.3572	0.3124	0.3504	0.3785	0.3968	0.4066	**0.4192**

3.4   Interpretation Results

By following Zhou *et al.* [4], we computed the SHAP values for all samples during the 10-fold cross-validation process. The Violin Plot was used to visualize the interpretation results of the edRVFL, lightGBM and the basic ensemble model. These plots serve the purpose of not only displaying the distribution of SHAP values for each feature but also indicating the importance of the features in the model, enabling a comparison of their impacts. The Violin plots of three models are shown in Fig. 4.

While it was observed that the focus on input features varies across different models, it can still be concluded that certain eye-tracking features such as 'backMirror', 'fMax', 'road', etc., generally have a more significant impact on the model outputs compared to other features. This finding aligns with previous studies demonstrating the crucial role of driver fixation activities in SA recognition [4, 17].

3.5   Feature Selection-Based Ensemble Modes

By using the SHAP values to assess the contributions of each feature to the model outputs, we selected high-importance features to enhance the performance of SA recognition. The number of selected features was determined using the validation datasets. Comparative results of the three proposed ensemble modes are depicted in Fig. 5. The feature selection-based ensemble modes exhibited improved performance compared to the basic ensemble mode. Particularly, the individual feature selection-based ensemble modes demonstrated the highest average performance among the three models. Analyzing the interpretation results illustrated in Fig. 5, it was evident that edRVFL and lightGBM models prioritized different input features. Therefore, opting for a separate selection of high-importance features allows us to capitalize on the strengths of both models, leading to superior SA recognition performance.

3.6   Statistical Analysis

The one-tail Wilcoxon paired signed-rank test was performed using three different evaluation metrics. The results were shown in Table 4. The basic ensemble modes exhibited significantly higher Correlation than the compared models. Notably, it was observed that both feature selection-based ensemble modes significantly outperformed the single models. These statistical findings further demonstrated the effectiveness of the proposed ensemble framework.

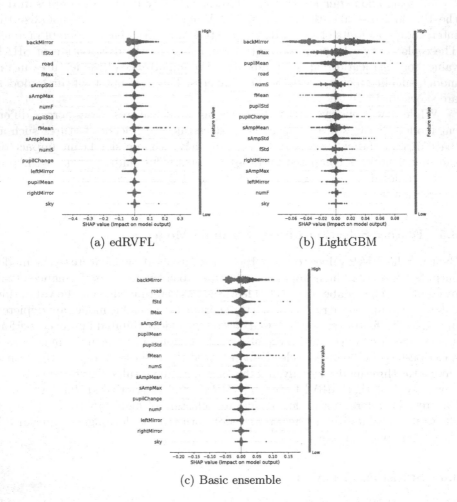

(a) edRVFL   (b) LightGBM

(c) Basic ensemble

Fig. 4. Violin Plot of three models. 1) The color ranging from blue (representing low values) to red (representing high values) visually represents the transition of the variable from a lower to a higher value. 2) The horizontal axis displays the SHAP values relative to the baseline value, providing insights into the impact of predictor variables on the target variable. Positive SHAP values indicate an increase in the target variable, while negative values suggest a decrease. 3) The vertical axis arranges the predictor variables based on their importance, with higher-ranked variables positioned at the top. This ordering emphasizes the relative significance of variables, highlighting those at the top as more influential than those lower down. (Color figure online)

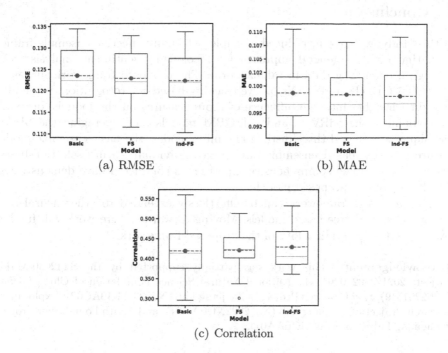

(a) RMSE   (b) MAE

(c) Correlation

Fig. 5. Comparison results of the proposed ensemble models in terms of three evaluation metrics. The term 'Basic' represents the basic ensemble model. The term 'FS' represents the feature selection-based ensemble model. The term 'Ind-FS' represents the individual feature selection-based ensemble model. The red lines and red dots represent the mean values, while the orange lines represent the median values. (Color figure online)

Table 4. Wilcoxon test results in terms of three metrics.

Model	Metric	LR	SVR	Ridge	ELM	RVFL	edRVFL	XGBoost	LightGBM
Basic	RMSE	+	+	+	+	+	+	−	−
	MAE	+	+	−	+	+	−	−	−
	Correlation	+	+	+	+	+	+	+	+
FS	RMSE	+	+	+	+	+	+	+	+
	MAE	+	+	−	+	+	+	+	+
	Correlation	+	+	+	+	+	+	+	+
IND-FS	RMSE	+	+	+	+	+	+	+	+
	MAE	+	+	+	+	+	+	+	+
	Correlation	+	+	+	+	+	+	+	+

'+' represents $p < 0.05$.
'−' represents $p > 0.05$.

4   Conclusion

In this study, we have introduced a simple yet highly effective ensemble framework that offers a general approach for recognizing sentiment analysis (SA) through eye-tracking data in tabular form. The proposed framework surpasses the SOTA GBDTs and RNNs on eye-tracking-based SA recognition, while also providing insights into the influence of input features on the models' outputs. By combining the edRVFL and lightGBM models and leveraging the SHAP technique, we devised three distinct ensemble modes: the basic ensemble mode, feature selection-based ensemble mode, and individual feature selection-based ensemble mode. The comprehensive experiments conducted have demonstrated the efficacy of our proposed models.

The proposed framework can be effortlessly extended to other neural networks (NNs) and tree-based models. Moving forward, future work will further investigate the generalizability of the proposed framework.

Acknowledgements. This work was partially supported by the STI2030-Major Projects 2021ZD0200201, the National Natural Science Foundation of China (Grant No. 62201519), Key Research Project of Zhejiang Lab (No. 2022KI0AC02), Exploratory Research Project of Zhejiang Lab (No. 2022ND0AN01), and Youth Foundation Project of Zhejiang Lab (No. K2023KI0AA01).

References

1. Endsley, M.R.: Measurement of situation awareness in dynamic systems. Hum. Factors **37**(1), 65–84 (1995)
2. de Winter, J.C., Eisma, Y.B., Cabrall, C., Hancock, P.A., Stanton, N.A.: Situation awareness based on eye movements in relation to the task environment. Cogn. Technol. Work **21**(1), 99–111 (2019). https://doi.org/10.1007/s10111-018-0527-6
3. Li, R., Wang, L., Sourina, O.: Subject matching for cross-subject EEG-based recognition of driver states related to situation awareness. Methods **202**, 136–143 (2022)
4. Zhou, F., Yang, X.J., de Winter, J.C.F.: Using eye-tracking data to predict situation awareness in real time during takeover transitions in conditionally automated driving. IEEE Trans. Intell. Transp. Syst. **23**(3), 2284–2295 (2022)
5. Li, F., Chen, C.-H., Lee, C.-H., Feng, S.: Artificial intelligence-enabled non-intrusive vigilance assessment approach to reducing traffic controller's human errors. Knowl.-Based Syst. **239**, 108047 (2022)
6. Kadra, A., Lindauer, M., Hutter, F., Grabocka, J.: Well-tuned simple nets excel on tabular datasets. In: Ranzato, M., Beygelzimer, A., Dauphin, Y., Liang, P., Vaughan, J.W. (eds.) Advances in Neural Information Processing Systems, vol. 34, pp. 23928–23941. Curran Associates Inc. (2021)
7. Gorishniy, Y., Rubachev, I., Khrulkov, V., Babenko, A.: Revisiting deep learning models for tabular data. In: Ranzato, M., Beygelzimer, A., Dauphin, Y., Liang, P., Vaughan, J.W. (eds.) Advances in Neural Information Processing Systems, vol. 34, pp. 18932–18943. Curran Associates Inc. (2021)
8. McElfresh, D., Khandagale, S., Valverde, J., Ramakrishnan, G., Goldblum, M., White, C., et al.: When do neural nets outperform boosted trees on tabular data? arXiv preprint arXiv:2305.02997 (2023)

9. Arik, S.O., Pfister, T.: TabNet: attentive interpretable tabular learning. In: Proceedings of the AAAI Conference on Artificial Intelligence, vol. 35, no. 8, pp. 6679–6687 (2021)
10. Shi, Q., Katuwal, R., Suganthan, P.N., Tanveer, M.: Random vector functional link neural network based ensemble deep learning. Pattern Recogn. **117**, 107978 (2021)
11. Malik, A.K., Ganaie, M.A., Tanveer, M., Suganthan, P.N., Alzheimer's Disease Neuroimaging Initiative Initiative: Alzheimer's disease diagnosis via intuitionistic fuzzy random vector functional link network. IEEE Trans. Comput. Soc. Syst. 1–12 (2022)
12. Hu, M., Chion, J.H., Suganthan, P.N., Katuwal, R.K.: Ensemble deep random vector functional link neural network for regression. IEEE Trans. Syst. Man Cybern. Syst. **53**(5), 2604–2615 (2023)
13. Malik, A., Gao, R., Ganaie, M., Tanveer, M., Suganthan, P.N.: Random vector functional link network: recent developments, applications, and future directions. Appl. Soft Comput. **143**, 110377 (2023)
14. Ke, G., et al.: LightGBM: a highly efficient gradient boosting decision tree. In: Guyon, I., et al. (eds.) Advances in Neural Information Processing Systems, vol. 30. Curran Associates Inc. (2017)
15. Lundberg, S.M., Lee, S.-I.: A unified approach to interpreting model predictions. In: Guyon, I., et al. (eds.) Advances in Neural Information Processing Systems, vol. 30, pp. 4765–4774. Curran Associates Inc. (2017)
16. Pelikan, M., Goldberg, D.E., Cantú-Paz, E.: BOA: the Bayesian optimization algorithm. In: Proceedings of the Genetic and Evolutionary Computation Conference, pp. 525–532 (1999)
17. Lu, Z., Happee, R., de Winter, J.C.: Take over! A video-clip study measuring attention, situation awareness, and decision-making in the face of an impending hazard. Transport. Res. F: Traffic Psychol. Behav. **72**, 211–225 (2020)
18. Galton, F.: Regression towards mediocrity in hereditary stature. J. Anthropol. Inst. Great Br. Irel. **15**, 246–263 (1886)
19. Platt, J., et al.: Probabilistic outputs for support vector machines and comparisons to regularized likelihood methods. Adv. Large Margin Classifiers **10**(3), 61–74 (1999)
20. Hoerl, A.E., Kennard, R.W.: Ridge regression: biased estimation for nonorthogonal problems. Technometrics **12**(1), 55–67 (1970)
21. Chen, T., Guestrin, C.: XGBoost: a scalable tree boosting system. In: Proceedings of the 22nd ACM SIGKDD International Conference on Knowledge Discovery and Data Mining, pp. 785–794 (2016)
22. Pao, Y.-H., Takefuji, Y.: Functional-link net computing: theory, system architecture, and functionalities. Computer **25**(5), 76–79 (1992)
23. Suganthan, P.N., Katuwal, R.: On the origins of randomization-based feedforward neural networks. Appl. Soft Comput. **105**, 107239 (2021)

Temporal Modeling Approach for Video Action Recognition Based on Vision-language Models

Yue Huang⬡ and Xiaodong Gu$^{(\boxtimes)}$⬡

Department of Electronic Engineering, Fudan University, Shanghai 200438, China
xdgu@fudan.edu.cn

Abstract. The usage of large-scale vision-language pre-training models plays an important role in reducing computational consumption and improving the accuracy of the video action recognition task. However, pre-training models trained by image data may ignore temporal information which is significant for video tasks. In this paper, we introduce a temporal modeling approach for the action recognition task based on large-scale pre-training models. We make the model capture the temporal information contained in frames by modeling the short-time local temporal information and the long-time global temporal information in videos separately. We introduce a multi-scale difference approach to getting the difference between adjacent frames, and employ a cross-frame attention approach to capturing semantic differences and details of temporal changes. In addition, we use residual attention blocks to implement the temporal Transformer and assign individual importance scores to each frame by computing the similarity of the frame to the clustering center, to obtain the overall temporal information of the video. Our model achieves 82.3% accuracy on the Kinetics400 dataset with just eight frames. Furthermore, zero-shot results on the HMDB51 dataset and UCF101 dataset demonstrate the strong transferability of our model.

Keywords: Action Recognition · CLIP · Multimodal Learning

1 Introduction

Action recognition has become an increasingly important component of video understanding and has attracted the attention of many researchers. Due to the development of the Internet, more and more large-scale datasets of action recognition like Kinetics400 [3] have been proposed. Thus, a large number of unique networks for action recognition sprang up. Unlike images, the existence of temporal dimension should be considered in the video model. For example, 3D convolution neural networks expand 2D image architectures into spacetime [4], and two-stream networks extract spatial and temporal features of video respectively [5]. However, conventional classifiers for video classification struggle to achieve

B. Luo et al. (Eds.): ICONIP 2023, LNCS 14449, pp. 512–523, 2024.
https://doi.org/10.1007/978-981-99-8067-3_38

good results due to the infinite label space for video, as well as the poor generalization and transfer abilities of previous classification models.

Transformer's dot-product attention provides a simple and effective fusion scheme for different modal information, and naturally builds bridges between visual and language. As a result, it has been applied to the video field with great success [7,11]. Compared with the CNN structure, Transformer has a less inductive bias, so more data or stronger regularization is required for training so training Transformer models for video recognition from scratch is highly inefficient, which is not friendly for academics with limited computing resources.

In recent years, there have been many large-scale vision-language models based on transformers, such as CLIP [8]. They can learn more powerful visual representations aligned with much richer language semantics and it is easy to do transfer learning and apply to downstream tasks by adapter [13] and prompt [14]. Therefore, instead of training the encoder from scratch, one can choose to use a model pre-trained on a large vision-language dataset to extract the desired features. But that paradigm faces a very serious problem: video has one more dimension than the image which contains temporal information. After extracting frame features with a pre-trained image encoder, we must find a way to extract and aggregate temporal information to obtain the final video features.

To solve this problem, we propose an effective framework that can realize both short-time and long-time temporal information aggregation with low computation cost and competitive results.

The main contributions of this paper can be summarized as follows.

- We propose a temporal modeling method that can effectively learn temporal information in videos for action recognition tasks. This method involves transferring a large model and applying it to the specific action recognition task.
- We use the difference method and the cross-frame attention method to extract short-time temporal information from video frames. Additionally, the temporal transformer method and frame importance assignment method are employed to process global temporal information.
- We do experiments on kinetics400 to demonstrate that the proposed temporal modeling approach significantly improves action recognition accuracy compared to direct frame averaging. Furthermore, it achieves improved zero-shot results on both the HMDB51 and UCF101 datasets.

2   Related Work

ConvNets-Based: When the input to the network is solely an RGB video, two types of models are typically considered: 2D ConvNets with LSTMs [17] and 3D ConvNets. The latter intuitively learns spatiotemporal features from RGB frames directly, extending the typical 2D CNN with an additional temporal dimension [4]. Because of the rising computational costs and parameters of 3D ConvNets, various efficient architectures have been designed [10] to find a balance between precision and speed. Furthermore, when pre-computed optical flow is included, other methods [5] model appearance and dynamics separately with two independent networks, fusing the two streams in the end.

Transformer-Based: Inspired by the success of Transformers in image recognition [6], video recognition has also begun to transition from 3D-CNN to Transformer-based architectures [11] that can jointly encode spatial and temporal features. Due to ViTs' high computational complexity, which restricts their practical applications, recent studies have concentrated on creating lightweight.

Transformer-based model architectures [12]. In comparison to CNN-based models, transformer-based models have demonstrated significant progress in video recognition tasks.

CLIP-Based: CLIP, which has shown considerable advantages in visual representation learning, has been transferred to video-related tasks. Recent works such as CLIP2Video [15] have achieved impressive results by considering temporal information in videos. Ju et al. [18] transferred the CLIP model to video recognition by learning cues and temporal modeling. Meanwhile, Wang et al. [9] utilized the CLIP model for video recognition through traditional end-to-end fine-tuning. However, as data for large-scale pre-training is often image-based rather than video-based, the main issue to address when processing video information is how to comprehensively process the vast amount of information in the time dimension.

3   Methodology

Inspired by ActionClip, we perform prompts on video and text separately, compared to their method, our method greatly enhances the utilization of temporal information in the video and focuses on the detail of temporal changes.

To make full use of the semantic information in the text, we expand texts to sentences through the text prompt proposed in ActionClip [9], and then use the pre-trained encoder of CLIP to extract features and get image representation I and text representation T. In order to make full use of the temporal information in the video, we add a video prompt to the image representation to get the video representation V.

After that, we calculate the softmax-normalized video-to-text and text-to-video similarity scores by contrasting the paired text against others in the sampled batch as:

$$p^{V2T} = \sum_i^N \log \frac{\exp\left(V_i^T T_i/\sigma\right)}{\sum_{j=1}^N \exp\left(V_i^T T_j/\sigma\right)} \tag{1}$$

$$p^{T2V} = \sum_i^N \log \frac{\exp\left(T_i^T V_i/\sigma\right)}{\sum_{j=1}^N \exp\left(T_i^T V_j/\sigma\right)} \tag{2}$$

where V_i and T_j are normalized embeddings of the video in the i-th pair and that of the text in the j-th pair. N is the batch size, and σ is a learnable temperature parameter to scale the logits. Finally, we can formulate the loss as:

$$L = -\frac{1}{N}\left(p^{V2T} + p^{V2T}\right) \tag{3}$$

To effectively utilize the temporal information in the video, some latest methods use late fusions of frame features as food representation, for example, ActionClip employs a temporal transformer on the output sequence, but it is easy to lose details of temporal changes.

For us, it is divided into long-term motion information and short-term motion information, the overall architecture is shown in Fig. 1. Just as the information in the image is divided into local information and global information, in the video, the short-term information describes the part of the whole video, and depicts the uniqueness of the action, that is, the movement difference per second, and the long-term information represents the whole of the video. It depicts the continuity of the action, that is, although the amplitude and direction of each frame are different, there is a complementary relationship between the actions of multiple frames.

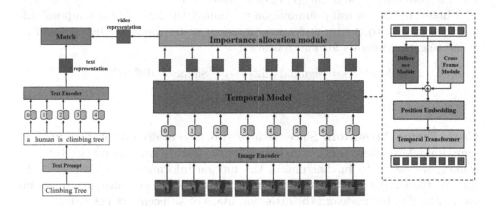

Fig. 1. Overview of the Network Structure

3.1   Short-Time Temporal Prompt

For local time information, we understand it as the information difference between frames, which is divided into local timing difference and local semantic difference. The inter-frame timing information depicts the difference of actions between frames, and the inter-frame semantic information depicts the similarity of actions between frames. The former can be achieved by modeling frame differences, and the latter can be achieved by cross-frame attention.

Multi-scale Temporal Difference. To use short-term information, a naive idea is to take the difference between adjacent frames, directly obtain the difference information between frames and use it. In order to avoid the large offset of object motion between adjacent frames, we adopt a form of inter-frame feature difference. Considering that direct subtraction of image representations may produce the problem of missing alignment of spatial location, we first use a convolution layer to the frame-level representation, which is because both spatial downsampling and convolution operations can increase the receptive field.

$$\text{diff}(t) = \text{Conv1}\,(d\,(t)) - \text{Conv1}\,(d\,(t-1)) \tag{4}$$

Then, We use pooling operations with different spatial scales for the results of inter-frame feature subtraction to create motion information with different receptive fields.

$$\text{diff}_1\,(t) = \text{Conv2}\,(\text{pool}_1\,(\text{diff}\,(t))) \tag{5}$$

$$\text{diff}_2\,(t) = \text{Conv3}\,(\text{pool}_2\,(\text{diff}\,(t))) \tag{6}$$

where pool_1 and pool_2 represent different space pooling sizes. Conv2 and Conv3 represent the convolution operations performed on the outputs of the two different pooling respectively, $\text{diff}_1\,(t)$ and $\text{diff}_2\,(t)$ represent the short time temporal difference with different receptive fields between adjacent frames, and the dimension for both $\text{diff}_1\,(t)$ and $\text{diff}_2\,(t)$ of each frame is 1.

Finally, instead of using summation to connect the inter-frame temporal differences directly to the input features in terms of residuals, we use sigmoid to generate gating weights for multiplication.

$$\text{Mark}\,(t) = \text{Sigmoid}\,(\text{diff}_1\,(t)) + \text{Sigmoid}\,(\text{diff}_2\,(t)) \tag{7}$$

$$I'\,(t) = I\,(t) + I\,(t) * \text{Mark}\,(t) \tag{8}$$

Cross Frame Attention. Since the attention map is also one of the sources of temporal information, because it can reflect the feature correspondence between frames, we use it to further obtain the motion information in the video. To achieve the short-term temporal information we need, we construct temporal cross-attention by modeling the attention maps of adjacent frames (Fig. 2).

We also calculate the query and key of adjacent frames to build the attention map. For the frame $I\,(t)$ corresponding to the current time t, we first calculate its Query. Then, we get the key of the previous frame and the next frame and obtain their semantic similarity by the following methods:

$$A_{prev}\,(t) = \text{SoftMax}\left((QI\,(t))^{T}\,(KI\,(t-1))\right) \tag{9}$$

$$A_{next}\,(t) = \text{SoftMax}\left((QI\,(t))^{T}\,(KI\,(t+1))\right) \tag{10}$$

For all frames in a segment, we obtain the relationship between the previous frame and the next frame of the current frame according to the above formula. After that, we make the model learn different weights for the frames before and after, multiply the attention map and the weights and then project them into the original feature space.

$$I''\,(t) = I\,(t) + I'\,(t) + W_{prev}A_{prev}\,(t) + W_{next}A_{next}\,(t) \tag{11}$$

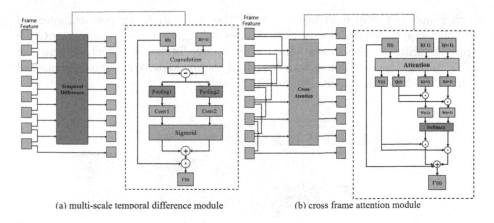

(a) multi-scale temporal difference module   (b) cross frame attention module

Fig. 2. Two short-time temporal modules

where W_{prev} represents the weight for the attention map with the previous frame, while W_{next} represents that with the next frame. Finally, the temporal features are mixed with the features of the original frames through the residual method, and this method is used to capture the semantic change information between frames in a short time.

3.2   Long-Time Temporal Prompt

For long-term information, it is also divided into long-term information and long-term semantic information. The former models the temporal relationship between all frames, and the latter models the semantic relationship between all frames. We use a long short-term memory block to model the temporal relationship of the frame sequence, spectral clustering to model the semantic distribution of the frame sequence, and finally use the residual to tie it to the original features.

Temporal Transformer. To strengthen the association of all frames, we still follow the commonly used Temporal transformer method, using residual attention block to collect the global information of the whole video and combine the global information with the original features, while maintaining the good properties of the original features. After the temporal attention module, we represent the current frame features as:

$$I''' (t) = \text{TempTransf} (I'' (t)) \tag{12}$$

Frame Importance Allocation. To provide the model with key frame information and help the model learn the key information in the video, we adopt the spectral clustering method which suits high-dimensional data points clustering. We regard the medoid obtained by spectral clustering of the frame embedding as the semantic center, and assign importance scores based on the similarity of each frame to the medoid like Fig. 3. In the application, we perform spectral clustering on the embedding of video frames belonging to the same segment. We apply k-means to these data points after doing dimension reduction, and use KKZ initialization combined with it to find the medoids. We regard the medoid frame $I'''(c)$ as the semantic center of the video.

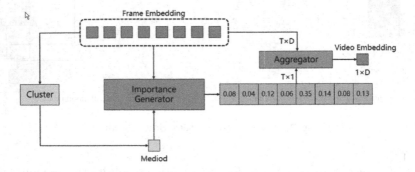

Fig. 3. Frame Importance Allocation Module

Then, we assign weights in the final embedding to the current frame by calculating the similarity between the semantic center frame and the current frame; frames with high similarity to the semantic center will occupy large weights, and frames with low similarity to the semantic center will be assigned small weights:

$$I''''(t) = \left(\text{SoftMax} \left(I'''(c)^T I'''(t) \right) \right) I'''(t) \tag{13}$$

In this way, the model can pay more attention to modules that are close to the semantic center, and ignore those that differ greatly from the semantic center. While using the time series Transformer to obtain the time series correlation, the key frame information is used to obtain better long-term temporal information utilization.

4   Experiments

In this section, we describe our implementation details and present the evaluation results with our method of action recognition benchmark. Besides, we report the zero-shot results of our model and show a visualization to demonstrate the effectiveness of the method.

4.1   Implementation

We give some implementation specifics here, including the datasets used, and the training parameters. The proposed approach is implemented in Pyhton utilizing Pytorch framework. The results in this part are obtained by using ViT-B/16 backbone, 8 input frames on Kinetics, and trained on only one NVIDIA GeForce RTX 3080Ti GPU with 12G video memory.

Datasets. We evaluate our proposed approach on four widely-used action recognition datasets, i.e., Kinetics-400, HMDB51 [2], and UCF101 [1]. Kinetics-400 contains around 240k training videos and 20k validation videos from 400 different action categories.

Due to the limitation of computational resources, it is too difficult for us to do ablation studies on the entire kinetics400 dataset. Therefore, we use two subsets of the kinetics400 dataset. The first split method is category split, we randomly selected 100 categories from the 400 categories data as the training set while the second split method is data split, we randomly selected 1/4 from all training data as the training set, and use all of val data.

HMDB51 dataset contains 6849 videos divided into 51 types of behavioral categories and the UCF101 dataset contains 101 action classes with a total of 13,320 video clips, these two datasets are used for zero-shot recognition in our paper.

Training Parameters. We apply the text encoder of CLIP which owns a Transformer of 12 layers with 8 attention heads and following ActionClip, we also use the ViT-B/16 visual encoder of CLIP which sets the input patch sizes to 16.

With a base learning rate of 5×10^{-6} for pre-trained parameters and 5×10^{-5} for learnable parameters that are newly added, we employ the AdamW optimizer. We train the model for 50 epochs and set the weight decay to 0.2. The learning rate is warmed up for 10% of the total training epochs and decayed to zero following a cosine schedule for the rest of the training. The shape of each extracted input frame is 224×224. Due to the limitation of computing resources, for each video clip, we just input 8 RGB frames.

4.2 Comparison with Previous Work

In this section, we compare our method with some previous methods. Comparisons with recent state-of-the-art models on Kinetics-400 are provided in Table 1. We find our models achieve competitive accuracy among regular video recognition methods, showing more efficient use of CLIP. Notice that the effect of the model tends to increase with the number of input frames, because the more input frames, the more information is collected, and our model achieves high results despite using only 8 frames for input. With an input of 8 frames, we have improved the result by almost 1.2 points over our baseline ActionClip, and only increase a few parameters.

4.3 Ablation Study

We realize the usage of the previous method did not make good use of the temporal information in video frames when applying CLIP to extract semantic information but only think of video frames as frame sets. To validate our approach, in this section, we investigate different factors affecting the performance of our method.

Table 1. Comparison with previous work on Kinetics-400

Method	Pre-train	Top1	Top5	Frames
I3D-NL	IN-1K	77.7	93.3	32
SlowFast-R50	No-extra-data	79.9	94.5	16+64
X3D-XXL	No-extra-data	80.4	94.7	16
TSM-R50	IN-1K	74.7	/	16
ViViT-L	JFT	82.8	95.3	32
Swin-B	IN-1K	80.6	94.6	32
MViT-B	No-extra-data	80.2	94.4	64
Uniformer-B	IN-1K	82.0	95.1	16
ActionClip	CLIP	81.1	95.5	8
ActionClip	CLIP	81.7	95.9	16
EVL-B [16]	CLIP	82.9	/	8
EVL-B	CLIP	83.6	/	16
Ours	CLIP	82.3	95.6	8

We examine the independent effect of the two short time block and the two long time block in Table 2, the 1 st row show the simplest baseline, where we just use the CLIP model on each frame separately, then perform a simple mean pooling to extract video feature. In this table, Difference and CrossAtt denote the multi-scale temporal difference module and the cross-frame attention module for capturing short-time temporal information while TempTrans and Allocation indicate the temporal Transformer module and the frame importance allocation module for aggregating long-time temporal information.

Table 2. Performance of temporal module on Kinetics-split dataset

Dataset	Difference	CrossAtt	TempTrans	Allocation	Top1	Top5
Kinetics400split1	✗	✗	✗	✗	69.33	91.59
	✓	✗	✗	✗	70.01	91.89
	✓	✓	✗	✗	70.37	92.04
	✓	✓	✓	✗	70.64	92.10
	✓	✓	✓	✓	70.83	92.09
Kinetics400split2	✗	✗	✗	✗	87.05	97.01
	✓	✗	✗	✗	87.35	97.08
	✓	✓	✗	✗	87.69	97.48
	✓	✓	✓	✗	88.04	97.78
	✓	✓	✓	✓	88.36	97.81

With the results on the first Kinetics split dataset in the first five lines in Table 2, we initially demonstrate the enhancement of the results by the short-time and long-time modules, where the model works best when these methods are used simultaneously.

In order to prevent data bias from splitting the dataset, we validate the results by training on the split2 in the last five lines in Table 2, the result further proves the validity and stackability of the short-time temporal module and the long-time aggregation module.

4.4 Visualization

To visualize the feasibility and effectiveness of the proposed model for the video recognition task, we employ t-SNE to cluster the video features before and after using the temporal modeling approach, and the clustering results are displayed on the 2D image.

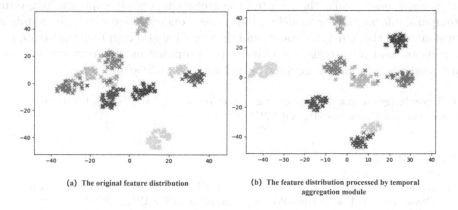

(a) The original feature distribution

(b) The feature distribution processed by temporal aggregation module

Fig. 4. Visualization of Video Feature

To visually compare the effectiveness of our proposed module, we randomly selected 9 classes of videos from the kinetics test list and show the clustering results before and after adding the module. Different classes are indicated by different colors. The results of Fig. 4 show that our proposed temporal aggregation module increases the differences between the features of video data from different classes while decreasing the differences between the features of video data from the same class. This demonstrates the usefulness of our proposed module for video classification tasks.

4.5 Zero-Shot Recognition

To verify the migration performance of the model, we test the zero-shot ability on the HMDB51 and UCF101 dataset, the results are shown in Table 3. It turns out that by temporal modeling, we allow the network to learn video features better and thus obtain better zero-shot results.

Table 3. Zero-shot performance

Dataset	Method	Pre-train	Average Score
HMDB51	ActionClip	CLIP+Kinetics400	50.2
	Ours	CLIP+Kinetics400	52.7
UCF101	ActionClip	CLIP+Kinetics400	69.7
	Ours	CLIP+Kinetics400	71.4

5   Conclusion

This paper proposes a novel and effective temporal modeling module for video action recognition. The module comprises two components: a short-term temporal module and a long-term aggregation module. The short-term temporal module captures short-term timing information by modeling the difference between adjacent frames, while the long-term aggregation module captures long-term temporal information by modeling the entire sequence. Moreover, the module is evaluated on the Kinetics dataset, and the experimental results demonstrate the significant and consistent superiority of the proposed module over state-of-the-art methods in terms of accuracy, robustness, and efficiency.

Acknowledgements. This work was supported in part by National Natural Science Foundation of China under grant 62176062.

References

1. Soomro, K., Zamir, A.R., Shah, M.: UCF101: a dataset of 101 human actions classes from videos in the wild. arXiv preprint arXiv:1212.0402 (2012)
2. Kuehne, H., Jhuang, H., Garrote, E., Poggio, T., Serre, T.: HMDB: a large video database for human motion recognition. In: 2011 International Conference on Computer Vision, pp. 2556–2563. IEEE (2011)
3. Carreira, J., Zisserman, A.: Quo vadis, action recognition? A new model and the kinetics dataset. In: Proceedings of the IEEE Conference on Computer Vision and Pattern Recognition, pp. 6299–6308 (2017)
4. Tran, D., Bourdev, L., Fergus, R., Torresani, L., Paluri, M.: Learning spatiotemporal features with 3D convolutional networks. In: Proceedings of the IEEE International Conference on Computer Vision, pp. 4489–4497 (2015)
5. Simonyan, K., Zisserman, A.: Two-stream convolutional networks for action recognition in videos. In: Advances in Neural Information Processing Systems, vol. 27 (2014)
6. Dosovitskiy, A., et al.: An image is worth 16x16 words: transformers for image recognition at scale. arXiv preprint arXiv:2010.11929 (2020)
7. Liu, Z., et al.: Video swin transformer. In: Proceedings of the IEEE/CVF Conference on Computer Vision and Pattern Recognition, pp. 3202–3211 (2022)
8. Radford, A., et al.: Learning transferable visual models from natural language supervision. In: International Conference on Machine Learning, pp. 8748–8763. PMLR (2021)

9. Wang, M., Xing, J., Liu, Y.: Actionclip: a new paradigm for video action recognition. arXiv preprint arXiv:2109.08472 (2021)
10. Feichtenhofer, C.: X3D: expanding architectures for efficient video recognition. In: Proceedings of the IEEE/CVF Conference on Computer Vision and Pattern Recognition, pp. 203–213 (2020)
11. Arnab, A., Dehghani, M., Heigold, G., Sun, C., Lučić, M., Schmid, C.: Vivit: a video vision transformer. In: Proceedings of the IEEE/CVF International Conference on Computer Vision, pp. 6836–6846 (2021)
12. Li, K., et al.: Uniformer: unifying convolution and self-attention for visual recognition. IEEE Trans. Pattern Anal. Mach. Intell. (2023)
13. Houlsby, N., et al.: Parameter-efficient transfer learning for NLP. In: International Conference on Machine Learning, pp. 2790–2799. PMLR (2019)
14. Zhou, K., Yang, J., Loy, C.C., Liu, Z.: Learning to prompt for vision-language models. Int. J. Comput. Vision **130**(9), 2337–2348 (2022)
15. Fang, H., Xiong, P., Xu, L., Chen, Y.: Clip2video: mastering video-text retrieval via image clip. arXiv preprint arXiv:2106.11097 (2021)
16. Lin, Z., et al.: Frozen clip models are efficient video learners. In: Avidan, S., Brostow, G., Cissé, M., Farinella, G.M., Hassner, T. (eds.) ECCV 2022. LNCS, vol. 13695, pp. 388–404. Springer, Cham (2022). https://doi.org/10.1007/978-3-031-19833-5_23
17. Shi, X., Chen, Z., Wang, H., Yeung, D.Y., Wong, W.K., Woo, W.C.: Convolutional LSTM network: a machine learning approach for precipitation nowcasting. In: Advances in Neural Information Processing Systems, vol. 28 (2015)
18. Ju, C., Han, T., Zheng, K., Zhang, Y., Xie, W.: Prompting visual-language models for efficient video understanding. In: Avidan, S., Brostow, G., Cissé, M., Farinella, G.M., Hassner, T. (eds.) ECCV 2022. LNCS, vol. 13695, pp. 105–124. Springer, Cham (2022). https://doi.org/10.1007/978-3-031-19833-5_7

A Deep Learning Framework with Pruning RoI Proposal for Dental Caries Detection in Panoramic X-ray Images

Xizhe Wang[1]([✉]), Jing Guo[2,3], Peng Zhang[4], Qilei Chen[1], Zhang Zhang[1], Yu Cao[1], Xinwen Fu[1], and Benyuan Liu[1]

[1] Miner School of Computer and Information Sciences,
University of Massachusetts Lowell, Lowell, USA
{xizhe_wang,zhang_zhang}@student.uml.edu,
{qilei_chen,yu_cao,xinwen_fu,benyuan_liu}@uml.edu
[2] Department of Dental General and Emergency, The Affiliated Stomatological
Hospital of Nanchang University, Nanchang, China
ndkqgj@ncu.edu.cn
[3] The Key Laboratory of Oral Biomedicine, Jiangxi Province Clinical Research
Center for Oral Diseases, Nanchang, China
[4] Department of Oral and Maxillofacial Imaging, The Affiliated Stomatological
Hospital of Nanchang University, Nanchang, China
ndfskqyy363@ncu.edu.cn

Abstract. Dental caries is a prevalent noncommunicable disease that affects over half of the global population. It can significantly diminish individuals' quality of life by impairing their eating and socializing abilities. Consistent dental check-ups and professional oral healthcare are crucial in preventing dental caries and other oral diseases. Deep learning based object detection provides an efficient approach to assist dentists in identifying and treating dental caries. In this paper, we present a deep learning framework with a lightweight pruning region of interest (P-RoI) proposal specifically designed for detecting dental caries in panoramic dental radiographic images. Moreover, this framework can be enhanced with an auxiliary head for label assignment during the training process. By utilizing the Cascade Mask R-CNN model with a ResNet-101 backbone as the baseline, our modified framework with the P-RoI proposal and auxiliary head achieves a notable 3.85 increase in Average Precision (AP) for the dental caries class within our dental dataset.

Keywords: Object detection · Pruning RoI proposal · Label assignment · Dental caries

1 Introduction

Oral health plays an important role in healthcare. Oral diseases, such as tooth decay, gum disease, and oral cancer [20], can have a seriously negative impact

B. Luo et al. (Eds.): ICONIP 2023, LNCS 14449, pp. 524–536, 2024.
https://doi.org/10.1007/978-981-99-8067-3_39

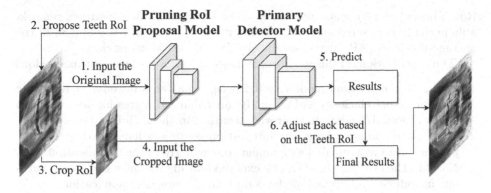

Fig. 1. The workflow of the proposed framework during the inference stage.

on the quality of life by impairing patients' abilities to eat and socialize. According to the "Global Oral Health Status Report" of World Health Organization (WHO), oral diseases are the most widespread noncommunicable diseases, affecting approximately half of the population worldwide. Particularly, tooth decay, also referred to as dental caries, is the most prevalent oral disease, affecting more than one third of the world's population throughout individuals' lifetime [1]. Regular dental check-ups and professional oral healthcare are essential preventive measures to combat dental caries and other oral diseases.

The detection of tooth and dental caries [5,11,17,18] in X-ray images has been proven to assist dentists in improving the identification and treatment of tooth decay. The choice of deep learning models and associated techniques plays a crucial role in achieving high performance in this domain. Among the state-of-the-art models for object detection, the You Only Look Once (YOLO) series stands out as a notable choice. These one-stage models have gained significant recognition for their performance. Notably, YOLO-X [3] and YOLOv7 [19] incorporate label assignment methods to improve classification accuracy. On the other hand, when segmentation labels are available in the dataset, the Cascade Mask R-CNN [2,6] emerges as one of the leading two-stage models for object detection.

In this study, we propose a deep learning framework for the detection of dental caries in panoramic dental X-ray images. Our framework integrates a lightweight pruning region of interest (P-RoI) proposal model, equipped with an enhanced regression loss function, to accurately identify the teeth RoI. The cropped RoI is then used for detecting dental caries. To enhance the label assignment, we introduce an auxiliary head that utilizes the AutoAssign model [22] during the training stage. The workflow of proposed framework is depicted in Fig. 1, illustrating the sequential steps during the inference stage. Initially, the P-RoI proposal module identifies the teeth RoI. Subsequently, the image is cropped according to the RoI, which serves as the input to the primary model for dental caries detection. Since the results are predicted within the cropped image, their coordinates are finally adjusted to align with the original image based on the

RoI. Through experimentation, our enhanced framework demonstrates remarkable performance, surpassing the baseline model with an increase of 3.85 in the average precision (AP) metric specifically for the dental caries class.

The contributions of this study encompass various aspects, as outlined below:

1. In the P-RoI proposal module, a lightweight model is introduced to ensure computational efficiency. Additionally, an enhanced regression loss function is incorporated, leading to improved accuracy in the P-RoI proposals.
2. During the training phase, the integration of an auxiliary head within the primary head aids in label assignment, particularly in the classification task. Notably, the auxiliary head can be excluded during the inference phase, without introducing additional model complexity in practical applications.
3. An alternative approach is explored, wherein the P-RoI proposal is utilized for cropping the extracted features generated by the primary detection model, instead of cropping the original image. While this exploration did not yield favorable outcomes, an extensive investigation was conducted to assess its potential for the specific task.

2   Related Work

Singh and Raza introduced TeethU^{2}Net [18], an innovative framework specifically designed for tooth saliency detection in panoramic X-ray images. Their work significantly improved upon the original U^2-Net by refining the loss function and training scheme, thereby enhancing the accuracy of tooth boundary segmentation. The proposed framework achieved exceptional performance with an accuracy of 0.9740, specificity of 0.9969, precision of 0.9880, recall of 0.8707, and an F_1-score of 0.9047. Moreover, in their study [17], the researchers also presented an optimized single-stage anchor-free deep learning model capable of detecting teeth and distinguishing their respective treatment types. Their proposed work demonstrated a remarkable AP of 91%.

Haghanifar *et al.* introduced PaXNet [5], a deep learning model designed for dental caries detection in panoramic X-ray images. The PaXNet pipeline involves a series of steps, including jaw extraction, jaw separation, tooth extraction, and classification of the cropped tooth images. According to the evaluation results, PaXNet achieved an accuracy of 86.05%, precision of 89.41%, and recall of 50.67% on their dataset. In addition, Imak *et al.* proposed an approach that employed a multi-input deep convolutional neural network ensemble model with a score fusion technique [15]. The model takes both raw and enhanced periapical images as inputs and fuses the scores in the Softmax layer. Evaluation on a dataset of 340 periapical images demonstrates the effectiveness of the proposed model, achieving an accuracy of 99.13% in diagnosing dental caries.

For the backbone architecture of the P-RoI proposal model, several lightweight models such as SqueezeNet [10], ShuffleNet series [13,21], and MobileNet series [8,9,16] can be considered as potential candidates. The neck architecture commonly employed is the Feature Pyramid Network (FPN) [12]. To effectively handle the RoI proposals, the Region Proposal Network (RPN) [14] is adopted.

These models are characterized by their compact and efficient design, making them suitable foundations for the framework due to their small-scale structure and computational efficiency. In terms of the regression loss function for the P-RoI proposal, our observations indicate that the Smooth L_1 Loss [4] outperforms the intersection of union (IOU) loss series.

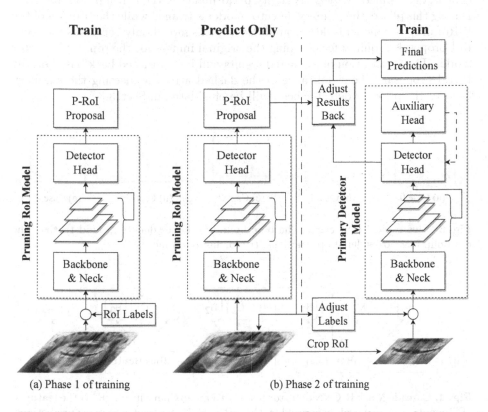

(a) Phase 1 of training   (b) Phase 2 of training

Fig. 2. The two training phases of our framework. "○" implies data combination.

3   Method

3.1   The Overview of Framework

The training of the framework is structured as a two-phase process, as illustrated in Fig. 2. The framework includes two detection models: a primary detector model specifically designed for dental caries detection and an additional detection model used for the pruning region of interest (P-RoI) proposal. The teeth RoI ground truth in each image can be obtained by finding the smallest top-y and left-x coordinates and the greatest bottom-y and right-x coordinates of all the bounding box labels. In phase 1, the P-RoI proposal model undergoes independent training for multiple epochs. Notably, the training of the primary detector model is temporarily deactivated during this phase. This precaution is taken to prevent the occurrence of loss exceptions that may arise from simultaneously training both models. The duration of phase 1 is empirically determined through experimentation to ensure effective training of the P-RoI proposal model.

Phase 2 of the training process initiates after reaching a pre-defined number of epochs, determined by the satisfactory performance of the P-RoI proposal model. During this phase, the primary detector model is trained while the training of the P-RoI proposal model is deactivated. Hence, this model only generates the teeth RoI proposals required for cropping the original images for the primary detector model. The localization of predicted results will be translated back based on the P-RoI proposal. The significance of the dashed arrow connecting the auxiliary head to the primary detector head will be elucidated in Sect. 3.3.

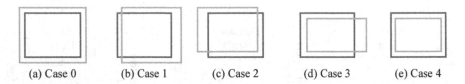

| (a) Case 0 | (b) Case 1 | (c) Case 2 | (d) Case 3 | (e) Case 4 |

Fig. 3. Five typical IOU cases. The blue bounding boxes denote ground truths. The red bounding boxes denote predictions. (Color figure online)

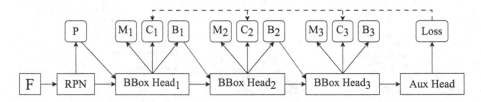

Fig. 4. Cascade Mask R-CNN detector head with an auxiliary head. "F" is the features, "P" denotes the proposals generated by the RPN, "M" represents the mask predictions, "C" signifies the classification results, and "B" indicates the bounding box predictions.

3.2   The Pruning RoI Proposal

In the second phase of the training process, the performance of the P-RoI proposal plays a vital role in influencing the final predictions of dental caries. Insufficient P-RoI proposals can lead to incomplete information being cropped, thereby deteriorating the performance of the primary detector model. In our implementation, we integrated the MobileNet-v3-small, FPN, and RPN with a single convolutional layer preceding the proposal head, which encompasses classification and regression branches, to form the P-RoI proposal model.

Figure 3 showcases several common scenarios of IOU cases, depicting the relationships between the bounding boxes of ground truths and predictions. The ground truths are represented by blue bounding boxes, while the predictions are denoted by red ones. The sub-figures demonstrate the following cases:

- (a) "Larger" prediction: This scenario displays a prediction with its top-y and left-x being equal or smaller than those of the ground truth, and bottom-y and right-x being equal or greater than the ground truth's. In other words, the ground truth bounding box is entirely contained within the prediction.
- (b) Insufficient on one side: In this case, only one side of the prediction bounding box is inadequate to achieve the "larger" prediction.
- (c) Insufficient on two sides: Here, two sides of the prediction bounding box are inadequate compared to the "larger" prediction.
- (d) Insufficient on three sides: This case showcases a prediction bounding box with three sides that are insufficient to reach the "larger" prediction.
- (e) Insufficient on all four sides: Finally, this scenario represents that a prediction bounding box is completely contained within the ground truth.

The performance of the P-RoI proposal is significantly influenced by the choice of the regression loss function. Through experimental observations, we find that the Smooth L_1 Loss function outperforms all variations within the IOU Loss series. While the selection of the Smooth L_1 Loss has yielded promising outcomes, there is still potential for further improvements by fine-tuning specific aspects of the loss function. When modifying the regression loss function, it is essential to adhere to several key principles that consider the characteristics and objectives of the RoI proposal process:

1. If an RoI proposal perfectly matches the ground truth, the loss should be 0.
2. When comparing multiple predictions, the loss associated with the more suitable prediction for detecting the ground truth should be smaller.
3. Assuming all the IOU values in the five cases of Fig. 3 are equal, the loss function should demonstrate a relative inclination to accommodate Case 0, the "larger" prediction. This inclination facilitates the preservation of valuable information within the RoI.

The Smooth L_1 Loss function satisfactorily fulfills principles 1 and 2. Thus, our modification primarily focuses on addressing principle 3 by introducing an additional loss penalty specifically for the "non-larger" predictions, which can be represented by cases 1 to 4 in Fig. 3. This modification aims to further refine the loss function and improve the overall performance of the P-RoI proposals. The original loss function of the P-RoI proposal model can be succinctly expressed as $\mathcal{L}_{P\text{-}RoI} = \mathcal{L}_{cls} + \mathcal{L}_{reg} = \mathcal{L}_{BCE} + \mathcal{L}_{SmoothL_1}$, where $\mathcal{L}_{cls}$ represents the classification loss implemented using Binary Cross Entropy Loss $\mathcal{L}_{BCE}$, $\mathcal{L}_{reg}$ denotes the regression loss achieved through Smooth L_1 Loss $\mathcal{L}_{SmoothL_1}$.

The modified regression loss function is defined as $\mathcal{L}_{reg} = \lambda_p^{\hat{n}} \cdot \mathcal{L}_{SmoothL_1}$, and the modified loss function of the P-RoI proposal model can be expressed as:

$$\mathcal{L}_{P\text{-}RoI} = \mathcal{L}_{BCE} + \lambda_p^{\hat{n}} \cdot \mathcal{L}_{SmoothL_1} \tag{1}$$

where λ_p is a weight used to fine-tune the regression loss, and $\lambda_p \in (0, +\infty)$, the exponent $\hat{n}$ indicates the number of sides of a prediction that are inadequate to reach the "larger" prediction case, and $\hat{n}$ can be 0, 1, 2, 3, or 4. Consequently, $\lambda_p > 1$ implies an additional penalty for "non-larger" predictions, and $0 < \lambda_p < 1$ indicates an additional penalty for "larger" predictions.

3.3 The Auxiliary Head for Label Assignment

The incorporation of an auxiliary head into the detector architecture plays a pivotal role in optimizing the label assignment process. There are two approaches to implementing the auxiliary head: one involves applying a label assignment algorithm to all of the extracted and cropped features provided by the proposed RoI, while the other involves globally performing the label assignment process on the entire set of features. In our framework, AutoAssign is selected as the auxiliary head to perform a global label assignment process, since its efficiency in distinguishing positive and negative targets with respect to the label assignment.

Figure 4 illustrates an example of the Cascade Mask R-CNN architecture with three cascaded bounding box regression stages, showcasing the integration of an auxiliary head at the conclusion of the primary head. The auxiliary head receives the complete set of features as input. The dashed arrow connecting the auxiliary head to the primary head signifies that the AutoAssign module influences the label assigning through exerting its impact on the loss function, rather than directly modifying the classification results within the primary head. Considering the loss function of Cascade Mask R-CNN, the loss function of the primary detector model can be expressed as follows:

$$\mathcal{L}_{Primary} = \mathcal{L}_{RPN} + \sum_{i=1,2,3} \lambda_i \cdot (\mathcal{L}_{cls,i} + \mathcal{L}_{reg,i} + \mathcal{L}_{mask,i}) + \mathcal{L}_{AutoAssign} \qquad (2)$$

where $\mathcal{L}_{RPN}$ is the loss of RPN, λ_i, $\mathcal{L}_{cls,i}$, $\mathcal{L}_{reg,i}$, and $\mathcal{L}_{mask,i}$ are the weight, classification loss, regression loss, and segmentation loss of the i-th stage of the cascaded bounding box heads, $\mathcal{L}_{AutoAssign}$ denotes the loss of the auxiliary head.

However, it is important to note that the final predictions are still obtained from the primary head, which is the last stage of the cascaded bounding box heads in the Cascade Mask R-CNN architecture, rather than directly from the auxiliary head itself. This design decision is driven by three main reasons. Firstly, the cascaded bounding box heads exhibit superior performance in terms of bounding box prediction. Secondly, the losses provided by AutoAssign are highly effective in fine-tuning the positive and negative of labels generated by the primary head. Lastly, the auxiliary head is solely utilized during the training stage to provide additional losses and can be safely omitted during the inference stage, thus not increasing the model's scale in practical applications.

4   Experiments

4.1   Dataset

A dataset consisting of 1,008 panoramic dental X-ray images, with each image exhibiting a resolution of 2,918 × 1,435 pixels, was created. The dataset incorporates a total of 31,296 annotations of teeth in both bounding box and segmentation, which can be classified into ten distinct dental categories, as specified in Table 1. Using a random and exclusive approach, we split the images, along with their corresponding labels, into training and test sets in a ratio of 3:1. Figure 5 illustrates representative samples of teeth within the ten classes. In each subfigure, the left one exhibits a tooth of a particular class, and the right one shows its related bounding box and segmentation annotations.

Table 1. Statistical information of our dental dataset.

	Class	Training	Test	Total
1	Normal	19,688	6,638	26,326
2	Caries	2,889	876	3,765
3	Implant	9	1	10
4	Inlay	4	4	8
5	Crown	183	49	232
6	Retainer	21	12	33
7	Pontic	21	5	26
8	Teeth Filling	355	118	473
9	Carie and Tooth Filling	163	55	218
10	Residual Root	146	59	205
	Others (Merged by Class 3–10)	902	303	1,205
	Sum (Class 1–10)	23,479	7,817	31,296
	Number of Images	756	252	1,008

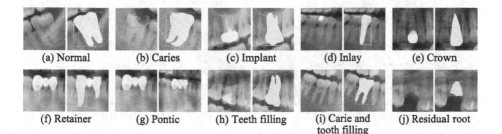

(a) Normal   (b) Caries   (c) Implant   (d) Inlay   (e) Crown

(f) Retainer   (g) Pontic   (h) Teeth filling   (i) Carie and tooth filling   (j) Residual root

Fig. 5. Tooth samples of ten dental classes.

Furthermore, we performed a class reorganization process on our dataset, resulting in four distinct scenarios, based on the usage of categories. In Scenario I (SCE I), we only employed the labels from the "Caries" class. By default, the AP metric refers to the evaluation for the "Caries" class only. Therefore, the mean average precision (mAP) and AP metrics for the test set of SCE I are identical. In the SCE I dataset, where only the "Caries" class is included, the mAP and AP values are identical. In Scenario II (SCE II), we utilized the labels of both the "Normal" and "Caries" classes. In Scenario III (SCE III), we reorganized all labels into a three-class scenario, consisting of "Normal," "Caries," and "Others," which is merged by labels from classes 3 to 10 in Table 1. Finally, in Scenario IV (SCE IV), we employed all labels present in the original ten classes.

4.2   Performance of the Pruning RoI Proposal

We have introduced a modified version of the regression loss function for the P-RoI proposal, as described in Eq. (1). In our implementation, λ_p is set to 1.05. To illustrate the impact of this modified loss function, we present the evaluation results on the training and test sets in Table 2. For the evaluation on the test set, we observe that the average IOU of the teeth RoI proposals increases by

Table 2. Evaluation of the P-RoI proposal.

	Modified $\mathcal{L}_{reg}$	Avg IOU	Best IOU	Worst IOU	IOU $\geqslant 0.7$	IOU $\geqslant 0.8$	IOU $\geqslant 0.9$	"Larger" RoI
Train		0.8790	0.9796	0.6489	99.34	94.18	40.78	32.28
Train	✓	0.9054	0.9843	0.6794	99.74	95.11	51.59	50.93
		+0.0264	-	-	+0.40	+0.93	+10.81	+18.65
Test		0.8623	0.9847	0.6517	99.60	92.46	32.94	34.52
Test	✓	0.8843	0.9823	0.6834	99.60	94.05	41.27	49.60
		+0.0220	-	-	+0.00	+1.59	+8.33	+15.08

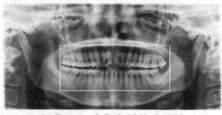
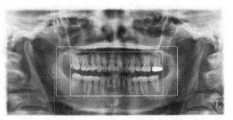

(a) Best teeth RoI: IOU=0.9823   (b) Worst teeth RoI: IOU=0.6834

Fig. 6. The best and worst cases of the P-RoI proposals in the test set. The RoI ground truths are labeled in yellow color, and the predictions are labeled in white color. (Color figure online)

0.0220 to 0.8843 when using the modified loss function. Furthermore, with the modified loss function, 99.60% of the RoI proposals achieve an IOU value greater than 0.7, 94.05% of them achieve an IOU value greater than 0.8, and 41.27% of them achieve an IOU value greater than 0.9. Notably, there is a significant increase of 15.08% points in the number of "larger" prediction cases benefiting from the modified regression loss. Figure 6 shows the best and the worst teeth RoI predictions in the test set with respect to IOU metric.

4.3 Fundamental Experiments

As part of our investigation, we conducted experiments using several contemporary models. In all of these models, we applied multi-scale image input with a ratio range of $[0.8, 1.2]$ and random flip with a probability of 0.5 as data augmentation techniques. All images underwent a keeping-ratio resize operation. Furthermore, for the YOLOv7 series, we employed the Mosaic data augmentation approach, and the input images were padded to achieve a square shape. In terms of the Cascade Mask R-CNN, λ_1, λ_2, and λ_3 are set to 1, 0.5, and 0.25.

A notable finding derived from the presented results in Table 3 is the remarkable performance of the YOLOv7-W model, particularly when trained and evaluated on the SCE IV dataset, surpassing all other models in terms of class-wise AP for the "Caries" class. In contrast, among the Cascade Mask R-CNN series, all models exhibit the highest class-wise AP for "Caries" when trained and evaluated on the SCE I dataset. The exceptional performance of YOLOv7-W can be attributed to its utilization of Mosaic data augmentation and the incorporation of an auxiliary head for label assignment. Consequently, the Cascade Mask R-CNN model with a ResNet-101 [7] backbone was selected as the baseline for subsequent improvements. The parameter scales and FLOPs denote the model configurations in the inference phase.

Table 3. Evaluations of selected models. The class-wise AP refers to "Caries" class.

Detector	Backbone	SCE I		SCE II		SCE III		SCE IV		Input	# of Param.	FLOPs
		mAP	AP	mAP	AP	mAP	AP	mAP	AP			
YOLOv7	L	53.83		66.46	**56.85**	63.41	55.18	30.70	56.26	640^2	36.50M	51.74G
	L	**57.28**		67.56	56.53	64.14	56.44	29.11	54.35	1280^2	36.50M	0.207T
	L	57.65		70.09	60.06	67.64	**60.61**	29.35	59.26	2560^2	36.50M	0.828T
	W	**56.89**		67.67	56.80	65.10	56.49	29.36	56.39	1280^2	79.39M	0.176T
	W	58.73		72.49	63.22	69.37	61.36	34.47	**63.51**	2560^2	79.39M	0.706T
Cascade Mask RCNN	R50	**60.86**		64.34	50.46	62.41	46.43	35.30	46.54	Orig.	75.42M	1.052T
	X50	**58.99**		63.98	49.93	64.10	50.86	38.75	48.16	Orig.	75.12M	1.090T
	R101	**60.94**		64.68	50.81	61.17	43.60	36.58	43.54	Orig.	94.41M	1.360T
	X101	**58.12**		64.91	52.05	63.99	51.06	39.23	49.46	Orig.	94.27M	1.408T

4.4 Performance of the Improved Models

Table 4 displays the AP and best F_1-score values for training and evaluation on the SCE I dataset. An additional experiment was conducted, referred to as the case of "baseline⊞", where the extracted features generated by the primary detection model, instead of the original image, were cropped using the P-RoI proposal. The results indicate a decrease of 3.01 in the "Caries" AP compared to the baseline. Alternatively, the framework proposed in this study, denoted by "baseline□", demonstrates an increase in AP by 2.62 when utilizing the P-RoI proposal to crop the original image. Moreover, incorporating the AutoAssign auxiliary head solely in the baseline model, signified by "baseline*", leads to an AP increase of 2.21. Ultimately, combining both the P-RoI proposal and the AutoAssign auxiliary head into the baseline model, indicated by "baseline□,*", results in a comprehensive improvement of 3.85 in the "Caries" class AP, reaching 64.79. This improvement also surpasses the performance of the YOLOv7-W model by 1.28.

It is worth highlighting that the parameter scales and FLOPs presented in Table 4 specifically pertain to the inference phase. Thus, the models equipped with the auxiliary head maintain the equivalent parameter scale and FLOPs as the models without the auxiliary head. When considering the training phase, the "baseline*" model is characterized by a parameter scale of 99.15M, while the "baseline□,*" model exhibits a parameter scale of 102.58M.

Table 4. Performance of Cascade Mask R-CNN R101 baseline and modified models.

	AP	AP$_{50}$	AP$_{75}$	conf.	TP	FP	FN	Pre	Rec	Best F_1	Param.	FLOPs
baseline	60.94	75.04	72.76	0.81	540	138	336	79.65	61.64	69.50	94.41M	1.360T
baseline⊞	57.93	72.82	69.63	0.85	568	253	308	69.18	64.84	66.94	97.84M	1.410T
	−3.01	−2.22	−3.13	-	-	-	-	-	-	−2.56	-	-
baseline□	63.56	75.87	73.86	0.86	578	192	298	75.06	65.98	70.23	97.84M	1.410T
	+2.62	+0.83	+1.10	-	-	-	-	-	-	+0.73	-	-
baseline*	63.15	76.14	74.12	0.87	566	173	310	76.59	64.61	70.09	94.41M	1.360T
	+2.21	+1.10	+1.35	-	-	-	-	-	-	+0.59	99.15M	-
baseline□,*	64.79	77.34	75.52	0.86	589	172	287	77.40	67.24	71.96	97.84M	1.410T
	+3.85	+2.30	+2.76	-	-	-	-	-	-	+2.46	102.58M	-

5   Conclusion and Future Work

In this paper, we propose a deep learning framework for the detection of dental caries in panoramic X-ray images. Our framework integrates a lightweight pruning region of interest proposal with an enhanced regression loss function to improve its accuracy. On the head of the primary detector, we introduce an auxiliary head, utilizing the AutoAssign to promote the accuracy through the label assignment during the training phase. Ultimately, our framework achieves a notable increase of 3.85 in the AP with respect to the dental caries class.

There are several aspects that can be further explored and improved upon. 1) The development and refinement of regression loss functions for the P-RoI proposal module could be investigated. 2) It would be beneficial to explore and experiment with additional label assignment methods. 3) The Mosaic data augmentation can be implemented with the mask to apply to the Cascade Mask R-CNN model.

References

1. Global oral health status report. World Health Organization
2. Cai, Z., Vasconcelos, N.: Cascade R-CNN: high quality object detection and instance segmentation. IEEE Trans. Pattern Anal. Mach. Intell. **43**(5), 1483–1498 (2019)
3. Ge, Z., Liu, S., Wang, F., Li, Z., Sun, J.: Yolox: exceeding yolo series in 2021. arXiv preprint arXiv:2107.08430 (2021)
4. Girshick, R.: Fast R-CNN. In: Proceedings of the IEEE International Conference on Computer Vision, pp. 1440–1448 (2015)
5. Haghanifar, A., Majdabadi, M.M., Ko, S.B.: Paxnet: dental caries detection in panoramic X-ray using ensemble transfer learning and capsule classifier. arXiv preprint arXiv:2012.13666 (2020)
6. He, K., Gkioxari, G., Dollár, P., Girshick, R.: Mask R-CNN. In: Proceedings of the IEEE International Conference on Computer Vision, pp. 2961–2969 (2017)
7. He, K., Zhang, X., Ren, S., Sun, J.: Deep residual learning for image recognition. In: Proceedings of the IEEE Conference on Computer Vision and Pattern Recognition, pp. 770–778 (2016)
8. Howard, A., et al.: Searching for mobilenetv3. In: Proceedings of the IEEE/CVF International Conference on Computer Vision, pp. 1314–1324 (2019)
9. Howard, A.G., et al.: Mobilenets: efficient convolutional neural networks for mobile vision applications. arXiv preprint arXiv:1704.04861 (2017)
10. Hu, J., Shen, L., Sun, G.: Squeeze-and-excitation networks. In: Proceedings of the IEEE Conference on Computer Vision and Pattern Recognition, pp. 7132–7141 (2018)
11. Imak, A., Celebi, A., Siddique, K., Turkoglu, M., Sengur, A., Salam, I.: Dental caries detection using score-based multi-input deep convolutional neural network. IEEE Access **10**, 18320–18329 (2022)
12. Lin, T.Y., Dollár, P., Girshick, R., He, K., Hariharan, B., Belongie, S.: Feature pyramid networks for object detection. In: Proceedings of the IEEE Conference on Computer Vision and Pattern Recognition, pp. 2117–2125 (2017)
13. Ma, N., Zhang, X., Zheng, H.T., Sun, J.: Shufflenet V2: practical guidelines for efficient CNN architecture design. In: Proceedings of the European Conference on Computer Vision (ECCV), pp. 116–131 (2018)
14. Ren, S., He, K., Girshick, R., Sun, J.: Faster R-CNN: towards real-time object detection with region proposal networks. In: Advances in Neural Information Processing Systems, vol. 28 (2015)
15. Saini, D., Jain, R., Thakur, A.: Dental caries early detection using convolutional neural network for tele dentistry. In: 2021 7th International Conference on Advanced Computing and Communication Systems (ICACCS), vol. 1, pp. 958–963. IEEE (2021)
16. Sandler, M., Howard, A., Zhu, M., Zhmoginov, A., Chen, L.C.: Mobilenetv 2: inverted residuals and linear bottlenecks. In: Proceedings of the IEEE Conference on Computer Vision and Pattern Recognition, pp. 4510–4520 (2018)

17. Singh, N.K., Faisal, M., Hasan, S., Goshwami, G., Raza, K.: Dental treatment type detection in panoramic X-rays using deep learning. In: Abraham, A., Pllana, S., Casalino, G., Ma, K., Bajaj, A. (eds.) ISDA 2022. LNNS, vol. 716, pp. 25–33. Springer, Cham (2023). https://doi.org/10.1007/978-3-031-35501-1_3

18. Singh, N.K., Raza, K.: TeethU^{2}Net: a deep learning-based approach for tooth saliency detection in dental panoramic radiographs. In: Tanveer, M., Agarwal, S., Ozawa, S., Ekbal, A., Jatowt, A. (eds.) ICONIP 2022. CCIS, vol. 1794, pp. 224–234. Springer, Cham (2023). https://doi.org/10.1007/978-981-99-1648-1_19

19. Wang, C.Y., Bochkovskiy, A., Liao, H.Y.M.: Yolov7: trainable bag-of-freebies sets new state-of-the-art for real-time object detectors. In: Proceedings of the IEEE/CVF Conference on Computer Vision and Pattern Recognition, pp. 7464–7475 (2023)

20. Welikala, R.A., et al.: Automated detection and classification of oral lesions using deep learning for early detection of oral cancer. IEEE Access 8, 132677–132693 (2020)

21. Zhang, X., Zhou, X., Lin, M., Sun, J.: Shufflenet: an extremely efficient convolutional neural network for mobile devices. In: Proceedings of the IEEE Conference on Computer Vision and Pattern Recognition, pp. 6848–6856 (2018)

22. Zhu, B., et al.: Autoassign: differentiable label assignment for dense object detection. arXiv preprint arXiv:2007.03496 (2020)

User Stance Aware Network for Rumor Detection Using Semantic Relation Inference and Temporal Graph Convolution

Danke Wu, Zhenhua Tan[✉], and Taotao Jiang

Software College, Northeastern University, Shenyang 110819, China
tanzh@mail.neu.edu.cn

Abstract. The massive propagation of rumor has impaired the credibility of online social networks while effective rumor detection remains a difficulty. Recent studies leverage stance inference to explore the semantic evidence in comments to improve detection performance. However, existing models only consider stance-relevant semantic features and ignore stance distribution and evolution, thus leaving room for improvement. Moreover, we argue that stance inference without considering the context in threads may lead to incorrect semantic features being accumulated and carried through to rumor detection. In this paper, we propose a user stance aware attention network (USAT), which learns the temporal features in semantic content, individual stance and collective stance for rumor detection. Specifically, a high-order graph convolutional operator is designed to aggregate the preceding posts of each post, ensuring a complete semantic context for stance inference. Two temporal graph convolutional networks work in parallel to model the evolution of stance distribution and semantic content respectively and share stance-based attention for de-nosing content aggregation. Extensive experiments demonstrate that our model outperforms the state-of-the-art baselines.

Keywords: Rumor detection · Stance Inference · GAT · Social network

1 Introduction

The social panic and economic losses brought by massive rumor spreading in online social networks have made rumor detection a focus of cyberspace security. Recently, significant progress has been made in rumor detection by utilizing the deep language models and Graph Neural network to learn features hidden in the content, structure and sequence [2–5].

This work is supported by the National Natural Science Foundation of China under Grants (61772125) and the Fundamental Research Funds for the Central Universities (N2217001).

Motivation. Among them, Some studies infer the implicit semantic relationship (i.e., users' stance) [10] to highlight the key evidence hidden in posts for rumor detection. Though achievements have been made, we argue that the neglect of the conversational nature of posts in social media may lead to incorrect semantics being accumulated in rumor detection. For instance, the P2 supports the claim P0 by refuting the P1, once ignoring the context P1, the P2 becomes the refutation to P0. Therefore, the semantic relationship must be inferred in the context of a thread or else cause the opposite effect. Moreover, we notice that the distribution and evolution of users' stances are informative for rumor detection but have not been explored. Analyses of users' responses to claims show that users' stances are various and evolving [1]. The non-rumor has a slightly higher support ratio (supporting to denying) than that of rumor at the early stage of diffusion, and the differences become widened, especially after the advent of resolving tweets [6]. Based on it, we attempt to model the stance distribution and evolution to provide temporal statistic features for rumor detection.

Our Approach. Hence, we proposed a user stance-aware attention (USAT) network, which extracts the temporal and statistical features in stance and content. Specifically, we design a high-order graph convolutional operator that prepares the post with its context for stance inference, and capture the temporal features of individual stance in different time stages. Two temporal graph attention networks are responsible for aggregating the semantic features and stance features following the chronological conversation threads respectively, generating the content representation and collective stance representation. Finally, the stance and content representation that integrated the semantic, textual, structural and temporal features are concatenated for identifying rumors.

To summarize, the main contributions of this work are as follows.

(1) We propose a novel user stance-aware attention (USAT) network. To the best of our knowledge, this is the first attempt to model the users' stance distribution and evolution with deep learning technologies for rumor detection.
(2) We design a high-order graph convolutional operator as a plug-in to complete the stance inference of posts with context, which provides more precise semantic evidence for rumor detection.
(3) Extensive experimental results have demonstrated the effectiveness of our approach in rumor detection and early rumor detection. Moreover, a case study is provided for the intuitive representation of stance inference and the user stance-aware attention.

2   Related Work

Rumor Detection. Recently, the positive outcome of applying deep learning technologies on the claim and its responsive posts has been reported. Besides the content of posts, some studies track temporal dependency between posts in threads. For instance, Ma et al. [13] used LSTM to learn the long-distance dependency of evidence hidden in posts. Some argue that the features hidden in diffusion structure are vital for rumor detection [13,14]. Bian et al. [15] proposed a bi-directional graph model to explore the characteristics of the top-down and bottom-up propagation of rumors. Furthermore, some recent works exploit the auxiliary features, such as image content, novelty and emotion of the claim [16], the credibility of publishers and users [8,17], debiased user preference [18] etc. The aforementioned studies achieved favorable results on rumor detection, however, they suffer from the same flaw: the feature of the claim is normalized with posts, thus some useless and even adverse semantic features could be mixed in the overall representation to disturb the classification.

Stance-Based Rumor Detection. Stance detection determines if each tweet relevant to a claim is supporting, denying, questioning or simply commenting on the claim, which has been identified as a crucial sub-step before rumor detection [19–21]. Existing stance-based rumor detection can be roughly classified into two categories, (1) stance detection based approaches for extracting the stance-relevant semantic features, and (2) stance inference based approaches for attending to the useful posts. The former typically resorted to multi-task learning that integrates stance classification into rumor detection as an auxiliary task [7–9]. Thus rumor detection could benefit from the task-shared semantic representation that is supervised by stance labels and rumor labels jointly. However, they require the stance annotation for each post, which is not always available in practice. Instead, stance inference based approaches implicitly infers the entailment relationships between the posts and the claim with learnable pairwise similarities [10,11]. Relying on the inference-based stance, the rumor detection models can attend to the semantically related posts of the claim while mitigating semantic conflicts. In this paper, we follow the implicit semantic inference and further explore the distribution and evolution of stance for rumor detection.

3 Preliminaries

This paper aims to verify whether the claim p_0^i is a rumor or not based on its social context c^i. We divide the survival times of claims into T stages and construct the sentence embedding tensor $P \in \mathbb{R}^{T \times m \times d}$, where each $P_t \in R^{m \times d}$ consists of $p_{j,t}$ if the user j replied at the stage $t \in T$, $p_{j,t-1}$ otherwise. Meanwhile, we construct the corresponding bi-direction graph g_t for each P_t to represent the conversation threads, where the posts are the nodes and the responsive relationships are the edges. Let $\overleftarrow{N}_{p_j,t}$ and $\overrightarrow{N}_{p_j,t}$ denote the 1-hop preceding and subsequent posts of the post $p_{j,t}$ in the graph g_t.

Here, we define the key conceptions: (1) user stance: the semantic relation between p_* and the claim p_0. (2) stance distribution: the collection of user's stance, (3) stance evolution: the change of individual and collective stance in different stages.

4   Methodology

As indicated in Fig. 1, our USAT framework mainly consists of (1) the user stance inference module, (2) two parallel branches that leverage temporal graph attention network (Tem-GAT) to model the users' stance and content interaction respectively. In the classification phase, the individual stance, collective stance, and content representation are concatenated to feed in a full connection layer for final prediction. We discuss each of these sub-modules as below.

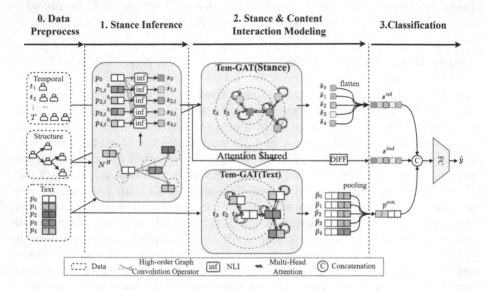

Fig. 1. USAT Framework.

4.1   User Stance Inference Module

In online social networks, a thorough understanding of the posts requires the preceding posts in their conversation threads. Inspired by GNN, we design a high-order graph convolution operator(HiGCO) for updating post representation with their context. To avoid the semantic features of posts over normalization during iterative aggregation in GNN, we construct the high-order neighbors set $\overrightarrow{N}_{p_{j,t}}^{H}$ for post $p_{j,t}$ based on the graph g_t, where $p_{k,t} \in \overrightarrow{N}_{p_{j,t}}^{H}$ is the preceding posts within H hop of $p_{j,t}$. The h as a hyper-parameter determines the receptive field for context aggregation (default value 3). The sentence vector of $p_{j,t}$ is then updated with contextual features using a simple gate mechanism, i.e., the high-order graph convolution operator is formulated as follow,

$$p_{j,t}{}^{H} = \left(\frac{1}{|\overrightarrow{N}_{p_j,t}^{H}|} \sum_{p_{k,t}\in\overrightarrow{N}_{p_j,t}^{H}} p_{k,t} \right) \odot W_u + p_{k,t} \odot (1 - W_u) \tag{1}$$

where $W_u \in R^{d \times d}$ is update gate parameters, $\odot$ refers to the Hadamard product.

After equipping the post with its context, we employ the element-wise product based matching scheme [10,22] to infer the stance of each post towards the claim $p_{0,t}$,

$$s_{p_{j,t}} = \tanh\left(\left(p_{0,t} \odot p_{j,t}{}^{H}\right) W_i\right) \tag{2}$$

Here, the activation function $\tanh(\cdot)$ maps the stance feature to a value $s_{p_{j,t}} \in [-1,1]$, $W_i \in R^{d \times 1}$ is inference parameters. Afterward, we can obtain the stance vector $s_t = [s_{p_{0,t}}, \ldots, s_{p_{m-1,t}}]$ for T stages. The evolution of individual stance is captured by the difference operation for users cross stages, which is formulated as,

$$s_{p_j}^{evo} = \frac{1}{T} \sum_{t=1}^{T} \left(s_{p_{j,t}} - s_{p_{j,t-1}}\right) \tag{3}$$

where $s_{p_j}^{evo}$ is the evolution of individual stance. To reflect the stance distribution, we flatten the $s^{evo} \in R^{m \times 1}$ rather than pooling operation. Therefore, we obtain the evolved individual stance representation is $s^{ind} \in R^{1 \times m}$.

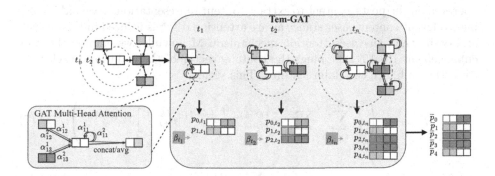

Fig. 2. The architecture of the Tem-GAT.

4.2   Stance Interaction Modeling Module

In this section, we propose a temporal graph attention network (Tem-GAT) to learn the collective stance by modeling the interaction and evolution of stance on conversation graph g_t. The core idea of GAT is to iteratively aggregate information from useful neighboring nodes to the target node via the attention mechanism. Based on this, we equip the GAT with a set of learnable coefficients to trace the temporal features of stance in different periods (As shown in Fig. 2). Formally, the attention mechanism and aggregation operation in GAT is formulated as,

$$\alpha_{p_i p_j, t} = \frac{\exp\left(\sigma\left(\left[s_{p_{i,t}} W_a \| s_{p_{j,t}} W_a\right] a\right)\right)}{\sum_{p_{k,t} \in N_{p_{i,t}}} \exp\left(\sigma\left(\left[s_{p_{i,t}} W_a \| s_{p_{j,t}} W_a\right] a\right)\right)} \tag{4}$$

$$s_{p_i,t}^L = \sigma\left(\frac{1}{k} \sum_{k=1}^{K} \sum_{p_j,t \in N_{p_{i,t}}} \alpha_{p_i p_j, t}^{(L-1),k'} s_{p_j,t}^{(L-1)} W_s^{(L-1),k}\right) \tag{5}$$

where W_* and a are parameters, L and K are the numbers of layers and heads in GAT. The output of the final layer is the stance interaction representations $s^{(L)} \in R^{T \times m \times 1}$ for T stages. Then, we fuse T stages through the temporal coefficients β_t, that is,

$$\bar{s}_{p_i} = \sum_{t=1}^{T} \beta_t s_{p_i,t}^{(L)} \tag{6}$$

The $\bar{s}$ is flattened to represent the collective stance representation $s^{col} \in R^{1 \times m}$.

4.3   Content Interaction Modeling Module

In addition to the implicit semantic relationships, another Tem-GAT is directly operated on the posts' content to extract content representation. To avoid involving irrelevant content, user stance aware attention $\alpha^{(l),k}$ is utilized in this branch and omits the pair-wise attention mechanism. Meanwhile, we employ the root enhancement [10,15] that plugs the content vector of the claim into each post to further enhance the claim features. This yields,

$$p_{i,t}^{(L)} = \sigma\left(\frac{1}{K} \sum_{k=1}^{K} \sum_{p_j,t \in N_{p_{i,t}}} \alpha_{p_i p_j, t}^{(L-1),k} \left[p_{j,t}^{(L-1)} \| p_{0,t}^{(L-1)}\right] W_c^{(L-1),k}\right) \tag{7}$$

$$\bar{p}_i = \sum_{t}^{T} \beta_t p_{i,t}^{(L)} \tag{8}$$

Here we employ mean-pooling over m post-level representations and generate the event-level representation, which is:

$$p^{evn} = \text{meanpooling}_i\left(\bar{p}_i\right) \tag{9}$$

4.4   Classification

We concatenate the evolved individual stance representations, collective stance representation, and event-level representation, and feed them into a fully connected layer to get a low-dimensional classification vector,

$$\hat{y} = \text{softmax}\left(\left(\left[s^{ind} \| s^{col} \| p^{evn}\right] W_{cls} + b_{cls}\right)\right) \tag{10}$$

where W_{cls}, b_{cls} are the parameters of the fully connected layer. The model is trained to minimize the cross-entropy of the predictions and the ground truth distributions, which is defined as,

$$\mathcal{L}_X = \sum_{i=1}^{|X|} y^t \log\left(\hat{y}^i\right) + \lambda\|\Theta\|_2^2 \tag{11}$$

where $\|\Theta\|_2^2$ is the L_2 regularizer over all the model parameters Θ, λ is a trade-off coefficient.

5 Experiments

In this section, we construct experiments to evaluate the effectiveness of USAT on two public datasets: WEIBO [12] and PHEME dataset [7]. Specifically, we aim to answer the following research questions:(RQ1) Can USAT improve rumor detection performance with the temporal and statistical feature in users' stance? (RQ2) How effective are stance inference and users' stance aware attention respectively, in improving the detection performance of USAT? (RQ3) Is USAT robust in early rumor detection tasks?

5.1 Implementation Details

All the baselines are implemented with Pytorch and use the sentence embedding vectors generated by a pre-trained Bert as input. The train, validation, and test set are split with a ratio of 3:1:1. Besides, we adopted a unified set of hyper-parameters throughout the WEIBO and PHEME datasets, where batch size = 16, $\lambda = 5e - 4$, the max number of posts is set to 50. The learning rate is initialized as $5e - 4$ and gradually decreases during the process of training. The training process is iterated 200 epochs, with early stopping implemented when the validation loss stops reducing after 30 iterations. The detailed code and case studies are available at https://github.com/TAN-OpenLab/USAT-RD.

5.2 Performance Comparison (RQ1)

To answer RQ1, we compare our approach with five social-context based baselines and two stance-aware baselines. The results are listed in Table 1. We find that our USAT yields consistently better results than other methods in terms of accuracy and F1 Score. Besides, some detailed observations are as follows. First, the overall performance of the PHEME dataset is lower than the WEIBO dataset. The reason is twofold, in the PHEME dataset, rumors differ slightly from non-rumors, and the average number of responsive posts is smaller, making classification more challenging. Second, the comparison between the RvNN series models further confirms that the bottom-up model suffers from larger information loss than the top-down one, All-RvNN has a slight advantage over TD-RvNN, which indicates that both directions in the diffusion carry useful information. Third, the results show that the ClaHi-GAT and our USAT outperform EBGCN in terms of accuracy and robustness. It can be attributed that the content aggregation guided by the claim reserves targeted evidence properly, thereby surpassing the EBGCN that relies on the semantic relations between posts. Fourth, our USAT achieves superior performance among all the baselines, reflecting the significance of stance distribution and evolution.

5.3   Ablation Study (RQ2)

To better understand the effect of the key modules in USAT, we make comparisons between the following variants of the USAT by removing each module (#module) from the overall framework. The performance is summarized in Table 2.

We see that each component contributes to the performance w.r.t. accuracy. Among them, inappropriate stance inference methods impair accuracy dramatically. The possible reasons are that the cosine similarity cares more about the direction of two posts and the element-wise difference is incapable of the distance metric between high-dimension vectors. Besides, the results show that

Table 1. Performance Comparison and *($p < 0.05$) indicates the pairwise t-test with the best baseline.

Method	WEIBO				PHEME			
	Acc.	Prec.	Rec.	F1	Acc.	Prec.	Rec.	F1
All-RvNN	0.9440	0.9452	0.9437	0.9402	0.8125	0.8162	0.8117	0.8004
TD-RvNN	0.9429	0.9414	0.9411	0.9368	0.8103	0.8251	0.7850	0.7889
BU-RvNN	0.8998	0.9020	0.9018	0.8961	0.7946	0.7991	0.7918	0.7832
EBGCN	0.9224	0.9255	0.9232	0.9190	0.7902	0.7915	0.7840	0.7784
BIGCN	0.9084	0.9170	0.9122	0.9053	0.8013	0.8039	0.7975	0.7883
SemGraph	0.9440	0.9436	0.9442	0.9413	0.8021	0.8113	0.7984	0.7935
ClaHi-GAT	0.9494	0.9473	0.9539	0.9463	0.8170	0.8156	0.8243	0.8099
USAT	**0.9571***	**0.9540**	**0.9560**	**0.9558**	**0.8237***	**0.8241**	**0.8263**	**0.8189**

Table 2. Ablation Studies, each row is the relative improvement over the baseline (%).

Method	WEOBO				PHEME			
	Acc.	Prec.	Rec.	F1	Acc.	Prec.	Rec.	F1
#Stance	−1.51	−2.01	−1.12	−1.73	−2.46	−2.94	−2.95	−3.17
#AT-shared	−0.65	−0.43	−1.58	−1.14	−3.80	−2.36	−3.34	−4.27
#HiGCO	−0.22	−0.74	0.10	−0.52	−0.67	−2.02	0.30	−1.65
#Text	−0.65	−0.76	−0.82	−0.77	−2.01	−0.28	−2.62	−2.91
ST-ED	−4.10	−3.82	−4.29	−4.51	−3.80	−4.24	−4.94	−4.63
ST-CosSim	−1.51	−2.16	−1.17	−1.87	−4.25	−4.04	−5.16	−4.86

#Text is superior to #Stance, which indicates stance interaction provides more information than textual content interaction. That also proves the performance improvement of our USAT was mainly achieved by finding evidence from the stance. Furthermore, the performance degradation in #AT-shared illustrates that the stance-aware attention favors USAT for de-nosing content aggregation while reducing the computational complexity. Meanwhile, the USAT benefits from high-order graph convolution operator, which demonstrates our semantic inference grasps the correct stance for rumor detection.

5.4   Early Detection (RQ3)

In this section, we examine all baselines by varying the number of posts to further demonstrate the advantages of our model in early rumor detection. For example, if the maximum number is 20, we only use the first 20 posts in the given claim for classification.

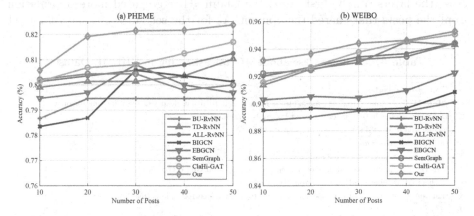

Fig. 3. Early rumor detection, social context-based methods are in cool colors and the stance aware methods are in warm colors.

On the subsets of the posts, the accuracy of each method is shown in Fig. 3. With more data involved in training, most of the methods perform better on the Weibo dataset. However, the results are slightly different on PHEME, the reason is that the average number of posts in PHEME is 16, and increasing the max number of posts may not actually input posts but zero noise. And the results demonstrate that USAT starts to beat the other models at the early stage, especially on the challenging PHEME dataset. Distinguishingly, there was no noticeable performance improvement under BU-RvNN, which further indicated the information loss when discarding context.

5.5   Case Study

In this section, we design an experiment to obtain deeper insights and detailed interpretability about the effectiveness of the stance inference and the attention mechanism. The social context of a rumor about the #ottawashooting is listed in Table 3. We visualize the corresponding stance value obtained from stance inference module and the corresponding attention score generated by Tem-GAT (stance) in Fig. 4.

From the results in Fig. 4, we give the detailed observations as follows: (1) In Fig. 4 the stance-value graph, despite the stance inference component is not as we expect that maps the semantic of denying to supporting in $[-1, 1]$ strictly, the stance is captured proportionately. Taking claim #1, Post #3, #7, #10 as examples, which support or is textual correlative to the Claim #1 be mapped to near -0.3. Fortunately, the stance inference component could distinguish the querying posts from the comments, such as Post #3, #4 and #5, but exists a deficiency in precision such phenomena are not surprising, as we do not provide further supervision on stance inference. Further exploration on implicit stance inference is needed. (2) Comparing the in-attention scores obtained from two Tem-GATs, it is obvious that the focuses of the Claim #1 were quite the opposite cross the layers. In the first layer, the Claim #1 aggregated more information from the posts that shared the opinion. As for the second layer, the Claim #1

Table 3. A Rumor Case about #Ottawashooting in PHEME Dataset

ID	Claim #1 and posts	Parent ID
1	#ISIS Media account posts pic claiming to be Michael Zehaf-Bibeau, dead #OttawaShooting terrorist	-
2	Did really anyone have any doubts about his 'roots' - no way	1
3	no one said it was ISIS. Stop your propoganda plz	1
4	can you share the source?	1
5	...and again...which acct? What's the source of this?	1
6	This thing is sub-human and deserves no publicity. Canada-On Guard For Thee!	1
7	Michael Zehaf-Bibeau: 5 Fast Facts You Need to Know	1
8	that you assume the zionist heretics of the USSA to be on the correct side is très drôle. #daBeast	1
9	???	5
10	Source? Anyone?	7

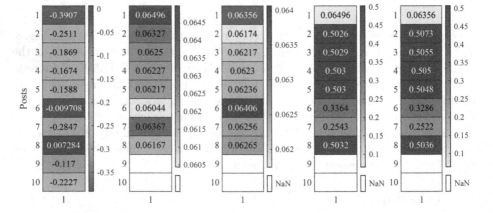

Fig. 4. The stance value of posts and the attention scores of the claim.

turned to update itself with more information from the contrary posts. Therefore, the interaction pattern, such as support-deny-deny, may be adaptively preserved in node representation during the iterative node aggregation. (3) For those posts directly linked to the Claim #1, more features of the Claim #1 were updated in the posts than themselves. The out-attention scores of Claim #1 exceed the self-attention of the posts in general, which indicates that the stance aware attention emphasizes the feature of the claim. That could work in concert with claim enhancement that concatenates the representations of posts with that of the claim in Tem-GAT(Text).

6   Conclusion and Discussion

In this paper, we propose a novel user stance aware attention (USAT) network that utilizes the inferred stance in content and temporal features for rumor detection. Two modules are vital for USAT, (1) the high-order graph convolutional operator to supply the posts with their context, ensuring the correct stance inference. (2) the parallel Tem-GATs that operate on the content and users' stance respectively, providing features that fuses spatiotemporal and statistic information. Extensive experiments conducted on real-world data demonstrate that our proposed method achieves state-of-the-art performance. In the future, we will go deeper in (1) the optimal semantic relation inference for rumor detection, (2) the generalization of evidence hidden in stance to emerging rumors.

References

1. Qazvinian, V., Rosengren, E., Radev, D.R., Mei, Q.: Rumor has it: identifying misinformation in microblogs. Proc. EMNLP **2011**, 1589–1599 (2011)
2. Song, C., Shu, K., Wu, B.: Temporally evolving graph neural network for fake news detection. Inf. Process. Manag. **58**(6), 102712 (2021). https://doi.org/10.1016/j.ipm.2021.102712
3. Wu, Z., Pi, D., Chen, J., Xie, M., Cao, J.: Rumor detection based on propagation graph neural network with attention mechanism. Expert Syst. Appl. **158**, 113595 (2020). https://doi.org/10.1016/j.eswa.2020.113595
4. He, Z., Li, C., Zhou, F., Yang, Y.: Rumor detection on social media with event augmentations. In: Proceedings of the 44th International ACM SIGIR Conference on Research and Development in Information Retrieval, pp. 2020–2024 (2021). https://doi.org/10.1145/3404835.3463001
5. Khoo, L.M.S., Chieu, H.L., Qian, Z., Jiang, J.: Interpretable rumor detection in microblogs by attending to user interactions. In: Proceedings of the AAAI Conference on Artificial Intelligence, pp. 8783–8790 (2020). https://doi.org/10.1609/aaai.v34i05.6405
6. Zubiaga, A., Liakata, M., Procter, R., Wong Sak Hoi, G., Tolmie, P.: Analysing how people orient to and spread rumours in social media by looking at conversational threads. PLoS ONE **11**(3), 1–29 (2016). https://doi.org/10.1371/journal.pone.0150989
7. Kochkina, E., Liakata, M., Zubiaga, A.: All-in-one: Multi-task Learning for Rumour Verification (2018)

8. Li, Q., Zhang, Q., Si, L.: Rumor detection by exploiting user credibility information, attention and multi-task learning. In: Proceedings of the ACL 2019, pp. 1173–1179. https://doi.org/10.18653/v1/p19-1113

9. Ma, J., Gao, W., Wong, K.-F.: Detect rumor and stance jointly by neural multi-task learning. In: Proceedings of the Web Conference 2018 - WWW 2018, vol. 2018, pp. 585–593 (2018). https://doi.org/10.1145/3184558.3188729

10. Lin, H., Ma, J., Cheng, M., Yang, Z., Chen, L., Chen, G.: Rumor detection on twitter with claim-guided hierarchical graph attention networks. In: Proceedings of the 2021 Conference on Empirical Methods in Natural Language, pp. 783–787 (2021)

11. Yuan, C., Qian, W., Ma, Q., Zhou, W., Hu, S.: SRLF: a stance-aware reinforcement learning framework for content-based rumor detection on social media. In: Proceedings of the International Joint Conference on Neural Networks 2021-July (2021)

12. Ma, J., et al.: Detecting rumors from microblogs with recurrent neural networks. In: Proceedings of IJCAI, pp. 3818–3824 (2016)

13. Huang, Q., Zhou, C., Wu, J., Wang, M., Wang, B.: Deep structure learning for rumor detection on twitter. In: Proceedings of the IJCNN, pp. 1–8 (2019). https://doi.org/10.1109/IJCNN.2019.8852468

14. Ma, J., Gao, W., Wong, K.F.: Rumor detection on twitter with tree-structured recursive neural networks. In: Proceedings of the ACL, pp. 1980–1989 (2018). https://doi.org/10.18653/v1/p18-1184

15. Bian, T., et al.: Rumor detection on social media with bi-directional graph convolutional networks. In: Proceedings of the AAAI Conference on Artificial Intelligence, vol. 34, no. 01, pp. 549–556 (2020). https://doi.org/10.1609/aaai.v34i01.5393

16. Kumari, R., Ashok, N., Ghosal, T., Ekbal, A.: What the fake? Probing misinformation detection standing on the shoulder of novelty and emotion. Inf. Process. Manage. 59(1), 102740 (2022). https://doi.org/10.1016/j.ipm.2021.102740

17. Yuan, C., Ma, Q., Zhou, W., Han, J., Hu, S.: Early detection of fake news by utilizing the credibility of news, publishers, and users based on weakly supervised learning. In: Proceedings of COLING, USA, pp. 5444–5454 (2020). https://doi.org/10.18653/v1/2020.coling-main.475

18. Zhang, W., Zhong, T., Li, C., Zhang, K., Zhou, F.: CausalRD: a causal view of rumor detection via eliminating popularity and conformity biases. In: IEEE INFOCOM 2022, pp. 1369–1378 (2022). https://doi.org/10.1109/INFOCOM48880.2022.9796678

19. Zubiaga, A., Kochkina, E., Liakata, M., Procter, R., Lukasik, M.: Stance classification in rumours as a sequential task exploiting the tree structure of social media conversations. In: Proceedings of COLING, pp. 2438–2448 (2016)

20. Zhao, Z., Resnick, P., Mei, Q.: Enquiring minds: early detection of rumors in social media from enquiry posts categories and subject descriptors detection problems in social media. In: Proceedings of the 24th International Conference on World Wide Web (WWW), pp. 1395–1405 (2015)

21. Jin, Z., Cao, J., Zhang, Y., Luo, J.: News verification by exploiting conflicting social viewpoints in microblogs. In: Proceedings of the 30th AAAI Conference on Artificial Intelligence(AAAI), pp. 2972–2978 (2016)

22. Mou, L., et al.: Natural language inference by tree-based convolution and heuristic matching. In: Proceedings of ACL- Short Papers, pp. 130–136 (2016). https://doi.org/10.18653/v1/p16-2022

IEEG-CT: A CNN and Transformer Based Method for Intracranial EEG Signal Classification

Mengxin Yu[1,2], Yuang Zhang[1,2(✉)], Haihui Liu[1,2], Xiaona Wu[1,2], Mingsen Du[1,2], and Xiaojie Liu[1,2]

[1] School of Information Science and Engineering, Shandong Normal University, Jinan 250358, China
zhangyuang@sdnu.edu.cn
[2] Shandong Provincial Key Laboratory for Distributed Computer Software Novel Technology, Shandong Normal University, Jinan 250358, China

Abstract. Intracranial electroencephalography (iEEG) is of great importance for the preoperative evaluation of drug-resistant epilepsy. Automatic classification of iEEG signals can speed up the process of epilepsy diagnosis. Existing deep learning-based approaches for iEEG signal classification usually rely on convolutional neural network (CNN) and long short-term memory network. However, these approaches have limitations in terms of classification accuracy. In this study, we propose a CNN and Transformer based method, which is named as IEEG-CT, for iEEG signal classification. Firstly, IEEG-CT utilizes deep one-dimensional CNN to extract the critical local features from the raw iEEG signals. Secondly, IEEG-CT combines a Transformer encoder, which employs a multi-head attention mechanism to capture long-range global information among the extracted features. In particular, we leverage a causal convolution multi-head attention instead of the standard Transformer block to efficiently capture the temporal dependencies within the input features. Finally, the obtained global features by the Transformer encoder are employed for the classification. We assess the performance of IEEG-CT on two publicly available multicenter iEEG datasets. According to the experimental results, IEEG-CT surpasses state-of-the-art techniques in terms of several evaluation metrics, i.e., accuracy, AUROC, and AUPRC.

Keywords: Drug-resistant epilepsy · Intracranial EEG · CNN · Transformer

1 Introduction

Epilepsy is a critical and persistent brain disease that has a significant impact on a large population of approximately 70 million people across the globe [1]. Drug-resistant epilepsy (DRE) accounts for approximately one-quarter of all

B. Luo et al. (Eds.): ICONIP 2023, LNCS 14449, pp. 549–563, 2024.
https://doi.org/10.1007/978-981-99-8067-3_41

patients with epilepsy. When the epilepsy is focal, patients with DRE may gain from epilepsy surgery [2]. For DRE patients, intracranial electroencephalography (iEEG) is considered the gold standard for assessing neuronal activity in epileptic. The iEEG signal has been widely adopted in preoperative planning to locate regions of epilepsy [3]. However, the analysis of iEEG signal by clinical experts is a highly subjective, time-consuming, and costly process [4]. Hence, there is a demand for an automated iEEG classification method.

Recently, researchers have developed a variety of algorithms for the automatic classification of iEEG signals. Existing approaches made remarkable advances under the frameworks of convolutional neural networks (CNN) and long short-term memory (LSTM) in iEEG signal classification. Specifically, CNN-based models employ convolutional computation to automatically extract features of physiological signals [5–7], while the LSTM approach is mainly utilized for capturing the temporal relationships of long time series iEEG data [8–10]. However, CNN-based methods usually perform poorly on capturing the global information, while LSTM-based approaches lack parallelization, and are less efficient [11].

The transformer model has achieved excellent results in natural language processing [12] and computer vision [13] due to its powerful attention mechanism, and some researchers have employed Transformer for iEEG decoding [14–16]. The self-attention mechanism in the Transformer model obtains long-range global information by performing similarity calculations and weighted averages for all vectors at all positions in the sequence.

The focus of this paper is to explore how the Transformer model can be leveraged to enhance the accuracy of iEEG signal classification. We propose an automated and end-to-end classification method for iEEG signals, which combines CNN and Transformer and is named IEEG-CT. IEEG-CT leverages CNN to extract local features from iEEG signals, and the Transformer module to mine long-range global information within iEEG signals. In summary, this paper's contributions can be outlined as follows:

- We propose IEEG-CT, which is a hybrid architecture for iEEG signal classification. IEEG-CT combines the strengths of CNN and Transformer to efficiently extract critical local features and long-range global information of raw iEEG segments, respectively.
- We take deep one-dimensional convolution (1D-CNN) to effectively capture local contextual information in a sequential manner.
- We perform the extensive experiment of proposed IEEG-CT on two publicly available multicentre iEEG datasets. The results of iEEG signal classification, i.e., artifacts, pathology activities, and physiology activities, demonstrate that our proposed IEEG-CT outperforms state-of-the-art methods.

Below is the organization of the rest of this paper. The related work of deep learning-based iEEG classification method is presented in Sect. 2. Section 3 illustrates the details of the IEEG-CT method. In Sect. 4, we present the experimental results of IEEG-CT on publicly available datasets. Section 5 provides the conclusion of this paper.

2   Related Work

This section discusses studies relevant to iEEG signal classification methods. Several deep learning backbones, i.e., CNN, LSTM, and Transformer, have been applied to iEEG signals.

2.1   CNN-Based Approaches

Approaches based on CNN are commonly employed to extract local features from signals. CNN can be applied for a variety of iEEG processing tasks, e.g., iEEG noise detection [5], localization of epileptic foci [6], and iEEG signal identification [7]. [5] proposes a CNN-based classification method for detecting iEEG artifacts. The architecture consists of two convolutional layers, and two max-pooling layers, and is designed to operate across five different frequency bands. Sui et al. [6] propose a method that utilizes short-time Fourier transform (STFT) and CNN to distinguish focal from non-focal iEEG signals. The time-frequency spectrogram obtained by STFT is fed into a 5-layer CNN. Guo et al. [7] propose a 1D temporal convolutional network for feature extraction, then use the attention layer to focus on the discriminative features of iEEG for classification.

Although the CNN model is effective in capturing local features of iEEG data, it is not good at capturing global features, which are also significant for the iEEG classification task.

2.2   LSTM-Based Approaches

Unlike CNN, which is unable to handle sequence-changing data, long short-term memory (LSTM) is specifically designed to handle longer sequences, which leads to improved performance [17]. Wang et al. [8] propose a deep learning model with two branches to identify epileptic signals. The first branch selects temporal features of the signal using an attention-based bidirectional LSTM, and the second branch automatically extracts features using 1D-CNN. Nejedly et al. [9] publicly share two iEEG multicentre datasets containing four types of signals in iEEG, namely physiological activity, pathological activity, artifacts, and power line noise. In [9], CNN and LSTM fusion models are adopted to perform the signal multi-classification. Wang et al. [10] propose a multi-scale CNN to learn the multi-frequency domain features of iEEG, while enhancing the sequence dependence of the features using an attention-based bidirectional LSTM.

Although LSTM can memorize long-range dependencies, the LSTM structure has inherent order in the temporal problem [18], which inevitably affects the efficiency of training.

2.3   Transformer-Based Approaches

Recently, Transformer has garnered significant attention in natural language processing [12] and computer vision [13]. The multi-head attention mechanism (MHA) in Transformer solves the problem of long dependencies [12].

Sun et al. [14] design an end-to-end model which fuses CNN and Transformer. In [14], Transformer is applied to increase the dependency between signal channels and experimentally demonstrated that Transformer can improve the performance of epilepsy detection classification. Xu et al. [15] propose a temporal cascaded channel network based on the Swin Transformer for motor imagery iEEG signal classification. The model in [15] leverages the global learning capability of the Transformer framework to extract temporal and spatial features that provide a global representation. Peh et al. [16] propose a multi-channel vision Transformer approach, which converts time-series iEEG signals into time-frequency images using the continuous wavelet transform. Each time-frequency image is segmented into fixed-size blocks and then fed as input into the model for iEEG signal classification.

3   Method

To improve the classification accuracy of iEEG signals, inspired by [19], we present the IEEG-CT model, which combines CNN and Transformer to capture the long-range global information of local features from raw iEEG signals.

3.1   Overview of IEEG-CT Model

The overall architecture diagram of the IEEG-CT model is depicted in Fig. 1. There are three modules in IEEG-CT, which are the convolution module, Transformer module, and classifier. Firstly, the convolution module receives the normalized iEEG signal as input and learns to extract local features from the original signal segments. After reshaping, the output vector from the convolution module is input into the Transformer module to further learn the global long-range information of the local features extracted from the iEEG signal. Finally, the classifier takes the input vectors extracted from the Transformer module to classify the iEEG signals.

3.2   Convolution Module

Successful training of deep learning algorithms requires adequate representation of data. Inspired by the VGG model [20], we adopt a deep 1D-CNN with a convolution kernel size of three for feature extraction, which increases the feature representation capability.

To extract critical local features of the iEEG signals, a deep CNN architecture is employed, as illustrated in Fig. 2. The module consists of three convolution blocks, where each block includes three 1D-CNN layers, three batch normalization layers, and three Gaussian error linear unit (GELU) activation functions. The 1D-CNN layer is utilized to process iEEG sequence data and capture local contextual information in a sequential manner. Our network that is stacked with small-size convolution reduces the parameters for network training and expands the perceptual field. Meanwhile, it also corresponds to more nonlinear mappings,

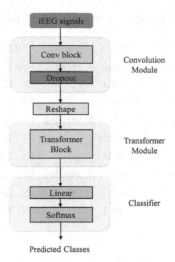

Fig. 1. The overall architecture diagram of IEEG-CT.

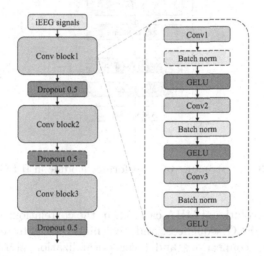

Fig. 2. The structure of convolution module in IEEG-CT.

which increases the fitting ability of the network. The detailed parameter can be found in Table 1. Here, S is the number of batch size and T means the length of the iEEG signal sequence.

3.3   Transformer Module

The Transformer module of IEEG-CT consists of three components, i.e., the causal convolutional MHA, skip connections and layer normalization, and two fully connected layers, as shown in Fig. 3.

Table 1. Detailed parameters of convolution module in IEEG-CT.

Name	Layer	Kernel	Output
Conv block1	Conv1	3, stride2	$(S, 128, T/2)$
	Conv2		$(S, 128, T/2)$
	Conv3		$(S, 128, T/2)$
Conv block2	Conv1	3, stride2	$(S, 128, T/2)$
	Conv2		$(S, 128, T/2)$
	Conv3		$(S, 128, T/2)$
Conv block3	Conv1	3, stride2	$(S, 128, T/2)$
	Conv2		$(S, 256, T/2)$
	Conv3		$(S, 256, T/2)$

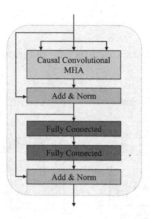

Fig. 3. The structure of Transformer module in IEEG-CT.

The causal convolutional MHA can obtain the correlations and dependencies between different iEEG sequences, and give higher weights to more significant features. The skip connections and layer normalization aim to reduce model complexity and alleviate overfitting. The fully connected layers are reinforced by means of an activation function for the representation of features. More details of the three components are presented as follows.

Causal Convolutional MHA. The success of MHA in deep learning applications can be attributed to its ability to learn long-range dependencies. As shown in Fig. 4, we develop a causal convolutional MHA, in which the causal convolution (Conv1D) is to encode the input feature vectors and capture temporal relationships. The causal convolution [21] is fast to process and significantly reduces the training time of the model compared with the traditional Transformer encoder.

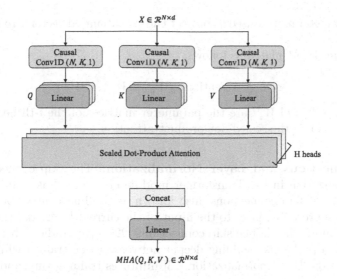

Fig. 4. The structure of causal convolutional MHA

The input of causal convolution MHA is from the convolution module, denoted as $X = \{x_1, ..., x_N\} \in R^{N \times d}$. Here, N denotes the total number of input features, while d represents the length of $x_i, 1 \leq i \leq N$. Specifically, for this purpose, MHA selects three repetitions of X as inputs.

Firstly, the causal convolution generates $\tilde{X}$ from X, which can be represented as $\tilde{X} = \psi(X)$, where $\psi(\cdot)$ is a causal convolution function. Specifically, causal convolution is a technique that ensures causality in convolutional operations, which is achieved by adding $K - 1$ zeros to the beginning of the input sequence and discarding $K - 1$ elements from the end of the output sequence. Here, K denotes the size of the convolution kernel, which is set to 7 in this case.

Secondly, the MHA module starts with a linear transform which transformers the input $\tilde{X} \in R^{N \times d}$ into $Q \in R^{N \times d_q}$, $K \in R^{N \times d_k}$ and $V \in R^{N \times d_v}$, d_q, d_k and d_v refer to the dimensions of q, k and v, respectively.

The dot product between Q and K allows computing the global positional dependencies, and V is used to add the weights of these long-range dependencies to the features. The scaled dot-product attention is calculated as follows:

$$Attention(Q, K, V) = Softmax(\frac{QK^T}{\sqrt{d_k}})V. \tag{1}$$

The MHA comprises H scaled dot-product attention layers, which enable the model to selectively attend to information from different representation subspaces at various locations. The MHA mechanism is calculated as follows:

$$MHA(Q, K, V) = Concat(head_1, ..., head_H)W^o, \tag{2}$$

where H represents the number of heads, $1 \leq i \leq H$. Concat is the concatenation operation and $head_i$ refers to the weighted representation vector of the i-th head.

W^o is a parameter matrix used to project the concatenated vector into the output space.

The $head_i$ is calculated as follows:

$$head_i = \text{Attention}(QW_i^Q, KW_i^K, VW_i^V), \tag{3}$$

where W_i^Q, W_i^K and W_i^V are the parameter matrices of the i-th head, respectively. Attention is the scaled dot-product attention.

Skip Connections and Layer Normalization. The skip connections and layer normalization in the Transformer module are marked as "Add & Norm" in Fig. 3. These skip connections, also known as residual connections, add the output of the preceding layer to the input of the current layer, and the resulting sum is then normalized. The skip connections [22] solve gradient disappearance and explosion problems, enabling deeper networks to be trained while preserving information. Layer normalization [23] improves training time and execution efficiency.

Fully Connected Layers. The output of the MHA sub-layer is passed to two fully connected layers that apply rectified linear unit activation functions to enhance the nonlinearity of the model. This structure employs 120 hidden layer units, and the output dimension is the same as the input dimension.

3.4   Classifier

In order to classify the extracted iEEG feature representations into the appropriate labels, a classifier module is employed, which comprises a linear layer and a softmax function. Initially, the iEEG features with global information representations are flattened into one-dimensional vectors, which are then input into the linear layer. The output layer of the linear layer comprises three neurons, corresponding to the number of categories. The softmax function is applied to transform the model's output into a probability distribution, yielding the predicted probability of each category. The category with the highest probability is selected as the classification result for the iEEG signal.

The cross-entropy function is employed as the loss function of IEEG-CT and is defined as follows:

$$J(\theta) = -\frac{1}{N} \sum_{i=1}^{N} \sum_{j=1}^{C} y_{ij} \log(\hat{y}_{ij}), \tag{4}$$

where N represents the total number of samples, C denotes the number of classes, and θ denotes the model parameters. Additionally, y_{ij} represents the true label of the i-th sample, and $\hat{y}_{ij}$ denotes the predicted probability of the i-th sample belonging to the j-th class.

4   Experiments

4.1   IEEG Datasets

To evaluate the performance of the proposed IEEG-CT method, two publicly available multicentre iEEG datasets, namely, the FNUSA and MAYO datasets, are utilized.

The FNUSA dataset [9] is collected at St. Anne's University Hospital, which records iEEG data of 14 patients with DRE who undergo standard pre-surgical monitoring in an awake resting state. The hospital records 30 min of iEEG signals with a sampling rate of 25 kHz. The raw data is filtered using a 2 kHz low-pass filter and downsampled to 5 kHz. For brevity, we use FNUSA to present the FNUSA dataset in the rest part of this paper.

The MAYO dataset [9] consists of iEEG recordings from 25 patients with DRE who are evaluated for epilepsy surgery during the nighttime. Specifically, the recordings are obtained between 1 AM and 3 AM. The MAYO Clinic acquires data at a sampling rate of 32 kHz and a hardware filter bandwidth of DC to 9 kHz. All data is filtered using an anti-aliasing filter with a cutoff frequency of 1 kHz. Subsequently, the data is downsampled to 5 kHz. We shorten the MAYO dataset as MAYO in the following part of this paper.

Each iEEG signal in FNUSA and MAYO is divided into 3 s data segments (15000 samples), and these segments can be classified into physiological activity, pathological activity, artifacts, and power line noise. Due to the difference in power line frequencies between FNUSA (50 Hz) and MAYO (60 Hz), our experiments exclude the power line noise set. The specific statistic information of the multicentre iEEG datasets utilized in this paper can be observed in Table 2.

Table 2. The classification category of multicentre iEEG dataset.

Category	FNUSA	MAYO
Physiological Activity	94560	56730
Pathological Activity	52470	15227
Artifacts	32599	41303
Total	179629	113260

4.2   Experimental Protocols

Evaluation Metrics. We use accuracy (ACC), the area under the receiver operating characteristic curve (AUROC), and the area under the precision-recall curve (AUPRC) as evaluation metrics for multi-classification tasks in this paper. In particular, AUROC and AUPRC are considered to be crucial and objective metrics for the unbiased assessment of the classification performance of the model on the unbalanced multicentre iEEG dataset [9].

The evaluation metrics are specified as follows:

$$ACC = \frac{TP + TN}{TP + TN + FP + FN} \times 100\%, \tag{5}$$

where TP, TN, FP, and FN represent true positive, true negative, false positive, and false negative, respectively. To calculate the area under the ROC [24] curve, we employ AUROC. The PRC is a curve that connects the points of precision and recall, and we use AUPRC to calculate the area under this curve.

Parameter Setting. We carry out all experiments on an RTX 3090 (24 GB) GPU with the CUDA 11.1 environment, using Python 3.8 and the PyTorch 1.8.1 framework. A batch size of 128 is selected, and the Adam optimizer with a learning rate of 1e−3 is utilized. The maximum epoch is 50. IEEG-CT uses eight MHA heads, has two Transformer module layers, and the number of hidden layer cells in the Transformer modules is 120. Table 3 lists the parameter settings of the IEEG-CT model.

Table 3. The parameter settings of IEEG-CT model.

IEEG-CT model parameters	Values
Model optimizer	Adam
Model learning rate	1e−3
The maximum epoch	50
The number of batch size (S)	128
The number of conv block	3
The number of MHA heads (H)	8
The number of Transformer module layer	2
The number of Transformer module hidden layer cells	120

Cross-Subject Experiments Protocols. In the cross-subject experiments, the hold-out method is employed to train the model, where two subjects are used for testing and the remaining subjects are utilized for training, as depicted in Fig. 5. We can observe the testing scheme in Table 4, which is the same as the one described in [7]. The test set comprises subjects 8 and 13 from FNUSA and subjects 1 and 19 from MAYO.

Out-of-Institution Experiments Protocols. Out-of-institution experiments are trained using data from one institution and tested using data from another institution, e.g., trained using FNUSA and tested using MAYO or vice versa.

Table 4. The sample distribution for each category in the training and testing set.

Institution	Category	Train	Test
FNUSA	Physiological Activity	59564	34996
	Pathological Activity	45256	7214
	Artifacts	32400	199
	Total	137220	42409
MAYO	Physiological Activity	48077	8653
	Pathological Activity	14344	883
	Artifacts	35690	5613
	Total	98111	15149

Training data   Testing data

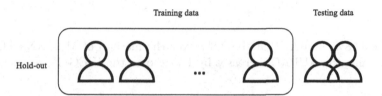

Hold-out

Fig. 5. The protocols of cross-subject experiments.

4.3 Cross-Subject Experiments

In the cross-subject experiments, three state-of-the-art deep learning approaches are evaluated alongside IEEG-CT, as presented in Table 5. The best results are highlighted in bold, and the second-best results are underlined.

From Table 5, we can find that for FNUSA, the proposed IEEG-CT achieves the best or second-best results. Only the AUROC of STFT+CNN+GRU [9] is higher than that of IEEG-CT. However, [9] has a much lower value of AUPRC (74.44%), which may lead to misclassification and is unacceptable in clinical practice. For MAYO, the proposed IEEG-CT exhibits the best performance. The AUROC of IEEG-CT demonstrates a noteworthy enhancement compared to STFT + CNN + GRU, with an increase of 9.5%, MRCNN + TCE, with an increase of 0.25%, and IEEG-TCN, with an increase of 1.5%.

In general, our model obtains good classification accuracy in both datasets, which demonstrates the effectiveness of our model in iEEG classification.

4.4 Out-of-Institution Experiments

Table 6 shows the out-of-institution experiment results, we can find that the proposed IEEG-CT model achieves better performance in all metrics. When FNUSA is used as the training set, IEEG-CT shows a significant improvement in ACC, with a 7.19% increase over STFT+CNN+GRU [9] and a 1.85% increase over MRCNN+TCE [19]. When MAYO is used as the training set, IEEG-CT increases

Table 5. Comparison between the proposed IEEG-CT and state-of-the-art models on the multicentre dataset: cross-subject.

Dataset	Method	ACC (%)	AUROC (%)	AUPRC (%)
FNUSA	STFT+CNN+GRU [9]	78.70	**94.13**	<u>74.44</u>
	MRCNN+TCE [19]	<u>92.78</u>	90.46	66.41
	IEEG-TCN [7]	**93.16**	84	67
	IEEG-CT (ours)	**93.16**	<u>93.44</u>	**82.97**
MAYO	STFT+CNN+GRU [9]	66.07	85	74
	MRCNN+TCE [19]	86.23	<u>94.25</u>	<u>84.59</u>
	IEEG-TCN [7]	<u>91.08</u>	93	78
	IEEG-CT (ours)	**91.83**	**94.50**	**84.91**

ACC by 3.18%, and AUROC by 2.8% over the method of MRCNN+TCE [19]. Additionally, the AUPRC increases by 1.71% over the IEEG-TCN.

Table 6. Comparison between the proposed IEEG-CT and state-of-the-art models on the multicentre dataset: out-of-institution.

Dataset	Method	ACC (%)	AUROC (%)	AUPRC (%)
Train: FNUSA, Test: MAYO	STFT+CNN+GRU [9]	69.95	84.70	75.79
	MRCNN+TCE [19]	<u>75.29</u>	88.47	79.92
	IEEG-TCN [7]	-	<u>89</u>	<u>81</u>
	IEEG-CT (ours)	**77.14**	**89.45**	**82.28**
Train: MAYO, Test: FNUSA	STFT+CNN+GRU [9]	66.65	84.41	72.29
	MRCNN+TCE [19]	<u>69.53</u>	<u>85.18</u>	72.26
	IEEG-TCN [7]	–	85	<u>76</u>
	IEEG-CT (ours)	**72.71**	**87.98**	**77.71**

The experimental results illustrate the good performance of our model on unbalanced datasets, which is important for real clinical scenarios.

4.5   Ablation Study

To study the contribution, validity, superior performance, and rationalization of the model architecture of each individual block in our proposed IEEG-CT, we perform cross-subject and out-of-institution ablation studies on the MAYO and FNUSA dataset. The process is outlined as follows:

(a) **CNN**: convolution module only.
(b) **CNN+Transformer**: the Transformer module is replaced with a traditional Transformer layer.

(c) **IEEG-CT (ours)**: combining convolution module Transformer module employed causal CNN together.

Tables 7 and 8 show the results of the cross-subject and out-of-institution ablation experiments, respectively. Based on the results of the ablation study, we can draw the following two conclusions. Firstly, the MHA mechanism of the Transformer improves the classification performance, which illustrates the necessity of capturing long-range global information. This can be derived by comparing CNN and CNN+Transformer comparisons. Secondly, by comparing CNN+Transformer and IEEG-CT, we conclude that applying the MHA of causal convolution to capture the correlation of time series is important for iEEG signal classification. As evidenced by the two tables, the IEEG-CT approach utilizing causal convolution of the MHA yields the best performance.

Table 7. Ablation experimental results for cross-subject experiments.

Dataset	Method	ACC (%)	AUROC (%)	AUPRC (%)
FNUSA	CNN	87.23	90.14	72.82
	CNN+Transformer	91.76	91.01	81.44
	IEEG-CT (ours)	**93.16**	**93.44**	**82.97**
MAYO	CNN	87.06	93.44	76.87
	CNN+Transformer	88.05	**95.54**	**89.97**
	IEEG-CT (ours)	**91.83**	94.50	84.91

Table 8. Ablation experimental results for out-of-institution experiments.

Dataset	Method	ACC (%)	AUROC (%)	AUPRC (%)
Train: FNUSA, Test: MAYO	CNN	70.51	87.92	78.45
	CNN+Transformer	75.63	89.16	81.62
	IEEG-CT (ours)	**77.14**	**89.45**	**82.28**
Train: MAYO, Test: FNUSA	CNN	64.19	83.47	71.86
	CNN+Transformer	71.89	83.66	71.91
	IEEG-CT (ours)	**72.71**	**87.98**	**77.71**

5   Conclusion

In this paper, we propose a deep learning model, IEEG-CT, for the classification of iEEG signal tasks. The proposed approach involves a CNN module for extracting physiological characteristics and a Transformer module to capture the long-range global information of local features from iEEG signals. Moreover, in both cross-subject experiments and out-of-institution experiments, our proposed method achieves superior results for the multicentre iEEG dataset. Our future work will focus on investigating the challenges of class imbalance and cross-individual variation in the iEEG dataset.

Acknowledgements. This work is supported by the Natural Science Foundation of Shandong Province, China, under Grant ZR2019MF071, and the Project of Shandong Province Higher Educational Science and Technology Program, China, under Grant J16LN05.

References

1. Thijs, R.D., Surges, R., O'Brien, T.J., Sander, J.W.: Epilepsy in adults. Lancet **393**(10172), 689–701 (2019)
2. Wiebe, S., Jette, N.: Pharmacoresistance and the role of surgery in difficult to treat epilepsy. Nat. Rev. Neurol. **8**(12), 669–677 (2012)
3. Wang, Y., Yan, J., Wen, J., Yu, T., Li, X.: An intracranial electroencephalography (iEEG) brain function mapping tool with an application to epilepsy surgery evaluation. Front. Neuroinform. **10**, 15 (2016)
4. Urrestarazu, E., Jirsch, J.D., LeVan, P., Hall, J.: High-frequency intracerebral EEG activity (100–500 Hz) following interictal spikes. Epilepsia **47**(9), 1465–1476 (2006)
5. Nejedly, P., Cimbalnik, J., Klimes, P., Plesinger, F., Halamek, J., et al.: Intracerebral EEG artifact identification using convolutional neural networks. Neuroinformatics **17**, 225–234 (2019)
6. Sui, L., Zhao, X., Zhao, Q., Tanaka, T., Cao, J.: Localization of epileptic foci by using convolutional neural network based on iEEG. In: MacIntyre, J., Maglogiannis, I., Iliadis, L., Pimenidis, E. (eds.) AIAI 2019. IAICT, vol. 559, pp. 331–339. Springer, Cham (2019). https://doi.org/10.1007/978-3-030-19823-7_27
7. Guo, J., Wang, Y., Yang, Y., Kang, G.: IEEG-TCN: a concise and robust temporal convolutional network for intracranial electroencephalogram signal identification. In: 2021 IEEE International Conference on Bioinformatics and Biomedicine (BIBM), pp. 668–673. IEEE (2021)
8. Wang, Y., et al.: Computer-aided intracranial EEG signal identification method based on a multi-branch deep learning fusion model and clinical validation. Brain Sci. **11**(5), 615 (2021)
9. Nejedly, P., et al.: Multicenter intracranial EEG dataset for classification of graphoelements and artifactual signals. Sci. data **7**(1), 179 (2020)
10. Wang, Y., et al.: SEEG-Net: an explainable and deep learning-based cross-subject pathological activity detection method for drug-resistant epilepsy. Comput. Biol. Med. **148**, 105703 (2022)
11. Jiang, W.B., Yan, X., Zheng, W.L., Lu, B.L.: Elastic graph transformer networks for EEG-based emotion recognition. In: 2023 IEEE International Conference on Acoustics, Speech and Signal Processing (ICASSP), pp. 1–5. IEEE (2023)
12. Vaswani, A., et al.: Attention is all you need. In: Advances in Neural Information Processing Systems, vol. 30 (2017)
13. Manzari, O.N., Ahmadabadi, H., Kashiani, H., Shokouhi, S.B., Ayatollahi, A.: MedViT: a robust vision transformer for generalized medical image classification. Comput. Biol. Med. **157**, 106791 (2023)

14. Sun, Y.: Continuous seizure detection based on transformer and long-term iEEG. IEEE J. Biomed. Health Inform. **26**(11), 5418–5427 (2022)

15. Xu, M., Zhou, W., Shen, X., Wang, Y., Mo, L., Qiu, J.: Swin-TCNet: swin-based temporal-channel cascade network for motor imagery iEEG signal recognition. Biomed. Signal Process. Control **85**, 104885 (2023)

16. Peh, W.Y., Thangavel, P., Yao, Y., Thomas, J., Tan, Y.L., Dauwels, J.: Multi-center assessment of CNN-transformer with belief matching loss for patient-independent seizure detection in scalp and intracranial EEG. arXiv preprint arXiv:2208.00025 (2022)

17. Yu, Y., Si, X., Hu, C., Zhang, J.: A review of recurrent neural networks: LSTM cells and network architectures. Neural Comput. **31**(7), 1235–1270 (2019)

18. Zhu, X., Li, L., Liu, J., Peng, H., Niu, X.: Captioning transformer with stacked attention modules. Appl. Sci. **8**(5), 739 (2018)

19. Eldele, E., et al.: An attention-based deep learning approach for sleep stage classification with single-channel EEG. IEEE Trans. Neural Syst. Rehabil. Eng. **29**, 809–818 (2021)

20. Simonyan, K., Zisserman, A.: Very deep convolutional networks for large-scale image recognition. In: 3rd International Conference on Learning Representations (ICLR 2015). Computational and Biological Learning Society (2015)

21. Van Den Oord, A., et al.: WaveNet: a generative model for raw audio. In: 9th ISCA Speech Synthesis Workshop, pp. 125–125 (2016)

22. He, K., Zhang, X., Ren, S., Sun, J.: Deep residual learning for image recognition. In: Proceedings of the IEEE Conference on Computer Vision and Pattern Recognition, pp. 770–778 (2016)

23. Ioffe, S., Szegedy, C.: Batch normalization: accelerating deep network training by reducing internal covariate shift. In: International Conference on Machine Learning, pp. 448–456. PMLR (2015)

24. Bandos, A.I., Rockette, H.E., Song, T., Gur, D.: Area under the free-response ROC curve (FROC) and a related summary index. Biometrics **65**(1), 247–256 (2009)

Multi-task Learning Network for Automatic Pancreatic Tumor Segmentation and Classification with Inter-Network Channel Feature Fusion

Kaiwen Chen[1], Chunyu Zhang[2], Chengjian Qiu[1], Yuqing Song[1], Anthony Miller[3], Lu Liu[3], Imran Ul Haq[1], and Zhe Liu[1(✉)]

[1] The School of Computer Science and Communication Engineering, Jiangsu University, Zhenjiang, Jiangsu, China
1000004088@ujs.edu.cn
[2] The Department of Radiology, The First Hospital of Jilin University, Changchun, Jilin, China
[3] The School of Computing and Mathematical Sciences, University of Leicester, Leicester, UK

Abstract. Pancreatic cancer is a malignant tumor with a high mortality rate. Therefore, accurately identifying pancreatic cancer is of great significance for early diagnosis and treatment. Currently, several methods have been developed using network structures based on multi-task learning to address tumor recognition issues. One common approach is to use the encoding part of a segmentation network as shared features for both segmentation and classification tasks. However, due to the focus on detailed features in segmentation tasks and the requirement for more global features in classification tasks, the shared features may not provide more discriminatory feature representation for the classification task. To address above challenges, we propose a novel multi-task learning network that leverages the correlation between the segmentation and classification networks to enhance the performance of both tasks. Specifically, the classification task takes the tumor region images extracted from the segmentation network's output as input, effectively capturing the shape and internal texture features of the tumor. Additionally, a feature fusion module is added between the networks to facilitate information exchange and fusion. We evaluated our model on 82 clinical CT image samples. Experimental results demonstrate that our proposed multi-task network achieves excellent performance with a Dice similarity coefficient (DSC) of 88.42% and a classification accuracy of 85.71%.

Keywords: Pancreatic Tumor · Segmentation · Classification · Multi-task · Channel Fusion

B. Luo et al. (Eds.): ICONIP 2023, LNCS 14449, pp. 564–577, 2024.
https://doi.org/10.1007/978-981-99-8067-3_42

1   Introduction

Pancreatic cancer is a common cancer of the abdominal organs and is associated with a high fatality rate [19]. With the current incidence and mortality of pancreatic cancer increasing year by year globally, this will constitute a significant public health burden. Due to the high imaging density resolution, convenient examination, fast speed and non-invasiveness, computed tomography (CT) scan has become a routine imaging modality for pancreatic cancer. However, designing an computer-aided diagnosis system(CAD) for pancreatic tumor is a challenging task, due to the small size, low contrast, significant deformation, and complex tumor features in abdominal images.

There are two important tasks in building a CAD of pancreatic cancer: tumor segmentation and tumor classification. The tumor segmentation task is used to achieve tumor location and boundary detection, the classification task is mainly aimed at achieving the diagnosis of benign and malignant tumors. In clinical diagnosis, tumor boundary information obtained by a segmentation network is helpful for tumor classification. Consequently, establishing an effective correlation between segmentation and classification in a multi-task learning manner is a promising research direction in medical image analysis.

In recent years, some work has also begun to explore the integration of segmentation and classification tasks [21,24,26]. In these multi-task learning methods, the segmentation network serves as the backbone network, leveraging the feature information from the segmentation network path for tumor classification. Although the above methods reduce model parameters and training difficulty, the tumor feature information extracted from the segmentation network encoding path is insufficient and unreliable for tumor classification. Figure 1 shows the feature maps extracted from the encoding path by the trained segmentation and classification networks, respectively. The input image of the classification network only contains the tumor part. Although the segmentation network focuses on identifying tumor features, it may also identify a significant amount of feature information that is not relevant to tumors. On the contrary, the classification network possesses stronger feature representation capability and can effectively capture the rich texture and boundary information of tumors. Therefore, relying solely on the feature information from the segmentation pathway is insufficient for accurate tumor classification, and it is also difficult to guarantee its effectiveness.

To tackle the above issue, we propose a novel 3D multi-task learning network with inter-network channel feature fusion, which consists of a preprocessing module to obtain tumor localization information and reduce the input image size, a multi-task learning network is utilized to generate tumor segmentation and classification predictions, and an inter-network channel feature fusion module to supplement the feature information in the classification network. Unlike previous mult-itask learning methods, we do not directly use the feature information from the segmentation network for the classification task. Instead, we leverage the tumor mask information generated by the segmentation network to extract images that contain only the tumor region as input for the classification network. This approach offers two key benefits: firstly, the classification network can recognize tumor boundary features; secondly, by excluding extraneous information outside the tumor region, the classification network can effectively focus on the intricate texture features within the tumor, leading to a marked enhancement in classification accuracy. The feature fusion module leverages the feature information from the segmentation network to enrich the feature information in the classification network and compensate for the loss of feature information resulting from low accuracy in tumor segmentation. Moreover, the feature fusion module not only evaluates the significance of each channel feature automatically before utilizing the segmentation network features but also eliminates redundant feature information that does not contribute to the classification network. Experiments on our self-collected CT images dataset demonstrate that the proposed method can significantly improve the classification task performance. Our main contributions are summarized as follows:

1. We propose a novel multi-task network framework for pancreatic tumor segmentation and classification, and establish an effective correlation between the networks, enabling the classification network to extract more effective feature representations.
2. We introduce an inter-network channel feature fusion module to enrich the tumor feature information of the classification network and compensate for the partial information loss caused by the low accuracy of tumor segmentation.
3. We demonstrate the effectiveness of our proposed method on our dataset, achieving the DSC value of 88.42% and classification accuracy of 85.71%. Furthermore, we qualitatively illustrate that the proposed feature fusion module effectively compensates for the information lost of tumors.

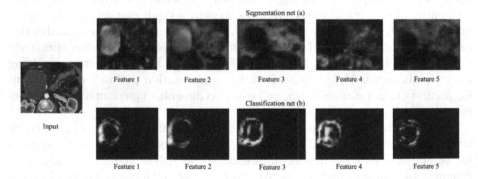

Fig. 1. The feature maps extracted from the encoding path by the trained segmentation and classification networks, respectively. The input image of the classification network only contains the tumor part.

2   Related Work

2.1   Pancreatic Tumor Segmentation

In recent years, with the rapid development of deep learning, many methods have emerged to address the challenging task of medical image. Among these methods, fully convolutional neural networks (FCN) [11], generative adversarial networks (GAN) [8], and the UNet segmentation network [16] as well as its variants [13,15,18] have played a critical role in medical image segmentation and demonstrated superior performance when compared to traditional techniques. In recent years, some methods using deep learning have also made great progress in the field of pancreatic tumor segmentation. For instance, Liang Y et al. [9] developed a CNN-based model and a framework that incorporates multi-modality images for the automatic segmentation of pancreatic tumors. A dual-path network was proposed by Zhou et al. [25] that fused dual-phase CT images of the arterial phase and the venous phase to segment pancreatic tumors. They added a matching loss function to encourage the commonality between high-level feature representations of different phases. Zhu et al. [27] cropped CT images into three different sizes of three-dimensional small blocks and sent them to three-dimensional U-Nets with a deep supervision mechanism. Finally, the three prediction results of the network were averaged to obtain the final pancreatic tumor segmentation.

2.2   Pancreatic Tumor Classification

In recent years, deep learning-based methods have made progress in some pancreatic tumor classification studies. Liu et al. [10] used a CNN classifier(based on the VGG network) to diagnose subjects with pancreatic cancer using contrast-enhanced CT images. Li et al. [7] proposed a computer-aided framework for early differential diagnosis of pancreatic cysts without pre-segmenting the lesions by using densely connected convolutional networks (DenseNet). Xuan W et al. [23] presented the prediction of the pancreatic tumor utilizing the hierarchical convolutional neural network method based on deep learning techniques. Ma et al. [12] developed a CNN classifier to automatically identify pancreas tumors in CT scans, which was evaluated using ternary and binary classifiers.

2.3   Multi-task Learning

Multi-task learning is a powerful approach that aims to solve multiple tasks concurrently. In the medical imaging domain, this approach has been used to perform segmentation and classification tasks simultaneously on the medical image. For instance, Wang et al. [21] modified the U-Net architecture by adding a classification branch for the segmentation and classification of bone surfaces in ultrasound images. Zhu et al. [26] adopted Segmentation for Classification strategy to make the classification result by a segmentation network more interpretable. Zhou et al. [24] designed a multi-task learning network for joint segmentation and classification of tumors in ABUS images.

3   Proposed Method

A novel multi-task learning approach for CT images of pancreatic tumors is proposed in this paper, as shown in Fig. 2. Our proposed framework comprises three main components: a preprocessing module, a multi-task learning network, and a feature fusion module. In addition, the Hounsfield Unit (HU) values of the input CT images are limited to [−200, 340] to increase contrast and prevent irrelevant noise.

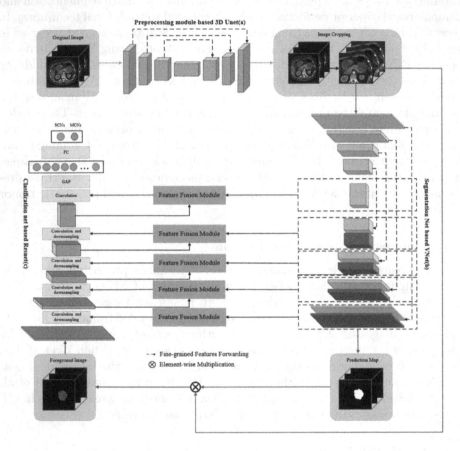

Fig. 2. The framework of the proposed network, which includes a preprocessing network, a multi-task learning network(segmentation and classification networks) and a feature fusion module.

3.1   Preprocessing Module

To overcome the challenge of training complex segmentation tasks on large 3D CT images, we employ a strategy that involves roughly locating the tumor locations and retaining only the images of the tumor region. This part is shown in

Fig. 2a, the input is sampled from the whole CT-scan volume denoted by X_1^i. We use medical images and their corresponding ground truth labels Y_1^i as our original input $\{X_1^i, Y_1^i\}$, where $i=\{0, 1, 2..., N\}$ and N is the total number of CT cases. Specifically, we predict the approximate region of the tumor using the 3D UNet [3] model with dice loss function (shown as in Eq. 1), written as $P_1^i = \mathcal{M}(X_1^i, \Theta_1)$, where Θ_1 is the model parameter and i is the sample token. Then, based on this approximate tumor volume, the size of each minimal 3D box covering the tumor prediction mask P_1^i is counted. Upon completion of the training, the largest rectangular cube among them is identified and extended outward by 10 pixels to obtain the final size $B \in R^{W,H,Z}$. Finally, during training and testing, we use the rectangular cube B to crop the original input $\{X_1^i, Y_1^i\}$ to obtain a new input $\{X_2^i, Y_2^i\}$, where X_2^i is consistent with Y_2^i.

3.2   Multi-task Learning Network

In this model, the classification network utilizes the tumor regions extracted from the segmentation network's predicted output as input. Additionally, we employ joint training to improve the robustness and accuracy of the model by incorporating mutual constraints between the tasks. The specific process of this part is as follows. For the cropped image $\{X_2^i, Y_2^i\}$ as the input of this part of the model. The VNet is used as the segmentation network for this part due to its excellent performance in 3D medical image segmentation (shown in Fig. 2b). This model aims to get accurate prediction results for tumor segmentation and is written as $P_2^i = \mathcal{S}(X_2^i, \Theta_2)$. Accurate tumor segmentation results mean that the classification network can focus on the feature information of the tumor itself, avoiding the interference of irrelevant information. Therefore, element-wise multiplication is performed between the prediction mask P_2^i and the input image X_2^i to obtain the image X_3^i containing only the tumor region corresponding to the prediction mask P_2^i. We then use X_3^i and the ground truth label for tumor categories Y_3^i as input $\{X_3^i, Y_3^i\}$ for the classification network. Regarding classification networks, Resnet and its variants have achieved excellent performance on natural images as well as medical images [1,4]. In this paper, ResNet [4] is applied to 3D medical image classification, as shown in Fig. 2c, and the model is written as $P_3^i = \mathcal{C}(X_3^i, \Theta_3)$. In this part, dice loss and cross-entropy loss(shown in Eq. 1, 2) are chosen as the loss functions for segmentation and classification networks respectively. The joint loss is shown in Eq. 3.

3.3   Feature Fusion Module

The module takes advantage of multi-task network feature extraction by using feature information from segmentation network paths to enrich feature information in classification network paths and compensate for any loss of feature information resulting from low accuracy in tumor segmentation. Additionally, a feature selection layer is incorporated to automatically evaluate the importance of channel features in the segmentation network and remove redundant feature information that does not contribute to the classification network. This fusion method follows a specific procedure. First, the corresponding feature information is extracted from the paths of the segmentation and classification networks. To ensure consistency in feature dimensions, the features obtained from the segmentation network's path are downsampled prior to concatenation. This is because our model for classification uses a ResNet variant in which the first convolutional block downsamples the input, while the segmentation network Vnet remains unchanged. Next, each channel's importance is automatically determined using Squeeze and Excitation (SE) [5] structure. The resulting features are then concatenated along the z-axis (horizontal plane) with features from the classification network and fed into a 3D convolutional layer, which reduces the number of channels. The process is illustrated in Fig. 3.

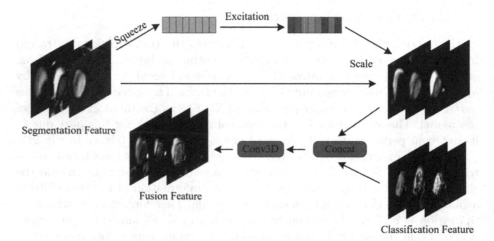

Fig. 3. The inter-network feature fusion module includes a Squeeze-Excitation structure, which is used to determine the importance of each channel. The Concat operation is used to combine the segmentation and classification networks along the channel dimension. Next, a Conv3D layer is applied to the concatenated feature maps to reduce the channel dimension.

3.4   Multi-task Loss Function

We choose the Dice coefficient loss and cross-entropy loss as loss functions for segmentation and classification tasks. In addition, classification loss and segmentation loss are linear combined as a multi-task loss by a hyperparameter α:

$$DiceLoss = 1 - \frac{2\left|P^i_{1,2} \cap Y^i_{1,2}\right|}{\left|P^i_{1,2}\right| + \left|Y^i_{1,2}\right|} \tag{1}$$

$$CELoss = -[Y^i_3 \times log P^i_3 + (1 - Y^i_3) \times log(1 - P^i_3)] \tag{2}$$

$$JointLoss = \alpha \times DiceLoss + (1 - \alpha) \times CELoss \qquad (3)$$

where P and Y represent the corresponding predictions and ground truth labels in the method described above. $\alpha \in [0, 1]$ is the weight of the segmentation task loss function.

4 Experiments and Results

4.1 Dataset

We conducted multi-task experiments on a private pancreatic tumor segmentation and classification dataset. It consists of 82 abdominal CT volumes, in which each sample was annotated by a professional abdominal radiologist and the tumors were classified as malignant or benign. The resolution of each CT scan is $512 \times 512 \times z$, where $z \in [153, 481]$ is the number of sampling slices along the horizontal plane. We randomly split the dataset into 4 fixed folds, randomly selecting three for training and the remaining one for testing. The classification task is a binary problem to distinguish serous cystic neoplasms (SCNs) and mucinous cystic neoplasms (MCNs). Among them, SCNs are considered to be benign tumors clinically, and MCNs are considered to be malignant tumors. Our dataset includes 40 SCNs and 42 MCNs samples.

4.2 Implementation Details

To verify the effectiveness of the proposed method, we compared it with other multi-task learning networks, including multi-feature guided CNN (cUNet) [21], Segmentation for Classification [26] and C_{MS}VNet [24]. Specifically, we extended all 2D operations in Multi-task CNN to 3D operations, and changed multi-class classification to two-class classification. In addition to the above comparison methods, we also compared our multi-task network with other single segmentation or classification networks. For single-task learning, the input images were obtained using the preprocessing module described above to ensure consistency with the multi-task experiments. For the segmentation task, we compared our proposed method with three segmentation networks: 3D SegNet [2], 3D ResUNet [22], and 3D Attention UNet [14]. We also compared our method with several classification methods, including 3D VGG16 [17], 3D SENet [5], 3D DenseNet [6], and 2D EfficientNet [20]. For EfficientNet, before model training, the original 3D CT image is converted into a 2D image in the form of slices, and only the cases where the number of tumor foreground pixels is greater than 100 are retained.

We also conducted several groups of ablation experiments. The VNet [13] and 3D ResNet [4] models were trained as the segmentation and classification baseline models, respectively. For multi-task learning, firstly, the classification network took the image containing only tumor regions as input (denoted as ResVNet$_{Tumor}$), which was obtained by the segmentation network. Secondly, we incorporated the proposed feature fusion module into the multi-task network (denoted as ResVNet$_{Fusion}$).

The model was developed by using the PyTorch framework and trained/tested on a single NVIDIA RTX 3090 GPU. During the training process in the preprocessing module and multi-task learning network, we all employed the Adam optimizer with a learning rate of 0.001 for 500 epochs and a batch size of two. During the training and testing process, the multi-task network takes the images and corresponding ground truth labels as inputs, which are obtained through the preprocessing module.

4.3 Performance Comparison

We evaluated the quality of tumor segmentation using three quantitative metrics: Dice similarity coefficient (DSC) and Jaccard similarity index (JI) are sensitive to tumor areas, and 95th percentile of the asymmetric Hausdorff distance (95HD) to shape. To demonstrate the superiority of the proposed network, we compared it with multiple existing segmentation methods including 3D SegNet, 3D ResUNet, and 3D Attention UNet. The segmentation performance of the proposed method and other segmentation methods are presented in Table 1. Experimental results show that our proposed network outperforms these existing single segmentation methods in pancreatic tumor segmentation. Specifically, without using additional

Table 1. Segmentation performance of different segmentation models(Mean ± Standard Deviation).

Model	DSC (%)	JI (%)	95HD (mm)
3D SegNet [2]	81.85± 2.31	52.78 ± 7.84	2.37 ± 2.63
3D ResUNet [22]	80.70±2.02	51.40 ± 8.18	2.23 ± 2.08
3D Attention UNet [14]	85.56±0.89	59.33 ± 4.47	2.31 ± 1.40
ResVNet$_{Fusion}$	**88.42±1.33**	**68.04±5.90**	**1.88 ± 2.85**

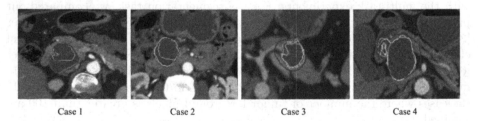

Case 1   Case 2   Case 3   Case 4

Fig. 4. This figure displays a 2D visual comparison of segmentation boundaries across different models, with each model represented by a different color. Specifically, VNet is represented in purple, 3D ResUnet in yellow, 3D Attention Unet in blue, ResVNet$_{Fusion}$($Color figure online$) in red, and the ground truth in green. (Color figure online)

data, our proposed network achieves an average Dice of 88.42%, JI of 68.04% and 95HD of 1.88 mm.

Figure 4 displays the segmented boundaries obtained by different 3D segmentation methods in 2D CT slices. Among all the methods, our proposed method (shown in red) has the most similar segmented boundaries to the ground truth (shown in green) in most slices.

Table 2 shows that our proposed method obtains quite remarkable classification performance in the classification task for pancreatic tumors, which outperforms the performance of other classification methods. Our proposed method achieves the best accuracy (ACC), precision (PRE), false positive rate (FPR), and F1-score (F1) arrive at 85.71%, 92.30%, 14.29% and 88.89%, respectively, except for the recall (REC). The performance improvement can be attributed to two factors: Firstly, the classification network focuses on the rich texture information within the tumor area. Secondly, the classification network incorporates effective feature information in the segmentation network as a supplement, improving the effectiveness of the feature. Additionally, we conducted a classification experiment on 2D pancreatic tumor images using the 2D EfficientNet. Our experiments validate that employing 3D methods for pancreatic tumors yields superior performance by capturing essential 3D spatial context information.

Table 2. Classification performance of different classification models.

Model	ACC (%)	REC (%)	PRE (%)	FPR (%)	F1 (%)
3D VGG16 [17]	66.67	78.57	73.33	57.14	75.86
3D SENet [5]	80.95	78.57	91.67	14.29	84.62
3D DenseNet [6]	71.43	**92.86**	72.22	71.43	81.25
2D EfficientNet [20]	79.40	84.62	74.92	24.45	81.45
ResVNet$_{Fusion}$	**85.71**	85.71	**92.30**	**14.29**	**88.89**

In order to further demonstrate the effectiveness of the feature fusion module in model ResVNet$_{Fusion}$, we analyzed a sample with poor segmentation prediction. Figure 5 shows that the segmentation prediction results are incomplete, with a portion of the tumor missing. Next, we extracted the feature maps from the first three layers of the classification network in model ResVNet$_{Fusion}$. In the proposed network framework, the fused features were able to effectively compensate for the lost tumor features. Additionally, redundant feature information that did not contribute to the classification network was suppressed through automatic evaluation of the importance of each segmentation network channel feature. Therefore, this confirms the effectiveness of our proposed method.

We compared the proposed method with the other three multi-task methods, which include multi-feature guided CNN (cUNet), Segmentation for Classification (S4C) and C$_{MS}$VNet. Table 3 displays the results of our method and the other three multi-task methods on our dataset. It shows that except for the

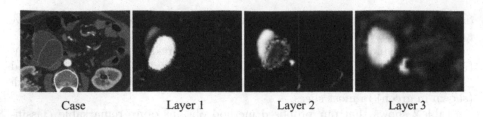

Case   Layer 1   Layer 2   Layer 3

Fig. 5. The first image is the segmentation prediction result of a case. The next few images are the feature maps obtained by the classification network after feature fusion. The masked red and green lines denote the predicted and ground truth labeled tumor boundaries, respectively. (Color figure online)

Table 3. Comparison with existing multi-task learning models.

Model	Segmentation (Mean ± SD)			Classification(%)				
	DSC (%)	JI (%)	95HD (mm)	ACC	REC	PRE	FPR	F1
cUNet [21]	77.92±3.39	48.25±11.32	2.09±2.36	76.19	**92.86**	76.47	57.14	83.87
S4C [26]	71.91±6.62	42.83±13.51	4.74±5.71	61.90	71.43	71.43	57.14	71.43
C_{MS}VNet [24]	87.14±1.34	64.16±5.95	**1.68±1.46**	71.43	71.43	83.33	28.57	76.92
ResVNet$_{Fusion}$	**88.42±1.33**	**68.04±5.90**	1.88±2.85	**85.71**	85.71	**92.30**	**14.29**	**88.89**

95HD and REC, our proposed model achieves the best segmentation and classification performance, and although the 95HD is not the best, it also reaches 1.88 mm, only second to 1.68 mm. As shown in Table 4, we list the quantitative results of the ablation experiments. For the segmentation task, the proposed model ResVNet$_{Fusion}$ outperforms other models in metrics with DSC, JI at 88.42%, 68.04%, except that the metric 95HD is slightly lower than the model ResVNet$_{Tumor}$. Comparison between ResVNet$_{Fusion}$ and VNet results shows that the segmentation task achieves better performance in ResVNet$_{Fusion}$, where the DSC, JI and 95HD increased by approximately 5.12%, 11.87%, and 0.41 mm, respectively. The result indicates that joint training of the segmentation and classification tasks can significantly improve the performance of the segmentation task.

Table 4. Segmentation and classification performance of ablation models.

Model	Segmentation (Mean ± SD)			Classification(%)				
	DSC (%)	JI (%)	95HD (mm)	ACC	REC	PRE	FPR	F1
VNet [13]	83.30±1.83	56.17±6.74	2.29±2.33	–	–	–	–	–
3D ResNet [4]	–	–	–	76.19	78.57	84.62	28.57	81.483
ResVNet$_{Tumor}$	86.30±1.41	64.92±6.68	**1.83±2.16**	80.95	78.57	91.67	14.29	84.62
ResVNet$_{Fusion}$	**88.42±1.33**	**68.04±5.90**	1.88±2.85	**85.71**	**85.71**	**92.30**	**14.29**	**88.89**

5   Conclusion

This paper proposes a multi-task network framework for pancreatic tumor, which consists of three modules: a preprocessing module, a multi-task learning network and a feature fusion module. Experimental results demonstrate that our proposed network framework achieves superior performance compared to other multi-task and single-task network models. This highlights the potential of multi-tasking computer-aided diagnosis systems to assist clinicians in developing optimal treatment plans. Therefore, we believe that further research in this direction is promising.

Acknowledgements. The authors would like to thank the Department of Radiology of First Hospital of Jilin University. This work was supported by the National Natural Science Foundation of China (62276116, 61976106); Six talent peaks project in Jiangsu Province (DZXX-122)

References

1. Abiwinanda, N., Hanif, M., Hesaputra, S.T., Handayani, A., Mengko, T.R.: Brain tumor classification using convolutional neural network. In: Lhotska, L., Sukupova, L., Lacković, I., Ibbott, G.S. (eds.) World Congress on Medical Physics and Biomedical Engineering 2018. IP, vol. 68/1, pp. 183–189. Springer, Singapore (2019). https://doi.org/10.1007/978-981-10-9035-6_33
2. Badrinarayanan, V., Kendall, A., Cipolla, R.: SegNet: a deep convolutional encoder-decoder architecture for image segmentation. IEEE Trans. Pattern Anal. Mach. Intell. **39**(12), 2481–2495 (2017)
3. Çiçek, Ö., Abdulkadir, A., Lienkamp, S.S., Brox, T., Ronneberger, O.: 3D U-Net: learning dense volumetric segmentation from sparse annotation. In: Ourselin, S., Joskowicz, L., Sabuncu, M.R., Unal, G., Wells, W. (eds.) MICCAI 2016. LNCS, vol. 9901, pp. 424–432. Springer, Cham (2016). https://doi.org/10.1007/978-3-319-46723-8_49
4. He, K., Zhang, X., Ren, S., Sun, J.: Deep residual learning for image recognition. In: Proceedings of the IEEE Conference on Computer Vision and Pattern Recognition, pp. 770–778 (2016)
5. Hu, J., Shen, L., Sun, G.: Squeeze-and-excitation networks. In: Proceedings of the IEEE Conference on Computer Vision and Pattern Recognition, pp. 7132–7141 (2018)
6. Huang, G., Liu, Z., Van Der Maaten, L., Weinberger, K.Q.: Densely connected convolutional networks. In: Proceedings of the IEEE Conference on Computer Vision and Pattern Recognition, pp. 4700–4708 (2017)
7. Li, H., et al.: Differential diagnosis for pancreatic cysts in CT scans using densely-connected convolutional networks. In: 2019 41st Annual International Conference of the IEEE Engineering in Medicine and Biology Society (EMBC), pp. 2095–2098. IEEE (2019)
8. Li, Z., Wang, Y., Yu, J.: Brain tumor segmentation using an adversarial network. In: Crimi, A., Bakas, S., Kuijf, H., Menze, B., Reyes, M. (eds.) BrainLes 2017. LNCS, vol. 10670, pp. 123–132. Springer, Cham (2018). https://doi.org/10.1007/978-3-319-75238-9_11

9. Liang, Y., et al.: Auto-segmentation of pancreatic tumor in multi-parametric MRI using deep convolutional neural networks. Radiother. Oncol. **145**, 193–200 (2020)

10. Liu, K.L., et al.: Deep learning to distinguish pancreatic cancer tissue from non-cancerous pancreatic tissue: a retrospective study with cross-racial external validation. Lancet Digital Health **2**(6), e303–e313 (2020)

11. Long, J., Shelhamer, E., Darrell, T.: Fully convolutional networks for semantic segmentation. In: Proceedings of the IEEE Conference on Computer Vision and Pattern Recognition, pp. 3431–3440 (2015)

12. Ma, H., et al.: Construction of a convolutional neural network classifier developed by computed tomography images for pancreatic cancer diagnosis. World J. Gastroenterol. **26**(34), 5156 (2020)

13. Milletari, F., Navab, N., Ahmadi, S.A.: V-Net: fully convolutional neural networks for volumetric medical image segmentation. In: 2016 Fourth International Conference on 3D Vision (3DV), pp. 565–571. IEEE (2016)

14. Oktay, O., et al.: Attention u-net: Learning where to look for the pancreas. arXiv preprint arXiv:1804.03999 (2018)

15. Qiu, C., et al.: RTUNet: residual transformer UNet specifically for pancreas segmentation. Biomed. Signal Process. Control **79**, 104173 (2023)

16. Ronneberger, O., Fischer, P., Brox, T.: U-Net: convolutional networks for biomedical image segmentation. In: Navab, N., Hornegger, J., Wells, W.M., Frangi, A.F. (eds.) MICCAI 2015. LNCS, vol. 9351, pp. 234–241. Springer, Cham (2015). https://doi.org/10.1007/978-3-319-24574-4_28

17. Simonyan, K., Zisserman, A.: Very deep convolutional networks for large-scale image recognition. arXiv preprint arXiv:1409.1556 (2014)

18. Song, L., Geoffrey, K., Kaijian, H.: Bottleneck feature supervised U-Net for pixel-wise liver and tumor segmentation. Expert Syst. Appl. **145**, 113131 (2020)

19. Sung, H., et al.: Global cancer statistics 2020: GLOBOCAN estimates of incidence and mortality worldwide for 36 cancers in 185 countries. CA Cancer J. Clin. **71**(3), 209–249 (2021)

20. Tan, M., Le, Q.: EfficientNet: rethinking model scaling for convolutional neural networks. In: International Conference on Machine Learning, pp. 6105–6114. PMLR (2019)

21. Wang, P., Patel, V.M., Hacihaliloglu, I.: Simultaneous segmentation and classification of bone surfaces from ultrasound using a multi-feature guided CNN. In: Frangi, A.F., Schnabel, J.A., Davatzikos, C., Alberola-López, C., Fichtinger, G. (eds.) MICCAI 2018. LNCS, vol. 11073, pp. 134–142. Springer, Cham (2018). https://doi.org/10.1007/978-3-030-00937-3_16

22. Xiao, X., Lian, S., Luo, Z., Li, S.: Weighted Res-UNet for high-quality retina vessel segmentation. In: 2018 9th International Conference on Information Technology in Medicine and Education (ITME), pp. 327–331. IEEE (2018)

23. Xuan, W., You, G.: Detection and diagnosis of pancreatic tumor using deep learning-based hierarchical convolutional neural network on the internet of medical things platform. Futur. Gener. Comput. Syst. **111**, 132–142 (2020)

24. Zhou, Y., et al.: Multi-task learning for segmentation and classification of tumors in 3D automated breast ultrasound images. Med. Image Anal. **70**, 101918 (2021)

25. Zhou, Y., et al.: Hyper-pairing network for multi-phase pancreatic ductal adenocarcinoma segmentation. In: Shen, D., et al. (eds.) MICCAI 2019. LNCS, vol. 11765, pp. 155–163. Springer, Cham (2019). https://doi.org/10.1007/978-3-030-32245-8_18

26. Zhu, Z., Lu, Y., Shen, W., Fishman, E.K., Yuille, A.L.: Segmentation for classification of screening pancreatic neuroendocrine tumors. In: Proceedings of the IEEE/CVF International Conference on Computer Vision, pp. 3402–3408 (2021)
27. Zhu, Z., Xia, Y., Xie, L., Fishman, E.K., Yuille, A.L.: Multi-scale coarse-to-fine segmentation for screening pancreatic ductal adenocarcinoma. In: Shen, D., et al. (eds.) MICCAI 2019. LNCS, vol. 11769, pp. 3–12. Springer, Cham (2019). https://doi.org/10.1007/978-3-030-32226-7_1

Fast and Efficient Brain Extraction with Recursive MLP Based 3D UNet

Guoqing Shangguan[1], Hao Xiong[2], Dong Liu[1,3,4], and Hualei Shen[1,3,4(✉)]

[1] College of Computer and Information Engineering, Henan Normal University, Xinxiang, China
2208283086@stu.htu.edu.cn
{liudong,shenhualei}@htu.edu.cn
[2] Centre for Health Informatics, Australian Institute of Health Innovation, Macquarie University, Sydney, Australia
hao.xiong@mq.edu.au
[3] Key Laboratory of Artificial Intelligence and Personalized Learning in Education of Henan Province, Xinxiang, China
[4] Big Data Engineering Lab of Teaching Resources and Assessment of Education Quality, Xinxiang, Henan, China

Abstract. Extracting brain from other non-brain tissues is an essential step in neuroimage analyses such as brain volume estimation. The transformers and 3D UNet based methods achieve strong performance using attention and 3D convolutions. They normally have complex architecture and are thus computationally slow. Consequently, they can hardly be deployed in computational resource-constrained environments. To achieve rapid segmentation, the most recent work UNeXt reduces convolution filters and also presents the Multilayer Perception (MLP) blocks that exploit simpler and linear MLP operations. To further boost performance, it shifts the feature channels in MLP block so as to focus on learning local dependencies. However, it performs segmentation on 2D medical images rather than 3D volumes. In this paper, we propose a recursive MLP based 3D UNet to efficiently extract brain from 3D head volume. Our network involves 3D convolution blocks and MLP blocks to capture both long range information and local dependencies. Meanwhile, we also leverage the simplicity of MLPs to enhance computational efficiency. Unlike UNeXt extracting one locality, we apply several shifts to capture multiple localities representing different local dependencies and then introduce a recursive design to aggregate them. To save computational cost, the shifts do not introduce any parameters and the parameters are also shared across recursions. Extensive experiments on two public datasets demonstrate the superiority of our approach against other state-of-the-art methods with respect to both accuracy and CPU inference time.

Keywords: MLP · 3D UNet · Brain Extraction · Resource-constrained Computing

G. Shangguan and H. Xiong—Equal contribution.

1 Introduction

Extracting brain region from head volume is clinically significant in the tasks including brain atrophy estimation and brain registration. Though a well-trained neurologist can manually produce brain masks, it is laborious, time consuming, and heavily relies on the well-trained personnel. Hence, the automatic brain extraction method FSL BET [12,21] was proposed. It utilized statistical machine learning method to estimate an optimized threshold differentiating between brain and non-brain areas in a head volume. However, the estimated threshold may be inaccurate and induces noises in the generated brain masks.

Due to the advancement of deep learning, many deep learning based approaches have been proposed for medical image segmentation. The most popular one should be UNet [19] which includes the encoder, decoder modules, and the skip connections aggregating information between them. Many UNet based methods [4,6,9–11,15,27] focused on volumetric data segmentation. For instance, 3D UNet [4,9] simply replaced 2D convolutional filters with 3D ones. The Modified 3D UNet [11] and V-Net [15] further added residual connection in 3D module to learn more representative features.

The transformer and attention based approaches also demonstrate strong performance in segmentation tasks [7,22,25]. Specifically, Non-local UNet [25] combined traditional convolution with self-attention to capture both local and non-local information. UNETR [7] first utilized transformer to encode long-distance dependencies across the whole volume, and then applied 3D convolutional neural network as decoder to represent multi-scale semantic features. MSMHA-CNN [22] used multiple kernels to obtain multi-scale local features which were then fused by a channel-wise attention module to effectively represent brain. All the above-mentioned methods invariably seek to enhance segmentation accuracy without considering computational complexity. In fact, their models are complex, parameters-heavy, and consequently cannot produce fast segmentation. However, fast and efficient brain extraction is pivotal in modern neuroimage analyses especially for those clinics or small-scale hospitals equipped with CPU or low-end GPU [5,14,20].

More recently, MLP-based networks [13,18,23,24,26] have also been utilized to accomplish vision tasks, which showed even better performance than transformers and had much less computations. Based on UNet architecture, Vala-narasu et al. [24] present UNeXt, an MLP based network, for rapid medical image segmentation. Its proposed MLP blocks include MLPs that are less complicated than convolution or attention. To attain better performance, they further introduced axial shift operations combined with MLPs to learn local dependency. However, it is a 2D medical image segmentation method that is not applicable to 3D image segmentation. In this work, we propose a recursive MLP based 3D UNet to extract brain from 3D head volumes. Our network first stacks a few 3D convolutions to extract long range features. These convolutional features will become much smaller in bottleneck where we apply a novel MLP block (shown in Fig. 1) to learn multiple local dependencies by different shifts. Unlike UNeXt focusing on learning one locality, we capture multiple localities to extract local features with different connections and introduce a recursive design to efficiently aggregate these localities. We call the aforementioned process as recursive MLP. Here, all shifts are parameters-free, MLPs are simpler than convolutions, and the parameters within all recursions are shared. Therefore, we are able to maintain rapid brain extraction. We show that our method surpasses other recent baselines on two public datasets: Internet Brain Segmentation Repository (IBSR) [1] and Neurofeedback Skull-stripped repository (NFBS) [2]. On these datasets, our method takes less than **7.5** s to extract brain from one head volume.

We summarize our contributions as follows: 1) We propose a brain extraction method that exploits both global information and local dependencies by combining 3D convolutions with MLPs. 2) Instead of focusing on one locality, we capture and aggregate multiple localities using an efficient recursive design for better representation extraction. 3) Our method shows superior performance against recent baselines with respect to both accuracy and inference speed. More importantly, it has great potential to be deployed in point-of-care circumstances.

2   Recursive MLP Based 3D UNet

2.1   Network Architecture

Figure 1 illustrates that our proposed network follows an UNet-like architecture with encoder and decoder. The 3D convolution blocks aim to extract long range information, while the MLP blocks focus on learning local dependencies.

3D Convolution Block. The 3D convolution block contains a 3D convolutional (Conv3d) layer and a batch normalization (BN) layer. For Conv3d, we set its kernel size as $3 \times 3 \times 3$, stride as 1, and padding as 1. In addition, the last 3D convolution block in decoder includes one more Conv3d with kernel size $1 \times 1 \times 1$, stride 1, and padding 1. Each 3D convolution block adjusts the feature resolutions using MaxPool3d or trilinear interpolation. Note that one 3D convolution block only extracts non-local features. However, stacking several 3D convolutions can gradually extract long range information.

MLP Block. Our MLP block starts with a Conv3d layer and the layer normalization (LN) [3,24]. For each MLP block, its Conv3d has same kernel size $3 \times 3 \times 3$ and padding 1. However, the Conv3d strides in first two and last two MLP blocks are 2 and 1, respectively. Afterwards, the 3D depth wise convolution (DwConv3d) layer encodes positional information [24], and a Gaussian error linear units (GELU) [8] layer smooths the activation maps. We then exploit the proposed recursive MLPs (details in Subsect. 2.2) to learn more robust feature representations. Finally, we apply another LN and add back its output to the input of MLP block.

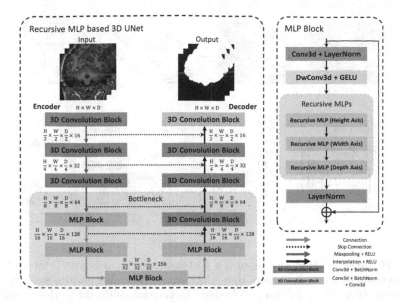

Fig. 1. The architecture of our network. It contains 3D convolution blocks and MLP blocks to capture both long range and local features. By stacking 3D convolutions, the features in bottleneck become much smaller and are already infused with global information. Instead, MLP blocks extract localities of these features in bottleneck.

2.2 Recursive MLP

UNeXt only extracts one locality. By contrast, our recursive MLP performs shifts several times and combines MLPs to capture multiple different local dependencies in a recursive manner. For efficient computation, the shifts do not introduce any parameters and the parameters in our recursive design are shared. In Fig. 2, our recursive MLP includes *group, shifts, recursive aggregation of localities*.

Group. The input feature of our recursive MLP is a 4D tensor $\mathbf{X} \in \mathbb{R}^{W \times H \times D \times C}$ where W, H, D, C respectively denotes width, height, depth and channel. Inspired by [13,24], we equally split $\mathbf{X}$ along channel into n_c thinner tensors $\{\boldsymbol{x}_i\}_{i=1}^{n_c}$ where $\boldsymbol{x}_i \in \mathbb{R}^{W \times H \times D \times \frac{C}{n_c}}$. As per [24], we set n_c as 5 in our case.

Shifts. Given one group $\boldsymbol{x}_i$ in $\mathbf{X}$, we first create a tensor $\hat{\boldsymbol{x}}_i \in \mathbb{R}^{W \times H \times D \times \frac{C}{n_c}}$ with all zeros, but having same size as $\boldsymbol{x}_i$. Suppose we shift $\boldsymbol{x}_i$ by l locations on width axes, then we have:

$$\hat{\boldsymbol{x}}_i\left[:, l : W, :, :\right] = \boldsymbol{x}_i\left[:, 0 : W - l, :, :\right], \tag{1}$$

$$\hat{\boldsymbol{x}}_i\left[:, 0 : W - l, :, :\right] = \boldsymbol{x}_i\left[:, l : W, :, :\right], \tag{2}$$

Here, the Eqs. 1–2 respectively correspond to shift right and left by l locations on width axis. After shifting, we have the shifted tensor $\hat{\mathbf{X}} = concat(\hat{\boldsymbol{x}}_i) \in$

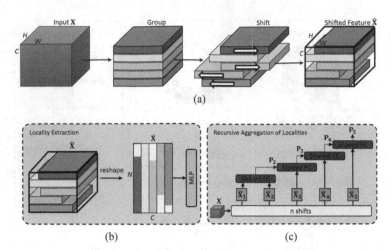

Fig. 2. Recursive MLP. Our input is a 4D tensor, yet we utilize a 3D tensor for ease of illustration. (a) Shift Operation. Here, the input **X** shifts along width axis, but it can shift along the other axes either. (b) Locality Extraction. With shifts, the generated $\bar{\mathbf{X}}$ contains zeros (blank area) that are not involved in MLP projection. In (c) Recursive Aggregation of Localities, it shows an example with 4 recursions. The **Shared FC** refers to the fully-connected layers in which their parameters are shared.

$\mathbb{R}^{W \times H \times D \times C}$ where *concat* concatenates all groups $\{\hat{\boldsymbol{x}}_i\}_{i=1}^{n_c}$ along channels. The *group* and *shift* operations are illustrated in top row of Fig. 2.

As per UNeXt [24], we subsequently reshape $\hat{\mathbf{X}} \in \mathbb{R}^{W \times H \times D \times C}$ to $\bar{\mathbf{X}} \in \mathbb{R}^{N \times C}$ where $N = W \times H \times D$, and then apply MLP projection as $\mathbf{U}\bar{\mathbf{X}}$ where $\mathbf{U} \in \mathbb{R}^{N \times N}$ is the MLP parameter. Due to shifts, some features in $\bar{\mathbf{X}}$ are vacant (zeros) and will not contribute to projection. Therefore, the MLP projection captures the connections among the remaining features to learn local dependency. (shown in Fig. 2(a)) We respectively shift along height, width, and depth to extract local information corresponding to different axial shifts.

Recursive Aggregation of Localities. We introduce a recursive design to effectively aggregate multiple localities that further enhance model capacity. To capture n different localities, we apply n shift operations (denoted as φ) to input feature **X**, generating n shifted feature tensors:

$$\left[\hat{\mathbf{X}}_1^{W \times H \times D \times C}, \cdots, \hat{\mathbf{X}}_n^{W \times H \times D \times C}\right] = \varphi\left(\mathbf{X}\right). \tag{3}$$

We then reshape $\{\hat{\mathbf{X}}_i\}_{i=1}^n \in \mathbb{R}^{W \times H \times D \times C}$ to $\{\bar{\mathbf{X}}_i\}_{i=1}^n \in \mathbb{R}^{N \times C}$ ($N = W \times H \times D$) and perform aggregation of localities recursively.

$$\mathbf{P}_{i+1} = \mathbf{W}[\mathbf{P}_i, \bar{\mathbf{X}}_{i+1}], \qquad 1 \leq i \leq n-1. \tag{4}$$

Here, we initialize $\mathbf{P}_1$ with $\bar{\mathbf{X}}_1$, $[\mathbf{P}_i, \bar{\mathbf{X}}_{i+1}] \in \mathbb{R}^{2N \times C}$ concatenates $\mathbf{P}_i$ and $\bar{\mathbf{X}}_{i+1}$ along rows, and $\mathbf{W} \in \mathbb{R}^{N \times 2N}$ is the parameter shared by all fully-connected layers to ensure efficient computation. (shown in Fig. 2 (b))

2.3   Loss Function

Our loss function $\mathcal{L}$ includes binary cross entropy (BCE) and dice loss (DL):

$$\mathcal{L}_{BCE} = -\frac{1}{N} \sum_{i=1}^{N} \left(y_i \log\left(p(\hat{y}_i)\right) + (1 - y_i) \log\left(1 - p(\hat{y}_i)\right)\right), \tag{5}$$

$$\mathcal{L}_{DL} = 1 - \frac{2 \sum_{i=1}^{N} \hat{y}_i y_i}{\sum_{i=1}^{N} \left(\hat{y}_i + y_i\right)}, \tag{6}$$

$$\mathcal{L} = \mathcal{L}_{BCE} + \alpha \mathcal{L}_{DL}, \tag{7}$$

where $\mathcal{L}_{BCE}$ is the binary cross-entropy loss, and $\mathcal{L}_{DL}$ is the dice loss. Meanwhile, $y_i, \hat{y}_i \in \{0, 1\}$, $p(\hat{y}_i)$ refer to the ground truth, predicted label and predicted probability of ith voxel. We empirically set the trade-off parameter α as 0.25.

3   Experiments

3.1   Setups

Data and Evaluation Metrics. We evaluate our method on two public datasets: IBSR [1] and NFBS [2]. The IBSR dataset contains 18 T1-weighted head volumes and corresponding brain masks. Each volume has size $256 \times 256 \times 128$ with resolution 1 mm $\times$ 1 mm $\times$ 2 mm. The NFBS dataset has 125 T1-weighted head volumes with corresponding masks. The size of each volume is $256 \times 256 \times 192$ and the resolution is 1 mm $\times$ 1 mm $\times$ 1 mm.

For computational complexity comparison, we demonstrate number of parameters, GFLOPs, and CPU inference time. To evaluate segmentation accuracy, we utilize the metrics of Dice coefficient (DICE), mean surface distance (MSD), 95th-percentile Hausdorff distance (HD95) and 99th-percentile Hausdorff distance (HD99).

Implementation Details. We implemented our network using PyTorch, and used Adam to optimize the model with learning rate $3e-3$ and weight decay $2e-6$. The total number of epochs for training was 2000 with batch size 4. At training stage, a large number of $128 \times 128 \times 128$ patches were randomly cropped from volumes for data augmentation. In inference, we followed [16] to equally partition each volume into $128 \times 128 \times 128$ patches with the overlapping set as 80. We performed 5-fold cross validation, for which each fold had training, validation and test sets with ratio 7 : 1 : 2. The best-performing model on validation set was used for inference on test set. Instead of using GPUs, our model was evaluated on a computer with Intel® Xeon® CPU E5-2620.

Table 1. Computing complexity comparison with other baselines.

	Params (M)	GFLOPs	CPU inference time (s)	
			IBSR	NFBS
3D UNet [4]	15.37	1730.87	241.47	239.23
Modified UNet [11]	7.13	57.57	32.89	33.11
UNETR [7]	92.62	195.64	53.50	55.06
MSMHA-CNN [22]	17.37	1943.74	210.32	207.55
UNeXt [24]	**1.47**	**0.31**	26.25	23.56
MLP-VNet [17]	14.45	14.41	8.30	8.36
Our Method	4.69	8.48	**7.01**	**6.75**

3.2   Results

We compare our method against several recent baselines including 3D convolutional networks - 3D UNet [4] and Modified UNet [11], transformer based methods - UNETR [7] and MSMHA-CNN [22], MLP based methods - UNeXt [24] and MLP-VNet [17].

Table 1 compares computational complexity of our model with other baselines. Other than UNeXt, our method has fewest number of parameters, since we reduce convolutional filters and involves simpler MLP operations. However, UNeXt exploits 2D convolutions and thus contains less parameters than ours. Despite that, it has to process 3D head volumes one by one, which leads to slower CPU inference time. By contrast, our method has fastest CPU inference time on both datasets. The transformers or all-convolutions based methods are parameters-heavy in general and hence inference much slower.

In terms of segmentation accuracy, we can see from Table 2 that our method outperforms the other baselines on both datasets. We simultaneously exploit long range information and localities, and show its superiority over those methods extracting either long range information or local features only. Besides, UNeXt is 2D image segmentation focused and can only process 3D volume slice by slice, which inevitably induces discontinuities across these slices. Figure 3 and 4 illustrate some examples to visually compare brain extraction effects. Compared to the other methods, our method produces more accurate brain segmentation.

3.3   Ablation Study

We study how the number of recursions affects performance. More recursions means more shifts, which extracts more local dependencies. In Fig. 5(a) and (c), we can see that our model reaches best performance with 4 recursions on IBSR and 3 recursions on NFBS. It is also shown that the performance degrades as the number of recursions increases. This is presumably because more recursions aggregate more information and accordingly the locality loses. Besides, more recursions lead to the increase of CPU inference time. (shown in Fig. 5(b), (d)) However, the increase is so minor that will not affect its practical usage.

Table 2. Performance comparison on the IBSR and NFBS datasets.

	IBSR				NFBS			
	DICE ↑	MSD ↓	HD95 ↓	HD99 ↓	DICE ↑	MSD ↓	HD95 ↓	HD99 ↓
3D UNet [4]	$0.978_{\pm0.002}$	$1.041_{\pm0.281}$	$2.707_{\pm0.497}$	$6.364_{+3.045}$	$0.988_{\pm0.001}$	$0.609_{\pm0.191}$	$1.578_{\pm0.841}$	$2.968_{\pm1.384}$
Modified UNet [11]	$0.949_{\pm0.030}$	$4.686_{\pm5.097}$	$14.100_{\pm15.686}$	$22.642_{\pm18.037}$	$0.918_{\pm0.034}$	$12.558_{\pm6.521}$	$38.050_{\pm18.885}$	$56.411_{\pm15.003}$
UNETR [7]	$0.971_{\pm0.012}$	$1.266_{\pm0.582}$	$3.485_{\pm1.520}$	$5.842_{\pm2.553}$	$0.987_{\pm0.000}$	$0.584_{\pm0.104}$	$1.589_{\pm0.688}$	$2.788_{\pm0.920}$
MSMHA-CNN [22]	$0.947_{\pm0.018}$	$5.011_{\pm2.137}$	$16.522_{\pm7.566}$	$30.292_{\pm11.766}$	$0.980_{\pm0.003}$	$5.133_{\pm1.663}$	$17.390_{\pm4.423}$	$32.219_{\pm5.554}$
UNeXt [24]	$0.865_{\pm0.083}$	$6.254_{\pm3.231}$	$16.316_{\pm8.033}$	$25.534_{\pm11.572}$	$0.931_{\pm0.044}$	$5.268_{\pm2.067}$	$16.327_{\pm5.593}$	$29.734_{\pm3.226}$
MLP-VNet [17]	$0.959_{\pm0.010}$	$2.786_{\pm0.879}$	$7.878_{\pm3.376}$	$19.157_{\pm6.413}$	$0.980_{\pm0.004}$	$1.763_{\pm0.550}$	$4.243_{\pm1.950}$	$15.371_{\pm5.426}$
Our Method	$\mathbf{0.981}_{\pm0.002}$	$\mathbf{0.898}_{\pm0.132}$	$\mathbf{2.104}_{\pm0.142}$	$\mathbf{5.512}_{\pm2.607}$	$\mathbf{0.988}_{\pm0.000}$	$\mathbf{0.573}_{\pm0.075}$	$\mathbf{1.359}_{\pm0.282}$	$\mathbf{2.693}_{\pm0.940}$

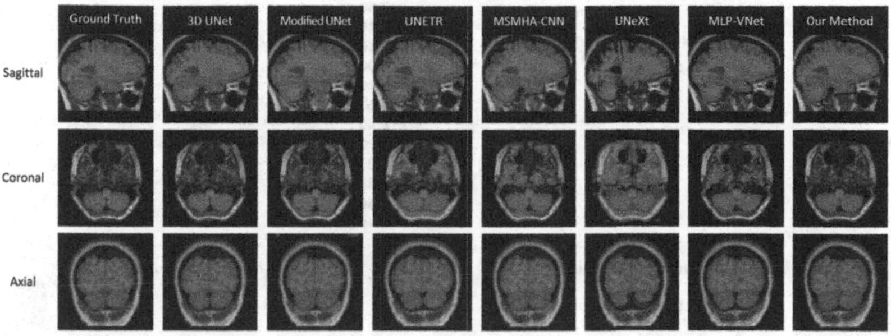

Fig. 3. Brain extraction examples from the IBSR dataset for visual comparison.

In addition, we also investigate the effects of shifting along different axes. We start with **no shifts**, and respectively shift along height axis (**H only**) and width axis (**W only**). Then, we do shifts along height and width axes in parallel (**H & W**) and finally shift along all three axes (namely our method). In Table 3, we can see that shifts along all three axes attain best results with respect to most metrics. Compared to those having shifts, **no shifts** has the lowest accuracy since it cannot capture locality without shifts.

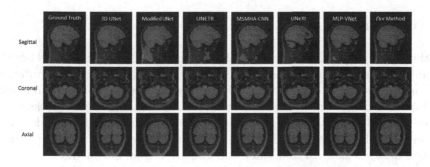

Fig. 4. Brain extraction examples from the NFBS dataset for visual comparison.

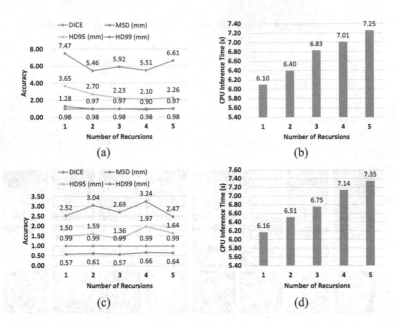

Fig. 5. Analysis on number of recursions. (a)–(b) accuracy and CPU inference time on IBSR dataset. (c)–(d) accuracy and CPU inference time on NFBS dataset.

Table 3. Ablation study. The effect of shift on the IBSR and NFBS datasets.

	IBSR				NFBS			
	DICE ↑	MSD ↓	HD95 ↓	HD99 ↓	DICE ↑	MSD ↓	HD95 ↓	HD99 ↓
no shifts	$0.978_{\pm0.003}$	$1.104_{\pm0.219}$	$2.737_{\pm0.652}$	$6.704_{\pm2.994}$	$0.987_{\pm0.001}$	$0.675_{\pm0.248}$	$1.656_{\pm0.834}$	$2.840_{\pm1.008}$
H only	$0.980_{\pm0.003}$	$0.964_{\pm0.249}$	$2.280_{\pm0.408}$	$5.595_{\pm3.022}$	$0.987_{\pm0.001}$	$0.626_{\pm0.183}$	$1.585_{\pm0.799}$	$2.715_{\pm1.087}$
W only	$0.979_{\pm0.005}$	$1.156_{\pm0.650}$	$3.363_{\pm2.650}$	$7.010_{\pm5.516}$	$0.987_{\pm0.001}$	$0.656_{\pm0.232}$	$1.640_{\pm0.864}$	$\mathbf{2.661}_{\pm1.028}$
H & W	$0.980_{\pm0.004}$	$1.167_{\pm0.556}$	$3.528_{\pm2.499}$	$6.563_{\pm3.519}$	$0.987_{\pm0.001}$	$0.591_{\pm0.079}$	$1.511_{\pm0.392}$	$3.402_{\pm0.853}$
Our method	$\mathbf{0.981}_{\pm0.002}$	$\mathbf{0.898}_{\pm0.132}$	$\mathbf{2.104}_{\pm0.142}$	$\mathbf{5.512}_{\pm2.607}$	$\mathbf{0.988}_{\pm0.000}$	$\mathbf{0.573}_{\pm0.075}$	$\mathbf{1.359}_{\pm0.282}$	$2.693_{\pm0.940}$

4   Conclusion

In this paper we propose a recursive MLP based 3D UNet for rapid and accurate brain extractions from 3D head volumes, aiming to deploy it in computational resources constrained environment. The network utilizes 3D convolution block to extract long range information. In parallel, it introduces the MLP blocks to capture localities and also enhance computational efficiency. We further design the recursive MLP to aggregate different localities in an efficient way. Experiments on two datasets show the efficiency and effectiveness of our method.

Acknowledgements. This work is supported by National Natural Science Foundation of China (No. 62072160) and Science and Technology Research Project of Henan Province (No. 232102211024).

References

1. The Internet Brain Segmentation Repository (IBSR). https://www.nitrc.org/projects/ibsr
2. The Neurofeedback Skull-stripped (NFBS) repository. https://preprocessed-connectomes-project.org/NFB_skullstripped/
3. Ba, J.L., Kiros, J.R., Hinton, G.E.: Layer normalization. arXiv preprint arXiv:1607.06450 (2016)
4. Çiçek, Ö., Abdulkadir, A., Lienkamp, S.S., Brox, T., Ronneberger, O.: 3D U-net: learning dense volumetric segmentation from sparse annotation. In: Ourselin, S., Joskowicz, L., Sabuncu, M.R., Unal, G., Wells, W. (eds.) MICCAI 2016. LNCS, vol. 9901, pp. 424–432. Springer, Cham (2016). https://doi.org/10.1007/978-3-319-46723-8_49
5. Cooley, C.Z., et al.: A portable scanner for magnetic resonance imaging of the brain. Nat. Biomed. Eng. 5(3), 229–239 (2021)
6. Fatima, A., Madni, T.M., Anwar, F., Janjua, U.I., Sultana, N.: Automated 2D slice-based skull stripping multi-view ensemble model on NFBS and IBSR datasets. J. Digit. Imaging 35(2), 374–384 (2022)
7. Hatamizadeh, A., et al.: UNETR: transformers for 3D medical image segmentation. In: Proceedings of the IEEE/CVF Winter Conference on Applications of Computer Vision, pp. 574–584 (2022)
8. Hendrycks, D., Gimpel, K.: Gaussian error linear units (GELUs). arXiv preprint arXiv:1606.08415 (2016)
9. Hwang, H., Rehman, H.Z.U., Lee, S.: 3D U-Net for skull stripping in brain MRI. Appl. Sci. 9(3), 569 (2019)
10. Ibtehaz, N., Rahman, M.S.: MultiResUNet: rethinking the U-Net architecture for multimodal biomedical image segmentation. Neural Netw. 121, 74–87 (2020)
11. Isensee, F., Kickingereder, P., Wick, W., Bendszus, M., Maier-Hein, K.H.: Brain tumor segmentation and radiomics survival prediction: contribution to the BRATS 2017 challenge. In: Crimi, A., Bakas, S., Kuijf, H., Menze, B., Reyes, M. (eds.) BrainLes 2017. LNCS, vol. 10670, pp. 287–297. Springer, Cham (2018). https://doi.org/10.1007/978-3-319-75238-9_25
12. Jenkinson, M., Beckmann, C.F., Behrens, T.E., Woolrich, M.W., Smith, S.M.: FSL. Neuroimage 62(2), 782–790 (2012)
13. Lian, D., Yu, Z., Sun, X., Gao, S.: AS-MLP: an axial shifted MLP architecture for vision. In: International Conference on Learning Representations (ICLR), pp. 1–19 (2022)
14. Mazurek, M.H., et al.: Portable, bedside, low-field magnetic resonance imaging for evaluation of intracerebral hemorrhage. Nat. Commun. 12(1), 1–11 (2021)
15. Milletari, F., Navab, N., Ahmadi, S.A.: V-net: fully convolutional neural networks for volumetric medical image segmentation. In: 2016 Fourth International Conference on 3D Vision (3DV), pp. 565–571 (2016)
16. Nie, D., Wang, L., Adeli, E., Lao, C., Lin, W., Shen, D.: 3-D fully convolutional networks for multimodal isointense infant brain image segmentation. IEEE Trans. Cybern. 49(3), 1123–1136 (2018)
17. Pan, S., et al.: Abdomen CT multi-organ segmentation using token-based MLP-mixer. Med. Phys. 50, 3027–3038 (2022)
18. Qiu, Z., Yao, T., Ngo, C.W., Mei, T.: MLP-3D: a MLP-Like 3D architecture with grouped time mixing. In: Proceedings of the IEEE/CVF Conference on Computer Vision and Pattern Recognition, pp. 3062–3072 (2022)

19. Ronneberger, O., Fischer, P., Brox, T.: U-net: convolutional networks for biomedical image segmentation. In: Navab, N., Hornegger, J., Wells, W.M., Frangi, A.F. (eds.) MICCAI 2015. LNCS, vol. 9351, pp. 234–241. Springer, Cham (2015). https://doi.org/10.1007/978-3-319-24574-4_28

20. Sheth, K.N., et al.: Assessment of brain injury using portable, low-field magnetic resonance imaging at the bedside of critically ill patients. JAMA Neurol. **78**(1), 41–47 (2021)

21. Smith, S.M.: Fast robust automated brain extraction. Hum. Brain Mapp. **17**(3), 143–155 (2002)

22. Sun, L., Shao, W., Zhu, Q., Wang, M., Li, G., Zhang, D.: Multi-scale multi-hierarchy attention convolutional neural network for fetal brain extraction. Pattern Recogn. **133**, 109029 (2023)

23. Tolstikhin, I.O., et al.: MLP-Mixer: an all-MLP architecture for vision. Adv. Neural. Inf. Process. Syst. **34**, 24261–24272 (2021)

24. Valanarasu, J.M.J., Patel, V.M.: UNeXt: MLP-based rapid medical image segmentation network. In: Wang, L., Dou, Q., Fletcher, P.T., Speidel, S., Li, S. (eds.) MICCAI 2022. LNCS, vol. 13435, pp. 23–33. Springer, Cham (2022). https://doi.org/10.1007/978-3-031-16443-9_3

25. Wang, Z., Zou, N., Shen, D., Ji, S.: Non-local U-nets for biomedical image segmentation. In: Proceedings of the AAAI Conference on Artificial Intelligence, vol. 34, pp. 6315–6322 (2020)

26. Yu, T., Li, X., Cai, Y., Sun, M., Li, P.: S^2-MLP spatial-shift MLP architecture for vision. In: the IEEE/CVF Winter Conference on Applications of Computer Vision, pp. 297–306 (2022)

27. Zhou, Z., Siddiquee, M.M.R., Tajbakhsh, N., Liang, J.: UNet++: redesigning skip connections to exploit multiscale features in image segmentation. IEEE Trans. Med. Imaging **39**(6), 1856–1867 (2020)

A Hip-Knee Joint Coordination Evaluation System in Hemiplegic Individuals Based on Cyclogram Analysis

Ningcun Xu[1], Chen Wang[2], Liang Peng[2(✉)], Jingyao Chen[1], Zhi Cheng[1], Zeng-Guang Hou[1,2(✉)], Pu Zhang[3], and Zejia He[3]

[1] Macau University of Science and Technology, Macao 999078, China
`2202853NMI30001@student.must.edu.mo`
[2] State Key Laboratory of Multimodal Artificial Intelligence Systems, Institute of Automation, Chinese Academy of Sciences, Beijing 100190, China
`{liang.peng,zengguang.hou}@ia.ac.cn`
[3] Beijing Bóai Hospital, China Rehabilitation Research Center, Beijing 100068, China

Abstract. Inter-joint coordination analysis can provide deep insights into assessing patients' walking ability. This paper developed a hip-knee joint coordination assessment system. Firstly, we introduced a hip-knee joint cyclogram generation model that takes into account walking speed. This model serves as a reference template for identifying abnormal patterns in the hip-knee joints when walking at different speeds. Secondly, we developed a portable motion capture platform based on stereovision technology. It uses near-infrared cameras and markers to accurately capture kinematic data of the human lower limb. Thirdly, we designed a hip-knee joint coordination assessment metric DTW-ED (Dynamic Time Wrapping - Euclidean Distance), which can score the subject's hip-knee joint coordination. Experimental results indicate that the hip-knee joint cyclogram generation model has an error range of $[0.78°, 1.08°]$. We conducted walking experiments with five hemiplegic subjects and five healthy subjects. The evaluation system successfully scored the hip-knee joint coordination of patients, allowing us to differentiate between healthy individuals and hemiplegic patients. This assessment system can also be used to distinguish between the affected and unaffected sides of hemiplegic subjects. In conclusion, the hip-knee joint coordination assessment system developed in this paper has significant potential for clinical disease diagnosis.

Keywords: Hip-Knee Cyclogram · Joint Coordination Evaluation Metric · B-spline Curve · Motion Capture System · Stereo Vision · Dynamic Time Warping

This work was supported in part by the National Key Research and Development Program of China under Grant 2022YFC3601200; in part by the National Natural Science Foundation of China under Grant 62203441 and Grant U21A20479, and in part by the Beijing Natural Science Foundation under Grant L222013.

B. Luo et al. (Eds.): ICONIP 2023, LNCS 14449, pp. 589–601, 2024.
https://doi.org/10.1007/978-981-99-8067-3_44

1   Introduction

Gait analysis is a valuable tool that systematically differentiates human walking patterns. It finds applications in various fields, including diagnosing the health status of patients with gait disorders for clinicians [1], providing comprehensive evaluations of the effects of surgery or lower limb rehabilitation aids on patients [2,3], aiding professional athletes in correcting their movement posture [4], and assessing the fall risk among the elderly [5], and more.

Traditionally, clinicians have conducted gait analysis by using motor ability evaluation scales, such as the Brunnstrom Recovery Stage (BRS) [6] and the Fugl-Meyer assessment (FMA) [7]. However, the assessment results depend heavily on the experience level of clinicians.

Cyclogram, also known as angle-angle plot, is a very useful technology to describe inter-joint coordination, which was first proposed by Grieve et al. [8]. Compared with the redundant gait parameters in the gait report, the cyclogram can capture the subtle variability of gait patterns during walking and provide a more insightful analysis of the abnormal gait.

Researchers have proposed many metrics to qualitatively and quantitatively evaluate the inter-joint coordination based on cyclogram. Goswami et al. [9] designed the hip-knee joint coordination evaluation metric "cyclogram moments" by taking the geometric moments of the cyclogram as the descriptor of the inter-joint coordination. Another cyclogram-based metric is vector coding [10], which was defined as the "coupling angle" between two adjacent points on the cyclogram [11]. In order to better understand the changes in the gait patterns, Chang et al. [12] divided the coupling angle into four bins named in-phase, anti-phase, rear-foot phase, and fore-foot phase. Needham et al. [13] also separated the coupling angle into four bins: in-phase with proximal dominance, in-phase with distal dominance, anti-phase with proximal dominance, and anti-phase with distal dominance. The improved vector coding techniques were called "coupling angle binning" or "phase binning" by Beitter et al. [14]. They proposed a more comprehensive binning-based metric by combining the previous two metrics.

In this paper, we developed an active lower limb motion capture system based on stereo vision technology to collect the walking kinematics data of the subjects. A hip-knee cyclogram generation model related to walking speed was designed by using the B-spline curve. The model's parameters are transparent and interpretable. The hip-knee cyclogram produced by the generation model can be used as the reference template for the hip-knee joint coordination analysis. We also proposed the hip-knee joint coordination assessment metric DTW-ED (Dynamic Time Wrapping - Euclidean Distance), which can score the subject's hip-knee joint coordination and distinguish between healthy individuals and hemiplegic patients. Compared with the existing gait analysis tools, the advantages of the evaluation system proposed by this paper are its low cost, high mobility, and convenient deployment.

2 Hip-Knee Joint Coordination Assessment System

2.1 Lower Limbs Motion Capture Platform

System Composition. In this study, an active lower limbs motion capture platform was developed to collect the kinematics data of hip and knee joints during walking. As shown in Fig. 1, the motion capture platform is used in a hand-held way for the moving scene, which mainly consists of a near-infrared binocular camera, an infrared light source, an RGB camera, a palm-sized computer, some magic cable ties, and passive reflective ball markers.

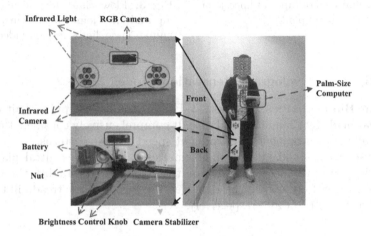

Fig. 1. Lower limbs motion capture platform in a hand-held way for mobile scene.

Hip and Knee Joint Angles Calculation. Firstly, the definitions of the hip and knee joints angle are illustrated in Fig. 2. The hip joint angle θ_{hip} was defined as the intersection angle of the body gravity vector relative to the connection line L_6 between marker M_1 and marker M_2. The knee angle θ_{knee} was the intersection angle of the extended line of L_6 relative to the connection line L_7 between reflective ball marker M_2 and marker M_3 [15]. In the walking experiment, these markers were placed bilaterally in the centerline of the thigh, the flexion-extension axis of the knee joint, and the lateral malleolus of the ankle joint [16].

Then, as shown in Fig. 3, the main steps of calculating the hip and knee joint angle include extracting frames from recording an individual's walking video, correcting frames based on the parameters of the binocular camera, detecting the reflective ball markers, matching the corresponding pixel points, calculating the position of markers based on stereo vision technology, and calculating the hip and knee joint angle.

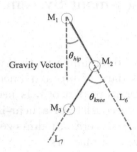

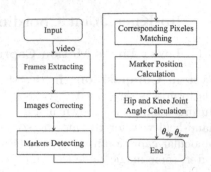

Fig. 2. Definitions of hip and knee joint angle in the sagittal plane.

Fig. 3. Flow chart for calculating the hip and knee joint angle from recording human lower limbs motion video.

2.2 Hip-Knee Cyclogram Generation Model

Walking Biomechanics Dataset. WBDS is the largest public gait dataset, which was made by Claudiane et al. [17]. It comprehensively considers the effects of the walking environment, age, and gait speed on gait patterns.

We use the hip and knee joint kinematics data in the sagittal plane from 24 healthy young volunteers walking on the treadmill to establish the hip-knee cyclogram generation model. The eight walking speeds in the treadmill trials are denoted as Speed1-Speed8, respectively.

B-Spline Curve. The spline curve is a continuous and smooth curve generated by a set of control points. The spline curve is divided into two types: the interpolation spline curve and the approximation spline curve. The former passes through the control points, while the latter does not. Therefore, we used the B-spline curve to establish the hip-knee cyclogram generation model. The definition of the B-spline curve is that

$$\mathbf{C}(u) = \sum_{i=0}^{n} N_{i,k}(u)\mathbf{cp}_i, u \in [0,1], \tag{1}$$

where, the two-dimensional vector $\mathbf{C}(u)$ represents the generating B-spline curve over the parameter u. The range of the parameter u is $[0, 1]$. We denote the $n+1$ control points as CP_0, CP_1, CP_2, ..., CP_n, and the vector $\mathbf{cp}_i$ is the coordinate of the i-th control point. The $N_{i,k}(u)$ is the basis function of the k-degree B-spline curve, which is over the parameter u.

The basis function $N_{i,k}(u)$ can be calculated by using the Cox-deBoor algorithm in recursive way, as shown in (2) [18].

$$
\begin{cases}
N_{i,0}(u) = \begin{cases} 1, \& if\ t_i \leq u \leq t_{i+1} \\ 0, \& other \end{cases} \\
N_{i,k}(u) = \dfrac{(u - t_i)N_{i,k-1}(u)}{t_{i+k} - t_i} + \\
\qquad\quad \dfrac{(t_{i+k+1} - u)N_{i+1,k-1}(u)}{t_{i+k+1} - t_{i+1}} \quad t_i \leq u \leq t_{i+1}, k \geq 1
\end{cases}
\tag{2}
$$

As shown in (2), when k is 0, the basis function $N_{i,0}(u)$ on the knot interval $[t_i, t_{i+1}]$ is constant 1; Otherwise, it is constant 0. And, when the degree $k \geq 1$, the basis functions $N_{i,k}(u)$ is calculated by the recursive formula. The t_i represents the knot value, and the $m + 1$ knot values form the knot vector $\mathbf{t} = \{t_0, t_1, t_2, ..., t_m\}$, which can divide the value range of parameter u into different bins, i.e. $[t_i, t_{i+1}]$. Thus, $N_{i,k}(u)$ can determine the influence of control points on the curve shape.

Above all, the three foundational elements of the B-spline curve are the degree k, $n + 1$ control points CP_0, CP_1, CP_2, ..., CP_n and the knot vector $\mathbf{t}$. There are following relationship: $m = n + k + 1$, among the three foundational elements [18].

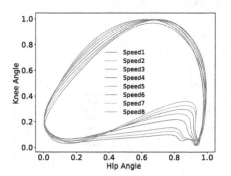

Fig. 4. The average hip-knee cyclograms from 24 healthy young subjects at each walking speed.

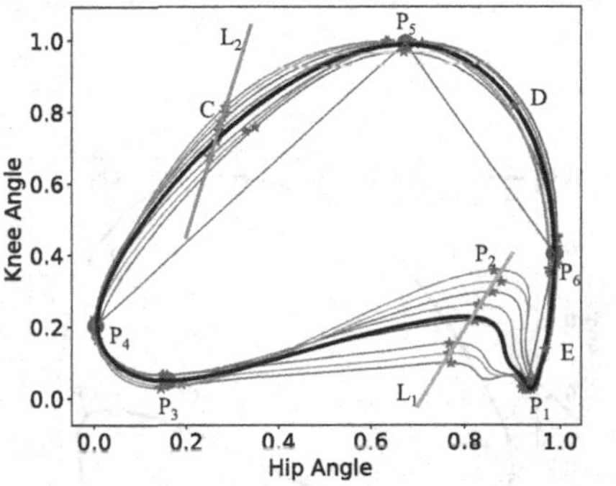

Fig. 5. Feature points of original hip-knee cyclograms.

Data Preprocess. The purpose of data preprocessing is to eliminate the amplitude difference of joint angle in different individuals and to extract general features on the hip-knee cyclograms of all volunteers. Firstly, all volunteers' hip joint angle and knee joint angle data were normalized, respectively. Secondly, we calculated the average hip-knee cyclogram of 24 healthy young volunteers at each walking speed, which was regarded as the original cyclograms, as shown in Fig. 4.

Feature Points. For determining the three foundational elements of the B-spline curve, we firstly extracted nine feature points and established their expression relative to walking speed s by analyzing the characteristic variety of the original hip-knee cyclograms at different walking speeds. Figure 5 shows nine feature points, P_1, P_2, P_3, P_4, P_5, P_6, C, D, E, on hip-knee cyclograms at the eight kinds of walking speeds. The black hip-knee cyclogram is the average curve of these cyclograms at all walking speeds. Feature points P_1, P_2, P_3, P_5 were located at the first valley, the first peak, the second valley, and the second peak of the knee joint angle curve, respectively. And the valley and the peak of the hip joint angle were defined as the feature points P_4 and P_6, separately. The feature point C was defined as the point farthest from the connection line between P_4 and P_5 on terminal stance and pre-swing phase of the hip-knee cyclogram. Feature points D and E were respectively defined as the points on the swing phase of the average angle cyclogram with the largest distance relative to the straight line between P_5 and P_6, and the straight line between P_6 and P_1.

In the next subsection, the nine feature points would be used to yield 12 control points of B-spline curve, and the Cox-deBoor algorithm would be used to generate the standard hip-knee cyclograms at different walking speeds.

Standard Hip-Knee Cyclogram Generating. Based on the nine feature points on hip-knee cyclograms, 12 control points of B-spline curve were defined as CP_i, $i = 0, 1, 2, 3, 4, 5, 6, 7, 8, 9, 10, 11$, as shown in Fig. 6.

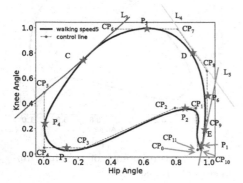

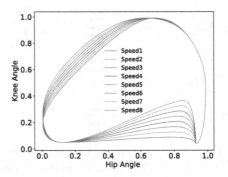

Fig. 6. Definitions of 12 control points. (Color figure online)

Fig. 7. The cyclograms produced by hip-knee cyclogram generating model at eight walking speeds.

The black curve represents the original hip-knee cyclogram at walking Speed5. The red stars represent the nine feature points. The blue circles represent the control points and the dotted lines represent the control line between two adjacent control points.

Control point CP_0 and control point CP_{11} were located at the feature point P_1 for generating the closed cyclogram. Control point CP_1 was defined as horizontal right of feature point P_2, the coordinate of which was $[P_{1x} - 0.01, P_{2y}]$. Control point CP_2 was defined as horizontal left of feature point P_2, the coordinate of which was $[P_{2x} -0.09, P_{2y}]$. Control point CP_3 was defined as the horizontal right of the feature point P_3, the coordinate of which was $[P_{3x} + 0.08, P_{3y}]$. The abscissa and ordinate of the control point CP_4 were respectively defined as the abscissa of the feature point P_4 and the ordinate of the feature point P_3, the coordinate of which was $[P_{4x}, P_{3y}]$. The control point CP_5 was defined as the intersection of the straight line $x = P_{4x}$ and the straight line L_3, which passed through point C and had the same slope as the connecting line between feature point P_4 and feature point P_5. The control point CP_6 was the intersection of the straight line $y = P_{5y}$ and the straight line L_3. The control point CP_7 was defined as the intersection of the straight line $y = P_{5y}$ and the straight line L_4, which passed through point D and had the same slope as the connecting line between feature point P_5 and feature point P_6. The control point CP_8 was defined as the intersection of the straight line $x = P_{6x}$ and the straight line L_4. The control point CP_9 was defined as the intersection of the straight line $x = P_{6x}$ and the straight line L_5, which passed through point E and had the same slope as the connecting line between feature point P_6 and feature point P_1. The control point CP_{10} was defined as the intersection of the straight line $y = P_{1y}$ and the straight line L_5.

Next, we used the cubic B-spline curve to establish the hip-knee cyclogram generation model, so that the order $k+1$ is 3. In light of the quantitative relationship between the three foundational elements of B-spline curve, $m = n + k + 1$, the number of knot values $m+1$ is 14. In order to generate a smooth and closed cyclogram, the knot vector t was defined as $t = \{0, 0, 0, 0.1, 0.2, 0.3, 0.4, 0.5, 0.6, 0.7, 0.8, 0.9, 1, 1, 1\}$.

The basis function $N_{i,k}(u)$ could be calculated by (2). Then, the hip-knee cyclogram can be generated by the formula (1), as shown in Fig. 7.

Above all, we indirectly established hip-knee cyclogram generation model related to walking speed s by using B-spline curve, which can produce the hip-knee cyclogram at some walking speed.

2.3 Hip-Knee Joint Coordination Assessment Metric

The hip-knee cyclograms of individuals with different severities of gait disorder have significant shape distinctions [15]. Clinicians can easily distinguish the walking ability of individuals by observing the shape features of their hip-knee cyclograms. Therefore, we regarded the hip-knee joint coordination evaluation of individuals as the shape similarity assessment of their hip-knee cyclograms relative to the standard cyclogram generated in former section.

We introduced a new metric called DTW-ED for evaluating hip-knee joint coordination. DTW-ED combines two key components: DTW-based corresponding points matching between the hip-knee cyclograms and ED-based mean matching error calculation. The resulting assessment value is referred to as the hip-knee joint coordination assessment value, denoted as *dtw-ed*.

Suppose we have two sets of sample points represented as $\mathbf{A}$ and $\mathbf{B}$ on the cyclograms, both with dimensions of $n \times 2$. Here, $\mathbf{A}$ refers to the standard hip-knee cyclogram generated in Sect. 2.2, while $\mathbf{B}$ represents the hip-knee cyclogram of an individual at a specific walking speed. The DTW-ED metric takes these two cyclograms, $\mathbf{A}$ and $\mathbf{B}$, as inputs and provides the hip-knee joint coordination evaluation value, *dtw-ed*, as the output. The main calculation steps involve determining the distance matrix $\mathbf{D_{AB}}$, computing the cumulative cost matrix $\mathbf{C}_{dtw}$, finding the shortest path $\mathbf{P}_{path}$, and calculating the evaluation value *dtw-ed*. The shortest matching path $\mathbf{P}_{path}$, with dimensions of $n \times 2$, plays a fundamental role in the quantitative assessment metric. Visualizing this path on the cyclograms can aid clinicians in qualitatively assessing hip-knee joint coordination. The *dtw-ed* value is defined as the average Euclidean distance between corresponding points on the shortest path $\mathbf{P}_{path}$.

3   Experiment and Results Analysis

3.1   Hip-Knee Cyclogram Modeling Results

As shown in Fig. 7, the standard hip-knee cyclograms at different walking speeds were produced by the hip-knee cyclogram generation model. The results show the hip-knee cyclogram generation model can accurately describe the characteristic variety of the cyclogram at different walking speeds.

Next, we employed Dynamic Time Warping (DTW) to extract corresponding points between the cyclogram generated by the hip-knee cyclogram generation model and the original cyclogram at the corresponding walking speed. The model's Mean Absolute Difference (MAD) was defined as the average Euclidean distance among all corresponding points.

Table 1. Hip-Knee cyclogram modeling error results

Walking Speed	Speed1	Speed2	Speed3	Speed4
MAD/degree[a]	0.96 ± 0.6	0.78 ± 0.54	1.02 ± 0.6	0.9 ± 0.42
Walking Speed	Speed5	Speed6	Speed7	Speed8
MAD/degree	0.84 ± 0.42	0.72 ± 0.42	0.96 ± 0.54	1.08 ± 0.54

[a] MAD: mean absolution difference

After inverse normalization, the modeling errors at different walking speeds are given in Table 1. The range of the modeling errors is $[0.72°, 1.08°]$[1]. Thus, the modeling errors are small enough. Figure 8 visualize the standard hip-knee cyclogram and the original hip-knee cyclogram at the walking speed8. The black stars

---

[1] The coefficient of inverse normalization was set as $60°$.

represent the cyclogram produced by the hip-knee cyclogram generation model. The red circles represent the original cyclogram. The black lines represent the connection line of corresponding points. It can be seen that the hip-knee cyclogram generated by the hip-knee cyclogram generating model is highly consistent with the original cyclogram.

3.2   Coordination Assessment Results

Data Collection To validate the effectiveness of the hip-knee joint coordination evaluation metric, this study conducted a data collection experiment using the lower limbs motion capture system. A total of ten subjects participated in the experiment, consisting of five hemiplegic individuals (5 males, age: 59.6 ± 8.1 years, height: 173.4 ± 3.1 cm, weight: 79.6 ± 10.4 kg) and five healthy individuals (4 males, 1 female, age: 27.6 ± 4.4 years, height: 172.2 ± 5.0 cm, weight: 64 ± 8.2 kg). All participants were recruited from the Chinese Rehabilitation Research Center.

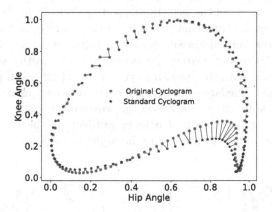

Fig. 8. The visualization of modeling error at the walking speed8. (Color figure online)

An experienced clinician assessed the walking ability of the hemiplegic subjects based on the Brunnstrom Recovery Stage Scale. This study obtained approval from the Ethics Committee of China Rehabilitation Research Center, and all participants provided written informed consent before taking part in the experiment. For the healthy subjects, they walked on a walking machine at a speed that felt comfortable to them while their hip and knee joint kinematics data were recorded for a duration of one minute. To ensure the safety of the hemiplegic subjects, a two-meter force-measuring platform was used for their walking experiment instead of the walking machine. They were instructed to walk at a comfortable speed. Given their limited walking ability, the hemiplegic subjects were asked to perform two walking trials on the force-measuring platform.

Assessment Performance. Clinicians can assess the hip-knee joint coordination of subjects by examining their hip-knee cyclograms. To aid clinicians in distinguishing abnormal gait, this study presents the matching results of corresponding points between each subject's hip-knee cyclogram and the cyclogram generated by the hip-knee cyclogram generation model at a comfortable walking speed. These results are illustrated in Fig. 9.

In Fig. 9, the first two rows show corresponding points matching results of the hip-knee cyclograms from the affected side and the unaffected side of the five hemiplegia subjects (i.e. AHS1-5, UHS1-5), respectively. The last row shows the corresponding point matching results from the right domain leg of the five healthy subjects (i.e. RHS1-5). The blue curves represent the cyclograms from subjects. The red curve represents the standard cyclogram at the normal walking speed. The black solid lines connect the matched corresponding points of the two cyclograms. The curves with △ on all cyclograms represent the stance phase of the gait cycle, and the curves with + on all cyclograms represent the swing phase of the gait cycle.

First of all, from the perspective of the hip-knee cyclogram shapes, the hip-knee cyclograms shapes of the healthy subjects are closer to the standard cyclogram and more regular than that of the hemiplegia subjects, especially than the affected side of the hemiplegia subjects. Secondly, from the view of the corresponding points matching degree, the cyclograms of healthy subjects have the best matching effect with the standard cyclograms at the peak points and valley points and also have the largest number of corresponding points. Conversely, the affected side of hemiplegia subjects has the worst matching effect. It shows that the hip-knee joint coordination of healthy subjects is the best, and the worst hip-knee joint coordinations are among hemiplegia subjects.

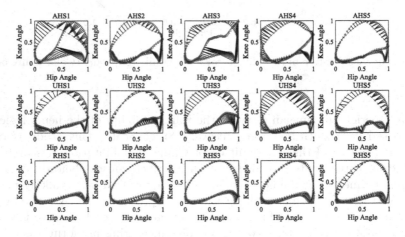

Fig. 9. The qualitative assessment results of the coordination evaluation metric DTW-ED. (Color figure online)

Table 2. Quantitative evaluation results for all subjects

Groups	Leg	Number of subject	DTW-ED	BRS
Hemiplegia	Affect Side	**Subject1**	**0.164 (0.621)**	III
		Subject2	0.161 (0.607)	II
		Subject3	**0.242 (1.000)**	III
		Subject4	0.186 (0.728)	II
		Subject5	0.090 (0.262)	II
Hemiplegia	Unaffect Side	**Subject1**	**0.103 (0.325)**	III
		Subject2	0.111 (0.364)	II
		Subject3	**0.096 (0.291)**	III
		Subject4	0.096 (0.291)	II
		Subject5	0.092 (0.272)	II
Healthy	Right Side	Subject1	0.036 (0.000)	–
		Subject2	0.062 (0.126)	–
		Subject3	0.038 (0.010)	–
		Subject4	0.050 (0.068)	–
		Subject5	0.038 (0.010)	–

[1] the raw result (the min-max normalization result)

Table 2 presents both the raw results and the min-max normalized results of the DTW-ED metric for all subjects. The min-max normalization is applied to facilitate the analysis of the coordination assessment metric's ability to identify abnormal gait. A result closer to zero indicates better hip-knee joint coordination. In Table 2, we can see that the hip-knee joint coordination assessment results of healthy subjects are smaller compared to those of both the affected and unaffected sides of the hemiplegic subjects. With the exception of Hemiplegic Subject 5, the hip-knee joint coordination assessment result for the unaffected side of each hemiplegic subject is smaller than that of their affected side. That is because the unaffected side of Hemiplegic Subject 5 compensates for the affected side, resulting in better hip-knee joint coordination on the affected side. Therefore, the DTW-ED metric for hip-knee joint coordination assessment can not only differentiate between healthy subjects and hemiplegic subjects but also distinguish between the affected and unaffected sides of hemiplegic subjects.

4   Conclusion

This paper introduces a hip-knee joint coordination assessment system. The lower limbs motion capture platform enhances the system's ease of deployment, portability, and cost-effectiveness. The hip-knee cyclogram generation model, utilizing B-spline curve, serves as a reference standard for analyzing abnormal gait patterns. This model provides a more comprehensive understanding of hip-knee joint coordination by considering walking speed. Additionally, the DTW-

ED metric is proposed, enabling both qualitative and quantitative evaluation of hip-knee joint coordination in hemiplegic subjects. This metric combines qualitative observations with quantitative data analysis, improving the comprehensiveness and reliability of hip-knee joint coordination assessment. In conclusion, the hip-knee joint coordination assessment system we proposed by us holds great potential for clinical disease diagnosis.

In future research, we will recruit more subjects with hemiplegia to evaluate the clinical value of the DTW-ED metric. Moreover, we intend to improve the lower limb motion capture system to acquire additional physiological signals, including plantar pressure signals and EMG signals, which will help us study the finer inter-joint coordination assessment model.

References

1. Cicirelli, G., Impedovo, D., Dentamaro, V., Marani, R., Pirlo, G., D'Orazio, T.R.: Human gait analysis in neurodegenerative diseases: a review. IEEE J. Biomed. Health Inf. **26**(1), 229–242 (2022)
2. Wang, W., Ackland, D.C., McClelland, J.A., Webster, K.E., Halgamuge, S.: Assessment of gait characteristics in total knee arthroplasty patients using a hierarchical partial least squares method. IEEE J. Biomed. Health Inf. **22**(1), 205–214 (2018)
3. Ma, H., Zhong, C., Chen, B., Chan, K.M., Liao, W.H.: User-adaptive assistance of assistive knee braces for gait rehabilitation. IEEE Trans. Biomed. Eng. **26**(10), 1994–2005 (2018)
4. Ramirez-Bautista, J.A., Huerta-Ruelas, J.A., Chaparro-Cárdenas, S.L., Hernández-Zavala, A.: A review in detection and monitoring gait disorders using in-shoe plantar measurement systems. IEEE Rev. Biomed. Eng. **10**, 299–309 (2017)
5. Tunca, C., Salur, G., Ersoy, C.: Deep learning for fall risk assessment with inertial sensors: utilizing domain knowledge in spatio-temporal gait parameters. IEEE J. Biomed. Health Inf. **24**(7), 1994–2005 (2020)
6. Saleh, M., Murdoch, G.: In defence of gait analysis observation and measurement in gait assessment. J. Bone Joint Surg. Brit. **67**(2), 237–241 (1985)
7. Gladstone, D.J., Danells, C.J., Black, S.E.: The Fugl-Meyer assessment of motor recovery after stroke: a critical review of its measurement properties. Neurorehabil. Neural Repair **16**(3), 232–240 (2002)
8. Grieve, D.W.: Gait patterns and the speed of walking. Biomed. Eng. **3**(3), 119–122 (1968)
9. Goswami, A.: A new gait parameterization technique by means of cyclogram moments: application to human slope walking. Gait Posture **8**(1), 15–36 (1998)
10. Tepavac, D., Field-Fote, E.C.: Vector coding: a technique for quantification of intersegmental coupling in multicyclic behaviors. J. Appl. Biomech. **17**(3), 259–270 (2001)
11. Hamill, J., Haddad, J.M., McDermott, W.J.: Issues in quantifying variability from a dynamical systems perspective. J. Appl. Biomech. **16**(4), 407–418 (2000)
12. Ryan, C., Emmerik, R.V., Hamill, J.: Quantifying rearfoot-forefoot coordination in human walking. J. Biomech. **41**(14), 3101–3105 (2008)
13. Needham, A.R., Roozbeh, N., Nachiappan, C.: A new coordination pattern classification to assess gait kinematics when utilising a modified vector coding technique. J. Biomech. **48**(12), 3506–3511 (2015)

14. Beitter, J., Kwon, Y.H., Kirsten, T.F.: A combined method for binning coupling angles to define coordination patterns. J. Biomech. **103**, 109598 (2020)
15. Sakoe, H., Chiba, S.: Dynamic programming algorithm optimization for spoken word recognition. IEEE Trans. Acoust. Speech Signal Process. **26**(1), 43–49 (1978)
16. Ren, S., Wang, W., Hou, Z.-G., Chen, B., Liang, X., Wang, J., et al.: Personalized gait trajectory generation based on anthropometric features using random forest. J. Ambient Intell. Hum. Comput. 1–12 (2019)
17. Claudiane, A.F., Reginaldo, F.K., Marcos, D.: A public dataset of overground and treadmill walking kinematics and kinetics in healthy individuals. PeerJ **6**, e4640 (2018)
18. Carl, D.B.: On calculating with B-splines. J. Approx. Theory **6**(1), 50–62 (1972)

Evaluation of Football Players' Performance Based on Multi-Criteria Decision Analysis Approach and Sensitivity Analysis

Jakub Więckowski[ID] and Wojciech Sałabun[✉][ID]

West Pomeranian University of Technology in Szczecin,
ul. Żołnierska 49, 71-210 Szczecin, Poland
{jakub-wieckowski,wojeciech.salabun}@zut.edu.pl

Abstract. The use of information systems and recommendation models in football has become a popular way to improve performance. With their help, it is possible to make more informed and effective decisions in terms of team management, the selection of training parameters, or the building of player line-ups. To this end, in this paper, we propose a decision model based on the Multi-Criteria Decision Analysis (MCDA) approach to assess defensive football players regarding overall and defense skills. The model was examined with selected objective weighting techniques and MCDA methods to comprehensively analyze the potential footballers' performance scores. A sensitivity analysis is performed to indicate what factors of the game the players should be focusing on throughout the season to increase the evaluation score of their performance and, thus, be a more attractive choice to clubs' managers. The results from the sensitivity analysis show that improving the performance regarding particular criteria can significantly improve the evaluation score of the players.

Keywords: Football players · Multi-Criteria Decision Analysis · Sport management · Sensitivity analysis

1 Introduction

Football is one of the most popular sports in the world, attracting fans, entrepreneurs, and politicians. Over the years, the sport has constantly grown, and alongside it, a range of tools used to ensure better results and more effective processes applied within the management of football clubs [8]. Many of the systems are directed towards developing recommendations for the selection of players to match the specifics of a particular team, the way the team plays, or the coach's idea of future development and tactics of the game [9]. To this end, evaluation systems are used to determine, based on a selected set of player characteristics and the preferences of the decision-maker, which of the players under consideration may prove to be the most rational choice [19]. To assess the quality of players' performance, a summary of game results and statistics is used to show how a particular footballer performs in a selected position [15]. Analyzing the quality of performance of these players allows the selection of footballers who potentially fit into a particular team formation or a specific style of play or development goal so that the team can expect to perform better in the future by filling an existing gap in the team.

A popular tool for developing evaluation and recommendation systems is to use the method from the group of Multi-Criteria Decision Analysis (MCDA), which analyzes the attractiveness of the decision variants under consideration based on a set of decision criteria [12]. In addition, they allow the decision-maker's preferences to be modeled using determining the relevance of the criteria weights based on his/her knowledge so that the system's operation can be adapted to current needs [13]. It is also possible to use methods based on measures of information so that the evaluation of decision-making variants is conducted objectively without taking into account potential biases in the decision-makers judgments [7]. Within the scope of the MCDA group of methods, there are many techniques for assessing the attractiveness of alternatives, and their performance often leads to different recommendation rankings [1]. To this end, it is useful to compare their results with each other to be able to compare potential differences depending on the method chosen [2]. In addition, sensitivity analysis is often used with MCDA methods to gain additional knowledge about potential changes in the attractiveness of decision alternatives that are affected by changing the value of the inputs in the decision model [10].

This paper proposes an MCDA evaluation approach of defensive football players from the Premier League and their performance in season 2020/2021. The decision model contains criteria assessing the overall skills and defensive skills of the players. In the first case, the evaluation was performed for only defensive skills and, in the second case, for all identified criteria. The statistical analysis of selected metrics calculated from the decision matrix was presented. Then, a sensitivity analysis was conducted to examine what factors of the game the players should be focusing on more to be a more attractive choice based on the determined decision model. The aim of the study is to propose an approach to how the recommendation system can be defined to evaluate the attractiveness of footballers and their performance throughout the season. The main contributions of the study are:

- performing the multi-criteria evaluation of football players' performance considering different areas of operation during the football game
- proposition of sensitivity analysis examination approach using potential performance values defined by strict bounds
- comparison of criteria weighting methods regarding the impact of different criteria relevance in the problem

The rest of the paper is organized as follows. Section 2 presents the preliminaries regarding the objective weighting methods. Section 3 includes a description of a case study directed to the evaluation of defensive football players' performance. In Sect. 4, the results obtained from the performed experiments are described. Finally, Sect. 5 includes the research summary and conclusions drawn from the work.

2 Preliminaries

2.1 Objective Weighting Methods

Equal Weights: This technique provides the same weight value for each criterion [16]. It translates into having equal importance of all criteria in the decision problem. For the problem with n criteria, the weights can be calculated as follows:

$$w_j = \frac{1}{n} \quad j \in \{1, 2, \dots, n\} \tag{1}$$

Entropy Weights: The entropy method considers the measure of uncertainty in the information formulated using probability theory [20]. The first step in the problem with m alternatives and n criteria is to normalize the decision matrix as below:

$$X_{ij}^* = \frac{X_{ij}}{\sum_{i=1}^m X_{ij}} \quad i \in \{1, 2, \dots, m\} \quad j \in \{1, 2, \dots, n\} \tag{2}$$

Then, the calculation of the information entropy of every given criterion should be performed as:

$$e_j = -\frac{\left(\sum_{i=1}^m X_{ij}^* \ln\left(X_{ij}^*\right)\right)}{\ln(m)} \tag{3}$$

Finally, calculating criteria weight using information entropy is done as follows:

$$w_j = \frac{1 - e_j}{\sum_{i=1}^n (1 - e_k)} \tag{4}$$

Standard Deviation Weights: The standard deviation method focuses on measuring the volatility of given values in the column in the decision matrix [17]. At first, in the problem with m alternatives and n criteria, calculating the standard deviation of a given decision matrix should be performed as below:

$$\sigma_j = \sqrt{\frac{\sum_{i=1}^m \left(X_{ij} - \overline{X_j}\right)^2}{m}} \quad i \in \{1, 2, \dots, m\} \quad j \in \{1, 2, \dots, n\} \tag{5}$$

With calculated standard deviation values, criteria weights can be calculated as:

$$w_j = \frac{\sigma_j}{\sum_{j=1}^n \sigma_j} \tag{6}$$

Variance Weights: Statistical Variance Procedure is a method in which objective weights are derived using a mathematical-statistical variance [4]. It describes the spread of variables from their mean value. Statistical variance is often used to describe the spread of variables in a data set. Initially, the decision matrix needs to be normalized as:

$$X_{ij}^* = normalize(X_{ij}) \quad i \in \{1, 2, \ldots, m\} \quad j \in \{1, 2, \ldots, n\} \tag{7}$$

With the obtained normalized decision matrix, a statistical variance of information is calculated as:

$$V_j = \left(\frac{1}{n}\right) \sum_{i=1}^{n} \left(X_{ij}^* - \overline{X_{ij}^*}\right)^2 \tag{8}$$

Then, based on the calculated values, the criteria weights values are determined as follows:

$$w_j = \frac{V_j}{\sum_{i=1}^{m} V_j} \tag{9}$$

2.2 Correlation Coefficients

Weighted Spearman Correlation Coefficient: The Weighted Spearman correlation coefficient allows for the comparison of two ranking vectors [6]. It uses the weights for determining the importance of differences occurring in the rankings coherence [3]. The formula for the mentioned coefficient is presented below.

$$r_w = 1 - \frac{6 \cdot \sum (x_i - y_i)^2 ((N - x_i + 1) + (N - y_i + 1))}{n \cdot (N^3 + N^2 - N - 1)} \tag{10}$$

The x_i means position in the reference ranking, y_i is the position in the second ranking, and N is the number of ranked elements.

WS Rank Similarity Coefficient: The WS rank similarity coefficient is based on assigning a greater significance to the elements in the top part of the ranking [14]. The discrepancies in higher positions will cause a lower similarity value than the same differences at the bottom of the ranking [11]. The calculation formula for the WS similarity coefficient stands is determined below.

$$WS = 1 - \sum \left(2^{-x_i} \frac{|x_i - y_i|}{max\{|x_i - 1|, |x_i - N|\}}\right) \tag{11}$$

The x_i means position in the reference ranking, y_i is the position in the second ranking, and N is the number of ranked elements.

3   Study Case

The evaluation of defensive football players' performance from the Premier League regarding their results from the season 2020/2021 was performed based on two groups of skills, namely the overall and defense skills. The criteria were identified based on the statistics presented on the website WhoScored.com, and the dataset of the players is available on the open repository. The footballers that were considered in the study were filtered based on the position in which they played. Only defenders were taken into account. The filtration was made based on the following positions: [D(C), D(R), D(L), D(CR), D(CL), D(LR)]. The D letter stands for **defensive** position, while the letters in brackets represent the specific position (C - central, R - right, L - left, CR- central right, CL - central left, LR - left or right). Moreover, based on the statistics gathered on WhoScored.com, 14 criteria were identified in the decision model. They are presented in Table 1, where the symbol, full name, and type of criteria are shown. The group of overall skills was determined with six criteria, and the defense skills had eight criteria.

Table 1. Criteria identified for the problem of defensive football players' performance evaluation.

Evaluation area	C_i	Symbol	Full name	Type
Overall skills	C_1	PS%	Pass success percentage	Profit
	C_2	ThrB	Through balls per game	Profit
	C_3	Crosses	Crosses per game	Profit
	C_4	LongB	Long balls per game	Profit
	C_5	KeyP	Key passes per game	Profit
	C_6	Assists	Total assists	Profit
Defense skills	C_7	Fouls	Fouls per game	Cost
	C_8	Drb	Dribbled past per game	Cost
	C_9	Tackles	Tackles per game	Profit
	C_{10}	Offsides	Offsides won per game	Profit
	C_{11}	Blocks	Outfielder block per game	Profit
	C_{12}	Inter	Interceptions per game	Profit
	C_{13}	Clear	Clearances per game	Profit
	C_{13}	OwnG	Own goals	Cost

The 108 footballers matched the filtering criterion with the defensive position. The decision matrix composed regarding the decision criteria was presented in Table 2. Data for the first eight and last two footballers in the list of 108 alternatives were presented, and the respective values corresponded to the players' performance for the given criteria. Based on the presented decision matrix, further evaluations were performed, as well as the sensitivity analysis of the results.

Figure 1 presents the visualizations of the data distributions in the decision matrix for the selected criteria. The number of bins in the visualization was set to 10. The visualizations present the distribution of pass success percentages (C_1), total assists (C_6), interceptions per game (C_{12}), and clearances per game

Table 2. Part of the decision matrix defined for the problem of defensive football players' performance evaluation.

A_i	C_1	C_2	C_3	C_4	C_5	C_6	C_7	C_8	C_9	C_{10}	C_{11}	C_{12}	C_{13}	C_{14}
A_1	87.1	0.1	0.4	2.4	1.6	3	1.3	1.4	2.4	0.3	0.1	1.5	1.0	0
A_2	68.2	0.0	0.1	0.5	1.0	5	1.1	0.2	0.3	0.0	0.1	0.3	0.3	0
A_3	84.9	0.0	0.4	0.8	0.9	4	0.9	0.4	2.6	0.2	0.5	1.8	1.8	0
A_4	77.5	0.0	1.5	1.3	1.5	7	1.1	0.8	2.1	0.1	0.1	1.0	2.4	0
A_5	90.9	0.0	0.0	3.8	0.2	0	0.7	0.2	0.7	0.5	0.4	1.2	3.8	0
A_6	85.6	0.0	0.0	3.8	0.0	0	0.9	0.9	2.7	0.5	1.3	2.4	2.7	0
A_7	87.3	0.0	0.0	4.5	0.2	1	1.2	0.3	0.9	0.6	0.7	1.8	3.6	0
A_8	93.3	0.0	0.0	2.8	0.1	0	0.3	0.2	1.0	0.7	0.7	0.8	2.5	0
$\vdots$	$\vdots$	$\vdots$	$\vdots$	$\vdots$	$\vdots$	$\vdots$	$\vdots$	$\vdots$	$\vdots$	$\vdots$	$\vdots$	$\vdots$	$\vdots$	$\vdots$
A_{107}	78.3	0.0	0.3	0.7	0.6	2	1.4	0.9	1.5	0.1	0.2	0.8	1.9	0
A_{108}	85.6	0.0	0.0	0.5	0.4	0	0.6	0.6	1.0	0.0	0.2	0.9	0.7	0

(C_{13}). It can be seen that most of the players had pass success percentages placed between 77% and 90% in the season. On the other hand, while taking into account the number of total assists, most of the defenders had zero assists, and only a few of them had more than four assists. Those criteria were marked as the overall skills factors. In the case of the defensive skills criteria, it can be seen that the distributions of the values are more focused in the middle part of the graphs. It can be caused by the fact that those factors are focused on the area of expertise of the defenders, making it more possible to occur than, for example, assists in the football game, which is more likely to occur for more offensive players.

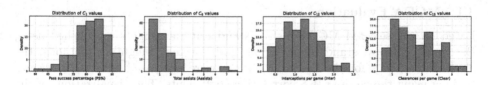

Fig. 1. Distribution of the values from the decision matrix for selected criteria C_1, C_6, C_{12}, and C_{13}.

Based on the two areas of criteria (defensive and overall), two assessment cases were planned. The first one concerns the assessment of the players regarding only defensive skills $(C_7–C_{14})$, while the second evaluation was based on all criteria $(C_1–C_{14})$. To provide an objective decision model, four criteria weighting methods that are based on the information measures in the decision matrix were used, namely the equal, entropy, standard deviation, and variance weights methods. Then, 12 MCDA methods were used to evaluate football players' performance with the combination of listed objective weighting methods. The obtained rankings were compared with the reference ranking established as a player rating in the WhoScored website. Moreover, the Weighted Spearman correlation coefficient (r_W) and WS rank similarity coefficient (WS) were used to indicate the coherence of the results. As the last phase of the evaluation, the sensitivity analysis was performed. The approach used a modified methodology presented by Wolters and Mareschal [18] and referred to modifications of the decision matrix to promote a given alternative in the ranking. The authors proposed to modify the single criterion value for one alternative until the changes allow for promotion to 1st place in the ranking. In this paper, the changes were limited by the bounds for each criterion, and the requirement of promoting an alternative to 1st position was not included.

4   Results

The football players' performances were evaluated by 12 MCDA methods. The Additive Ratio ASsessment (ARAS), COmbined COmpromise SOlution (COCOSO), COmbinative Distance-based ASsessment (CODAS), COmplex PRoportional ASsessment (COPRAS), Evaluation based on Distance from Average Solution (EDAS), Multi-Attributive Border Approximation area Comparison (MABAC), MultiAttributive Ideal-Real Comparative Analysis (MAIRCA), Measurement Alternatives and Ranking according to COmpromise Solution (MARCOS), Multi-Objective Optimization Method by Ratio Analysis (MOORA), Operational Competitiveness Ratings (OCRA), Preference Ranking Organization METHod for Enrichment of Evaluations II (PROMETHEE II), and Technique for the Order of Prioritisation by Similarity to Ideal Solution (TOPSIS) methods were used for this purpose. The pymcdm package written in Python was applied to perform all evaluations [5].

4.1   MCDA Evaluation

Based on the selected MCDA methods and criteria weights established using the equal, entropy, standard deviation, and variance weighting methods, the obtained rankings for the defensive criteria were compared with the reference ranking using the r_W and WS correlation coefficients. The visualizations of the coherence of the results were presented in Fig. 2. It can be seen that despite using different combinations of the weighting techniques and MCDA methods, the obtained correlations with the reference ranking indicated low similarity. In the case of the Weighted Spearman correlation coefficient, the highest correlation value was for the PROMETHEE II method with equal weights and for the CODAS method with the same weighting technique. The WS rank similarity coefficient values were slightly higher. However, the obtained similarities indicated significant divergence between the compared rankings. The EDAS and COPRAS methods combined with the entropy weighting methods produce the highest correlation values.

Based on the identified approaches that guaranteed the most correlated results, the visualizations of the rankings comparison for the selected techniques were presented in Fig. 3. The comparisons of reference ranking and the rankings obtained from the MCDA assessment for defensive skills criteria show that the proposed decision model produces a significantly different order of players' performance than the rating presented on the WhoScored website.

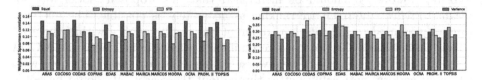

Fig. 2. Rankings correlation for selected objective weighting methods for players' performance assessment regarding defensive criteria.

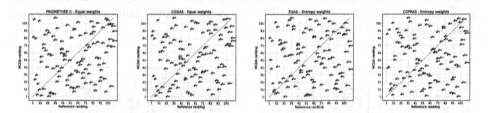

Fig. 3. Rankings comparison for selected objective weighting and MCDA methods for players' performance assessment regarding defensive criteria.

The second phase of the assessment was based on considering all identified criteria in the evaluation model. To this end, all 14 criteria presented in Table 1 were used to assess the players' performance. Since multiple clubs aim to have players who perform well in their position and other areas during the match, six overall skills criteria were also considered. It contributes to possessing players that can be relied on in defensive and offensive challenges, making the squad more complete. To this end, the players' performances were evaluated again with all identified decision criteria with four weighting techniques and 12 MCDA methods. The obtained rankings were compared with the reference ranking with two selected correlation coefficients. Figure 4 presents the visualizations of the coherence of the rankings. It can be seen that while considering all criteria from Table 1, the correlation values were significantly higher. In the case of the Weighted Spearman correlation coefficient, the highest correlation values were obtained for the standard deviation weights with a combination of MOORA and ARAS methods. On the other hand, for the WS rank similarity, the entropy weights with TOPSIS and COPRAS methods guaranteed the most similar rankings.

The visualization of the comparison of reference ranking and rankings obtained based on the determined decision model is presented in Fig. 5. It can be seen that the positions of the alternatives from the reference ranking and the given ranking from the MCDA assessment are less diverse and focused near the line that refers to the complete consistency of the rankings. Based on the obtained results, it can be concluded that taking into account all criteria, the obtained rankings were more similar to the ones presented on the WhoScored website. However, it should be mentioned that the proposed model was not meant to reflect the reference ranking as accurately as possible. The reference ranking was used to indicate how different the results can be within the evaluation approaches. The proposed recommendations obtained from the MCDA assessment show that players with worse rating scores on the WhoScored website may perform noticeably better throughout the season. It should be noted that the evaluation approach used in the study is based on objective weighting techniques and the determined criteria weights may differ from the ones used in the WhoScored website. Since different decision-makers could assign different importance to the considered criteria, the scope of the study was limited to the analysis of the results obtained with objective techniques, and expert knowledge and preferences were not considered.

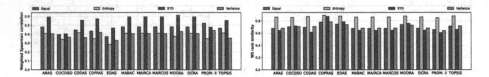

Fig. 4. Rankings correlation for selected objective weighting methods for players' performance assessment regarding all criteria.

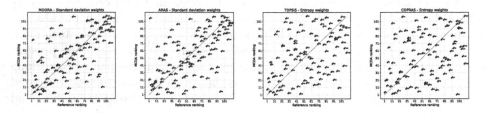

Fig. 5. Rankings comparison for selected objective weighting and MCDA methods for players' performance assessment regarding all criteria.

4.2 Sensitivity Analysis

Based on the approach proposed by Walters and Mareschal [18] to perform a sensitivity analysis of the results obtained with the MCDA assessment, an experiment toward examining changes in the decision matrix was performed. Walters and Mareschal described an approach of changing the single criterion value for a single alternative and re-assessing the decision variants until the change promotes the given alternative to 1st position in the ranking. This approach was modified in the following experiment. The changes for single criterion and single alternative were modeled in the decision matrix considering the maximum bounds of changes. The minimum and maximum values from the decision matrix were presented in Table 3, where the bounds refer to the maximum change that can be made within the given criterion values. For the profit-type criteria, the values in the decision matrix were increased, while for the cost-type criteria, these values were decreased with the step that equals 0.1 for all criteria.

Table 3. Ranges for subsequent criteria values used in the sensitivity analysis of players assessments.

	C_1	C_2	C_3	C_4	C_5	C_6	C_7	C_8	C_9	C_{10}	C_{11}	C_{12}	C_{13}	C_{14}
Min	59.4	0.0	0.0	0.4	0.0	0.0	0.3	0.0	0.3	0.0	0.0	0.3	0.3	0
Max	93.3	0.1	1.9	6.7	2.3	8.0	1.6	1.5	3.4	1.1	1.5	2.4	6.0	2
Bound	100.0	0.3	3.0	9.0	4.0	12.0	0.0	0.0	5.0	2.0	2.5	4.0	9.0	0

The sensitivity analysis was performed for the determined model based on the COPRAS method and entropy weights technique, which produced the most similar rankings in the second phase of the experiment. Table 4 presents the matrix of changes that caused a promotion of a given player regarding other footballers. It can be seen that some values in the matrix are filled with −. It translates into a situation in which the maximum change was reached, and it did not cause a promotion in the ranking. Moreover, the value of the threshold presented refers to the lowest possible value that promotes the given alternative to the highest possible in the ranking. It translates into not taking into account greater changes in values while they do not guarantee further improvements in the overall score.

For a single criterion and a single alternative, the initial value of change was set to the value from the decision matrix. Then, with subsequent iterations, this value was modified with the step of 0.01 toward direction compliant with the objective of the criterion (profit or cost). Thus, from Table 4, it can be seen that for the alternative A_1 and criterion C_2, the change of 0.2 representing the increased through balls per game statistics by 0.1 with respect to initial value could guarantee this player a higher position in the ranking. On the other hand, for alternative A_3 and criterion C_4, the change from 0.8 to 8.5 with respect to long balls per game could also provide this player a higher position in the ranking. Based on the presented data, the players and the coaches could get additional knowledge about which factors of the game should be improved to make them more efficient and provide them with better evaluation scores.

Table 4. Changes in decision matrix regarding given criterion and alternative that cause the most significant promotion in ranking.

A_i	C_1	C_2	C_3	C_4	C_5	C_6	C_7	C_8	C_9	C_{10}	C_{11}	C_{12}	C_{13}	C_{14}
A_1	–	0.2	0.7	–	2.0	4.0	0.0	0.5	–	0.6	0.5	–	–	0.0
A_2	–	0.1	2.6	–	3.6	11.6	0.1	0.1	–	1.9	2.4	–	–	0.0
A_3	–	0.1	2.9	8.5	3.9	11.8	0.0	0.1	–	1.8	2.3	–	–	0.0
A_4	–	0.2	2.8	–	3.7	10.8	0.0	0.1	–	1.6	2.1	–	–	0.0
A_5	–	0.2	2.8	4.3	3.8	10.9	0.6	0.1	1.3	1.9	2.1	1.7	4.4	0.0
A_6	–	0.2	2.8	8.2	3.3	11.6	0.0	0.1	4.7	1.9	2.4	3.9	7.5	0.0
A_7	–	0.2	2.9	–	3.8	11.2	0.0	0.1	–	1.8	2.2	–	–	0.0
A_8	–	0.2	2.9	7.1	3.8	11.6	0.2	0.1	2.6	1.9	2.3	2.0	7.2	0.0
$\vdots$	$\vdots$	$\vdots$	$\vdots$	$\vdots$	$\vdots$	$\vdots$	$\vdots$	$\vdots$	$\vdots$	$\vdots$	$\vdots$	$\vdots$	$\vdots$	$\vdots$
A_{107}	–	0.2	2.8	–	3.8	11.6	0.0	0.1	–	1.9	2.4	–	–	0.0
A_{108}	–	0.2	2.8	1.6	3.2	11.8	0.0	0.1	1.9	1.8	2.4	0.8	1.9	0.0

5   Conclusions

The proposed decision model aimed to evaluate the performance of football players from the Premier League based on their statistics from the season 2020/2021. Defenders were taken into consideration for the assessment process. Besides defensive skills, overall skills were also considered to propose players who can engage in offensive actions and defend the position on the pitch. To determine the decision model, MCDA methods were used with a combination of selected objective weighting techniques. The sensitivity analysis was performed to show what factors of the game should be improved to increase the assessment score from the model. The examination of the robustness of the results showed that the through balls per game factor could have the biggest impact on the score, and making progress in this area by defenders could make them more attractive to other clubs as their evaluation score and position in the ranking improve.

For future directions, it is worth considering using a vector of criteria weights that represent the preferences of specific decision-makers. It would allow to identify which players are meeting the requirements of specific coaches and clubs thus the most important criteria in the assessment process could be identified. Moreover, it would be meaningful to provide a similar evaluation for footballers playing in other positions as forwards or midfielders, to indicate the most attractive players for the clubs.

References

1. Ceballos, B., Lamata, M.T., Pelta, D.A.: A comparative analysis of multi-criteria decision-making methods. Prog. Artif. Intell. **5**, 315–322 (2016)

2. Chen, T.Y.: A comparative analysis of score functions for multiple criteria decision making in intuitionistic fuzzy settings. Inf. Sci. **181**(17), 3652–3676 (2011)
3. Costa, J.F.: Weighted correlation. In: Lovric, M. (ed.) International Encyclopedia of Statistical Science, pp. 1653–1655. Springer, Heidelberg (2011). https://doi.org/10.1007/978-3-642-04898-2_612
4. Hongjiu, L., Yanrong, H.: An evaluating method with combined assigning-weight based on maximizing variance. Sci. Program. **2015**, 3 (2015)
5. Kizielewicz, B., Shekhovtsov, A., Sałabun, W.: pymcdm-the universal library for solving multi-criteria decision-making problems. SoftwareX **22**, 101368 (2023)
6. Kizielewicz, B., Wątróbski, J., Sałabun, W.: Identification of relevant criteria set in the MCDA process-wind farm location case study. Energies **13**(24), 6548 (2020)
7. Koksalmis, E., Kabak, Ö.: Deriving decision makers' weights in group decision making: an overview of objective methods. Inf. Fusion **49**, 146–160 (2019)
8. Li, D., Zhang, J.: Computer aided teaching system based on artificial intelligence in football teaching and training. Mob. Inf. Syst. **2021**, 1–10 (2021)
9. Meng, X., et al.: A video information driven football recommendation system. Comput. Electr. Eng. **85**, 106699 (2020)
10. Pamučar, D.S., Božanić, D., Ranđelović, A.: Multi-criteria decision making: an example of sensitivity analysis. Serb. J. Manage. **12**(1), 1–27 (2017)
11. Paradowski, B., Shekhovtsov, A., Bączkiewicz, A., Kizielewicz, B., Sałabun, W.: Similarity analysis of methods for objective determination of weights in multi-criteria decision support systems. Symmetry **13**(10), 1874 (2021)
12. Pelissari, R., Alencar, P.S., Amor, S.B., Duarte, L.T.: The use of multiple criteria decision aiding methods in recommender systems: a literature review. In: Xavier-Junior, J.C., Rios, R.A. (eds.) BRACIS 2022. LNCS, vol. 13653, pp. 535–549. Springer, Cham (2022). https://doi.org/10.1007/978-3-031-21686-2_37
13. Roszkowska, E., Wachowicz, T., Bajwa, D., Koeszegi, S., Vetschera, R.: Analyzing the applicability of selected MCDA methods for determining the reliable scoring systems. In: Bajwa, D.S., Koeszegi, S., Vetschera, R. (eds.) Proceedings of the 16th International Conference on Group Decision and Negotiation Bellingham, pp. 180–187. Western Washington University (2016)
14. Sałabun, W., Urbaniak, K.: A new coefficient of rankings similarity in decision-making problems. In: Krzhizhanovskaya, V.V., et al. (eds.) ICCS 2020, Part II. LNCS, vol. 12138, pp. 632–645. Springer, Cham (2020). https://doi.org/10.1007/978-3-030-50417-5_47
15. Stanojevic, R., Gyarmati, L.: Towards data-driven football player assessment. In: 2016 IEEE 16th International Conference on Data Mining Workshops (ICDMW), pp. 167–172. IEEE (2016)
16. Wang, J.J., Jing, Y.Y., Zhang, C.F., Zhao, J.H.: Review on multi-criteria decision analysis aid in sustainable energy decision-making. Renew. Sustain. Energy Rev. **13**(9), 2263–2278 (2009)
17. Wang, Y.M., Luo, Y.: Integration of correlations with standard deviations for determining attribute weights in multiple attribute decision making. Math. Comput. Model. **51**(1–2), 1–12 (2010)
18. Wolters, W., Mareschal, B.: Novel types of sensitivity analysis for additive MCDM methods. Eur. J. Oper. Res. **81**(2), 281–290 (1995)
19. Yılmaz, Ö.İ., Öğüdücü, Ş.G.: Learning football player features using graph embeddings for player recommendation system. In: Proceedings of the 37th ACM/SIGAPP Symposium on Applied Computing, pp. 577–584 (2022)
20. Zhu, Y., Tian, D., Yan, F.: Effectiveness of entropy weight method in decision-making. Math. Probl. Eng. **2020**, 1–5 (2020)

Author Index

Printed in the United States
by Baker & Taylor Publisher Services

Printed in the United States
by Baker & Taylor Publisher Services